Theory of Ridge Regression Estimation with Applications

WILEY SERIES IN PROBABILITY AND STATISTICS

Established by *Walter A. Shewhart and Samuel S. Wilks*

The ***Wiley Series in Probability and Statistics*** is well established and authoritative. It covers many topics of current research interest in both pure and applied statistics and probability theory. Written by leading statisticians and institutions, the titles span both state-of-the-art developments in the field and classical methods.

Reflecting the wide range of current research in statistics, the series encompasses applied, methodological and theoretical statistics, ranging from applications and new techniques made possible by advances in computerized practice to rigorous treatment of theoretical approaches. This series provides essential and invaluable reading for all statisticians, whether in academia, industry, government, or research.

A complete list of titles in this series can be found at http://www.wiley.com/go/wsps

Theory of Ridge Regression Estimation with Applications

A.K. Md. Ehsanes Saleh
Carleton University
Ottawa, Canada

Mohammad Arashi
Shahrood University of Technology, Shahrood, Iran
University of Pretoria, Pretoria, South Africa

B.M. Golam Kibria
Florida International University
Miami, Florida, USA

Registered Office
John Wiley & Sons, Inc., 111 River Street, Hoboken, NJ 07030, USA

Editorial Office
111 River Street, Hoboken, NJ 07030, USA

For details of our global editorial offices, customer services, and more information about Wiley products visit us at www.wiley.com.

Wiley also publishes its books in a variety of electronic formats and by print-on-demand. Some content that appears in standard print versions of this book may not be available in other formats.

Library of Congress Cataloging-in-Publication Data Applied For
ISBN: 9781118644614 (hardback)

Cover design: Wiley
Cover image: © hrui/Shutterstock, Courtesy of B. M. Golam Kibria

Set in 10/12pt WarnockPro by SPi Global, Chennai, India

Printed in the United States of America

V10007111_011519

To our wives
Shahidara Saleh
Reihaneh Arashi
Farhana Kibria
(Orchi)

Contents in Brief

Contents

List of Figures

List of Tables

Preface

Regression analysis is the most useful statistical technique for analyzing multifaceted data in numerous fields of science, engineering, and social sciences. The estimation of regression parameters is a major concern for researchers and practitioners alike. It is well known that the least-squares estimators (LSEs) are popular for linear models because they are unbiased with minimum variance characteristics. But data analysts point out some deficiencies of the LSE with respect to prediction accuracy and interpretation. Further, the LSE may not exist if the design matrix is singular. Hoerl and Kennard (1970) introduced "ridge regression," which opened the door for "penalty estimators" based on the Tikhonov (1963) regularization. This methodology is the minimization of the least squares subject to an L_2 penalty. This methodology now impacts the development of data analysis for low- and high-dimensional cases, as well as applications of neural networks and big data analytics. However, this procedure does not produce a sparse solution. Toward this end, Tibshirani (1996) proposed the least absolute shrinkage and selection operator (LASSO) to overcome the deficiencies of LSE such as prediction and interpretation of the reduced model. LASSO is applicable in high- and low-dimensional cases as well as in big data analysis. LASSO simultaneously estimates and selects the parameters of a given model. This methodology minimizes the least squares criteria subject to an L_1 penalty, retaining the good properties of "subset selection" and "ridge regression."

There are many other shrinkage estimators in the literature such as the preliminary test and Stein-type estimators originally developed by Bancroft (1944), and Stein (1956), and James and Stein (1961), respectively. They do not select coefficients but only shrink them toward a predecided target value. There is extensive literature on a parametric approach to preliminary test and Stein-type estimators. The topic has been expanded toward robust rank-based, M-based, and quantile-based preliminary test and Stein-type estimation of regression coefficients by Saleh and Sen (1978–1985) and Sen and Saleh (1979, 1985, 1987). There is extensive literature focused only on this topic, and most recently documented by Saleh (2006). Due to the immense impact of Stein's approach on point estimation, scores of technical papers appeared in various areas of application.

The objective of this book is to provide a clear and balanced introduction of the theory of ridge regression, LASSO, preliminary test, and Stein-type estimators for graduate students and research-oriented statisticians, postdoctoral, and researchers. We start with the simplest models like the location model, simple linear model, and analysis of variance (ANOVA). Then we introduce the seemingly unrelated simple linear models. Next, we consider multiple regression, logistic regression, robust ridge regression, and high dimensional models. And, finally, as applications, we consider neural networks and big data to demonstrate the importance of ridge and logistic regression in these applications.

This book has 12 chapters, according to the given description of materials covered. Chapter 1 presents an introduction to ridge regression and different aspects of it, stressing the multicollinearity problem and its application to high-dimensional problems. Chapter 2 considers the simple linear model and location model, and provides theoretical developments of it. Chapters 3 and 4 deal with the ANOVA model and the seemingly unrelated simple linear models, respectively. Chapter 5 considers ridge regression and LASSO for multiple regression together with preliminary test and Stein-type estimators and a comparison thereof when the design matrix is nonorthogonal. Chapter 6 considers the ridge regression estimator and its relation with LASSO. Further, we study the properties of the preliminary test and Stein-type estimators with low dimension in detail. In Chapter 7, we cover the partially linear model and the properties of LASSO, ridge, preliminary test, and the Stein-type estimators. Chapter 8 contains the discussion of the logistic regression model and the related estimators of diverse kinds as described before in other chapters. Chapter 9 discusses the multiple regression model with autoregressive errors. In Chapter 10, we provide a comparative study of LASSO, ridge, preliminary test, and Stein-type estimators using rank-based theory. In Chapter 11, we discuss the estimation of parameters of a regression model with high dimensions. Finally, we conclude the book with Chapter 12 to illustrate recent applications of ridge, LASSO, and logistic regression to neural networks and big data analysis.

We appreciate the immense work done by Dr Mina Norouzirad with respect to the expert typing, editing, numerical computations, and technical management. Without her help, this book could not have been completed. Furthermore, Chapters 5 and 10 are the results of her joint work with Professor Saleh while she visited Carleton University, Ottawa during 2017. The authors thank Professor Resve Saleh for his assistance in preparing Chapter 12, along with reviewing the chapters and editing for English. We express our sincere thanks to Professor Mahdi Roozbeh (Semnan University, Iran) and Professor

Fikri Akdeniz (Cag University, Turkey) for their kind help with some numerical computations and providing references.

Professor A.K. Md. Ehsanes Saleh is grateful to NSERC for supporting his research for more than four decades. He is grateful to his loving wife, Shahidara Saleh, for her support over 67 years of marriage. M. Arashi wishes to thank his family in Iran, specifically his wife, Reihaneh Arashi (maiden name: Soleimani) for her everlasting love and support. This research is supported in part by the National Research Foundation of South Africa (Grant Numbers 109214 and 105840) and DST-NRF Centre of Excellence in Mathematical and Statistical Sciences (CoE-MaSS). Finally, B.M. Golam Kibria is thankful to Florida International University for excellent research facilities for 18 years. He spent most of his sabbatical leave (Fall, 2017) to read, correct, and edit all the chapters (1–11) collaborating with Professor Saleh. He is also grateful to his wife, Farhana Kibria, daughter, Narmeen Kibria, and son, Jinan Kibria and his family in Bangladesh for their support and encouragement during the tenure of writing this work.

August, 2018

Carleton University, Canada
Shahrood University of Technology,
Iran – University of Pretoria, South Africa
Florida International University, USA
A.K. Md. Ehsanes Saleh
Mohammad Arashi
B.M. Golam Kibria

Abbreviations and Acronyms

ADB	asymptotic distributional bias
ADR	asymptotic distributional risk
ANOVA	analysis of variance
ARRE	adaptive ridge regression estimator
BLUE	best linear unbiased estimator
c.d.f	cumulative distribution function
D.F.	degrees of freedom
DP	diagonal projection
Eff	Efficiency
GCV	generalized cross validation
GRRE	generalized ridge regression estimator
HTE	hard threshold estimator
LASSO	least absolute shrinkage and selection operator
LSE	least squares estimator
MLASSO	modified least absolute shrinkage and selection operator
MLE	maximum likelihood estimator
MSE	mean squared error
p.d.f.	probability density function
PLM	partially linear regression
PLS	penalized least squares
PRSE	positive-rule Stein-type estimator
PTE	preliminary test estimator
REff	relative efficiency
RHS	right-hand side
RLSE	restricted least squares estimator
RRE	ridge regression estimator

RSS	residual sum of squares
SE	Stein-type estimator
STE	soft threshold estimator
SVD	singular value decomposition
VIF	variance inflation factor
w.r.t.	with respect to

List of Symbols

Symbol	Description
$\boldsymbol{\alpha}, \boldsymbol{A}, \ldots$	matrices and vectors
$^\top$	transpose of a vector or a matrix
$\mathbb{E}[\cdot]$	expectation of a random variable
$\mathrm{Var}(\cdot)$	variance of a random variable
Diag	diagonal matrix
$\mathrm{tr}(\cdot)$	trace of a matrix
$\mathbb{R}$	real numbers
$\boldsymbol{Y}$	response vector
$\boldsymbol{X}$	design matrix
$\boldsymbol{\beta}$	vector of unknown regression parameters
$\tilde{\boldsymbol{\beta}}_n$	least squares estimator
$\hat{\boldsymbol{\beta}}_n$	restricted least squares estimator
$\hat{\boldsymbol{\beta}}_n^{\mathrm{PT}}$	preliminary test least squares estimator
$\hat{\boldsymbol{\beta}}_n^{\mathrm{S}}$	James–Stein-type least squares estimator
$\hat{\boldsymbol{\beta}}_n^{\mathrm{S+}}$	positive-rule James–Stein-type least squares estimator
$\boldsymbol{\epsilon}$	error term
$\tilde{\boldsymbol{\beta}}_n(k)$	ridge regression estimator
$\hat{\boldsymbol{\beta}}_n(k)$	restricted ridge regression estimator
$\hat{\boldsymbol{\beta}}_n^{\mathrm{PT}}(k)$	preliminary test ridge regression estimator
$\hat{\boldsymbol{\beta}}_n^{\mathrm{S}}(k)$	James–Stein-type ridge regression estimator
$\hat{\boldsymbol{\beta}}_n^{\mathrm{S+}}(k)$	positive-rule James–Stein-type ridge regression estimator
$\hat{\boldsymbol{\beta}}_n^{\mathrm{GRR}}(\boldsymbol{K})$	generalized ridge regression estimator
$\hat{\theta}_n^{\mathrm{Shrinkage}}(c)$	shrinkage estimator of the location parameter
$\hat{\theta}_n^{\mathrm{LASSO}}(c)$	LASSO estimator of the location parameter
$\hat{\theta}_n^{\mathrm{MLASSO}}(c)$	modified LASSO estimator of the location parameter
$\tilde{\boldsymbol{\beta}}_n^{(\mathrm{G})}$	generalized least squares estimator
$\hat{\boldsymbol{\beta}}_n^{(\mathrm{G})}$	restricted generalized least squares estimator
$\hat{\boldsymbol{\beta}}_n^{\mathrm{PT(G)}}$	preliminary test generalized least squares estimator

$\hat{\boldsymbol{\beta}}_n^{\mathrm{S(G)}}$	James–Stein-type generalized least squares estimator
$\hat{\boldsymbol{\beta}}_n^{\mathrm{S(G)+}}$	positive-rule James–Stein-type generalized least squares estimator
$\tilde{\boldsymbol{\beta}}_n^{\mathrm{(R)}}$	R-estimator
$\hat{\boldsymbol{\beta}}_n^{\mathrm{(R)}}$	restricted R-estimator
$\hat{\boldsymbol{\beta}}_n^{\mathrm{PT(R)}}$	preliminary test R-estimator
$\hat{\boldsymbol{\beta}}_n^{\mathrm{S(R)}}$	James–Stein-type R-estimator
$\hat{\boldsymbol{\beta}}_n^{\mathrm{S(R)+}}$	positive-rule James–Stein-type R-estimator
$\tilde{\boldsymbol{\beta}}_n^{\mathrm{(M)}}$	M-estimator
$\hat{\boldsymbol{\beta}}_n^{\mathrm{(M)}}$	restricted M-estimator
$\hat{\boldsymbol{\beta}}_n^{\mathrm{PT(M)}}$	preliminary test M-estimator
$\hat{\boldsymbol{\beta}}_n^{\mathrm{S(M)}}$	James–Stein-type M-estimator
$\hat{\boldsymbol{\beta}}_n^{\mathrm{S(M)+}}$	positive-rule James–Stein-type M-estimator
$b(\cdot)$	bias expression of an estimator
$\mathrm{R}(\cdot)$	L_2-risk function of an estimator
$\Phi(\cdot)$	c.d.f. of a standard normal distribution
$H_m(\cdot;\Delta^2)$	c.d.f of a chi-square distribution with m d.f. and noncentrality parameter Δ^2
$G_{m_1,m_2}(\cdot;\Delta^2)$	c.d.f. of a noncentral F distribution with (m_1,m_2) d.f. and noncentrality parameter $\Delta^2/2$
$I(\cdot)$	indicator function
$\boldsymbol{I}_n$	identity matrix of order n
$\mathcal{N}(\mu,\sigma^2)$	normal distribution with mean μ and variance σ^2
$\mathcal{N}_p(\boldsymbol{\mu},\boldsymbol{\Sigma})$	p-variate normal distribution with mean $\boldsymbol{\mu}$ and covariance $\boldsymbol{\Sigma}^2$
Δ^2	noncentrality parameter
Cov	covariance matrix of an estimator
sgn	sign function

1

Introduction to Ridge Regression

This chapter reviews the developments of ridge regression, starting with the definition of ridge regression together with the covariance matrix. We discuss the multicollinearity problem and ridge notion and present the preliminary test and Stein-type estimators. In addition, we discuss the high-dimensional problem. In conclusion, we include detailed notes, references, and organization of the book.

1.1 Introduction

Consider the common multiple linear regression model with the vector of coefficients, $\boldsymbol{\beta} = (\beta_1, \dots, \beta_p)^\top$ given by

$$Y = X\boldsymbol{\beta} + \boldsymbol{\epsilon}, \tag{1.1}$$

where $Y = (y_1, \dots, y_n)^\top$ is a vector of n responses, $X = (\boldsymbol{x}_1, \dots, \boldsymbol{x}_n)^\top$ is an $n \times p$ design matrix of rank p $(\leq n)$, $\boldsymbol{x}_i \in \mathbb{R}^p$ is the vector of covariates, and $\boldsymbol{\epsilon}$ is an n-vector of independently and identically distributed (i.i.d.) random variables (r.v.).

The least squares estimator (LSE) of $\boldsymbol{\beta}$, denoted by $\tilde{\boldsymbol{\beta}}_n$, can be obtained by minimizing the residual sum of squares (RSS), the convex optimization problem,

$$\min_{\boldsymbol{\beta}}\{(Y - X\boldsymbol{\beta})^\top (Y - X\boldsymbol{\beta})\} = \min_{\boldsymbol{\beta}}\{S(\boldsymbol{\beta})\},$$

where $S(\boldsymbol{\beta}) = Y^\top Y - 2\boldsymbol{\beta}^\top X^\top Y + \boldsymbol{\beta}^\top X^\top X\boldsymbol{\beta}$ is the RSS. Solving

$$\frac{\partial S(\boldsymbol{\beta})}{\partial \boldsymbol{\beta}} = -2X^\top Y + 2X^\top X\boldsymbol{\beta} = \mathbf{0}$$

Theory of Ridge Regression Estimation with Applications, First Edition.
A.K. Md. Ehsanes Saleh, Mohammad Arashi, and B.M. Golam Kibria.

with respect to (w.r.t.) $\boldsymbol{\beta}$ gives

$$\tilde{\boldsymbol{\beta}}_n = (\boldsymbol{X}^\top \boldsymbol{X})^{-1}\boldsymbol{X}^\top \boldsymbol{Y}. \tag{1.2}$$

Suppose that $\mathbb{E}(\boldsymbol{\epsilon}) = 0$ and $\mathbb{E}(\boldsymbol{\epsilon}\boldsymbol{\epsilon}^\top) = \sigma^2 \boldsymbol{I}_n$ for some $\sigma^2 \in \mathbb{R}^+$. Then, the variance–covariance matrix of LSE is given by

$$\mathrm{Var}(\tilde{\boldsymbol{\beta}}_n) = \sigma^2(\boldsymbol{X}^\top \boldsymbol{X})^{-1}. \tag{1.3}$$

Now, we consider the canonical form of the multiple linear regression model to illustrate how large eigenvalues of the design matrix $\boldsymbol{X}^\top \boldsymbol{X}$ may affect the efficiency of estimation.

Write the spectral decomposition of the positive definite design matrix $\boldsymbol{X}^\top \boldsymbol{X}$ to get $\boldsymbol{X}^\top \boldsymbol{X} = \boldsymbol{\Gamma}\boldsymbol{\Lambda}\boldsymbol{\Gamma}^\top$, where $\boldsymbol{\Gamma}(p \times p)$ is a column orthogonal matrix of eigenvectors and $\boldsymbol{\Lambda} = \mathrm{Diag}(\lambda_1, \dots, \lambda_p)$, where $\lambda_j > 0$, $j = 1, \dots, p$ is the ordered eigenvalue matrix corresponding to $\boldsymbol{X}^\top \boldsymbol{X}$. Then,

$$\boldsymbol{Y} = \boldsymbol{T}\boldsymbol{\xi} + \boldsymbol{\epsilon}, \quad \boldsymbol{T} = \boldsymbol{X}\boldsymbol{\Gamma}, \quad \boldsymbol{\xi} = \boldsymbol{\Gamma}^\top \boldsymbol{\beta}. \tag{1.4}$$

The LSE of $\boldsymbol{\xi}$ has the form,

$$\begin{aligned}\tilde{\boldsymbol{\xi}}_n &= (\boldsymbol{T}^\top \boldsymbol{T})^{-1}\boldsymbol{T}^\top \boldsymbol{Y} \\ &= \boldsymbol{\Lambda}^{-1}\boldsymbol{T}^\top \boldsymbol{Y}.\end{aligned} \tag{1.5}$$

The variance–covariance matrix of $\tilde{\boldsymbol{\xi}}_n$ is given by

$$\mathrm{Var}(\tilde{\boldsymbol{\xi}}_n) = \sigma^2 \boldsymbol{\Lambda}^{-1} = \sigma^2 \begin{bmatrix} \frac{1}{\lambda_1} & 0 & \cdots & 0 \\ 0 & \frac{1}{\lambda_2} & 0 & 0 \\ \vdots & \cdots & \ddots & \vdots \\ 0 & 0 & \cdots & \frac{1}{\lambda_p} \end{bmatrix}. \tag{1.6}$$

Summation of the diagonal elements of the variance–covariance matrix of $\tilde{\boldsymbol{\xi}}_n$ is equal to $\mathrm{tr}(\mathrm{Var}(\tilde{\boldsymbol{\xi}}_n)) = \sigma^2 \sum_{j=1}^{p} \lambda_j^{-1}$. Apparently, small eigenvalues inflate the total variance of estimate or energy of $\boldsymbol{X}^\top \boldsymbol{X}$. Specifically, since the eigenvalues are ordered, if the first eigenvalue is small, it causes the variance to explode. If this happens, what must one do? In the following section, we consider this problem. Therefore, it is of interest to realize when the eigenvalues become small.

Before discussing this problem, a very primitive understanding is that if we enlarge the eigenvalues from λ_j to $\lambda_j + k$, for some positive value, say, k, then we can prevent the total variance from exploding. Of course, the amount of recovery depends on the correct choice of the parameter, k.

An artificial remedy is to have $\mathrm{tr}(\mathrm{Var}(\tilde{\xi}_n)) = \sigma^2 \sum_{j=1}^{p} (\lambda_j + k)^{-1}$ based on the variance matrix given by

$$\mathrm{Var}(\tilde{\xi}_n) = \sigma^2 \begin{pmatrix} \frac{1}{\lambda_1 + k} & 0 & \cdots & 0 \\ 0 & \frac{1}{\lambda_2 + k} & 0 & 0 \\ \vdots & \cdots & \ddots & \vdots \\ 0 & 0 & \cdots & \frac{1}{\lambda_p + k} \end{pmatrix} = \sigma^2(\mathbf{\Lambda} + k\boldsymbol{I}_p)^{-1}, \quad k \in \mathbb{R}^+. \tag{1.7}$$

Replacing the eigenvector matrix $\mathbf{\Gamma}$ in (1.5) by this matrix (1.7), we get the $\tilde{\xi}_n = (\mathbf{\Lambda} + k\boldsymbol{I}_p)^{-1}\boldsymbol{T}^\top \boldsymbol{Y}$ and the variance as in (1.8).

$$\begin{aligned} \mathrm{Var}(\tilde{\xi}_n) &= (\mathbf{\Lambda} + k\boldsymbol{I}_p)^{-1}\boldsymbol{T}^\top \mathrm{Var}(\boldsymbol{Y})\boldsymbol{T}(\mathbf{\Lambda} + k\boldsymbol{I}_p)^{-1} \\ &= \sigma^2(\mathbf{\Lambda} + k\boldsymbol{I}_p)^{-1}\mathbf{\Lambda}(\mathbf{\Lambda} + k\boldsymbol{I}_p)^{-1}, \end{aligned} \tag{1.8}$$

which shows

$$\begin{aligned} \mathrm{tr}(\mathrm{Var}(\tilde{\xi}_n)) &= \sigma^2 \mathrm{tr}((\mathbf{\Lambda} + k\boldsymbol{I}_p)^{-1}\mathbf{\Lambda}(\mathbf{\Lambda} + k\boldsymbol{I}_p)^{-1}) \\ &= \sigma^2 \sum_{j=1}^{p} \frac{\lambda_j}{(\lambda_j + k)^2} \end{aligned} \tag{1.9}$$

Further, we show that achieving the total variance of $\sigma^2 \sum_{j=1}^{p} \lambda_j/(\lambda_j + k)^2$ is the target.

1.1.1 Multicollinearity Problem

Multicollinearity or collinearity is the existence of near-linear relationships among the regressors, predictors, or input/exogenous variables. There are terms such as exact, complete and severe, or supercollinearity and moderate collinearity. Supercollinearity indicates that two (or multiple) covariates are linearly dependent, and moderate occurs when covariates are moderately correlated. In the complete collinearity case, the design matrix is not invertible. This case mostly occurs in a high-dimensional situation (e.g. microarray measure) in which the number of covariates (p) exceeds the number of samples (n).

Moderation occurs when the relationship between two variables depends on a third variable, namely, the moderator. This case mostly happens in structural equation modeling. Although moderate multicollinearity does not cause the mathematical problems of complete multicollinearity, it does affect the interpretation of model parameter estimates. According to Montgomery et al. (2012), if there is no linear relationship between the regressors, they are said to be orthogonal.

Multicollinearity or ill-conditioning can create inaccurate estimates of the regression coefficients, inflate the standard errors of the regression coefficients, deflate the partial t-tests for the regression coefficients, give false and nonsignificant p-values, and degrade the predictability of the model. It also causes changes in the direction of signs of the coefficient estimates. According to Montgomery et al. (2012), there are five sources for multicollinearity: (i) data collection, (ii) physical constraints, (iii) overdefined model, (iv) model choice or specification, and (v) outliers.

There are many studies that well explain the problem of multicollinearity. Since theoretical aspects of ridge regression and related issues are our goal, we refer the reader to Montgomery et al. (2012) for illustrative examples and comprehensive study on the multicollinearity and diagnostic measures such as correlation matrix, eigen system analysis of $\boldsymbol{X}^{\mathrm{T}}\boldsymbol{X}$, known as condition number, or variance decomposition proportion and variance inflation factor (VIF). To end this section, we consider a frequently used example in a ridge regression, namely, the Portland cement data introduced by Woods et al. (1932) from Najarian et al. (2013). This data set has been analyzed by many authors, e.g. Kaciranlar et al. (1999), Kibria (2003), and Arashi et al. (2015). We assemble the data as follows:

$$\boldsymbol{X} = \begin{bmatrix} 7 & 26 & 6 & 60 \\ 1 & 29 & 15 & 52 \\ 11 & 56 & 8 & 20 \\ 11 & 31 & 8 & 47 \\ 7 & 52 & 6 & 33 \\ 11 & 55 & 9 & 22 \\ 3 & 71 & 17 & 6 \\ 1 & 31 & 22 & 44 \\ 2 & 54 & 18 & 22 \\ 21 & 47 & 4 & 26 \\ 1 & 40 & 23 & 34 \\ 11 & 66 & 9 & 12 \\ 10 & 68 & 8 & 12 \end{bmatrix}, \quad \boldsymbol{Y} = \begin{bmatrix} 78.5 \\ 74.3 \\ 104.3 \\ 87.6 \\ 95.9 \\ 109.2 \\ 102.7 \\ 72.5 \\ 93.1 \\ 115.9 \\ 83.8 \\ 113.3 \\ 109.4 \end{bmatrix}. \tag{1.10}$$

The following (see Tables 1.1 and 1.2) is the partial output of linear regression fit to this data using the software SPSS, where we selected Enter as the method.

We display the VIF values to diagnose the multicollinearity problem in this data set. Values greater than 10 are a sign of multicollinearity. Many softwares can be used.

Table 1.1 Model fit indices for Portland cement data.

Model	R	R^2	Adjacent R^2	Standard error of the estimate
1	0.991	0.982	0.974	2.446

Table 1.2 Coefficient estimates for Portland cement data.

Coefficient	Standardized coefficient	t	Significance	VIF
(Constant)		0.891	0.399	
x_1	0.607	2.083	0.071	38.496
x_2	0.528	0.705	0.501	254.423
x_3	0.043	0.135	0.896	46.868
x_4	−0.160	−0.203	0.844	282.513

1.2 Ridge Regression Estimator: Ridge Notion

If the regression coefficients β_js are unconstrained, then they can explode (become large); this results in high variance. Hence, in order to control the variance, one may regularize the regression coefficients and determine how large the coefficient grows. In other words, one may impose a constraint on them so as not to get unboundedly large or penalized large regression coefficients. One type of constraint is the ridge constraint given by $\sum_{j=1}^{p} \beta_j^2 \leq t$ for some positive value t. Hence, the minimization of the penalized residual sum of squares (PRSS) is equivalent to solving the following convex optimization problem,

$$\min_{\beta}\{(\boldsymbol{Y}-\boldsymbol{X\beta})^\top(\boldsymbol{Y}-\boldsymbol{X\beta})\} \quad \text{such that} \quad \sum_{j=1}^{p} \beta_j^2 \leq t, \tag{1.11}$$

for some positive value t.

In general, the PRSS is defined by

$$(\boldsymbol{Y}-\boldsymbol{X\beta})^\top(\boldsymbol{Y}-\boldsymbol{X\beta}) + k\|\boldsymbol{\beta}\|^2, \quad \|\boldsymbol{\beta}\|^2 = \sum_{j=1}^{p} \beta_j^2. \tag{1.12}$$

Since the PRSS is a convex function w.r.t. $\boldsymbol{\beta}$, it has a unique solution. Because of the ridge constraint, the solution is termed as the ridge regression estimator (RRE).

To derive the RRE, we solve the following convex optimization problem

$$\min_{\beta}\{(Y - X\beta)^\top(Y - X\beta) + k\|\beta\|^2\} = \min_{\beta}\{\mathrm{PS}(\beta)\}, \tag{1.13}$$

where $\mathrm{PS}(\beta) = Y^\top Y - 2\beta^\top X^\top Y + \beta^\top X^\top X\beta + k\beta^\top\beta$ is the PRSS. Solving

$$\frac{\partial \mathrm{PS}(\beta)}{\partial \beta} = -2X^\top Y + 2X^\top X\beta + 2k\beta = \mathbf{0}$$

w.r.t. β gives the RRE,

$$\hat{\beta}_n^{\mathrm{RR}}(k) = (X^\top X + kI_p)^{-1}X^\top Y. \tag{1.14}$$

Here, k is the shrinkage (tuning) parameter. Indeed, k tunes (controls) the size of the coefficients, and hence regularizes them. As $k \to 0$, the RRE simplifies to the LSE. Also, as $k \to \infty$, the RREs approach zero. Hence, the optimal shrinkage parameter k is of interest.

One must note that solving the optimization problem (1.13) is not the only way of yielding the RRE. It can also be obtained by solving a RSS of another data, say augmented data. To be specific, consider the following augmentation approach. Let

$$X^* = \begin{bmatrix} X \\ \sqrt{k}I_p \end{bmatrix}, \quad Y^* = \begin{bmatrix} Y \\ \mathbf{0} \end{bmatrix}. \tag{1.15}$$

Assume the following multiple linear model,

$$Y^* = X^*\beta + \epsilon^*, \tag{1.16}$$

where ϵ^* is an $(n + p)$-vector of i.i.d. random variables. Then, the LSE of β is obtained as

$$\begin{aligned}\beta_n^* &= \min_{\beta}\{(Y^* - X^*\beta)^\top(Y^* - X^*\beta)\} \\ &= (X^{*\top}X^*)^{-1}X^{*\top}Y^* \\ &= (X^\top X + kI_p)^{-1}X^\top Y \\ &= \hat{\beta}_n^{\mathrm{RR}}(k). \end{aligned} \tag{1.17}$$

Thus, the LSE of the augmented data is indeed the RRE of the normal data.

1.3 LSE vs. RRE

Indeed, the RRE is proportional to the LSE. Under the orthonormal case, i.e. $X^\top X = I_p$, the RRE simplifies to

$$\begin{aligned}\hat{\beta}_n^{\mathrm{RR}}(k) &= (X^\top X + kI_p)^{-1}X^\top Y \\ &= (1 + k)^{-1}I_p X^\top Y \\ &= (1 + k)^{-1}(X^\top X)^{-1}X^\top Y \\ &= \frac{1}{1 + k}\tilde{\beta}_n. \end{aligned} \tag{1.18}$$

Theorem 1.1 *The ridge penalty shrinks the eigenvalues of the design matrix.*

Proof: Write the singular value decomposition (SVD) of the design matrix $\boldsymbol{X}$ to get $\boldsymbol{X} = \boldsymbol{U}\boldsymbol{\Lambda}\boldsymbol{V}^\top$, where $\boldsymbol{U}$ $(n \times p)$ and $\boldsymbol{V}$ $(p \times p)$ are column orthogonal matrices, and $\boldsymbol{\Lambda} = \mathrm{Diag}(\lambda_1, \dots, \lambda_p)$ $\lambda_j > 0, j = 1, \dots, p$ is the eigenvalue matrix corresponding to $\boldsymbol{X}^\top\boldsymbol{X}$. Then,

$$\begin{aligned}
\tilde{\boldsymbol{\beta}}_n &= (\boldsymbol{X}^\top\boldsymbol{X})^{-1}\boldsymbol{X}^\top\boldsymbol{Y} \\
&= (\boldsymbol{V}\boldsymbol{\Lambda}\boldsymbol{U}^\top\boldsymbol{U}\boldsymbol{\Lambda}\boldsymbol{V}^\top)^{-1}\boldsymbol{V}\boldsymbol{\Lambda}\boldsymbol{U}^\top\boldsymbol{Y} \\
&= (\boldsymbol{V}\boldsymbol{\Lambda}^2\boldsymbol{V}^\top)^{-1}\boldsymbol{V}\boldsymbol{\Lambda}\boldsymbol{U}^\top\boldsymbol{Y} \\
&= \boldsymbol{V}\boldsymbol{\Lambda}^{-2}\boldsymbol{V}^\top\boldsymbol{V}\boldsymbol{\Lambda}\boldsymbol{U}^\top\boldsymbol{Y} \\
&= \boldsymbol{V}\{(\boldsymbol{\Lambda}^2)^{-1}\boldsymbol{\Lambda}\}\boldsymbol{U}^\top\boldsymbol{Y}. \qquad (1.19)
\end{aligned}$$

In a similar manner, one obtains

$$\begin{aligned}
\hat{\boldsymbol{\beta}}_n^{\mathrm{RR}}(k) &= (\boldsymbol{X}^\top\boldsymbol{X} + k\boldsymbol{I}_p)^{-1}\boldsymbol{X}^\top\boldsymbol{Y} \\
&= (\boldsymbol{V}\boldsymbol{\Lambda}\boldsymbol{U}^\top\boldsymbol{U}\boldsymbol{\Lambda}\boldsymbol{V}^\top + k\boldsymbol{I}_p)^{-1}\boldsymbol{V}\boldsymbol{\Lambda}\boldsymbol{U}^\top\boldsymbol{Y} \\
&= (\boldsymbol{V}\boldsymbol{\Lambda}^2\boldsymbol{V}^\top + k\boldsymbol{V}\boldsymbol{V}^\top)^{-1}\boldsymbol{V}\boldsymbol{\Lambda}\boldsymbol{U}^\top\boldsymbol{Y} \\
&= \boldsymbol{V}(\boldsymbol{\Lambda}^2 + k\boldsymbol{I}_p)^{-1}\boldsymbol{V}^\top\boldsymbol{V}\boldsymbol{\Lambda}\boldsymbol{U}^\top\boldsymbol{Y} \\
&= \boldsymbol{V}\{(\boldsymbol{\Lambda}^2 + k\boldsymbol{I}_p)^{-1}\boldsymbol{\Lambda}\}\boldsymbol{U}^\top\boldsymbol{Y}. \qquad (1.20)
\end{aligned}$$

The difference between the two estimators w.r.t. the SVD is in the bracket terms. Let

$$\boldsymbol{\Lambda}^{-2}\boldsymbol{\Lambda} = \boldsymbol{\Lambda}^{-1} = \mathrm{Diag}\left(\frac{1}{\lambda_1}, \dots, \frac{1}{\lambda_p}\right),$$

$$(\boldsymbol{\Lambda}^2 + k\boldsymbol{I}_p)^{-1}\boldsymbol{\Lambda} = \mathrm{Diag}\left(\frac{\lambda_1}{\lambda_1^2 + k}, \dots, \frac{\lambda_p}{\lambda_p^2 + k}\right).$$

Since $\lambda_j/(\lambda_j^2 + k) \le \lambda_j^{-1}$, the ridge penalty shrinks the λ_js. □

1.4 Estimation of Ridge Parameter

We can observe from Eq. (1.18) that the RRE heavily depends on the ridge parameter k. Many authors at different times worked in this area of research and developed and proposed different estimators for k. They considered various models such as linear regression, Poisson regression, and logistic regression models. To mention a few, Hoerl and Kennard (1970), Hoerl et al. (1975), McDonald and Galarneau (1975), Lawless and Wang (1976), Dempster et al. (1977), Gibbons (1981), Kibria (2003), Khalaf and Shukur (2005), Alkhamisi and Shukur (2008), Muniz and Kibria (2009), Gruber et al. (2010), Muniz et al. (2012), Mansson et al. (2010), Hefnawy and Farag (2013), Aslam (2014), and Arashi and Valizadeh (2015), and Kibria and Banik (2016), among others.

1.5 Preliminary Test and Stein-Type Ridge Estimators

In previous sections, we discussed the notion of RRE and how it shrinks the elements of the ordinary LSE. Sometimes, it is needed to shrink the LSE to a subspace defined by $\boldsymbol{H\beta} = \boldsymbol{h}$, where $\boldsymbol{H}$ is a $q \times p$ known matrix of full row rank q $(q \leq p)$ and $\boldsymbol{h}$ is a q vector of known constants. It is also termed as constraint or restriction. Such a configuration of the subspace is frequently used in the design of experiments, known as contrasts. Therefore, sometimes shrinking is for two purposes. We refer to this as *double shrinking*.

In general, unlike the Bayesian paradigm, correctness of the prior information $\boldsymbol{H\beta} = \boldsymbol{h}$ can be tested on the basis of samples through testing $\mathcal{H}_o : \boldsymbol{H\beta} = \boldsymbol{h}$ vs. a set of alternatives. Following Fisher's recipe, we use the non-sample information $\boldsymbol{H\beta} = \boldsymbol{h}$; if based on the given sample, we accept $\mathcal{H}_o$. In situations where this prior information is correct, an efficient estimator is the one which satisfies this restriction, called the *restricted estimator*.

To derive the restricted estimator under a multicollinear situation, satisfying the condition $\boldsymbol{H\beta} = \boldsymbol{h}$, one solves the following convex optimization problem,

$$\min_{\boldsymbol{\beta}}\{\text{PS}_{\lambda}(\boldsymbol{\beta})\}, \quad \text{PS}_{\lambda}(\boldsymbol{\beta}) = (\boldsymbol{Y} - \boldsymbol{X\beta})^{\top}(\boldsymbol{Y} - \boldsymbol{X\beta}) + k\|\boldsymbol{\beta}\|^2 + \boldsymbol{\lambda}^{\top}(\boldsymbol{H\beta} - \boldsymbol{h}), \tag{1.21}$$

where $\boldsymbol{\lambda} = (\lambda_1, \ldots, \lambda_q)^{\top}$ is the vector of Lagrangian multipliers. Grob (2003) proposed the restricted RRE, under a multicollinear situation, by correcting the restricted RRE of Sarkar (1992).

In our case, we consider prior information with the form $\boldsymbol{\beta} = \boldsymbol{0}$, which is a test used for checking goodness of fit. Here, the restricted RRE is simply given by $\hat{\boldsymbol{\beta}}_n(k) = \boldsymbol{0}$, where $\boldsymbol{0}$ is the restricted estimator of $\boldsymbol{\beta}$. Therefore, one uses $\hat{\boldsymbol{\beta}}_n^{\text{RR}}(k)$ if $\mathcal{H}_o : \boldsymbol{\beta} = \boldsymbol{0}$ rejects and $\hat{\boldsymbol{\beta}}_n^{\text{RR(R)}}(k)$ if $\mathcal{H}_o : \boldsymbol{\beta} = \boldsymbol{0}$ accepts. Combining the information existing in both estimators, one may follow an approach by Bancroft (1964), to propose the preliminary test RRE given by

$$\begin{aligned}
\hat{\boldsymbol{\beta}}_n^{\text{RR(PT)}}(k, \alpha) &= \begin{cases} \hat{\boldsymbol{\beta}}_n^{\text{RR}}(k); & \mathcal{H}_o \text{ is rejected} \\ \hat{\boldsymbol{\beta}}_n^{\text{RR(R)}}(k); & \mathcal{H}_o \text{ is accepted} \end{cases} \\
&= \hat{\boldsymbol{\beta}}_n^{\text{RR}}(k) I(\mathcal{H}_o \text{ is rejected}) + \hat{\boldsymbol{\beta}}_n^{\text{RR(R)}}(k) I(\mathcal{H}_o \text{ is accepted}) \\
&= \hat{\boldsymbol{\beta}}_n^{\text{RR}}(k) I(\mathcal{L}_n > F_{p,m}(\alpha)) + \hat{\boldsymbol{\beta}}_n^{\text{RR(R)}}(k) I(\mathcal{L}_n \leq F_{p,m}(\alpha)) \\
&= \hat{\boldsymbol{\beta}}_n^{\text{RR}}(k) - \hat{\boldsymbol{\beta}}_n^{\text{RR}}(k) I(\mathcal{L}_n \leq F_{p,m}(\alpha)),
\end{aligned} \tag{1.22}$$

where $I(A)$ is the indicator function of the set A and $\mathcal{L}_n$ is the test statistic for testing $\mathcal{H}_o : \boldsymbol{\beta} = \boldsymbol{0}$, and $F_{q,m}(\alpha)$ is the upper α-level critical value from the F-distribution with (q, m) degrees of freedom (D.F.) See Judge and Bock (1978) and Saleh (2006) for the test statistic and details.

After some algebra, it can be shown that

$$\hat{\boldsymbol{\beta}}_n^{\mathrm{RR(R)}}(k) = \boldsymbol{R}(k)\hat{\boldsymbol{\beta}}_n, \quad \hat{\boldsymbol{\beta}}_n = \mathbf{0}, \tag{1.23}$$

where

$$\boldsymbol{R}(k) = (\boldsymbol{I}_p + k(\boldsymbol{X}^\top\boldsymbol{X})^{-1})^{-1}.$$

Then, it is easy to show that

$$\hat{\boldsymbol{\beta}}_n^{\mathrm{RR(PT)}}(k,\alpha) = \boldsymbol{R}(k)\hat{\boldsymbol{\beta}}_n^{\mathrm{PT}}, \quad \hat{\boldsymbol{\beta}}_n^{\mathrm{PT}} = \tilde{\boldsymbol{\beta}}_n - \tilde{\boldsymbol{\beta}}_n I(\mathcal{L}_n \leq F_{p,m}(\alpha)). \tag{1.24}$$

The preliminary test RRE is discrete in nature and heavily dependent on α, the level of significance. Hence, a continuous α-free estimator is desired. Following James and Stein (1961), the Stein-type RRE is given by

$$\hat{\boldsymbol{\beta}}_n^{\mathrm{RR(S)}}(k) = \boldsymbol{R}(k)\hat{\boldsymbol{\beta}}_n^{\mathrm{S}}; \quad \hat{\boldsymbol{\beta}}_n^{\mathrm{S}} = \tilde{\boldsymbol{\beta}}_n - d\mathcal{L}_n^{-1}\tilde{\boldsymbol{\beta}}_n, \tag{1.25}$$

for some $d > 0$. It is shown in Saleh (2006) that the optimal choice of d is equal to $m(p+2)/p(m+2)$, $m = n - p$.

Note that if $\mathcal{L}_n \to \infty$, then $\hat{\boldsymbol{\beta}}_n^{\mathrm{RR(S)}}(k) = \hat{\boldsymbol{\beta}}_n^{\mathrm{RR}}(k)$, which matches with the estimator $\hat{\boldsymbol{\beta}}_n^{\mathrm{RR(PT)}}(k)$. However, if $\mathcal{L}_n \to 0$, then $\hat{\boldsymbol{\beta}}_n^{\mathrm{RR(S)}}(k)$ becomes a negative estimator. In order to obtain the true value, one has to restrict ($\mathcal{L}_n > d$). Hence, one may define the positive-rule Stein-type RRE given by

$$\hat{\boldsymbol{\beta}}_n^{\mathrm{RR(S+)}}(k) = \boldsymbol{R}(k)\hat{\boldsymbol{\beta}}_n^{\mathrm{S+}}; \quad \hat{\boldsymbol{\beta}}_n^{\mathrm{S+}} = \hat{\boldsymbol{\beta}}_n^{\mathrm{S}} - (1 - d\mathcal{L}_n^{-1})I(\mathcal{L}_n \leq d)\tilde{\boldsymbol{\beta}}_n. \tag{1.26}$$

Although these shrinkage RREs are biased, they outperform the RRE, $\hat{\boldsymbol{\beta}}_n^{\mathrm{RR}}(k)$ in the mean squared error (MSE) sense. One must note that this superiority is not uniform over the parameter space $\mathbb{R}^p$. According to how much $\boldsymbol{\beta}$ deviates from the origin, the superiority changes. We refer to Saleh (2006) for an extensive overview on this topic and statistical properties of the shrinkage estimators. This type of estimator will be studied in more detail and compared with the RRE in the following chapters.

1.6 High-Dimensional Setting

In high-dimensional analysis, the number of variables p is greater than the number of observations, n. In such situations, $\boldsymbol{X}^\top\boldsymbol{X}$ is not invertible and, hence, to derive the LSE of $\boldsymbol{\beta}$, one may use the generalized inverse of $\boldsymbol{X}^\top\boldsymbol{X}$, which does not give a unique solution. However, the RRE can be obtained, regardless of the relationship between p and n. In this section, we briefly mention some related endeavors in manipulating the RRE to adapt with high-dimensional setting. However, in Chapter 11, we discuss the growing dimension, i.e. $p \to \infty$ when the sample size is fixed.

Wang et al. (2016) used this fact and proposed a new approach in high-dimensional regression by considering

$$(\boldsymbol{X}^\top\boldsymbol{X})^{-1}\boldsymbol{X}^\top\boldsymbol{Y} = \lim_{k\to 0}(\boldsymbol{X}^\top\boldsymbol{X} + k\boldsymbol{I}_p)^{-1}\boldsymbol{X}^\top\boldsymbol{Y}. \tag{1.27}$$

They used an orthogonal projection of $\boldsymbol{\beta}$ onto the row space of $\boldsymbol{X}$ and proposed the following high-dimensional version of the LSE for dimension reduction:

$$\hat{\boldsymbol{\beta}}_n^{\text{HD}} = \lim_{k\to 0}\boldsymbol{X}^\top(\boldsymbol{X}\boldsymbol{X}^\top + k\boldsymbol{I}_n)^{-1}\boldsymbol{Y} = \boldsymbol{X}^\top(\boldsymbol{X}\boldsymbol{X}^\top)^{-1}\boldsymbol{Y}. \tag{1.28}$$

They applied the following identity to obtain the estimator:

$$(\boldsymbol{X}^\top\boldsymbol{X} + k\boldsymbol{I}_p)^{-1}\boldsymbol{X}^\top\boldsymbol{Y} = \boldsymbol{X}^\top(\boldsymbol{X}\boldsymbol{X}^\top + k\boldsymbol{I}_n)^{-1}\boldsymbol{Y} \tag{1.29}$$

for every $n, p, k > 0$. Buhlmann et al. (2014) also used the projection of $\boldsymbol{\beta}$ onto the row space of $\boldsymbol{X}$ and developed a bias correction in the RRE to propose a bias-corrected RRE for the high-dimensional setting. Shao and Deng (2012) considered the RRE of the projection vector and proposed to threshold the RRE when the projection vector is sparse, in the sense that many of its components are small.

Specifically, write the SVD of $\boldsymbol{X}$ as $\boldsymbol{X} = \boldsymbol{U}\boldsymbol{\Lambda}\boldsymbol{V}^\top$ where $\boldsymbol{U}(n\times p)$ and $\boldsymbol{V}(p\times p)$ are column orthogonal matrices, and let $\boldsymbol{\Lambda} = \text{Diag}(\lambda_1, \dots, \lambda_p)$, $\lambda_j > 0$, $j = 1, \dots, p$ be the eigenvalue matrix corresponding to $\boldsymbol{X}^\top\boldsymbol{X}$. Let

$$\boldsymbol{\theta} = \boldsymbol{X}^\top(\boldsymbol{X}\boldsymbol{X}^\top)^{-}\boldsymbol{X}\boldsymbol{\beta} = \boldsymbol{V}\boldsymbol{V}^\top\boldsymbol{\beta}, \quad (\boldsymbol{X}\boldsymbol{X}^\top)^{-} = \boldsymbol{U}\boldsymbol{\Lambda}^{-2}\boldsymbol{U}^\top, \tag{1.30}$$

where $\boldsymbol{\theta}$ is the projection of $\boldsymbol{\beta}$ onto the row space of $\boldsymbol{X}$.

Since $\boldsymbol{X}\boldsymbol{X}^\top(\boldsymbol{X}\boldsymbol{X}^\top)^{-}\boldsymbol{X} = \boldsymbol{X}$, we have $\boldsymbol{X}\boldsymbol{\theta} = \boldsymbol{X}\boldsymbol{\beta}$. Obviously, $\boldsymbol{X}\boldsymbol{\theta} = \boldsymbol{X}\boldsymbol{\beta}$ yields $\boldsymbol{Y} = \boldsymbol{X}\boldsymbol{\theta} + \boldsymbol{\epsilon}$ as the underlying model instead of (1.1). Then, the RRE of the projection is simply

$$\hat{\boldsymbol{\theta}}_n^{\text{RR}}(k_n) = (\boldsymbol{X}^\top\boldsymbol{X} + k_n\boldsymbol{I}_p)^{-1}\boldsymbol{X}^\top\boldsymbol{Y}, \tag{1.31}$$

where $k_n > 0$ is an appropriately chosen regularization parameter.

Shao and Deng (2012) proved that the RRE $\hat{\boldsymbol{\theta}}_n(k_n)$ is consistent for the estimation of any linear combination of $\boldsymbol{\theta}$; however, it is not L_2 consistent, i.e. $n^{-1}\mathbb{E}\|\boldsymbol{X}\hat{\boldsymbol{\theta}}_n(k_n) - \boldsymbol{X}\boldsymbol{\theta}\|^2$ may not converge to zero. Then, they proposed an L_2-consistent RRE by thresholding.

Another approach to deal with high-dimensional data in the case $p > n$ is to partition the vector of regression parameters to main and nuisance effects, where we have p_1 (say) less than n, main and active coefficients that must be estimated and $p_2 = p - p_1$, redundant coefficients that can be eliminated from the analysis by a variable selection method. In fact, we assume some level of sparsity in our inference. Hence, we partition $\boldsymbol{\beta} = (\boldsymbol{\beta}_1^\top, \boldsymbol{\beta}_2^\top)^\top$ in such a way in which $\boldsymbol{\beta}_1$ is a p_1 vector of main effects and $\boldsymbol{\beta}_2$ is a p_2 vector of nuisance effects. We further assume that $p_1 < n$. Accordingly, we have the partition $\boldsymbol{X} = (\boldsymbol{X}_1, \boldsymbol{X}_2)$ for the design matrix. Then, the multiple linear model (1.1) is rewritten as

$$\boldsymbol{Y} = \boldsymbol{X}_1\boldsymbol{\beta}_1 + \boldsymbol{X}_2\boldsymbol{\beta}_2 + \boldsymbol{\epsilon}. \tag{1.32}$$

This model may be termed as the full model. As soon as a variable selection method is used and p_1 main effects are selected, the sub model has the form

$$Y = X_1\beta_1 + \epsilon. \tag{1.33}$$

Then, the interest is to estimate the main effects, β_1 (this technique is discussed in more detail in Chapter 11). Therefore, one needs a variable selection method. Unlike the RRE that only shrinks the coefficients, there are many estimators which simultaneously shrink and select variables in large p and small n problems. The most well-known one is the least absolute shrinkage and selection operator (LASSO) proposed by Tibshirani (1996). He suggested using an absolute-type constraint with form $\sum_{j=1}^{p} |\beta_j| \leq t$ for some positive values t in the minimization of the PRSS rather than the ridge constraint. Specifically, he defined the PRSS by

$$(Y - X\beta)^{\top}(Y - X\beta) + k\|\beta\|^{\top}\mathbf{1}_p, \quad \|\beta\|^{\top}\mathbf{1}_p = \sum_{j=1}^{p} |\beta_j|, \tag{1.34}$$

where $\|\beta\| = (|\beta_1|, \ldots, |\beta_p|)^{\top}$ and $\mathbf{1}_p$ is an n-tuple of 1's. The LASSO estimator, denoted by $\hat{\beta}_n^{\text{LASSO}}$, is the solution to the following convex optimization problem,

$$\min_{\beta} (Y - X\beta)^{\top}(Y - X\beta) + k\|\beta\|^{\top}\mathbf{1}_p. \tag{1.35}$$

To see some recent related endeavors in the context of ridge regression, we refer to Yuzbasi and Ahmed (2015) and Aydin et al. (2016) among others.

In the following section, some of the most recent important references about ridge regression and related topics are listed for more studies.

1.7 Notes and References

The first paper on ridge analysis was by Hoerl (1962); however, the first paper on multicollinearity appeared five years later, roughly speaking, by Farrar and Glauber (1967). Marquardt and Snee (1975) reviewed the theory of ridge regression and its relation to generalized inverse regression. Their study includes several illustrative examples about ridge regression. For the geometry of multicollinearity, see Akdeniz and Ozturk (1981). We also suggest that Gunst (1983) and and Sakallioglu and Akdeniz (1998) not be missed. Gruber (1998) in his monograph motivates the need for using ridge regression and allocated a large portion to the analysis of ridge regression and its generalizations. For historical survey up to 1998, we refer to Gruber (1998).

Beginning from 2000, a comprehensive study in ridge regression is the work of Ozturk and Akdeniz (2000), where the authors provide some solutions for ill-posed inverse problems. Wan (2002) incorporated measure of goodness of fit in evaluating the RRE and proposed a feasible generalized RRE. Kibria

(2003) gave a comprehensive analysis about the estimation of ridge parameter k for the linear regression model. For application of ridge regression in agriculture, see Jamal and Rind (2007). Maronna (2011) proposed an RRE based on repeated M-estimation in robust regression. Saleh et al. (2014) extensively studied the performance of preliminary test and Stein-type ridge estimators in the multivariate-t regression model. Huang et al. (2016) defined a weighted VIF for collinearity diagnostic in generalized linear models.

Arashi et al. (2017) studied the performance of several ridge parameter estimators in a restricted ridge regression model with stochastic constraints. Asar et al. (2017) defined a restricted RRE in the logistic regression model and derived its statistical properties. Roozbeh and Arashi (2016a) developed a new ridge estimator in partial linear models. Roozbeh and Arashi (2016b) used difference methodology to study the performance of an RRE in a partial linear model. Arashi and Valizadeh (2015) compared several estimators for estimating the biasing parameter in the study of partial linear models in the presence of multicollinearity. Roozbeh and Arashi (2013) proposed a feasible RRE in partial linear models and studied its properties in details. Roozbeh et al. (2012) developed RREs in seemingly partial linear models. Recently, Chandrasekhar et al. (2016) proposed the concept of partial ridge regression, which involves selectively adjusting the ridge constants associated with highly collinear variables to control instability in the variances of coefficient estimates. Norouzirad and Arashi (2017) developed shrinkage ridge estimators in the context of robust regression. Fallah et al. (2017) studied the asymptotic performance of a general form of shrinkage ridge estimator. Recently, Norouzirad et al. (2017) proposed improved robust ridge M-estimators and studied their asymptotic behavior.

1.8 Organization of the Book

This book has 12 chapters including this one. In this light we consider the chapter with location and simple linear model first. Chapter 1 presents an introduction to ridge regression and different aspects of it, stressing the multicollinearity problem and its application to high-dimensional problems. Chapter 2 considers simple linear and location models, and provides theoretical developments. Chapters 3 and 4 deal with the analysis of variance (ANOVA) model and seemingly unrelated simple linear models, respectively. Chapter 5 considers ridge regression and LASSO for multiple regression together with preliminary test and Stein-type estimator and a comparison thereof when the design matrix is non-orthogonal. Chapter 6 considers the RRE and its relation to LASSO. Further, we study in detail the properties of the preliminary test and Stein-type estimator with low dimension. In Chapter 7, we cover the partially linear model and the properties of LASSO, ridge, preliminary test, and the Stein-type estimators. Chapter 8 contains the discussion on the logistic regression model and the related estimators of diverse kinds as described in earlier chapters. Chapter 9 discusses the multiple regression model with autoregressive errors. In Chapter 10, we provide a comparative study of LASSO, ridge, preliminary test, and Stein-type estimators using rank-based theory. In Chapter 11, we discuss the estimation of parameters of a regression model with high dimensions. Finally, we conclude the book with Chapter 12 to illustrate recent applications of ridge, LASSO, and logistic regression to neural networks and big data analysis.

Problems

1.1 Derive the estimates given in (1.12) and (1.13).

1.2 Find the optimum value $k > 0$ for which the MSE of RRE in (1.18) becomes the smallest.

1.3 Derive the bias and MSE functions for the shrinkage RREs given by (1.24)–(1.26).

1.4 Verify that

$$\hat{\boldsymbol{\beta}}_n^{\mathrm{RR}}(k) = (\boldsymbol{I}_p + k(\boldsymbol{X}^\top\boldsymbol{X})^{-1})^{-1}\tilde{\boldsymbol{\beta}}_n.$$

1.5 Verify the identity (1.29).

1.6 Find the solution of the optimization problem (1.35) for the orthogonal case $\boldsymbol{X}^\top\boldsymbol{X} = n\boldsymbol{I}_p$.

2

Location and Simple Linear Models

In statistical literature, various methods of estimation of the parameters of a given model are available, primarily based on the least squares estimator (LSE) and maximum likelihood estimator (MLE) principle. However, when uncertain prior information of the parameters is known, the estimation technique changes. The methods of circumvention, including the uncertain prior information, are of immense importance in the current statistical literature. In the area of classical approach, preliminary test (PT) and the Stein-type (S) estimation methods dominate the modern statistical literature, side by side with the Bayesian methods.

In this chapter, we consider the simple linear model and the estimation of the parameters of the model along with their shrinkage version and study their properties when the errors are normally distributed.

2.1 Introduction

Consider the simple linear model with slope β and intercept θ, given by

$$Y = \theta \mathbf{1}_n + \beta \boldsymbol{x} + \epsilon. \tag{2.1}$$

If $\boldsymbol{x} = (x_1, \dots, x_n)^\top = (0, \dots, 0)^\top$, the model (2.1) reduces to

$$Y = \theta \mathbf{1}_n + \epsilon, \tag{2.2}$$

where θ is the location parameter of a distribution.

In the following sections, we consider the estimation and test of the location model, i.e. the model of (2.2), followed by the estimation and test of the simple linear model.

Theory of Ridge Regression Estimation with Applications, First Edition.
A.K. Md. Ehsanes Saleh, Mohammad Arashi, and B.M. Golam Kibria.

2.2 Location Model

In this section, we introduce two basic penalty estimators, namely, the *ridge regression* estimator (RRE) and the *least absolute shrinkage and selection operator* (LASSO) estimator for the location parameter of a distribution. The penalty estimators have become viral in statistical literature. The subject evolved as the solution to *ill-posed* problems raised by Tikhonov (1963) in mathematics. In 1970, Hoerl and Kennard applied the Tikhonov method of solution to obtain the *RRE* for linear models. Further, we compare the estimators with the LSE in terms of L_2-risk function.

2.2.1 Location Model: Estimation

Consider the simple location model,

$$Y = \theta \mathbf{1}_n + \epsilon, \quad n > 1, \tag{2.3}$$

where $Y = (Y_1, \dots, Y_n)^\top$, $\mathbf{1}_n = (1, \dots, 1)^\top$- n-tuple of 1's, and $\epsilon = (\epsilon_1, \dots, \epsilon_n)^\top$ n-vector of i.i.d. random errors such that $\mathbb{E}(\epsilon) = \mathbf{0}$ and $\mathbb{E}(\epsilon\epsilon^\top) = \sigma^2 I_n$, I_n is the identity matrix of rank n ($n \geq 2$), θ is the location parameter, and, in this case, σ^2 may be unknown.

The LSE of θ is obtained by

$$\min_\theta \{(Y - \theta\mathbf{1}_n)^\top (Y - \theta\mathbf{1}_n)\} = \tilde{\theta}_n = \frac{1}{n}\mathbf{1}_n^\top Y = \bar{y}. \tag{2.4}$$

Alternatively, it is possible to minimize the log-likelihood function when the errors are normally distributed:

$$l(\theta) = -n \log \sigma + \frac{n}{2} \log(2\pi) - \frac{1}{2}(Y - \theta\mathbf{1}_n)^\top (Y - \theta\mathbf{1}_n),$$

giving the same solution (2.4) as in the case of LSE. It is known that the $\tilde{\theta}_n$ is unbiased, i.e. $\mathbb{E}(\tilde{\theta}_n) = \theta$ and the variance of $\tilde{\theta}_n$ is given by

$$Var(\tilde{\theta}_n) = \frac{\sigma^2}{n}.$$

The unbiased estimator of σ^2 is given by

$$s_n^2 = (n-1)^{-1}(Y - \tilde{\theta}_n\mathbf{1}_n)^\top (Y - \tilde{\theta}_n\mathbf{1}_n). \tag{2.5}$$

The mean squared error (MSE) of θ_n^*, any estimator of θ, is defined as

$$\text{MSE}(\theta_n^*) = \mathbb{E}(\theta_n^* - \theta)^2.$$

Test for $\theta = 0$ when σ^2 is known:

For the test of null-hypothesis $\mathcal{H}_o : \theta = 0$ vs. $\mathcal{H}_A : \theta \neq 0$, we use the test statistic

$$Z_n = \frac{\sqrt{n}\tilde{\theta}_n}{\sigma}. \tag{2.6}$$

Under the assumption of normality of the errors, $Z_n \sim \mathcal{N}(\Delta, 1)$, where $\Delta = \frac{\sqrt{n}\theta}{\sigma}$. Hence, we reject $\mathcal{H}_o$ whenever $|Z_n|$ exceeds the threshold value from the null distribution. An interesting threshold value is $\sqrt{2\log 2}$.

For large samples, when the distribution of errors has zero mean and finite variance σ^2, under a sequence of local alternatives,

$$\mathcal{K}_{(n)} : \theta_{(n)} = n^{-\frac{1}{2}}\delta, \quad \delta \neq 0, \tag{2.7}$$

and assuming $\mathbb{E}(\epsilon_j) = 0$ and $\mathbb{E}(\epsilon_j^2) = \sigma^2\ (< \infty), j = 1, \ldots, n$, the asymptotic distribution of $\sqrt{n}\tilde{\theta}_n/s_n$ is $\mathcal{N}(\Delta, 1)$. Then the test procedure remains the same as before.

2.2.2 Shrinkage Estimation of Location

In this section, we consider a shrinkage estimator of the location parameter θ of the form

$$\hat{\theta}_n^{\text{Shrinkage}}(c) = c\,\tilde{\theta}_n, \quad 0 \leq c \leq 1, \tag{2.8}$$

where $\tilde{\theta}_n \sim \mathcal{N}(\theta, \sigma^2/n)$. The bias and the MSE of $\hat{\theta}_n^{\text{Shrinkage}}(c)$ are given by

$$\begin{aligned} \mathrm{b}(\hat{\theta}_n^{\text{Shrinkage}}(c)) &= \mathbb{E}[c\tilde{\theta}_n] - \theta = c\theta - \theta = -(1-c)\theta \\ \mathrm{MSE}(\hat{\theta}_n^{\text{Shrinkage}}(c)) &= c^2\frac{\sigma^2}{n} + (1-c)^2\theta^2. \end{aligned} \tag{2.9}$$

Minimizing $\mathrm{MSE}(\hat{\theta}_n^{\text{Shrinkage}}(c))$ w.r.t. c, we obtain

$$c^* = \frac{\Delta^2}{1+\Delta^2}, \qquad \Delta^2 = \frac{n\theta^2}{\sigma^2}. \tag{2.10}$$

So that

$$\mathrm{MSE}(\hat{\theta}_n^{\text{Shrinkage}}(c^*)) = \frac{\sigma^2}{n}\frac{\Delta^2}{1+\Delta^2}. \tag{2.11}$$

Thus, $\mathrm{MSE}(\hat{\theta}_n^{\text{Shrinkage}}(c^*))$ is an increasing function of $\Delta^2 \geq 0$ and the relative efficiency (REff) of $\hat{\theta}_n^{\text{Shrinkage}}(c^*)$ compared to $\tilde{\theta}_n$ is

$$\mathrm{REff}(\hat{\theta}_n^{\text{Shrinkage}}(c^*); \tilde{\theta}_n) = 1 + \frac{1}{\Delta^2}, \quad \Delta^2 \geq 0. \tag{2.12}$$

Further, the MSE difference is

$$\frac{\sigma^2}{n} - \frac{\sigma^2}{n}\frac{\Delta^2}{1+\Delta^2} = \frac{\sigma^2}{n}\frac{1}{1+\Delta^2} \geq 0, \quad \forall \Delta^2 \geq 0. \tag{2.13}$$

Hence, $\hat{\theta}_n^{\text{Shrinkage}}(c^*)$ outperforms the $\tilde{\theta}_n$ uniformly.

2.2.3 Ridge Regression–Type Estimation of Location Parameter

Consider the problem of estimating θ when one suspects that θ may be 0. Then following Hoerl and Kennard (1970), if we define

$$\tilde{\theta}_n^{\mathrm{R}}(k) = \mathrm{argmin}_\theta\{\|\boldsymbol{Y} - \theta\mathbf{1}_n\|^2 + nk\theta^2\}, \quad k \geq 0. \tag{2.14}$$

Then, we obtain the ridge regression–type estimate of θ as

$$n\theta + nk\theta = n\tilde{\theta}_n \tag{2.15}$$

or

$$\tilde{\theta}_n^{\mathrm{R}}(k) = \frac{\tilde{\theta}_n}{1+k}. \tag{2.16}$$

Note that it is the same as taking $c = 1/(1+k)$ in (2.8).

Hence, the bias and MSE of $\tilde{\theta}_n^{\mathrm{R}}(k)$ are given by

$$b(\tilde{\theta}_n^{\mathrm{R}}(k)) = -\frac{k}{1+k}\theta \tag{2.17}$$

and

$$\begin{aligned}\mathrm{MSE}(\tilde{\theta}_n^{\mathrm{R}}(k)) &= \frac{\sigma^2}{n(1+k)^2} + \frac{k^2\theta^2}{(1+k)^2}\\ &= \frac{\sigma^2}{n(1+k)^2}(1+k^2\Delta^2).\end{aligned} \tag{2.18}$$

It may be seen that the optimum value of k is Δ^{-2} and MSE at (2.18) equals

$$\frac{\sigma^2}{n}\left(\frac{\Delta^2}{1+\Delta^2}\right). \tag{2.19}$$

Further, the MSE difference equals

$$\frac{\sigma^2}{n} - \frac{\sigma^2}{n}\frac{\Delta^2}{1+\Delta^2} = \frac{\sigma^2}{n}\frac{1}{1+\Delta^2} \geq 0, \tag{2.20}$$

which shows $\tilde{\theta}_n^{\mathrm{R}}(\Delta^{-2})$ uniformly dominates $\tilde{\theta}_n$.

The REff of $\tilde{\theta}_n^{\mathrm{R}}(\Delta^{-2})$ is given by

$$\mathrm{REff}(\tilde{\theta}_n^{\mathrm{R}}(\Delta^{-2}) : \tilde{\theta}_n) = 1 + \frac{1}{\Delta^2}.$$

2.2.4 LASSO for Location Parameter

In this section, we define the LASSO estimator of θ introduced by Tibshirani (1996) in connection with the regression model.

Theorem 2.1 *The LASSO estimator of θ is defined by*

$$\begin{aligned}\hat{\theta}_n^{\mathrm{LASSO}}(\lambda) &= \mathrm{argmin}_\theta\{\|\boldsymbol{Y} - \theta\boldsymbol{1}_n\|^2 + 2\lambda\sqrt{n}\sigma|\theta|\}\\ &= \mathrm{sgn}\,(\tilde{\theta}_n)\left(|\tilde{\theta}_n| - \lambda\frac{\sigma}{\sqrt{n}}\right)^+\end{aligned} \tag{2.21}$$

(when one suspects that θ may be zero).

Proof: The derivative of objective function inside $\{\|Y-\theta\mathbf{1}_n\|^2+2\lambda\sqrt{n}\sigma|\theta|\}$ is given by

$$-2\mathbf{1}_n^\top(Y-\hat{\theta}_n^{\mathrm{LASSO}}\mathbf{1}_n)+2\lambda\sqrt{n}\sigma\mathrm{sgn}(\hat{\theta}_n^{\mathrm{LASSO}})=0 \tag{2.22}$$

or

$$\hat{\theta}_n^{\mathrm{LASSO}}(\lambda)-\tilde{\theta}_n+\lambda\frac{\sigma}{\sqrt{n}}\,\mathrm{sgn}(\hat{\theta}_n^{\mathrm{LASSO}})=0,\quad \tilde{\theta}_n=\overline{Y} \tag{2.23}$$

or

$$\frac{\sqrt{n}\hat{\theta}_n^{\mathrm{LASSO}}(\lambda)}{\sigma}-\frac{\sqrt{n}\tilde{\theta}_n}{\sigma}+\lambda\mathrm{sgn}(\hat{\theta}_n^{\mathrm{LASSO}}(\lambda))=0. \tag{2.24}$$

(i) If $\mathrm{sgn}(\hat{\theta}_n^{\mathrm{LASSO}}(\lambda))=1$, then Eq. (2.24) reduces to

$$\frac{\sqrt{n}}{\sigma}\hat{\theta}_n^{\mathrm{LASSO}}(\lambda)-\frac{\sqrt{n}\tilde{\theta}_n}{\sigma}+\lambda=0.$$

Hence, using (2.21)

$$0<\hat{\theta}_n^{\mathrm{LASSO}}(\lambda)=\frac{\sigma}{\sqrt{n}}(Z_n-\lambda)=\frac{\sigma}{\sqrt{n}}(|Z_n|-\lambda),$$

with $Z_n>0$ and $|Z_n|>\lambda$.

(ii) If $\mathrm{sgn}(\hat{\theta}_n^{\mathrm{LASSO}}(k))=-1$, then we have

$$0>\hat{\theta}_n^{\mathrm{LASSO}}(\lambda)=\frac{\sigma}{\sqrt{n}}(Z_n+\lambda)=-\frac{\sigma}{\sqrt{n}}(|Z_n|-\lambda).$$

(iii) If $\mathrm{sgn}(\hat{\theta}_n^{\mathrm{LASSO}}(\lambda))=0$, we have $-Z_n+\lambda\gamma=0, \gamma\in(-1,1)$. Hence, we obtain $Z_n=\lambda\gamma$, which implies $|Z_n|<\lambda$.

Combining (i)–(iii), we obtain

$$\hat{\theta}_n^{\mathrm{LASSO}}(\lambda)=\frac{\sigma}{\sqrt{n}}\,\mathrm{sgn}(Z_n)(|Z_n|-\lambda)^+. \tag{2.25}$$

□

Donoho and Johnstone (1994) defined this estimator as the "soft threshold estimator" (STE).

2.2.5 Bias and MSE Expression for the LASSO of Location Parameter

In order to derive the bias and MSE of LASSO estimators, we need the following lemma.

Lemma 2.1 *If* $Z\sim\mathcal{N}(\Delta,1)$, *then*

(i) $\mathbb{E}[|Z|]=\Delta[2\Phi(\Delta)-1]+\sqrt{\frac{2}{\pi}}\exp\left\{-\frac{\Delta^2}{2}\right\}$

(ii) $\mathbb{E}[I(|Z|<\lambda)]=\Phi(\lambda-\Delta)-\Phi(-\lambda-\Delta)$

(iii) $\mathbb{E}[ZI(|Z| < \lambda)] = \Delta[\Phi(\lambda - \Delta) - \Phi(-\lambda - \Delta)] - [\phi(\lambda - \Delta) - \phi(\lambda + \Delta)]$

(iv) $\mathbb{E}[\text{sgn}(Z)] = 2\Phi(\Delta) - 1$

(v) $\mathbb{E}[\text{sgn}(Z)I(|Z| < \lambda)] = [\Phi(\lambda - \Delta) - \Phi(\lambda + \Delta)] + [2\Phi(\Delta) - 1]$

(vi) $\mathbb{E}[|Z|I(|Z| < \lambda)] = [\phi(\Delta) - \phi(\lambda + \Delta)] + [\phi(\Delta) - \phi(\lambda - \Delta)] - \Delta[\Phi(\lambda + \Delta) - \Phi(\lambda - \Delta) + 1] + 2\Delta\Phi(\Delta)$

(vii) $\mathbb{E}[Z^2 I(|Z| < \lambda)] = 2\Delta[\phi(\lambda + \Delta) - \phi(\lambda - \Delta)] - [(\lambda - \Delta)\phi(\lambda - \Delta) + (\lambda + \Delta)\phi(\lambda + \Delta)] + (\Delta^2 + 1)[\Phi(\lambda - \Delta) + \Phi(\lambda + \Delta) - 1].$

Proof: In the proof, we use the facts that $\Phi(-\Delta) = 1 - \Phi(\Delta)$, $\phi'(x) = -x\phi(x)$ (first derivative), $\phi''(x) = (x^2 - 1)\phi(x)$ (second derivative).

(i) By changing variable $z - \Delta = u$, we have

$$\begin{aligned}
\mathbb{E}[|Z|] &= \int_{-\infty}^{\infty} |z|\phi(z - \Delta)\mathrm{d}z \\
&= \int_{-\infty}^{-\Delta} -(u + \Delta)\phi(u)\mathrm{d}u + \int_{-\Delta}^{\infty} (u + \Delta)\phi(u)\mathrm{d}u \\
&= -\int_{\Delta}^{\infty} \phi'(u)\mathrm{d}u - \Delta\int_{\Delta}^{\infty} \phi(u)\mathrm{d}u - \int_{-\Delta}^{\infty} \phi'(u)\mathrm{d}u + \Delta\int_{-\Delta}^{\infty} \phi(u)\mathrm{d}u \\
&= 2\phi(\Delta) + \Delta[2\phi(\Delta) - 1].
\end{aligned}$$

For (ii), by definition of standard expectation

$$\begin{aligned}
\mathbb{E}[I(|Z| < \lambda)] &= \int_{-\infty}^{\infty} I(|z| < \lambda)\phi(z - \Delta)\mathrm{d}z \\
&= \int_{-\lambda}^{\lambda} \phi(z - \Delta)\mathrm{d}z \\
&= \Phi(\lambda - \Delta) - \Phi(-\lambda - \Delta).
\end{aligned}$$

To prove (iii), using (ii) we can write

$$\begin{aligned}
\mathbb{E}[ZI(|Z| < \lambda)] &= \int_{-\lambda}^{\lambda} (z - \Delta + \Delta)\phi(z - \Delta)\mathrm{d}z \\
&= \int_{-\lambda-\Delta}^{\lambda-\Delta} u\phi(u)\mathrm{d}u + \Delta\mathbb{E}[I(|Z| < \lambda)] \\
&= -\int_{-\lambda-\Delta}^{\lambda-\Delta} \phi'(u)\mathrm{d}u + \Delta\mathbb{E}[I(|Z| < \lambda)],
\end{aligned}$$

which gives the result.

(iv) By definition of sgn function and changing variable $z - \Delta = u$, we readily obtain

$$\begin{aligned}
\mathbb{E}[\text{sgn}(Z)] &= -\int_{-\infty}^{0} \phi(z - \Delta)\mathrm{d}z + \int_{0}^{\infty} \phi(z - \Delta)\mathrm{d}z \\
&= -\int_{-\infty}^{-\Delta} \phi(u)\mathrm{d}u + \int_{-\Delta}^{\infty} \phi(u)\mathrm{d}u,
\end{aligned}$$

which gives the desired result.

To prove (v), by the same strategies used in (ii) and (iv), we get

$$\begin{aligned}\mathbb{E}[\text{sgn}(Z)I(|Z| < \lambda)] &= \int_{-\lambda}^{\lambda} \text{sgn}(z)\phi(z-\Delta)\text{d}z \\ &= -\int_{-\lambda}^{0} \phi(z-\Delta)\text{d}z + \int_{0}^{\lambda} \phi(z-\Delta)\text{d}z.\end{aligned}$$

By making a change of variable $z - \Delta = u$, the required result follows directly.

To prove (vi), the probability density function (p.d.f.) of $|Z|$ is

$$\phi_{|Z|}(z) = \phi(z-\Delta) + \phi(z+\Delta).$$

Thus,

$$\begin{aligned}\mathbb{E}[|Z|I(|Z| < \lambda)] &= \int_{0}^{\lambda} z\phi(z-\Delta)\text{d}z + \int_{0}^{\lambda} z\phi(z+\Delta)\text{d}z \\ &= \int_{-\Delta}^{\lambda-\Delta} (u+\Delta)\phi(u)\text{d}u + \int_{\Delta}^{\lambda+\Delta} (u-\Delta)\phi(u)\text{d}u \\ &= -\int_{-\Delta}^{\lambda-\Delta} \phi'(u)\text{d}u + \Delta\int_{-\Delta}^{\lambda-\Delta} \phi(u)\text{d}u - \int_{\Delta}^{\lambda+\Delta} \phi'(u)\text{d}u \\ &\quad -\Delta\int_{\Delta}^{\lambda+\Delta} \phi(u)\text{d}u \\ &= -[\phi(\lambda-\Delta) - \phi(-\Delta)] + \Delta[\Phi(\lambda-\Delta) - \Phi(-\Delta)] \\ &\quad -[\phi(\lambda+\Delta) - \phi(\Delta)] - \Delta[\Phi(\lambda+\Delta) - \Phi(\Delta)] \\ &= [\phi(\Delta) - \phi(\lambda+\Delta)] + [\phi(\Delta) - \phi(\lambda-\Delta)] \\ &\quad -\Delta[\Phi(\lambda+\Delta) - \Phi(\lambda-\Delta) + 1] + 2\Delta\Phi(\Delta).\end{aligned}$$

Finally, for (vii), making use of the property of indicator function,

$$\begin{aligned}\mathbb{E}[Z^2 I(|Z| < \lambda)] &= \mathbb{E}[|Z|^2 I(|Z| < \lambda)] \\ &= \int_{0}^{\lambda} z^2\phi(z-\Delta)\text{d}z + \int_{0}^{\lambda} z^2\phi(z+\Delta)\text{d}z \\ &= \int_{-\Delta}^{\lambda-\Delta} (u+\Delta)^2\phi(u)\text{d}u + \int_{\Delta}^{\lambda+\Delta} (u-\Delta)^2\phi(u)\text{d}u \\ &= \int_{-\Delta}^{\lambda-\Delta} u^2\phi(u)\text{d}u + \Delta^2\int_{-\Delta}^{\lambda-\Delta} \phi(u)\text{d}u \\ &\quad +2\Delta\int_{-\Delta}^{\lambda-\Delta} u\phi(u)\text{d}u + \int_{\Delta}^{\lambda+\Delta} u^2\phi(u)\text{d}u \\ &\quad +\Delta^2\int_{\Delta}^{\lambda+\Delta} \phi(u)\text{d}u - 2\Delta\int_{\Delta}^{\lambda+\Delta} u\phi(u)\text{d}u\end{aligned}$$

$$
\begin{aligned}
&= \int_{-\Delta}^{\lambda-\Delta} (u^2-1)\phi(u)\mathrm{d}u + (\Delta^2+1)\int_{-\Delta}^{\lambda-\Delta} \phi(u)\mathrm{d}u \\
&\quad -2\Delta \int_{-\Delta}^{\lambda-\Delta} \phi'(u)\mathrm{d}u + \int_{\Delta}^{\lambda+\Delta} (u^2-1)\phi(u)\mathrm{d}u \\
&\quad +(\Delta^2+1)\int_{\Delta}^{\lambda+\Delta} \phi(u)\mathrm{d}u + 2\Delta \int_{\Delta}^{\lambda+\Delta} \phi'(u)\mathrm{d}u \\
&= \int_{-\Delta}^{\lambda-\Delta} \phi''(u)\mathrm{d}u + (\Delta^2+1)\int_{-\Delta}^{\lambda-\Delta} \phi(u)\mathrm{d}u \\
&\quad -2\Delta \int_{-\Delta}^{\lambda-\Delta} \phi'(u)\mathrm{d}u + \int_{\Delta}^{\lambda+\Delta} \phi''(u)\mathrm{d}u \\
&\quad +(\Delta^2+1)\int_{\Delta}^{\lambda+\Delta} \phi(u)\mathrm{d}u + 2\Delta \int_{\Delta}^{\lambda+\Delta} \phi'(u)\mathrm{d}u \\
&= [\phi'(\lambda-\Delta)-\phi'(-\Delta)] + (\Delta^2+1)[\Phi(\lambda-\Delta)-\Phi(-\Delta)] \\
&\quad -2\Delta[\phi(\lambda-\Delta)-\phi(-\Delta)] + [\phi'(\lambda+\Delta)-\phi'(\Delta)] \\
&\quad +(\Delta^2+1)[\Phi(\lambda+\Delta)-\Phi(\Delta)] + 2\Delta[\phi(\lambda+\Delta)-\phi(\Delta)] \\
&= [-(\lambda-\Delta)\phi(\lambda-\Delta)-\Delta\phi(\Delta)] \\
&\quad +(\Delta^2+1)[\Phi(\lambda-\Delta)+\Phi(\lambda+\Delta)-1] \\
&\quad +2\Delta[\phi(\lambda+\Delta)-\phi(\lambda-\Delta)] \\
&\quad +[-(\lambda+\Delta)\phi(\lambda+\Delta)+\Delta\phi(\Delta)] \\
&= 2\Delta[\phi(\lambda+\Delta)-\phi(\lambda-\Delta)] \\
&\quad -[(\lambda-\Delta)\phi(\lambda-\Delta)+(\lambda+\Delta)\phi(\lambda+\Delta)] \\
&\quad +(\Delta^2+1)[\Phi(\lambda-\Delta)+\Phi(\lambda+\Delta)-1].
\end{aligned}
$$

The proof is complete. □

Using Lemma 2.1, we can find the bias and MSE expressions of $\hat{\theta}_n^{\text{LASSO}}(\lambda)$.

Theorem 2.2 *The bias and MSE expressions of the LASSO estimator in Theorem 2.1 are given by*

$$
\begin{aligned}
b(\hat{\theta}_n^{\text{LASSO}}(\lambda)) &= \frac{\sigma}{\sqrt{n}}\{\lambda[\Phi(\lambda-\Delta)-\Phi(\lambda+\Delta)] \\
&\quad +\Delta[\Phi(-\lambda-\Delta)-\Phi(\lambda-\Delta)] \\
&\quad +[\phi(\lambda-\Delta)-\phi(\lambda+\Delta)]\} \\
\text{MSE}(\hat{\theta}_n^{\text{LASSO}}(\lambda)) &= \frac{\sigma^2}{n}\rho_{\text{ST}}(\lambda,\Delta),
\end{aligned}
$$

where

$$
\begin{aligned}
\rho_{\text{ST}}(\lambda,\Delta) &= (1+\lambda^2) + (\Delta^2-\lambda^2-1)[\Phi(\lambda-\Delta)-\Phi(-\lambda-\Delta)] \\
&\quad -[(\lambda-\Delta)\phi(\lambda+\Delta)+(\lambda+\Delta)\phi(\lambda-\Delta)].
\end{aligned}
$$

Proof: We give the proof for the bias expression here and the MSE is left as an exercise. Using (2.21), the expectation of $\hat{\theta}_n^{\text{LASSO}}(\lambda)$ is

$$\begin{aligned}
\mathbb{E}(\hat{\theta}_n^{\text{LASSO}}(\lambda)) &= \frac{\sigma}{\sqrt{n}}\mathbb{E}[\text{sgn}(Z_n)(|Z_n| - \lambda)^+] \\
&= \frac{\sigma}{\sqrt{n}}\{\mathbb{E}[\text{sgn}(Z_n)(|Z_n| - \lambda)I(|Z_n| > \lambda)]\} \\
&= \frac{\sigma}{\sqrt{n}}\{\mathbb{E}[\text{sgn}(Z_n)(|Z_n| - \lambda)\{1 - I(|Z_n| < \lambda)\}]\} \\
&= \frac{\sigma}{\sqrt{n}}\mathbb{E}[Z_n] - \frac{\sigma\lambda}{\sqrt{n}}\mathbb{E}[\text{sgn}(Z_n)] - \frac{\sigma}{\sqrt{n}}\mathbb{E}[Z_n I(|Z_n| < \lambda)] \\
&\quad + \frac{\sigma\lambda}{\sqrt{n}}\mathbb{E}[\text{sgn}(Z_n)I(|Z_n| < \lambda)]. \qquad (2.26)
\end{aligned}$$

Applying Lemma 2.1 to (2.26) gives

$$\begin{aligned}
\mathbb{E}(\hat{\theta}_n^{\text{LASSO}}(\lambda)) = \theta + \frac{\sigma}{\sqrt{n}}\{&\lambda[\Phi(\lambda - \Delta) - \Phi(\lambda + \Delta)] \\
&+\Delta[\Phi(-\lambda - \Delta) - \Phi(\lambda - \Delta)] \\
&+[\phi(\lambda - \Delta) - \phi(\lambda + \Delta)]\},
\end{aligned}$$

which gives the desired result. □

2.2.6 Preliminary Test Estimator, Bias, and MSE

Based on Saleh (2006), the preliminary test estimators (PTEs) of θ under normality assumption of the errors are given by

$$\begin{aligned}
\hat{\theta}_n^{\text{PT}}(\lambda) = \tilde{\theta}_n I\left(|\tilde{\theta}_n| > \lambda\frac{\sigma}{\sqrt{n}}\right) &= \frac{\sigma}{\sqrt{n}} Z_n I(|Z_n| > \lambda) \\
&= \frac{\sigma}{\sqrt{n}}[Z_n - Z_n I(|Z_n| < \lambda)], \quad \text{if } \sigma^2 \text{ is known.} \qquad (2.27)
\end{aligned}$$

Thus, we have the following theorem about bias and MSE.

Theorem 2.3 *The bias and MSE functions of the PTE, under normality of errors when σ^2 is known (Donoho and Johnstone 1994), are given as*

$$\begin{aligned}
b(\hat{\theta}_n^{\text{PT}}(\lambda)) &= \Delta[\Phi(\lambda - \Delta) - \Phi(-\lambda - \Delta)] - [\phi(\lambda - \Delta) - \phi(-\lambda + \Delta)] \\
\text{MSE}(\hat{\theta}_n^{\text{PT}}(\lambda)) &= \frac{\sigma^2}{n}\{[\tilde{\Phi}(\lambda - \Delta) + \tilde{\Phi}(\lambda + \Delta)] + (\lambda - \Delta)\phi(\lambda - \Delta) \\
&\quad +(\lambda + \Delta)\phi(\lambda + \Delta) + \Delta^2[\Phi(\lambda - \Delta) - \Phi(-\lambda - \Delta)]\}
\end{aligned}$$

with $\tilde{\Phi}(x) = 1 - \Phi(x)$.

2.2.7 Stein-Type Estimation of Location Parameter

The PT heavily depends on the critical value of the test that θ may be zero. Thus, due to down effect of discreteness of the PTE, we define the Stein-type estimator of θ as given here assuming σ is known

$$\begin{aligned}
\hat{\theta}_n^{\rm S} &= \tilde{\theta}_n(1-\lambda|Z_n|^{-1}), \quad Z_n = \frac{\sqrt{n}\tilde{\theta}_n}{\sigma} \\
&= \tilde{\theta}_n\left(1-\frac{\lambda\sigma}{\sqrt{n}|\tilde{\theta}_n|}\right) \\
&= \frac{\sigma}{\sqrt{n}}(Z_n - \lambda \mathrm{sgn}(Z_n)), \quad Z_n \sim \mathcal{N}(\Delta, 1), \quad \Delta = \frac{\sqrt{n}\theta}{\sigma}.
\end{aligned} \tag{2.28}$$

The bias of $\hat{\theta}_n^{\rm S}$ is $-\frac{\lambda\sigma}{n}[2\Phi(\Delta)-1]$, and the MSE of $\hat{\theta}_n^{\rm S}$ is given by

$$\begin{aligned}
\mathrm{MSE}(\hat{\theta}_n^{\rm S}) = \mathbb{E}(\hat{\theta}_n^{\rm S}-\theta)^2 &= \mathbb{E}[\tilde{\theta}_n-\theta]^2 + \frac{\lambda^2\sigma^2}{n} - 2\frac{\lambda\sigma}{\sqrt{n}}\mathbb{E}[Z_n\ \mathrm{sgn}(Z_n)] \\
&= \frac{\sigma^2}{n}\left[1+\lambda^2 - 2\lambda\sqrt{\frac{2}{\pi}}\exp\{-\Delta^2/2\}\right].
\end{aligned} \tag{2.29}$$

The value of λ that minimizes $\mathrm{MSE}(\hat{\theta}_n^{\rm S}, \theta)$ is $\lambda_0 = \sqrt{\frac{2}{\pi}}\exp\{-\Delta^2/2\}$, which is a decreasing function of Δ^2 with a maximum at $\Delta = 0$ and maximum value $\sqrt{\frac{2}{\pi}}$. Hence, the optimum value of MSE is

$$\mathrm{MSE}_{\rm opt}(\hat{\theta}_n^{\rm S}) = \frac{\sigma^2}{n}\left[1-\frac{2}{\pi}(2\exp\{-\Delta^2/2\}-1)\right]. \tag{2.30}$$

The REff compared to LSE, $\tilde{\theta}_n$ is

$$\mathrm{REff}(\hat{\theta}_n^{\rm S}:\tilde{\theta}_n) = \left[1-\frac{2}{\pi}(2\exp\{-\Delta^2/2\}-1)\right]^{-1}. \tag{2.31}$$

In general, the $\mathrm{REff}(\hat{\theta}_n^{\rm S}:\tilde{\theta}_n)$ decreases from $\left(1-\frac{2}{\pi}\right)^{-1}$ at $\Delta^2 = 0$, then it crosses the 1-line at $\Delta^2 = 2\log 2$, and for $0 < \Delta^2 = 2\log 2$, $\hat{\theta}_n^{\rm S}$ performs better than $\tilde{\theta}_n$.

2.2.8 Comparison of LSE, PTE, Ridge, SE, and LASSO

We know the following MSE from previous sections:

$$\begin{aligned}
\mathrm{MSE}(\tilde{\theta}_n) &= \frac{\sigma^2}{n} \\
\mathrm{MSE}(\hat{\theta}_n^{\rm PT}(\lambda)) &= \frac{\sigma^2}{n}\{[\tilde{\Phi}(\lambda-\Delta)+\tilde{\Phi}(\lambda+\Delta)] \\
&\quad +(\lambda-\Delta)\phi(\lambda-\Delta)+(\lambda+\Delta)\phi(\lambda+\Delta) \\
&\quad +\Delta^2[\Phi(\lambda-\Delta)-\Phi(-\lambda-\Delta)]\}
\end{aligned}$$

$$\begin{aligned}\text{MSE}(\hat{\theta}_n^{\text{RR}}(k)) &= \frac{\sigma^2}{n(1+k)^2}(1+k^2\Delta^2)\\ &= \frac{\sigma^2}{n}\frac{\Delta^2}{1+\Delta^2}, \quad \text{for } k=\Delta^{-2}\end{aligned}$$

$$\text{MSE}(\hat{\theta}_n^{\text{S}}) = \frac{\sigma^2}{n}\left[1-\frac{2}{\pi}(2\exp\{-\Delta^2/2\}-1)\right]$$

$$\text{MSE}(\hat{\theta}_n^{\text{LASSO}}(\lambda)) = \frac{\sigma^2}{n}\rho_{\text{ST}}(\lambda,\Delta), \quad \lambda=\sqrt{2\log 2}.$$

Hence, the REff expressions are given by

$$\begin{aligned}\text{REff}(\hat{\theta}_n^{\text{PT}}(\alpha):\tilde{\theta}_n) = \{&[\tilde{\Phi}(\lambda-\Delta)+\tilde{\Phi}(\lambda+\Delta)]\\ &+(\lambda-\Delta)\phi(\lambda-\Delta)+(\lambda+\Delta)\phi(\lambda+\Delta)\\ &+\Delta^2[\Phi(\lambda-\Delta)-\Phi(-\lambda-\Delta)]\}^{-1}\end{aligned}$$

$$\text{REff}(\hat{\theta}_n^{\text{RR}}(\Delta^2):\tilde{\theta}_n) = 1+\frac{1}{\Delta^2}$$

$$\text{REff}(\hat{\theta}_n^{\text{S}},\tilde{\theta}_n) = \left[1-\frac{2}{\pi}(2\exp\{-\Delta^2/2\}-1)\right]^{-1}$$

$$\text{REff}(\hat{\theta}_n^{\text{LASSO}}(\lambda):\tilde{\theta}_n) = \rho_{\text{ST}}^{-1}(\lambda,\Delta) \quad \lambda=\sqrt{2\log 2}.$$

It is seen from Table 2.1 that the RRE dominates all other estimators uniformly and LASSO dominates UE, PTE, and $\hat{\theta}_n^{\text{PT}} > \hat{\theta}_n^{\text{S}}$ in an interval near 0.

Table 2.1 Table of relative efficiency.

Δ	$\tilde{\theta}_n$	$\hat{\theta}_n^{\text{RR}}(k)$	$\hat{\theta}_n^{\text{PT}}(\lambda)$	$\hat{\theta}_n^{\text{S}}$	$\hat{\theta}_n^{\text{LASSO}}(\lambda)$
0.000	1.000	∞	4.184	2.752	9.932
0.316	1.000	11.000	2.647	2.350	5.694
0.548	1.000	4.333	1.769	1.849	3.138
0.707	1.000	3.000	1.398	1.550	2.207
1.000	1.000	2.000	1.012	1.157	1.326
1.177	1.000	1.721	0.884	1.000	1.046
1.414	1.000	1.500	0.785	0.856	0.814
2.236	1.000	1.200	0.750	0.653	0.503
3.162	1.000	1.100	0.908	0.614	0.430
3.873	1.000	1.067	0.980	0.611	0.421
4.472	1.000	1.050	0.996	0.611	0.419
5.000	1.000	1.040	0.999	0.611	0.419
5.477	1.000	1.033	1.000	0.611	0.419
6.325	1.000	1.025	1.000	0.611	0.419
7.071	1.000	1.020	1.000	0.611	0.419

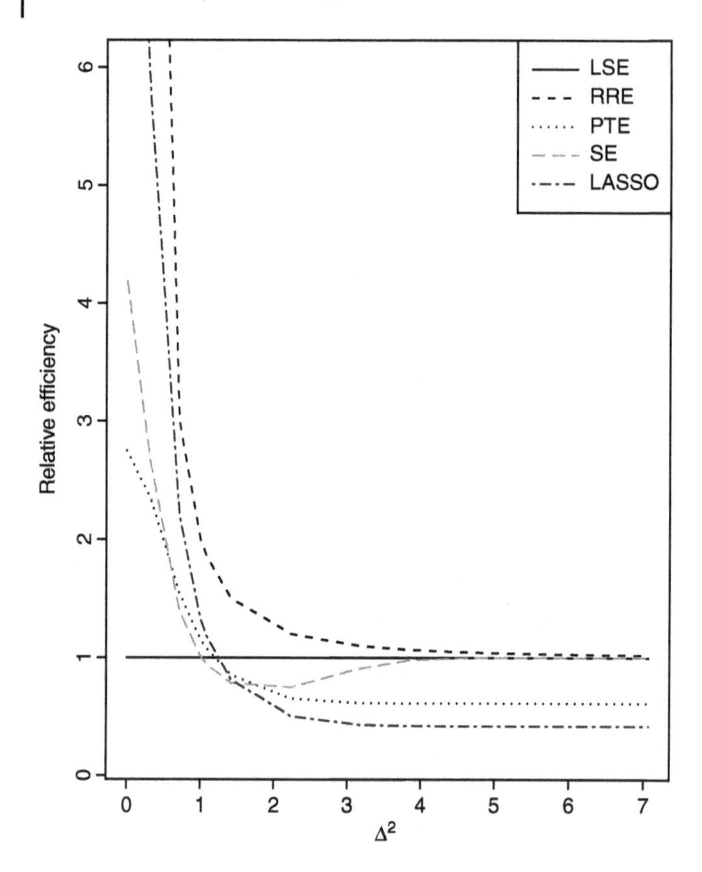

Figure 2.1 Relative efficiencies of the estimators.

From Table 2.1, we find $\hat{\theta}_n^{\mathrm{PT}}(\lambda) > \hat{\theta}_n^{\mathrm{S}}$ in the interval $[0, 0.316] \cup [2.236, \infty)$ while outside this interval $\hat{\theta}_n^{\mathrm{S}} > \hat{\theta}_n^{\mathrm{PT}}(\lambda)$. Figure 2.1 confirms that.

2.3 Simple Linear Model

In this section, we consider the model (2.1) and define the PT, ridge, and LASSO-type estimators when it is suspected that the slope may be zero.

2.3.1 Estimation of the Intercept and Slope Parameters

First, we consider the LSE of the parameters. Using the model (2.1) and the sample information from the normal distribution, we obtain the LSEs of $(\theta, \beta)^\top$ as

$$\begin{pmatrix} \tilde{\theta}_n \\ \tilde{\beta}_n \end{pmatrix} = \begin{pmatrix} \bar{y} - \tilde{\beta}_n \bar{x} \\ \frac{1}{Q}\left[\boldsymbol{x}^\top \boldsymbol{Y} - \frac{1}{n}(\mathbf{1}_n^\top \boldsymbol{x})(\mathbf{1}_n^\top \boldsymbol{Y}) \right] \end{pmatrix}, \tag{2.32}$$

where

$$\bar{x} = \frac{1}{n}\mathbf{1}_n^\top \boldsymbol{x}, \quad \bar{y} = \frac{1}{n}\mathbf{1}_n^\top \boldsymbol{Y}, \quad Q = \boldsymbol{x}^\top \boldsymbol{x} - \frac{1}{n}(\mathbf{1}_n^\top \boldsymbol{x})^2. \tag{2.33}$$

The exact distribution of $(\tilde{\theta}_n, \tilde{\beta}_n)^\top$ is a bivariate normal with mean $(\theta, \beta)^\top$ and covariance matrix

$$\frac{\sigma^2}{n}\begin{pmatrix} 1 + \dfrac{n\bar{x}^2}{Q} & -\dfrac{n\bar{x}}{Q} \\ \dfrac{n\bar{x}}{Q} & \dfrac{n}{Q} \end{pmatrix}. \tag{2.34}$$

An unbiased estimator of the variance σ^2 is given by

$$s_n^2 = (n-2)^{-1}(\boldsymbol{Y} - \tilde{\theta}_n \mathbf{1}_n - \tilde{\beta}_n \boldsymbol{x})^\top (\boldsymbol{Y} - \tilde{\theta}_n \mathbf{1}_n - \tilde{\beta}_n \boldsymbol{x}), \tag{2.35}$$

which is independent of $(\tilde{\theta}_n, \tilde{\beta}_n)$, and $(n-2)s_n^2/\sigma^2$ follows a central chi-square distribution with $(n-2)$ degrees of freedom (DF)

2.3.2 Test for Slope Parameter

Suppose that we want to test the null-hypothesis $\mathcal{H}_o : \beta = \beta_o$ vs. $\mathcal{H}_A : \beta \neq \beta_o$. Then, we use the likelihood ratio (LR) test statistic

$$\mathcal{L}_n^{(\sigma)} = \frac{(\tilde{\beta}_n - \beta_o)^2 Q}{\sigma^2}, \qquad \text{if } \sigma^2 \text{ is known}$$

$$\mathcal{L}_n^{(s)} = \frac{(\tilde{\beta}_n - \beta_o)^2 Q}{s_n^2}, \qquad \text{if } \sigma^2 \text{ is unknown} \tag{2.36}$$

where $\mathcal{L}_n^{(\sigma)}$ follows a noncentral chi-square distribution with 1 DF and noncentrality parameter $\Delta^2/2$ and $\mathcal{L}_n^{(s)}$ follows a noncentral F-distribution with $(1, m)$, where $m = n - 2$ is DF and also the noncentral parameter is

$$\Delta^2 = \frac{(\beta - \beta_o)^2 Q}{\sigma^2}.$$

Under $\mathcal{H}_o$, $\mathcal{L}_n^{(\sigma)}$ follows a central chi-square distribution and $\mathcal{L}_n^{(s)}$ follows a central F-distribution. At the α-level of significance, we obtain the critical value $\chi_1^2(\alpha)$ or $F_{1,m}(\alpha)$ from the distribution and reject $\mathcal{H}_o$ if $\mathcal{L}_n^{(\sigma)} > \chi_1^2(\alpha)$ or $\mathcal{L}_n^{(s)} > F_{1,m}(\alpha)$; otherwise, we accept $\mathcal{H}_o$.

2.3.3 PTE of the Intercept and Slope Parameters

This section deals with the problem of estimation of the intercept and slope parameters (θ, β) when it is suspected that the slope parameter β may be β_o.

From (2.30), we know that the LSE of θ is given by

$$\tilde{\theta}_n = \bar{y} - \tilde{\beta}_n \bar{x}. \tag{2.37}$$

If we know β to be β_o exactly, then the restricted least squares estimator (RLSE) of θ is given by

$$\hat{\theta}_n = \bar{y} - \beta_o\bar{x}. \tag{2.38}$$

In practice, the prior information that $\beta = \beta_o$ is uncertain. The doubt regarding this prior information can be removed using *Fisher's recipe* of testing the null-hypothesis $\mathcal{H}_o : \beta = \beta_o$ against the alternative $\mathcal{H}_A : \beta \neq \beta_o$. As a result of this test, we choose $\tilde{\theta}_n$ or $\hat{\theta}_n$ based on the rejection or acceptance of $\mathcal{H}_o$. Accordingly, in case of the unknown variance, we write the estimator as

$$\hat{\theta}_n^{\rm PT}(\alpha) = \hat{\theta}_n I(\mathcal{L}_n^{(s)} \leq F_{1,m}(\alpha)) + \tilde{\theta}_n I(\mathcal{L}_n^{(s)} > F_{1,m}(\alpha)), \quad m = n-2 \tag{2.39}$$

called the PTE, where $F_{1,m}(\alpha)$ is the α-level upper critical value of a central F-distribution with $(1, m)$ DF and $I(A)$ is the indicator function of the set A. For more details on PTE, see Saleh (2006), Ahmed and Saleh (1988), Ahsanullah and Saleh (1972), Kibria and Saleh (2012) and, recently Saleh et al. (2014), among others. We can write PTE of θ as

$$\hat{\theta}_n^{\rm PT}(\alpha) = \tilde{\theta}_n + (\tilde{\beta}_n - \beta_o)\bar{x}I(\mathcal{L}_n^{(s)} \leq F_{1,m}(\alpha)), \quad m = n-2. \tag{2.40}$$

If $\alpha = 1$, $\tilde{\theta}_n$ is always chosen; and if $\alpha = 0$, $\hat{\theta}_n$ is chosen. Since $0 < \alpha < 1$, $\hat{\theta}_n^{\rm PT}(\alpha)$ in repeated samples, this will result in a combination of $\tilde{\theta}_n$ and $\hat{\theta}_n$. Note that the PTE procedure leads to the choice of one of the two values, namely, either $\tilde{\theta}_n$ or $\hat{\theta}_n$. Also, the PTE procedure depends on the level of significance α.

Clearly, $\tilde{\beta}_n$ is the unrestricted estimator of β, while β_o is the restricted estimator. Thus, the PTE of β is given by

$$\hat{\beta}_n^{\rm PT}(\alpha) = \tilde{\beta}_n - (\tilde{\beta}_n - \beta_o)I(\mathcal{L}_n^{(s)} \leq F_{1,m}(\alpha)), \quad m = n-2. \tag{2.41}$$

Now, if $\alpha = 1$, $\tilde{\beta}_n$ is always chosen; and if $\alpha = 0$, β_o is always chosen.

Since our interest is to compare the LSE, RLSE, and PTE of θ and β with respect to bias and the MSE, we obtain the expression of these quantities in the following theorem. First we consider the bias expressions of the estimators.

Theorem 2.4 *The bias expressions of the estimators are given here:*

$$\begin{aligned}
&b(\tilde{\theta}_n) = 0, \quad \text{and} \quad b(\tilde{\beta}_n) = 0\\
&b(\hat{\theta}_n) = (\beta - \beta_o)\bar{x}, \quad \text{and} \quad b(\hat{\beta}_n) = -(\beta - \beta_o)\\
&b(\hat{\theta}_n^{\rm PT}) = -(\beta - \beta_o)\bar{x}H_3(\chi_1^2(\alpha); \Delta^2), \ \text{when } \sigma^2 \text{ is known}\\
&b(\hat{\theta}_n^{\rm PT}) = -(\beta - \beta_o)\bar{x}G_{3,m}\left(\frac{1}{3}F_{1,m}(\alpha); \Delta^2\right), \ \text{when } \sigma^2 \text{ is unknown}\\
&b(\hat{\beta}_n^{\rm PT}) = (\beta - \beta_o)H_3(\chi_1^2(\alpha); \Delta^2), \ \text{when } \sigma^2 \text{ is known}\\
&b(\hat{\beta}_n^{\rm PT}) = (\beta - \beta_o)G_{3,m}\left(\frac{1}{3}F_{1,m}(\alpha; \Delta^2)\right), \ \text{when } \sigma^2 \text{ is unknown}
\end{aligned}$$

where $G_{m_1,m_2}(\cdot; \Delta^2)$ is the cumulative distributional function (c.d.f.) of a noncentral F-distribution with (m_1, m_2) DF and non-centrality parameter $\Delta^2/2$.

Next, we consider the expressions for the MSEs of $\tilde{\theta}_n$, $\hat{\theta}_n$, and $\hat{\theta}_n^{\rm PT}$ along with the $\tilde{\beta}_n$, $\hat{\beta}_n$, and $\hat{\beta}_n^{\rm PT}(\alpha)$.

Theorem 2.5 *The MSE expressions of the estimators are given here:*

$$\text{MSE}(\tilde{\theta}_n) = \frac{\sigma^2}{n}\left(1 + \frac{n\bar{x}^2}{Q}\right), \quad \text{and} \quad \text{MSE}(\tilde{\beta}_n) = \frac{\sigma^2}{Q}$$

$$\text{MSE}(\hat{\theta}_n) = \frac{\sigma^2}{n}\left(1 + \frac{n\bar{x}^2}{Q}\Delta^2\right), \quad \text{and} \quad \text{MSE}(\hat{\beta}_n) = (\beta - \beta_o)^2 = \frac{\sigma^2}{Q}\Delta^2$$

$$\begin{aligned}\text{MSE}(\hat{\theta}_n^{\rm PT}(\alpha)) = {} & \frac{\sigma^2}{n} + \frac{\sigma^2\bar{x}^2}{Q}[1 - H_3(\chi_1^2(\alpha);\Delta^2) \\ & + \Delta^2\{2H_3(\chi_1^2(\alpha);\Delta^2) - H_5(\chi_1^2(\alpha);\Delta^2)\}] \\ & \text{when } \sigma^2 \text{ is known}\end{aligned}$$

$$\begin{aligned}\text{MSE}(\hat{\theta}_n^{\rm PT}(\alpha)) = {} & \frac{\sigma^2}{n} + \frac{\sigma^2\bar{x}^2}{Q}\left[1 - G_{3,m}\left(\frac{1}{3}F_{1,m}(\alpha);\Delta^2\right)\right. \\ & \left. + \Delta^2\left\{2G_{3,m}\left(\frac{1}{3}F_{1,m}(\alpha);\Delta^2\right) - G_{5,m}\left(\frac{1}{5}F_{1,m}(\alpha);\Delta^2\right)\right\}\right] \\ & \text{when } \sigma^2 \text{ is unknown}\end{aligned}$$

$$\begin{aligned}\text{MSE}(\hat{\beta}_n^{\rm PT}(\alpha)) = {} & \frac{\sigma^2}{Q}[1 - H_3(\chi_1^2(\alpha);\Delta^2) \\ & + \Delta^2\{2H_3(\chi_1^2(\alpha);\Delta^2) - H_5(\chi_1^2(\alpha);\Delta^2)\}] \\ & \text{when } \sigma^2 \text{ is known}\end{aligned}$$

$$\begin{aligned}\text{MSE}(\hat{\beta}_n^{\rm PT}(\alpha)) = {} & \frac{\sigma^2}{Q}\left[1 - G_{3,m}\left(\frac{1}{3}F_{1,m}(\alpha);\Delta^2\right)\right. \\ & \left. + \Delta^2\left\{2G_{3,m}\left(\frac{1}{3}F_{1,m}(\alpha);\Delta^2\right) - G_{5,m}\left(\frac{1}{5}F_{1,m}(\alpha);\Delta^2\right)\right\}\right] \\ & \text{when } \sigma^2 \text{ is unknown.}\end{aligned}$$

2.3.4 Comparison of Bias and MSE Functions

Since the bias and MSE expressions are known to us, we may compare them for the three estimators, namely, $\tilde{\theta}_n$, $\hat{\theta}_n$, and $\hat{\theta}_n^{\rm PT}(\alpha)$ as well as $\tilde{\beta}_n$, $\hat{\beta}_n$, and $\hat{\beta}_n^{\rm PT}(\alpha)$. Note that all the expressions are functions of Δ^2, which is the noncentrality parameter of the noncentral F-distribution. Also, Δ^2 is the standardized distance between β and β_o. First, we compare the bias functions as in Theorem 2.4, when σ^2 is unknown.

For $\bar{x} = 0$ or under $\mathcal{H}_o$,

$$\begin{aligned} b(\tilde{\theta}_n) &= b(\hat{\theta}_n) = b(\hat{\theta}_n^{\rm PT}(\alpha)) = 0 \\ b(\tilde{\beta}_n) &= b(\hat{\beta}_n) = b(\hat{\beta}_n^{\rm PT}(\alpha)) = 0. \end{aligned}$$

Otherwise, for all Δ^2 and $\overline{x} \neq 0$,

$$0 = b(\tilde{\theta}_n) \leq |b(\hat{\theta}_n^{\mathrm{PT}}(\alpha))| = |\beta - \beta_o|\overline{x}G_{3,m}\left(\frac{1}{3}F_{1,m}(\alpha); \Delta^2\right) \leq |b(\hat{\theta}_n)|$$
$$0 = b(\tilde{\beta}_n) \leq |b(\hat{\beta}_n^{\mathrm{PT}}(\alpha))| = |\beta - \beta_o|G_{3,m}\left(\frac{1}{3}F_{1,m}(\alpha); \Delta^2\right) \leq |b(\hat{\beta}_n)|.$$

The absolute bias of $\hat{\theta}_n$ is linear in Δ^2, while the absolute bias of $\hat{\theta}_n^{\mathrm{PT}}(\alpha)$ increases to the maximum as Δ^2 moves away from the origin, and then decreases toward zero as $\Delta^2 \to \infty$. Similar conclusions hold for $\hat{\beta}_n^{\mathrm{PT}}(\alpha)$.

Now, we compare the MSE functions of the restricted estimators and PTEs with respect to the traditional estimator, $\tilde{\theta}_n$ and $\tilde{\beta}_n$, respectively. The REff of $\hat{\theta}_n$ compared to $\tilde{\theta}_n$ may be written as

$$\mathrm{REff}(\hat{\theta}_n : \tilde{\theta}_n) = \left(1 + \frac{n\overline{x}^2}{Q}\right)\left[1 + \frac{n\overline{x}^2}{Q}\Delta^2\right]^{-1}. \tag{2.42}$$

The efficiency is a decreasing function of Δ^2. Under $\mathcal{H}_o$ (i.e. $\Delta^2 = 0$), it has the maximum value

$$\mathrm{REff}(\hat{\theta}_n; \tilde{\theta}_n) = \left(1 + \frac{n\overline{x}^2}{Q}\right) \geq 1, \tag{2.43}$$

and $\mathrm{REff}(\hat{\theta}_n; \tilde{\theta}_n) \geq 1$, accordingly, as $\Delta^2 \geq 1$. Thus, $\hat{\theta}_n$ performs better than $\tilde{\theta}_n$ whenever $\Delta^2 < 1$; otherwise, $\tilde{\theta}_n$ performs better $\hat{\theta}_n$.

The REff of $\hat{\theta}_n^{\mathrm{PT}}(\alpha)$ compared to $\tilde{\theta}_n$ may be written as

$$\mathrm{REff}(\hat{\theta}_n^{\mathrm{PT}}(\alpha); \tilde{\theta}_n) = [1 + g(\Delta^2)]^{-1}, \tag{2.44}$$

where

$$g(\Delta^2) = -\frac{\overline{x}^2}{Q}\left(\frac{1}{n} + \frac{\overline{x}^2}{Q}\right)^{-1}\left[G_{3,m}\left(\frac{1}{3}F_{1,m}(\alpha); \Delta^2\right) - \Delta^2\left\{2G_{3,m}\left(\frac{1}{3}F_{1,m}(\alpha); \Delta^2\right) - G_{5,m}\left(\frac{1}{5}F_{1,m}(\alpha); \Delta^2\right)\right\}\right]. \tag{2.45}$$

Under the $\mathcal{H}_o$, it has the maximum value

$$\mathrm{REff}(\hat{\theta}_n^{\mathrm{PT}}(\alpha); \tilde{\theta}_n) = \left\{1 - \frac{\overline{x}^2}{Q}\left(\frac{1}{n} + \frac{\overline{x}^2}{Q}\right)^{-1} G_{3,m}\left(\frac{1}{3}F_{1,m}(\alpha); 0\right)\right\}^{-1} \geq 1 \tag{2.46}$$

and $\mathrm{REff}(\hat{\theta}_n^{\mathrm{PT}}(\alpha); \tilde{\theta}_n)$ according as

$$\Delta^2 \lesseqgtr \Delta^2(\alpha) = \frac{G_{3,m}\left(\frac{1}{3}F_{1,m}(\alpha); \Delta^2\right)}{2G_{3,m}\left(\frac{1}{3}F_{1,m}(\alpha); \Delta^2\right) - G_{5,m}\left(\frac{1}{5}F_{1,m}(\alpha); \Delta^2\right)}. \tag{2.47}$$

Hence, $\hat{\theta}_n^{\text{PT}}$ performs better than $\tilde{\theta}_n$ if $\Delta^2 \leq \Delta^2(\alpha)$; otherwise, $\tilde{\theta}_n$ is better than $\hat{\theta}_n^{\text{PT}}$. Since

$$2G_{3,m}\left(\frac{1}{3}F_{1,m}(\alpha);\Delta^2\right) - G_{5,m}\left(\frac{1}{5}F_{1,m}(\alpha);\Delta^2\right) > 0, \tag{2.48}$$

we obtain

$$\Delta^2(\alpha) \leq 1. \tag{2.49}$$

As for the PTE of β, it is better than $\tilde{\beta}_n$, if

$$\Delta^2 \lesseqgtr \Delta^2(\alpha) = \frac{G_{3,m}\left(\frac{1}{3}F_{1,m}(\alpha);\Delta^2\right)}{2G_{3,m}\left(\frac{1}{3}F_{1,m}(\alpha);\Delta^2\right) - G_{5,m}\left(\frac{1}{5}F_{1,m}(\alpha);\Delta^2\right)}. \tag{2.50}$$

Otherwise, $\tilde{\beta}_n$ is better than $\hat{\beta}_n^{\text{PT}}(\alpha)$. The

$$\begin{aligned}\text{REff}(\hat{\beta}_n^{\text{PT}}(\alpha):\tilde{\beta}_n) = \Bigg[1 - G_{3,m}\left(\frac{1}{3}F_{1,m}(\alpha);\Delta^2\right)\\ +\Delta^2\left\{2G_{3,m}\left(\frac{1}{3}F_{1,m}(\alpha);\Delta^2\right) - G_{5,m}\left(\frac{1}{5}F_{1,m}(\alpha);\Delta^2\right)\right\}\Bigg]^{-1}.\end{aligned} \tag{2.51}$$

Under $\mathcal{H}_o$,

$$\text{REff}(\hat{\beta}_n^{\text{PT}}(\alpha):\tilde{\beta}_n) = \left[1 - G_{3,m}\left(\frac{1}{3}F_{1,m}(\alpha);0\right)\right]^{-1} \geq 1. \tag{2.52}$$

See Figure 2.2 for visual comparison between estimators.

2.3.5 Alternative PTE

In this subsection, we provide the alternative expressions for the estimator of PT and its bias and MSE. To test the hypothesis $\mathcal{H}_0 : \beta = 0$ vs. $\mathcal{H}_A : \beta \neq 0$, we use the following test statistic:

$$Z_n = \frac{\sqrt{Q}\tilde{\beta}_n}{\sigma}. \tag{2.53}$$

The PTE of β is given by

$$\begin{aligned}\hat{\beta}_n^{\text{PT}}(\alpha) &= \tilde{\beta}_n - \tilde{\beta}_n I\left(|\tilde{\beta}_n| < \frac{\lambda\sigma}{\sqrt{Q}}\right)\\ &= \frac{\sigma}{\sqrt{Q}}[Z_n - Z_n I(|Z_n| < \lambda)],\end{aligned} \tag{2.54}$$

where $\lambda = \sqrt{2\log 2}$.

Hence, the bias of $\tilde{\beta}_n$ equals $\beta[\Phi(\lambda-\Delta) - \Phi(-\lambda-\Delta)] - [\phi(\lambda-\Delta) - \phi(\lambda+\Delta)]$, and the MSE is given by

$$\text{MSE}(\hat{\beta}_n^{\text{PT}}) = \frac{\sigma^2}{Q}\rho_{\text{PT}}(\lambda,\Delta). \tag{2.55}$$

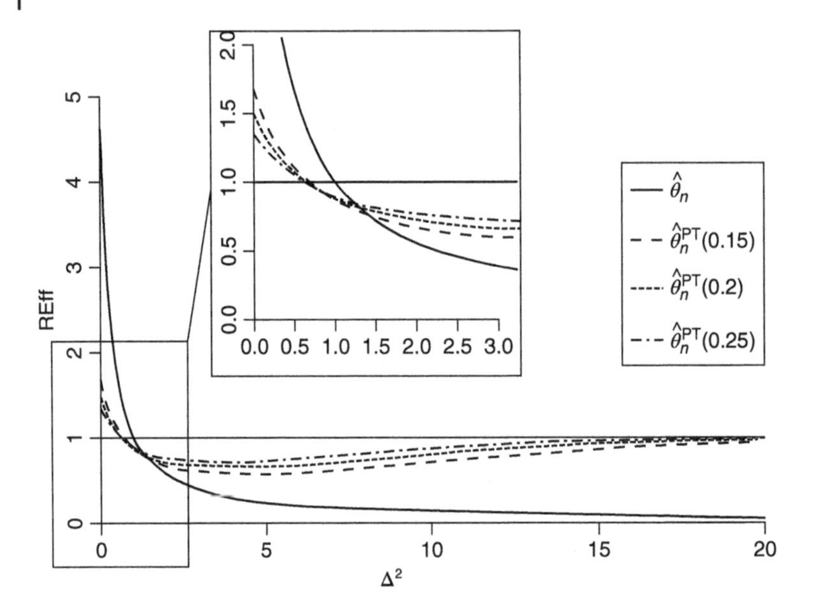

Figure 2.2 Graph of $\mathrm{REff}(\hat{\theta}_n; \tilde{\theta}_n)$ and $\mathrm{REff}(\hat{\theta}_n^{\mathrm{PT}}(\alpha); \tilde{\theta}_n)$ for $n = 8$ and $\overline{x}^2/Q = 0.5$.

Next, we consider the Stein-type estimator of β as

$$\hat{\beta}_n^{\mathrm{S}} = \tilde{\beta}_n - \lambda \frac{\tilde{\beta}_n}{|\tilde{\beta}_n|} = \frac{\sigma}{\sqrt{Q}} \left[Z_n - \lambda \frac{Z_n}{|Z_n|} \right]. \tag{2.56}$$

The bias and MSE expressions are given respectively by

$$b(\hat{\beta}_n^{\mathrm{S}}) = -\frac{\sigma}{\sqrt{Q}} \lambda [2\Phi(\Delta) - 1]$$

$$\mathrm{MSE}(\hat{\beta}_n^{\mathrm{S}}) = \frac{\sigma^2}{Q} \left[1 - \frac{2}{\pi} (2 \exp\{-\Delta^2/2\} - 1) \right]. \tag{2.57}$$

As a consequence, we may define the PT and Stein-type estimators of θ given by

$$\begin{aligned} \hat{\theta}_n^{\mathrm{PT}}(\alpha) &= \overline{y} - \hat{\beta}_n^{\mathrm{PT}}(\alpha)\overline{x} \\ &= (\overline{y} - \tilde{\beta}_n \overline{x}) - \tilde{\beta}_n \overline{x} I \left(|\tilde{\beta}_n| < \frac{\lambda \sigma}{\sqrt{Q}} \right) \\ &= \tilde{\theta}_n - \frac{\sigma}{\sqrt{Q}} \overline{x} Z_n I(|Z_n| < \lambda). \end{aligned} \tag{2.58}$$

Then, the bias and MSE expressions of $\hat{\theta}_n^{\mathrm{PT}}(\alpha)$ are

$$b(\hat{\theta}_n^{\mathrm{PT}}(\alpha)) = -\frac{\sigma}{\sqrt{Q}} [\Delta(\Phi(\lambda - \Delta) - \Phi(-\lambda - \Delta)) - [\phi(\lambda - \Delta) - \phi(\lambda + \Delta)]]$$

$$\mathrm{MSE}(\hat{\theta}_n^{\mathrm{PT}}(\alpha)) = \frac{\sigma^2}{Q} \left[1 + \frac{n\overline{x}^2}{Q} \rho_{\mathrm{PT}}(\lambda, \Delta) \right], \tag{2.59}$$

where

$$\rho_{PT}(\lambda,\Delta) = (1+\lambda^2) + (\Delta^2 - \lambda^2 - 1)[\Phi(\lambda-\Delta) - \Phi(-\lambda-\Delta)] \\ -[(\lambda-\Delta)\phi(\lambda+\Delta) + (\lambda+\Delta)\phi(\lambda-\Delta)].$$

Similarly, the bias and MSE expressions for $\hat{\theta}_n^{PT}(\alpha)$ are given by

$$b(\hat{\theta}_n^{PT}(\alpha)) = -b(\tilde{\beta}_n)\overline{x}$$

$$\text{MSE}(\hat{\theta}_n^{PT}(\alpha)) = \frac{\sigma^2}{n} + \frac{\sigma^2\overline{x}^2}{Q}\left[1 - G_{3,m}\left(\frac{1}{3}F_{1,m}(\alpha);\Delta^2\right)\right. \\ \left. + \Delta^2\left\{2G_{3,m}\left(\frac{1}{3}F_{1,m}(\alpha);\Delta^2\right) - G_{5,m}\left(\frac{1}{5}F_{1,m}(\alpha);\Delta^2\right)\right\}\right]. \tag{2.60}$$

2.3.6 Optimum Level of Significance of Preliminary Test

Consider the REff of $\hat{\theta}_n^{PT}(\alpha)$ compared to $\tilde{\theta}_n$. Denoting it by REff$(\alpha;\Delta^2)$, we have

$$\text{REff}(\alpha,\Delta^2) = [1 + g(\Delta^2)]^{-1}, \tag{2.61}$$

where

$$g(\Delta^2) = -\frac{\overline{x}^2}{Q}\left(\frac{1}{n} + \frac{\overline{x}^2}{Q}\right)^{-1}\left[G_{3,m}\left(\frac{1}{3}F_{1,m}(\alpha);\Delta^2\right)\right. \\ \left. -\Delta^2\left\{2G_{3,m}\left(\frac{1}{3}F_{1,m}(\alpha);\Delta^2\right) - G_{5,m}\left(\frac{1}{5}F_{1,m}(\alpha);\Delta^2\right)\right\}\right]. \tag{2.62}$$

The graph of REff(α,Δ^2), as a function of Δ^2 for fixed α, is decreasing crossing the 1-line to a minimum at $\Delta^2 = \Delta_0^2(\alpha)$ (say); then it increases toward the 1-line as $\Delta^2 \to \infty$. The maximum value of REff(α,Δ^2) occurs at $\Delta^2 = 0$ with the value

$$\text{REff}(\alpha;0) = \left\{1 - \frac{\overline{x}^2}{Q}\left(\frac{1}{n} + \frac{\overline{x}^2}{Q}\right)^{-1} G_{3,m}\left(\frac{1}{3}F_{1,m}(\alpha);0\right)\right\}^{-1} \geq 1,$$

for all $\alpha \in A$, the set of possible values of α. The value of REff$(\alpha;0)$ decreases as α-values increase. On the other hand, if $\alpha = 0$ and Δ^2 vary, the graphs of REff$(0,\Delta^2)$ and REff$(1,\Delta^2)$ intersect at $\Delta^2 = 1$. In general, REff(α_1,Δ^2) and REff(α_2,Δ^2) intersect within the interval $0 \leq \Delta^2 \leq 1$; the value of Δ^2 at the intersection increases as α-values increase. Therefore, for two different α-values, REff(α_1,Δ^2) and REff(α_2,Δ^2) will always intersect below the 1-line.

In order to obtain a PTE with a minimum guaranteed efficiency, E_0, we adopt the following procedure: If $0 \leq \Delta^2 \leq 1$, we always choose $\tilde{\theta}_n$, since REff$(\alpha,\Delta^2) \geq 1$ in this interval. However, since in general Δ^2 is unknown, there is no way to choose an estimate that is uniformly best. For this reason, we select an estimator with minimum guaranteed efficiency, such as E_0, and look for a suitable α from the set, $A = \{\alpha | \text{REff}(\alpha,\Delta^2) \geq E_0\}$. The estimator chosen

Table 2.2 Maximum and minimum guaranteed relative efficiency.

	α					
	0.05	0.10	0.15	0.20	0.25	0.50
			$p = 5$			
E_{max}	4.825	2.792	2.086	1.726	1.510	1.101
E_{min}	0.245	0.379	0.491	0.588	0.670	0.916
Δ^2	8.333	6.031	5.005	4.429	4.004	3.028
			$p = 6$			
E_{max}	4.599	2.700	2.034	1.693	1.487	1.097
E_{min}	0.268	0.403	0.513	0.607	0.686	0.920
Δ^2	7.533	5.631	4.755	4.229	3.879	3.028
			$p = 8$			
E_{max}	4.325	2.587	1.970	1.652	1.459	1.091
E_{min}	0.268	0.403	0.513	0.607	0.686	0.920
Δ^2	6.657	5.180	4.454	4.004	3.704	2.978
			$p = 10$			
E_{max}	4.165	2.521	1.933	1.628	1.443	1.088
E_{min}	0.319	0.452	0.557	0.644	0.717	0.928
Δ^2	6.206	4.955	4.304	3.904	3.629	2.953

maximizes $\text{REff}(\alpha, \Delta^2)$ over all $\alpha \in A$ and Δ^2. Thus, we solve the following equation for the optimum α^*:

$$\min_{\Delta^2} \text{REff}(\alpha, \Delta^2) = E(\alpha, \Delta_0^2(\alpha)) = E_0. \tag{2.63}$$

The solution α^* obtained this way gives the PTE with minimum guaranteed efficiency E_0, which may increase toward $\text{REff}(\alpha^*, 0)$ given by (2.61), and Table 2.2. For the following given data, we have computed the maximum and minimum guaranteed REff for the estimators of θ and provided them in Table 2.2.

$$\begin{aligned} x = (&19.383, 21.117, 18.99, 19.415, 20.394, 20.212, 20.163, 20.521, 20.125,\\ &19.944, 18.345, 21.45, 19.479, 20.199, 20.677, 19.661, 20.114, 19.724,\\ &18.225, 20.669)^\top. \end{aligned}$$

2.3.7 Ridge-Type Estimation of Intercept and Slope

In this section, we consider the ridge-type shrinkage estimation of (θ, β) when it is suspected that the slope β may be 0. In this case, we minimize the objective function with a solution as given here:

$$(\hat{\theta}_n^{\mathrm{RR}^\top}, \tilde{\beta}_n^{\mathrm{RR}^\top})^\top = \mathrm{argmin}_{(\theta,\beta)}\{\|\boldsymbol{Y} - \theta\mathbf{1}_n - \beta\boldsymbol{x}\|^2 + k\beta^2\}, \tag{2.64}$$

which yields two equations

$$n\theta + n\bar{x}\beta = n\bar{y}$$

$$n\bar{x}\theta + (Q + n\bar{x}^2)\beta + k\beta = \boldsymbol{x}^\top\boldsymbol{Y}, \quad \beta(Q+k) = \boldsymbol{x}^\top\boldsymbol{Y} - n\overline{xy}.$$

Hence,

$$\begin{pmatrix}\hat{\theta}_n^{\mathrm{RR}}(k)\\ \hat{\beta}_n^{\mathrm{RR}}(k)\end{pmatrix} = \begin{pmatrix}\bar{y} - \dfrac{Q}{Q+k}\tilde{\beta}_n\\ \dfrac{Q}{Q+k}\tilde{\beta}_n\end{pmatrix}. \tag{2.65}$$

2.3.7.1 Bias and MSE Expressions

From (2.65), it is easy to see that the bias expression of $\hat{\theta}_n^{\mathrm{RR}}(k)$ and $\hat{\beta}_n^{\mathrm{RR}}(k)$, respectively, are given by

$$b(\hat{\theta}_n^{\mathrm{RR}}(k)) = -\frac{k}{Q+k}\beta\bar{x}$$

$$b(\hat{\beta}_n^{\mathrm{RR}}(k)) = -\frac{k}{Q+k}\beta. \tag{2.66}$$

Similarly, MSE expressions of the estimators are given by

$$\begin{aligned}\mathrm{MSE}(\hat{\theta}_n^{\mathrm{RR}}(k)) &= \frac{\sigma^2}{n}\left(1 + \frac{n\bar{x}^2 Q}{(Q+k)^2}\right) + \frac{k^2\beta^2\bar{x}^2}{(Q+k)^2}\\ &= \frac{\sigma^2}{n}\left\{\left(1 + \frac{n\bar{x}^2 Q}{(Q+k)^2}\right) + \frac{n\bar{x}^2 k^2\beta^2}{(Q+k)^2\sigma^2}\right\}\\ &= \frac{\sigma^2}{n}\left\{1 + \frac{n\bar{x}^2}{Q(Q+k)^2}(Q^2 + k^2\Delta^2)\right\},\end{aligned} \tag{2.67}$$

where $\Delta^2 = \frac{Q\beta^2}{\sigma^2}$ and

$$\begin{aligned}\mathrm{MSE}(\hat{\beta}_n^{\mathrm{RR}}(k)) &= \frac{\sigma^2 Q}{(Q+k)^2} + \frac{k^2\beta^2}{(Q+k)^2}\\ &= \frac{\sigma^2}{Q(Q+k)^2}(Q^2 + k^2\Delta^2).\end{aligned} \tag{2.68}$$

Hence, the REff of these estimators are given by

$$\begin{aligned}\mathrm{REff}(\hat{\theta}_n^{\mathrm{RR}}(k):\tilde{\theta}_n) &= \left(1 + \frac{n\bar{x}^2}{Q}\right)\left\{1 + \frac{n\bar{x}^2}{Q(Q+k)^2}(Q^2 + k^2\Delta^2)\right\}^{-1}\\ \mathrm{REff}(\hat{\beta}_n^{\mathrm{RR}}(k):\tilde{\beta}_n) &= (Q+k)^2(Q^2 + k\Delta^2)^{-1}\\ &= \left(1 + \frac{k}{Q}\right)\left\{1 + \frac{k\Delta^2}{Q^2}\right\}^{-1}.\end{aligned} \tag{2.69}$$

Note that the optimum value of k is $Q\Delta^{-2}$. Hence,

$$\mathrm{MSE}(\hat{\theta}_n^{\mathrm{RR}}(Q\Delta^{-2})) = \frac{\sigma^2}{n}\left(1 + \frac{n\overline{x}^2\Delta^2}{Q(1+\Delta^2)}\right),$$

$$\mathrm{REff}(\hat{\theta}_n^{\mathrm{RR}}(Q\Delta^{-2}) : \tilde{\theta}_n) = \left(1 + \frac{n\overline{x}^2}{Q}\right)\left\{1 + \frac{n\overline{x}^2\Delta^2}{Q(1+\Delta^2)}\right\}^{-1},$$

$$\mathrm{MSE}(\hat{\beta}_n^{\mathrm{RR}}(Q\Delta^{-2})) = \frac{\sigma^2}{Q}\left(\frac{1+\Delta^2}{\Delta^2}\right)\left(\frac{Q}{1+Q}\right),$$

$$\mathrm{REff}(\hat{\beta}_n^{\mathrm{RR}}(Q\Delta^{-2}) : \tilde{\beta}_n) = \left(\frac{1+\Delta^2}{\Delta^2}\right)\left(\frac{Q}{1+Q}\right). \tag{2.70}$$

2.3.8 LASSO Estimation of Intercept and Slope

In this section, we consider the LASSO estimation of (θ, β) when it is suspected that β may be 0. For this case, the solution is given by

$$(\hat{\theta}_n^{\mathrm{LASSO}}(\lambda), \hat{\beta}_n^{\mathrm{LASSO}}(\lambda))^{\top} = \mathrm{argmin}_{(\theta),\beta)}\{\|\boldsymbol{Y} - \theta\boldsymbol{1}_n - \beta\boldsymbol{x}\|^2 + \sqrt{Q}\lambda\sigma|\beta|\}.$$

Explicitly, we find

$$\hat{\theta}_n^{\mathrm{LASSO}}(\lambda) = \overline{y} - \hat{\beta}_n^{\mathrm{LASSO}}\overline{x}$$

$$\begin{aligned}\hat{\beta}_n^{\mathrm{LASSO}} &= \mathrm{sgn}(\tilde{\beta})\left(|\tilde{\beta}| - \lambda\frac{\sigma}{\sqrt{Q}}\right)^{+} \\ &= \frac{\sigma}{\sqrt{Q}}\,\mathrm{sgn}(Z_n)(|Z_n| - \lambda)^{+},\end{aligned}$$

where $Z_n = \sqrt{Q}\tilde{\beta}_n/\sigma \sim \mathcal{N}(\Delta, 1)$ and $\Delta = \sqrt{Q}\beta/\sigma$.

According to Donoho and Johnstone (1994), and results of Section 2.2.5, the bias and MSE expressions for $\hat{\beta}_n^{\mathrm{LASSO}}(\lambda)$ are given by

$$\begin{aligned}b(\hat{\beta}_n^{\mathrm{LASSO}}(\lambda)) &= \frac{\sigma}{\sqrt{Q}}\{\lambda[\Phi(\lambda-\Delta) - \Phi(\lambda+\Delta)] \\ &\quad +\Delta[\Phi(-\lambda-\Delta) - \Phi(\lambda-\Delta)] \\ &\quad +[\phi(\lambda-\Delta) - \phi(\lambda+\Delta)]\}, \\ \mathrm{MSE}(\hat{\beta}_n^{\mathrm{LASSO}}(\lambda)) &= \frac{\sigma^2}{Q}\rho_{\mathrm{ST}}(\lambda, \Delta),\end{aligned} \tag{2.71}$$

where

$$\begin{aligned}\rho_{\mathrm{ST}}(\lambda, \Delta) = 1 + \lambda^2 &+ (1 - \Delta^2 - \lambda^2)\{\Phi(\lambda-\Delta) - \Phi(-\lambda-\Delta)\} \\ &-(\lambda-\Delta)\phi(\lambda+\Delta) - (\lambda+\Delta)\phi(\lambda-\Delta).\end{aligned} \tag{2.72}$$

Similarly, the bias and MSE expressions for $\hat{\theta}_n^{\text{LASSO}}(\lambda)$ are given by

$$\begin{aligned} b(\hat{\theta}_n^{\text{LASSO}}(\lambda)) &= b(\hat{\beta}_n^{\text{LASSO}}(\lambda))\overline{x} \\ \text{MSE}(\hat{\theta}_n^{\text{LASSO}}(\lambda)) &= \frac{\sigma^2}{n}\left[1 + \frac{n\overline{x}^2}{Q}\rho_{\text{ST}}(\lambda, \Delta)\right]. \end{aligned} \tag{2.73}$$

Then the REff is obtained as

$$\text{REff}(\hat{\theta}_n^{\text{LASSO}} : \tilde{\theta}_n) = \left(1 + \frac{n\overline{x}^2}{Q}\right)\left[1 + \frac{n\overline{x}^2}{Q}\rho_{\text{ST}}(\lambda, \Delta)\right]^{-1}. \tag{2.74}$$

For the following given data, we have computed the REff for the estimators of θ and β and provided them in Tables 2.3 and 2.4 and in Figures 2.3 and 2.4, respectively.

$$\begin{aligned} \boldsymbol{x} = (&19.9956, 35.3042, 7.293, 18.3734, 19.5403, 28.8734, 17.5883, \\ &15.9005, 16.2462, 23.6399, 14.4378, 25.4821, 30.1329, 24.6978, \\ &32.1264, 8.9395, 28.4798, 31.5587, 16.8541, 27.9487)^\top. \end{aligned}$$

It is seen from Tables 2.3 and 2.4 and Figures 2.3 and 2.4 that the RRE dominates all other estimators but the restricted estimator uniformly and that LASSO dominates LSE, PTE, and SE uniformly except RRE and RLSE in a subinterval $[0, \sqrt{2\log 2}]$.

Table 2.3 Relative efficiency of the estimators for θ.

Delta	LSE	RLSE	PTE	RRE	LASSO	SE
0.000	1.000	∞	2.987	9.426	5.100	2.321
0.100	1.000	10.000	2.131	5.337	3.801	2.056
0.300	1.000	3.333	1.378	3.201	2.558	1.696
0.500	1.000	2.000	1.034	2.475	1.957	1.465
1.000	1.000	1.000	0.666	1.808	1.282	1.138
1.177	1.000	0.849	0.599	1.696	1.155	1.067
2.000	1.000	0.500	0.435	1.424	0.830	0.869
5.000	1.000	0.200	0.320	1.175	0.531	0.678
10.000	1.000	0.100	0.422	1.088	0.458	0.640
15.000	1.000	0.067	0.641	1.059	0.448	0.638
20.000	1.000	0.050	0.843	1.044	0.447	0.637
25.000	1.000	0.040	0.949	1.036	0.447	0.637
30.000	1.000	0.033	0.986	1.030	0.447	0.637
40.000	1.000	0.025	0.999	1.022	0.447	0.637
50.000	1.000	0.020	1.000	1.018	0.447	0.637

Table 2.4 Relative efficiency of the estimators for β.

Delta	LSE	RLSE	PTE	RRE	LASSO	SE
0.000	1.000	∞	3.909	∞	9.932	2.752
0.100	1.000	10.000	2.462	10.991	5.694	2.350
0.300	1.000	3.333	1.442	4.330	3.138	1.849
0.500	1.000	2.000	1.039	2.997	2.207	1.550
1.000	1.000	1.000	0.641	1.998	1.326	1.157
1.177	1.000	0.849	0.572	1.848	1.176	1.075
2.000	1.000	0.500	0.407	1.499	0.814	0.856
5.000	1.000	0.200	0.296	1.199	0.503	0.653
10.000	1.000	0.100	0.395	1.099	0.430	0.614
15.000	1.000	0.067	0.615	1.066	0.421	0.611
20.000	1.000	0.050	0.828	1.049	0.419	0.611
25.000	1.000	0.040	0.943	1.039	0.419	0.611
30.000	1.000	0.033	0.984	1.032	0.419	0.611
40.000	1.000	0.025	0.999	1.024	0.419	0.611
50.000	1.000	0.020	1.000	1.019	0.419	0.611

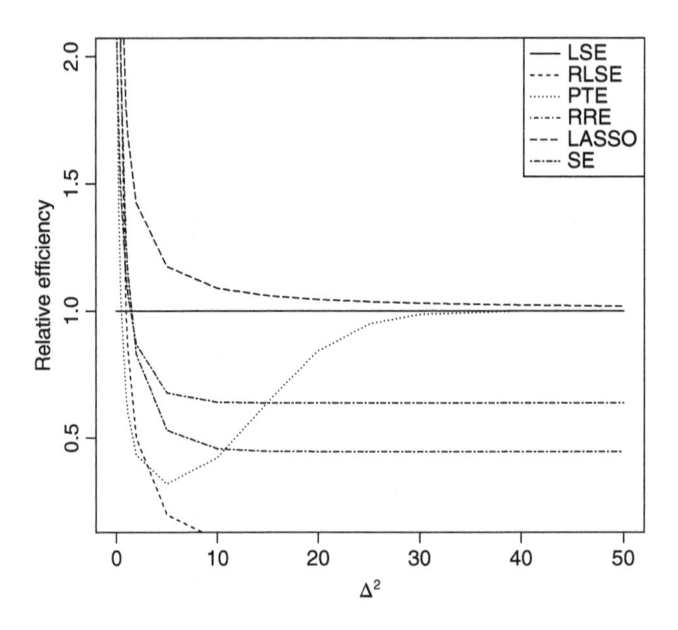

Figure 2.3 Relative efficiency of the estimators for θ.

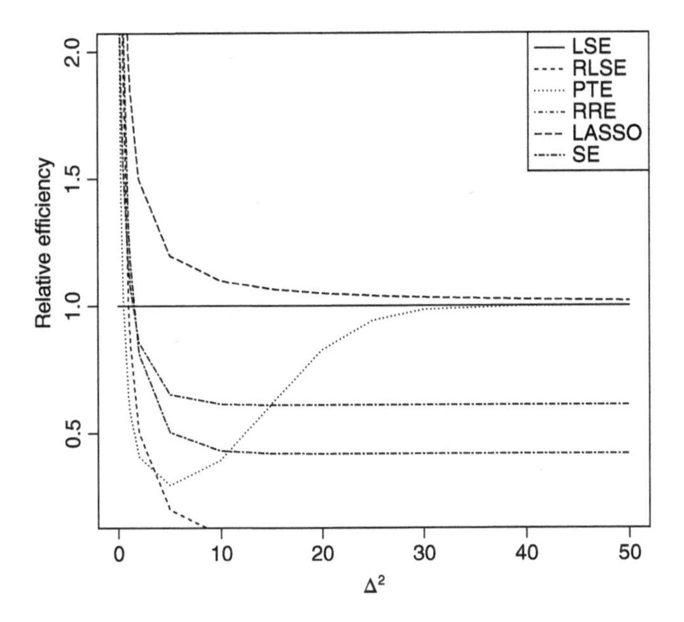

Figure 2.4 Relative efficiency of the estimators for β.

2.4 Summary and Concluding Remarks

This chapter considers the location model and the simple linear regression model when errors of the models are normally distributed. We consider LSE, RLSE, PTE, SE and two penalty estimators, namely, the RRE and the LASSO estimator for the location parameter for the location model and the intercept and slope parameter for the simple linear regression model. We found that the RRE uniformly dominates LSE, PTE, SE, and LASSO. However, RLSE dominates all estimators near the null hypothesis. LASSO dominates LSE, PTE, and SE uniformly.

Problems

2.1 Derive the estimate in (2.4) using the least squares method.

2.2 (a) Consider the simple location model $\mathbf{Y} = \theta 1_n + \xi$ and show that for testing the null-hypothesis $\mathcal{H}_o : \theta = 0$ against of $\mathcal{H}_A : \theta \neq 0$, the test statistic is

$$\mathcal{Z}_n = \frac{n\tilde{\theta}_n^2}{\sigma^2} \quad \text{or} \quad \mathcal{Z}_n = \frac{\sqrt{n}\tilde{\theta}_n}{\sigma}, \quad \text{for known } \sigma^2$$

and

$$\mathcal{T}_n = \frac{n\tilde{\theta}_n^2}{s_n^2} \quad \text{or} \quad \mathcal{T}_n = \frac{\sqrt{n}\tilde{\theta}_n}{s_n}, \quad \text{for unknown } \sigma^2,$$

where s_n^2 is the unbiased estimator of σ^2.

(b) What will be the distribution for $\mathcal{Z}_n$ and $\mathcal{T}_n$ under null and alternative hypotheses?

2.3 Show that the optimum value of ridge parameter k is Δ^{-2} and

$$\text{MSE}(\tilde{\theta}_n^R(k)) = \frac{\sigma^2}{n}\left(\frac{\Delta^2}{1+\Delta^2}\right).$$

2.4 Consider LASSO for location parameter θ and show that

$$\begin{aligned}\text{MSE}(\hat{\theta}_n^{\text{LASSO}}(\lambda)) = \frac{\sigma^2}{n}\{&1+\lambda^2+(\lambda+\Delta)\phi(\lambda-\Delta)\\ &+(\Delta^2-\lambda^2-1)[\Phi(\lambda-\Delta)-\Phi(-\lambda-\Delta)]\}.\end{aligned}$$

2.5 Prove Theorem 2.3.

2.6 Consider the simple linear model and derive the estimates given in (2.32) using the least squares method.

2.7 Consider the simple linear model and show that to test the null-hypothesis $\mathcal{H}_o : \beta = \beta_o$ vs. $\mathcal{H}_A : \beta \neq \beta_o$. Then, we use the LR test statistic

$$\mathcal{L}_n^{(\sigma)} = \frac{(\tilde{\beta}_n - \beta_o)^2 Q}{\sigma^2}, \qquad \text{if } \sigma^2 \text{ is known}$$

$$\mathcal{L}_n^{(s)} = \frac{(\tilde{\beta}_n - \beta_o)^2 Q}{s_n^2}, \qquad \text{if } \sigma^2 \text{ is unknown,}$$

where $\mathcal{L}_n^{(\sigma)}$ follows a noncentral chi-square distribution with 1 DF and noncentrality parameter $\Delta^2/2$ and $\mathcal{L}_n^{(s)}$ follows a noncentral F-distribution with $(1, m)$, $m = n-2$ DF and noncentral parameter,

2.8 Prove Theorems 2.4 and 2.5.

2.9 Consider the LASSO estimation of the intercept and slope models and show that the bias and MSE of $\hat{\theta}_n^{\text{LASSO}}$ are, respectively,

$$b(\hat{\theta}_n^{\text{LASSO}}(\lambda)) = \beta\overline{x} - \mathbb{E}[\hat{\beta}_n^{\text{LASSO}}(\lambda)]\overline{x}$$

$$\text{MSE}(\hat{\theta}_n^{\text{LASSO}}(\lambda)) = \frac{\sigma^2}{n}\left(1 + \frac{\overline{x}}{Q}\rho_{\text{ST}}(\lambda, \Delta)\right).$$

2.10 Show that when σ^2 is known, LASSO outperforms the PTE whenever

$$0 \leq \Delta^2 \leq \frac{1 - H_3\left(Z^2_{\frac{\alpha}{2}}; \Delta^2\right)}{1 - 2H_3\left(Z^2_{\frac{\alpha}{2}}; \Delta^2\right) + H_5\left(Z^2_{\frac{\alpha}{2}}; \Delta^2\right)}.$$

3

ANOVA Model

3.1 Introduction

An important model belonging to the class of general linear hypothesis is the analysis of variance (ANOVA) model. In this model, we consider the assessment of p treatment effects by considering sample experiments of sizes n_1, $n_2, \ldots, n_p$, respectively, with the responses $\{(y_{i1}, \ldots, y_{in_i})^\top; i = 1, 2, \ldots, p\}$ which satisfy the model, $y_{ij} = \theta_i + e_{ij}$ ($j = 1, \ldots, n_i$, $i = 1, \ldots, p$). The main objective of the chapter is the selection of the treatments which would yield best results. Accordingly, we consider the penalty estimators, namely, ridge, subset selection rule, and least absolute shrinkage and selection operator (LASSO) together with the classical shrinkage estimators, namely, the preliminary test estimator (PTE), the Stein-type estimators (SE), and positive-rule Stein-type estimator (PRSE) of $\boldsymbol{\theta} = (\theta_1, \ldots, \theta_p)^\top$. For LASSO and related methods, see Breiman (1996), Fan and Li (2001), Zou and Hastie (2005), and Zou (2006), among others; and for PTE and SE, see Judge and Bock (1978) and Saleh (2006), among others.

The chapter points to the useful "selection" aspect of LASSO and ridge estimators as well as limitations found in other papers. Our conclusions are based on the ideal L_2 risk of LASSO of an oracle which would supply optimal coefficients in a diagonal projection scheme given by Donoho and Johnstone (1994, p. 437). The comparison of the estimators considered here are based on mathematical analysis as well as by tables of L_2-risk efficiencies and graphs and not by simulation.

In his pioneering paper, Tibshirani (1996) examined the relative performance of the subset selection, ridge regression, and LASSO in three different scenarios, under orthogonal design matrix in a linear regression model:

(a) *Small number of large coefficients*: subset selection does the best here, the LASSO not quite as well, ridge regression does quite poorly.
(b) *Small to moderate numbers of moderate-size coefficients*: LASSO does the best, followed by ridge regression and then subset selection.

Theory of Ridge Regression Estimation with Applications, First Edition.
A.K. Md. Ehsanes Saleh, Mohammad Arashi, and B.M. Golam Kibria.

(c) *Large number of small coefficients*: ridge regression does the best by a good margin, followed by LASSO and then subset selection.

These results refer to *prediction accuracy*.

Recently, Hansen (2016) considered the comparison of LASSO, SE, and subset selection based on the upper bounds of L_2 risk under an infeasible condition: all parameters may be zero. His findings may be summarized as follows:

(i) The abovementioned condition yielded an infeasible estimate equal to the $\mathbf{0}$ vector with the L_2 risk equal to the divergence parameter, Δ^2. This led to a doubt of the "oracle properties" of LASSO.
(ii) Neither LASSO nor SE uniformly dominate one other.
(iii) Via simulation studies, he concludes that LASSO estimation is particularly sensitive to coefficient parametrization; and that for a significant portion of the parameter space, LASSO has higher L_2 risk than the least squares estimator (LSE).

He did not specify the regions where one estimator or the other has lower L_2 risk. In his analysis, he used the normalized L_2-risk bounds (NRB) to arrive at his conclusion with a total sparse model.

Hansen (2016) did not specify the regions where one estimator or the other has lower L_2 risk. In his analysis, he used the NRB to arrive at his conclusion with a total sparse model.

3.2 Model, Estimation, and Tests

Consider the ANOVA model

$$Y = B\theta + \epsilon = B_1\theta_1 + B_2\theta_2 + \epsilon, \tag{3.1}$$

where $\boldsymbol{Y} = (y_{11}, \dots, y_{1n_1}, \dots, y_{p_1}, \dots, y_{pn_p})^\top, \boldsymbol{\theta} = (\theta_1, \dots, \theta_{p_1}, \theta_{p_1+1}, \dots, \theta_p)^\top$ is the unknown vector that can be partitioned as $\boldsymbol{\theta} = (\boldsymbol{\theta}_1^\top, \boldsymbol{\theta}_2^\top)^\top$, where $\boldsymbol{\theta}_1 = (\theta_1, \dots, \theta_{p_1})^\top$, and $\boldsymbol{\theta}_2 = (\theta_{p_1+1}, \dots, \theta_p)^\top$.

The error vector $\boldsymbol{\epsilon}$ is $(\epsilon_{11}, \dots, \epsilon_{1n_1}, \dots, \epsilon_{p_1}, \dots, \epsilon_{pn_p})^\top$ with $\boldsymbol{\epsilon} \sim \mathcal{N}_n(\mathbf{0}, \sigma^2 \boldsymbol{I}_n)$. The notation $\boldsymbol{B}$ stands for a block-diagonal vector of $(\mathbf{1}_{n_1}, \dots, \mathbf{1}_{n_p})$ which can subdivide into two matrices $\boldsymbol{B}_1$ and $\boldsymbol{B}_2$ as $(\boldsymbol{B}_1\ ,\ \boldsymbol{B}_2)$, where $\mathbf{1}_{n_i} = (1, \dots, 1)^\top$ is an n_i-tuples of 1s, $\boldsymbol{I}_n$ is the n-dimensional identity matrix where $n = n_1 + \cdots + n_p$, and σ^2 is the known variance of the errors.

Our objective is to estimate and select the treatments $\boldsymbol{\theta} = (\theta_1, \dots, \theta_p)^\top$ when we suspect that the subset $\boldsymbol{\theta}_2 = (\theta_{p_1+1}, \dots, \theta_p)^\top$ may be $\mathbf{0}$, i.e. ineffective. Thus, we consider the model (3.1) and discuss the LSE of $\boldsymbol{\theta}$ in Section 3.2.1.

3.2.1 Estimation of Treatment Effects

First, we consider the unrestricted LSE of $\boldsymbol{\theta} = (\boldsymbol{\theta}_1^\top, \boldsymbol{\theta}_2^\top)^\top$ given by

$$\begin{aligned}
\tilde{\boldsymbol{\theta}}_n &= \text{argmin}_{\boldsymbol{\theta}}\{(\boldsymbol{Y} - \boldsymbol{B}_1\boldsymbol{\theta}_1 - \boldsymbol{B}_2\boldsymbol{\theta}_2)^\top(\boldsymbol{Y} - \boldsymbol{B}_1\boldsymbol{\theta}_1 - \boldsymbol{B}_2\boldsymbol{\theta}_2)\} \\
&= \begin{pmatrix} \boldsymbol{B}_1^\top\boldsymbol{B}_1 & \boldsymbol{B}_1^\top\boldsymbol{B}_2 \\ \boldsymbol{B}_2^\top\boldsymbol{B}_1 & \boldsymbol{B}_2^\top\boldsymbol{B}_2 \end{pmatrix}^{-1} \begin{pmatrix} \boldsymbol{B}_1^\top\boldsymbol{Y} \\ \boldsymbol{B}_2^\top\boldsymbol{Y} \end{pmatrix} = \begin{pmatrix} \boldsymbol{N}_1 & \boldsymbol{0} \\ \boldsymbol{0} & \boldsymbol{N}_2 \end{pmatrix}^{-1} \begin{pmatrix} \boldsymbol{B}_1^\top\boldsymbol{Y} \\ \boldsymbol{B}_2^\top\boldsymbol{Y} \end{pmatrix} \\
&= \begin{pmatrix} \boldsymbol{N}_1^{-1}\boldsymbol{B}_1^\top\boldsymbol{Y} \\ \boldsymbol{N}_2^{-1}\boldsymbol{B}_2^\top\boldsymbol{Y} \end{pmatrix} = \begin{pmatrix} \tilde{\boldsymbol{\theta}}_{1n} \\ \tilde{\boldsymbol{\theta}}_{2n} \end{pmatrix},
\end{aligned}$$

where $\boldsymbol{N} = \boldsymbol{B}^\top\boldsymbol{B} = \text{Diag}(n_1, \dots, n_p)$, $\boldsymbol{N}_1 = \text{Diag}(n_1, \dots, n_{p_1})$, and $\boldsymbol{N}_2 = \text{Diag}(n_{p_1+1}, \dots, n_p)$.

In case σ^2 is unknown, the best linear unbiased estimator (BLUE) of σ^2 is given by

$$s_n^2 = (n-p)^{-1}(\boldsymbol{Y} - \boldsymbol{B}_1\tilde{\boldsymbol{\theta}}_{1n} - \boldsymbol{B}_2\tilde{\boldsymbol{\theta}}_{2n})^\top(\boldsymbol{Y} - \boldsymbol{B}_1\tilde{\boldsymbol{\theta}}_{1n} - \boldsymbol{B}_2\tilde{\boldsymbol{\theta}}_{2n}).$$

Clearly, $\tilde{\boldsymbol{\theta}}_n \sim \mathcal{N}_p(\boldsymbol{\theta}, \sigma^2\boldsymbol{N}^{-1})$ is independent of ms_n^2/σ^2 $(m = n - p)$, which follows a central χ^2 distribution with m degrees of freedom (DF).

When $\boldsymbol{\theta}_2 = \boldsymbol{0}$, the restricted least squares estimator (RLSE) of $\boldsymbol{\theta}_{\text{R}} = (\boldsymbol{\theta}_1^\top, \boldsymbol{0}^\top)^\top$ is given by $\hat{\boldsymbol{\theta}}_{\text{R}} = (\tilde{\boldsymbol{\theta}}_{1n}^\top, \boldsymbol{0}^\top)^\top$, where $\tilde{\boldsymbol{\theta}}_{1n} = \boldsymbol{N}_1^{-1}\boldsymbol{B}_1^\top\boldsymbol{Y}$.

3.2.2 Test of Significance

For the test of $\mathcal{H}_o : \boldsymbol{\theta}_2 = \boldsymbol{0}$ vs. $\mathcal{H}_A : \boldsymbol{\theta}_2 \neq \boldsymbol{0}$, we consider the statistic $\mathcal{L}_n$ given by

$$\begin{aligned}
\mathcal{L}_n &= \frac{1}{\sigma^2}\tilde{\boldsymbol{\theta}}_{2n}^\top\boldsymbol{N}_2\tilde{\boldsymbol{\theta}}_{2n}, \qquad \text{if } \sigma^2 \text{ is known} \\
&= \frac{1}{p_2 s_n^2}\tilde{\boldsymbol{\theta}}_{2n}^\top\boldsymbol{N}_2\tilde{\boldsymbol{\theta}}_{2n}, \quad \text{if } \sigma^2 \text{ is unknown.}
\end{aligned} \tag{3.2}$$

Under a null-hypothesis $\mathcal{H}_o$, the null distribution of $\mathcal{L}_n$ is the central χ^2 distribution with p_2 DF. when σ^2 is known and the central F-distribution with (p_2, m) DF. in the case of σ^2 being unknown, respectively. Under the alternative hypothesis, $\mathcal{H}_A$, the test statistics $\mathcal{L}_n$ follows the noncentral version of the mentioned densities. In both cases, the noncentrality parameter is $\Delta^2 = \boldsymbol{\theta}_2^\top\boldsymbol{N}_2\boldsymbol{\theta}_2/\sigma^2$. In this paper, we always assume that σ^2 is known, then $\mathcal{L}_n$ follows a chi-square distribution with p_2 DF.

Further, we note that

$$\tilde{\theta}_{jn} \sim \mathcal{N}(\theta_j, \sigma^2 n_j^{-1}), \quad j = 1, \dots, p \tag{3.3}$$

so that $\mathcal{Z}_j = \sqrt{n_j}\tilde{\theta}_{jn}/\sigma \sim \mathcal{N}(\Delta_j, 1)$, where $\Delta_j = \sqrt{n_j}\theta_j/\sigma$. Thus, one may use $\mathcal{Z}_j$ to test the null-hypothesis $\mathcal{H}_o^{(j)} : \theta_j = 0$ vs. $\mathcal{H}_A^{(j)} : \theta_j \neq 0, j = p_1 + 1, \dots, p$.

In this chapter, we are interested in studying three penalty estimators, namely,

(i) the subset rule called "hard threshold estimator" (HTE),
(ii) LASSO or the "soft threshold estimator" (STE),
(iii) the "ridge regression estimator" (RRE),
(iv) the classical PTE and shrinkage estimators such as "Stein estimator" (SE) and "positive-rule Stein-type estimator" (PRSE).

3.2.3 Penalty Estimators

In this section, we discuss the penalty estimators. Define the HTE as

$$\begin{aligned}\hat{\boldsymbol{\theta}}_n^{\mathrm{HT}}(\kappa) &= \left(\tilde{\theta}_{jn} I\left(|\tilde{\theta}_{jn}| > \kappa\sigma n_j^{-\frac{1}{2}}\right) | j=1,\ldots,p\right)^{\top} \\ &= \left(\sigma n_j^{-\frac{1}{2}} \mathcal{Z}_j I(|\mathcal{Z}_j| > \kappa) | j=1,\ldots,p\right)^{\top},\end{aligned} \tag{3.4}$$

where κ is a positive threshold parameter.

This estimator is discrete in nature and may be extremely variable and unstable due to the fact that small changes in the data can result in very different models and can reduce the prediction accuracy. As such we obtain the continuous version of $\hat{\boldsymbol{\theta}}_n^{\mathrm{HT}}(\kappa)$, and the LASSO is defined by

$$\hat{\boldsymbol{\theta}}_n^{\mathrm{L}}(\lambda) = \mathrm{argmin}_{\boldsymbol{\theta}}(\boldsymbol{Y} - \boldsymbol{B}\boldsymbol{\theta})^{\top}(\boldsymbol{Y} - \boldsymbol{B}\boldsymbol{\theta}) + 2\lambda\sigma \sum_{j=1}^{p} \sqrt{n_j}\kappa|\theta_j|,$$

where $|\boldsymbol{\theta}| = (|\theta_1|, \ldots, |\theta_p|)^{\top}$, yielding the equation

$$\boldsymbol{B}^{\top}\boldsymbol{B}\boldsymbol{\theta} - \boldsymbol{B}^{\top}\boldsymbol{Y} + \lambda\sigma\boldsymbol{N}^{\frac{1}{2}} \operatorname{sgn}(\boldsymbol{\theta}) = \boldsymbol{0}$$

or

$$\hat{\boldsymbol{\theta}}_n^{\mathrm{L}}(\lambda) - \tilde{\boldsymbol{\theta}}_n + \frac{1}{2}\lambda\sigma\boldsymbol{N}^{-\frac{1}{2}} \operatorname{sgn}(\hat{\boldsymbol{\theta}}_n^{\mathrm{L}}(\lambda)) = \boldsymbol{0}. \tag{3.5}$$

Now, the jth component of (3.5) is given by

$$\hat{\theta}_{jn}^{\mathrm{L}}(\lambda) - \tilde{\theta}_{jn} + \lambda\sigma n_j^{-\frac{1}{2}} \operatorname{sgn}(\hat{\theta}_{jn}^{\mathrm{L}}(\lambda)) = 0. \tag{3.6}$$

Then, we consider three cases:

(i) $\operatorname{sgn}(\hat{\theta}_{jn}^{\mathrm{L}}(\lambda)) = +1$, then, (3.6) reduces to

$$0 < \frac{\hat{\theta}_{jn}^{\mathrm{L}}(\lambda)}{\sigma n_j^{-\frac{1}{2}}} - \frac{\tilde{\theta}_{jn}}{\sigma n_j^{-\frac{1}{2}}} + \lambda = 0.$$

Hence,

$$0 < \hat{\theta}_{jn}^{\mathrm{L}}(\lambda) = \sigma n_j^{-\frac{1}{2}}(\mathcal{Z}_j - \lambda) = \sigma n_j^{-\frac{1}{2}}(|\mathcal{Z}_j| - \lambda), \tag{3.7}$$

with, clearly, $\mathcal{Z}_j > 0$ and $|\mathcal{Z}_j| > \lambda$.

(ii) $\text{sgn}(\hat{\theta}^{\text{L}}_{jn}(\lambda)) = -1$; then we have

$$0 > \frac{\hat{\theta}^{\text{L}}_{jn}(\lambda)}{\sigma n_j^{-\frac{1}{2}}} = \mathcal{Z}_j + \lambda = -(|\mathcal{Z}_j| - \lambda), \tag{3.8}$$

with, clearly, $\mathcal{Z}_j < 0$ and $|\mathcal{Z}_j| > \lambda$, and

(iii) For $\hat{\theta}^{\text{L}}_{jn}(\lambda) = 0$, we have $-\mathcal{Z}_j + \lambda\gamma = 0$ for some $\gamma \in (-1, 1)$. Hence, we obtain $\mathcal{Z}_j = \lambda\gamma$, which implies $|\mathcal{Z}_j| < \lambda$.

Combining (3.6)–(3.8) and (iii), we obtain

$$\hat{\theta}^{\text{L}}_{jn}(\lambda) = \sigma n_j^{-\frac{1}{2}} \text{sgn}(\mathcal{Z}_j)(|\mathcal{Z}_j| - \lambda)^+, \quad j = 1, \dots, p,$$

where $a^+ = \max(0, a)$. Hence, the LASSO is given by

$$\hat{\boldsymbol{\theta}}^{\text{L}}_n(\lambda) = \left(\sigma n_j^{-\frac{1}{2}} \text{sgn}(\mathcal{Z}_j)(|\mathcal{Z}_j| - \lambda)^+ | j = 1, \dots, p\right)^\top. \tag{3.9}$$

Next, we consider the RRE given by

$$\hat{\boldsymbol{\theta}}^{\text{RR}}_n(k) = \begin{pmatrix} \tilde{\boldsymbol{\theta}}_{1n} \\ \frac{1}{1+k}\tilde{\boldsymbol{\theta}}_{2n} \end{pmatrix}, \quad \kappa \in \mathbb{R}^+ \tag{3.10}$$

to accommodate a sparse condition.

We may obtain $\tilde{\boldsymbol{\theta}}^{\text{RR}}_n(k)$ equal to $\tilde{\boldsymbol{\theta}}_{1n}$ when $\boldsymbol{\theta}_2 = \mathbf{0}$ and $\tilde{\boldsymbol{\theta}}_{2n}(k) = \frac{1}{1+k}\tilde{\boldsymbol{\theta}}_{2n}$ by minimizing the objective function,

$$(\boldsymbol{Y} - \boldsymbol{B}_1\tilde{\boldsymbol{\theta}}_{1n} - \boldsymbol{B}_2\boldsymbol{\theta}_2)^\top(\boldsymbol{Y} - \boldsymbol{B}_1\tilde{\boldsymbol{\theta}}_{1n} - \boldsymbol{B}_2\boldsymbol{\theta}_2) + k\boldsymbol{\theta}_2^\top\boldsymbol{N}_2\boldsymbol{\theta}_2$$

with respect to $\boldsymbol{\theta}_2$. Thus, Eq. (3.10) is a "feasible estimator of $\boldsymbol{\theta}$" when $\boldsymbol{\theta}_2$ consists of small-sized parameters.

3.2.4 Preliminary Test and Stein-Type Estimators

We recall that the unrestricted estimator of $\boldsymbol{\theta} = (\boldsymbol{\theta}_1^\top, \boldsymbol{\theta}_2^\top)^\top$ is given by $(\tilde{\boldsymbol{\theta}}_{1n}^\top, \tilde{\boldsymbol{\theta}}_{2n}^\top)^\top$ with marginal distribution $\tilde{\boldsymbol{\theta}}_{1n} \sim \mathcal{N}_{p_1}(\boldsymbol{\theta}_1, \sigma^2\boldsymbol{N}_1^{-1})$ and $\tilde{\boldsymbol{\theta}}_{2n} \sim \mathcal{N}_{p_2}(\boldsymbol{\theta}_2, \sigma^2\boldsymbol{N}_2^{-1})$, respectively. The restricted estimator of $(\boldsymbol{\theta}_1^\top, \mathbf{0}^\top)^\top$ is $(\tilde{\boldsymbol{\theta}}_{1n}^\top, \mathbf{0}^\top)^\top$. Similarly, the PTE of $\boldsymbol{\theta}$ is given by

$$\hat{\boldsymbol{\theta}}^{\text{PT}}_n(\alpha) = \begin{pmatrix} \tilde{\boldsymbol{\theta}}_{1n} \\ \tilde{\boldsymbol{\theta}}_{2n} I(\mathcal{L}_n > c_\alpha) \end{pmatrix},$$

where $I(A)$ is the indicator function of the set A, $\mathcal{L}_n$ is the test statistic given in Section 2.2, and c_α is the α-level critical value.

Similarly, the Stein estimator (SE) is given by

$$\hat{\boldsymbol{\theta}}^{\text{S}}_n = \begin{pmatrix} \tilde{\boldsymbol{\theta}}_{1n} \\ \tilde{\boldsymbol{\theta}}_{2n}(1 - (p_2 - 2)\mathcal{L}_n^{-1}) \end{pmatrix}, \quad p_2 \geq 3$$

and the positive-rule Stein-type estimator (PRSE) is given by

$$\hat{\boldsymbol{\theta}}_n^{\mathrm{S+}} = \begin{pmatrix} \tilde{\boldsymbol{\theta}}_{1n} \\ \hat{\boldsymbol{\theta}}_{2n}^{\mathrm{S}} I(\mathcal{L}_n > p_2 - 2) \end{pmatrix}.$$

3.3 Bias and Weighted L$_2$ Risks of Estimators

This section contains the bias and the weighted L$_2$-risk expressions of the estimators. We study the comparative performance of the seven estimators defined on the basis of the weighted L$_2$ risks defined by

$$\begin{aligned} R(\boldsymbol{\theta}_n^* : \boldsymbol{W}_1, \boldsymbol{W}_2) = & \mathbb{E}[(\boldsymbol{\theta}_{1n}^* - \boldsymbol{\theta}_1)^\top \boldsymbol{W}_1 (\boldsymbol{\theta}_{1n}^* - \boldsymbol{\theta}_1)] \\ & + \mathbb{E}[(\boldsymbol{\theta}_{2n}^* - \boldsymbol{\theta}_2)^\top \boldsymbol{W}_2 (\boldsymbol{\theta}_{2n}^* - \boldsymbol{\theta}_2)], \end{aligned} \tag{3.11}$$

where $\boldsymbol{\theta}_n^* = (\boldsymbol{\theta}_{1n}^{*\top}, \boldsymbol{\theta}_{2n}^{*\top})^\top$ is any estimator of $\boldsymbol{\theta} = (\boldsymbol{\theta}_1^\top, \boldsymbol{0}_2^\top)^\top$, and $\boldsymbol{W}_1$ and $\boldsymbol{W}_2$ are weight matrices. For convenience, when $\boldsymbol{W}_1 = \boldsymbol{I}_{p_1}$ and $\boldsymbol{W}_2 = \boldsymbol{I}_{p_2}$, we get the mean squared error (MSE) and write $R(\boldsymbol{\theta}_n^* : \boldsymbol{I}_p) = \mathbb{E}[\| \boldsymbol{\theta}_n^* - \boldsymbol{\theta}\|^2]$.

First, we note that for LSE,

$$\begin{aligned} b(\tilde{\boldsymbol{\theta}}_n) &= \boldsymbol{0} \\ R(\tilde{\boldsymbol{\theta}}_n : \boldsymbol{N}_1, \boldsymbol{N}_2) &= \sigma^2 (p_1 + p_2), \end{aligned}$$

and for RLSE, $\hat{\boldsymbol{\theta}}_{\mathrm{R}} = (\tilde{\boldsymbol{\theta}}_{1n}^\top, \boldsymbol{0}^\top)^\top$, we have

$$\begin{aligned} b(\hat{\boldsymbol{\theta}}_{\mathrm{R}}) &= (\boldsymbol{0}^\top, \boldsymbol{\theta}_2^\top) \\ R(\hat{\boldsymbol{\theta}}_{\mathrm{R}}; \boldsymbol{N}_1, \boldsymbol{N}_2) &= \sigma^2 (p_1 + \Delta^2). \end{aligned}$$

3.3.1 Hard Threshold Estimator (Subset Selection Rule)

The bias of this estimator is given by

$$b(\hat{\boldsymbol{\theta}}_n^{\mathrm{HT}}(\kappa)) = \left(-\sigma n_j^{-\frac{1}{2}} \Delta_j H_3(\kappa^2; \Delta_j^2) | j = 1, \dots, p \right)^\top,$$

where $H_\nu(\cdot; \Delta_j^2)$ is the cumulative distribution function (c.d.f.) of a noncentral χ^2 distribution with ν DF. and noncentrality parameter Δ_j^2 $(j = 1, \dots, p)$.

The MSE of $\hat{\boldsymbol{\theta}}_n^{\mathrm{HT}}(\kappa)$ is given by

$$\begin{aligned} R(\hat{\boldsymbol{\theta}}_n^{\mathrm{HT}}(\kappa) : \boldsymbol{I}_p) &= \sum_{j=1}^p \mathbb{E}\left[\tilde{\theta}_{jn} I\left(|\tilde{\theta}_{jn}| > \kappa \sigma n_j^{-\frac{1}{2}} \right) - \theta_j \right]^2 \\ &= \sigma^2 \sum_{j=1}^p n_j^{-1} \{ (1 - H_3(\kappa^2; \Delta_j^2)) \\ &\quad + \Delta_j^2 (2 H_3(\kappa^2; \Delta_j^2) - H_5(\kappa^2; \Delta_j^2)) \}. \end{aligned} \tag{3.12}$$

Since $\left[\tilde{\theta}_{jn} I\left(|\tilde{\theta}_{jn}| > \kappa\sigma n_j^{-\frac{1}{2}}\right) - \theta_j\right]^2 \leq (\tilde{\theta}_{jn} - \theta_j)^2 + \theta_j^2$, we obtain

$$R(\hat{\boldsymbol{\theta}}_n^{\mathrm{HT}}(\kappa) : \boldsymbol{I}_p) \leq \sigma^2 \operatorname{tr} \boldsymbol{N}^{-1} + \boldsymbol{\theta}^\top\boldsymbol{\theta} \quad \text{(free of } \kappa).$$

Following Donoho and Johnstone (1994), one can show that what follows holds:

$$R(\hat{\boldsymbol{\theta}}_n^{\mathrm{HT}}(\kappa) : \boldsymbol{I}_p) \leq \begin{cases} \text{(i)} \;\; \sigma^2(1+\kappa^2) \operatorname{tr} \boldsymbol{N}^{-1} & \forall\, \boldsymbol{\theta} \in \mathbb{R}^p,\; \kappa > 1, \\ \text{(ii)} \;\; \sigma^2 \operatorname{tr} \boldsymbol{N}^{-1} + \boldsymbol{\theta}^\top\boldsymbol{\theta} & \forall\, \boldsymbol{\theta} \in \mathbb{R}^p, \\ \text{(iii)} \;\; \sigma^2 \rho_{\mathrm{HT}}(\kappa, 0) \operatorname{tr} \boldsymbol{N}^{-1} & \\ \qquad +1.2\, \boldsymbol{\theta}^\top\boldsymbol{\theta} & \boldsymbol{0} < \boldsymbol{\theta} < \kappa \boldsymbol{1}_p^\top, \end{cases}$$

where $\rho_{\mathrm{HT}}(\kappa, 0) = 2[(1 - \Phi(\kappa)) + \kappa\varphi(\kappa)]$, and $\varphi(\cdot)$ and $\Phi(\cdot)$ are the probability density function (p.d.f.) and c.d.f. of standard normal distribution, respectively.

Theorem 3.1 *Under the assumed regularity conditions, the weighted* L_2*-risk bounds are given by*

$$R(\hat{\boldsymbol{\theta}}_n^{\mathrm{HT}}(\kappa) : \boldsymbol{N}_1, \boldsymbol{N}_2) \leq \begin{cases} \text{(i)} \;\; \sigma^2(1+\kappa^2)(p_1+p_2) & \kappa > 1, \\ \text{(ii)} \;\; \sigma^2(p_1+p_2) + \boldsymbol{\theta}_1^\top \boldsymbol{N}_1 \boldsymbol{\theta}_1 & \\ \qquad +\boldsymbol{\theta}_2^\top \boldsymbol{N}_2 \boldsymbol{\theta}_2 & \forall \boldsymbol{\theta} \in \mathbb{R}^p, \\ \text{(iii)} \;\; \sigma^2 \rho_{\mathrm{HT}}(\kappa, 0)(p_1+p_2) & \\ \qquad +1.2\{\boldsymbol{\theta}_1^\top \boldsymbol{N}_1 \boldsymbol{\theta}_1 + \boldsymbol{\theta}_2^\top \boldsymbol{N}_2 \boldsymbol{\theta}_2\} & 0 < \boldsymbol{\theta} < k\boldsymbol{1}_p^\top. \end{cases}$$

If the solution of $\hat{\boldsymbol{\theta}}_n^{\mathrm{HT}}(\kappa)$ has the configuration $(\tilde{\boldsymbol{\theta}}_{1n}^\top, \boldsymbol{0}^\top)^\top$, then the L_2 risk of $\hat{\boldsymbol{\theta}}_n^{\mathrm{HT}}(\kappa)$ is given by

$$R(\hat{\boldsymbol{\theta}}_n^{\mathrm{HT}}(\kappa) : \boldsymbol{N}_1, \boldsymbol{N}_2) = \sigma^2[p_1 + \Delta^2],$$

independent of κ.

3.3.2 LASSO Estimator

The bias expression of the LASSO estimator is given by

$$b(\boldsymbol{\theta}_n^{\mathrm{L}}(\lambda)) = \left(\sigma n_j^{-\frac{1}{2}}[\lambda(2\Phi(\Delta_j) - 1); j = 1, \ldots, p_1; -\Delta_{p_1+1}, \ldots, \Delta_p\right)^\top.$$

The MSE of the LASSO estimator has the form

$$R(\hat{\boldsymbol{\theta}}_n^{\mathrm{L}}(\lambda) : \boldsymbol{I}_p) = \sigma^2 \sum_{j=1}^{p_1} n_j^{-1} \rho_{\mathrm{ST}}(\lambda, \Delta_j) + \Delta^2,$$

where

$$\rho_{\mathrm{ST}}(\lambda,\Delta_j) = (1+\lambda^2)\{1-\Phi(\lambda-\Delta_j)+\Phi(-\lambda-\Delta_j)\}$$
$$+\Delta_j^2\{\Phi(\lambda-\Delta_j)-\Phi(-\lambda-\Delta_j)\}$$
$$-\{(\lambda-\Delta_j)\varphi(\lambda+\Delta_j)+(\lambda+\Delta_j)\varphi(\lambda-\Delta_j)\}.$$

Thus, according to Donoho and Johnstone (1994, Appendix 2), we have the following result.

Under the assumed regularity conditions,

$$R(\hat{\boldsymbol{\theta}}_n^{\mathrm{L}}(\lambda) : \boldsymbol{I}_p) \le \begin{cases} \text{(i)} \quad \sigma^2(1+\lambda^2) \operatorname{tr} \boldsymbol{N}^{-1} & \forall\, \boldsymbol{\theta} \in \mathbb{R}^p,\ \kappa > 1, \\ \text{(ii)} \quad \sigma^2 \operatorname{tr} \boldsymbol{N}^{-1} + \boldsymbol{\theta}^\top\boldsymbol{\theta} & \forall\, \boldsymbol{\theta} \in \mathbb{R}^p, \\ \text{(iii)} \quad \sigma^2\rho_{\mathrm{ST}}(\lambda,0) \operatorname{tr} \boldsymbol{N}^{-1} + 1.2\boldsymbol{\theta}^\top\boldsymbol{\theta} & \forall \boldsymbol{\theta} \in \mathbb{R}^p, \end{cases} \tag{3.13}$$

where $\rho_{\mathrm{ST}}(\lambda,0) = 2[(1+\lambda^2)(1-\Phi(\lambda)) - \kappa\phi(\lambda)]$.

If the solution of $\hat{\boldsymbol{\theta}}_n^{\mathrm{L}}(\lambda)$ has the configuration $(\hat{\boldsymbol{\theta}}_{1n}^\top, \boldsymbol{0}^\top)$, then the L_2 risk of $\hat{\boldsymbol{\theta}}_n^{\mathrm{L}}(\lambda)$ is given by

$$R(\hat{\boldsymbol{\theta}}_n^{\mathrm{L}}(\lambda) : \boldsymbol{N}_1, \boldsymbol{N}_2) = \sigma^2(p_1 + \Delta^2). \tag{3.14}$$

Thus, we note that

$$R(\hat{\boldsymbol{\theta}}_n; \boldsymbol{N}_1, \boldsymbol{N}_2) = R(\hat{\boldsymbol{\theta}}_n^{\mathrm{HT}}(\kappa); \boldsymbol{N}_1, \boldsymbol{N}_2) = R(\hat{\boldsymbol{\theta}}_n^{\mathrm{L}}(\lambda); \boldsymbol{N}_1, \boldsymbol{N}_2) = \sigma^2(p_1 + \Delta^2).$$

To prove the L_2 risk of LASSO, we consider the multivariate decision theory. We are given the LSE of $\boldsymbol{\theta}$ as $\tilde{\boldsymbol{\theta}}_n = (\tilde{\theta}_{1n}, \dots, \tilde{\theta}_{pn})^\top$ according to

$$\tilde{\theta}_{jn} = \theta_j + \sigma n_j^{-\frac{1}{2}} \mathcal{Z}_j, \quad \mathcal{Z}_j \sim \mathcal{N}(0,1),$$

where $\sigma n_j^{-\frac{1}{2}}$ is the marginal variance of $\tilde{\theta}_{jn}$ and noise level, and $\{\theta_j\}_{j=1,\dots,p}$ are the treatment effects of interest. We measure the quality of the estimators based on the L_2 risk, $R(\tilde{\boldsymbol{\theta}}_n : \boldsymbol{I}_p) = \mathbb{E}[\| \tilde{\boldsymbol{\theta}}_n - \boldsymbol{\theta}\|^2]$. Note that for a sparse solution, we use (3.11).

Consider the family of diagonal linear projections,

$$\boldsymbol{T}_{\mathrm{DP}}(\hat{\boldsymbol{\theta}}_n^{\mathrm{L}}(\lambda) : \boldsymbol{\delta}) = (\delta_1\hat{\theta}_{1n}^{\mathrm{L}}(\lambda), \dots, \delta_p\hat{\theta}_{pn}^{\mathrm{L}}(\lambda))^\top, \tag{3.15}$$

where $\boldsymbol{\delta} = (\delta_1, \dots, \delta_p)^\top$, $\delta_j \in (0,1), j = 1, \dots, p$. Such estimators "kill" or "keep" the coordinates.

Suppose we had available an oracle which would supply for us the coefficients δ_j optimal for use in the diagonal projection scheme (3.15). These "ideal" coefficients are $\delta_j = I\left(|\theta_j| > \sigma n_j^{-\frac{1}{2}}\right)$. Ideal diagonal projections consist of estimating only those θ_j, which are larger than its noise, $\sigma n_j^{-\frac{1}{2}}$ $(j = 1, \dots, p)$. These yield the "ideal" L_2 risk given by (3.16).

Then, the ideal diagonal coordinates in our study are $I\left(|\theta_j| > \sigma n_j^{-\frac{1}{2}}\right)$. These coordinates estimate those treatment effects θ_j which are larger than the noise level $\sigma n_j^{-1/2}$, yielding the "ideal" L_2 risk as

$$R(\hat{\boldsymbol{\theta}}_n^{\mathrm{L}}(\lambda) : \boldsymbol{I}_p) = \sum_{j=1}^{p} \min(\theta_j^2, \sigma^2 n_j^{-1})$$
$$= \begin{cases} \sigma^2 \operatorname{tr} \boldsymbol{N}^{-1} & \forall\ |\theta_j| > \sigma n_j^{-\frac{1}{2}},\ j = 1, \dots, p, \\ \boldsymbol{\theta}^\top \boldsymbol{\theta} & \forall\ |\theta_j| < \sigma n_j^{-\frac{1}{2}},\ j = 1, \dots, p. \end{cases} \tag{3.16}$$

Thus, we find that the ideal lower bound of the L_2 risk of $\hat{\boldsymbol{\theta}}_n^{\mathrm{L}}(\lambda)$ leading to a "keep" or "kill" solution is $(\tilde{\boldsymbol{\theta}}_{1n}^\top, \boldsymbol{0}^\top)^\top$. Thus, $R(\boldsymbol{T}_{\mathrm{DP}}; \boldsymbol{N}_1, \boldsymbol{N}_2) = \sigma^2(p_1 + \Delta^2)$, which is the same as (3.16).

In general, the L_2 risk given by (3.16) cannot be achieved for all $\boldsymbol{\theta}$ by any linear or nonlinear estimator of treatment effects. However, in the sparse case, if p_1 treatment effects $|\theta_j|$ exceed $\sigma n_j^{-1/2}$ and p_2 coefficients are null, then we obtain the ideal L_2 risk given by

$$R(\hat{\boldsymbol{\theta}}_n^{\mathrm{L}}(\lambda); \boldsymbol{I}_p) = \sigma^2 \operatorname{tr} \boldsymbol{N}_1^{-1} + \boldsymbol{\theta}_2^\top \boldsymbol{\theta}_2.$$

This ideal L_2 risk happens to be the lower bound of the L_2 risk given by (3.17). We shall use this ideal L_2 risk to compare with the L_2 risk of other estimators. Consequently, the lower bound of the weighted L_2 risk is given by

$$R(\hat{\boldsymbol{\theta}}_n^{\mathrm{L}}(\lambda); \boldsymbol{N}_1, \boldsymbol{N}_2) = \sigma^2(p_1 + \Delta^2). \tag{3.17}$$

3.3.3 Ridge Regression Estimator

Recall that the RRE is given by

$$\hat{\boldsymbol{\theta}}_n^{\mathrm{RR}}(k) = \begin{pmatrix} \tilde{\boldsymbol{\theta}}_{1n} \\ \frac{1}{1+k} \tilde{\boldsymbol{\theta}}_{2n} \end{pmatrix}, \quad k \in \mathbb{R}^+. \tag{3.18}$$

The bias and MSE of $\hat{\boldsymbol{\theta}}_n^{\mathrm{RR}}(k)$ have forms

$$b(\hat{\boldsymbol{\theta}}_n^{\mathrm{RR}}(k)) = \begin{pmatrix} \boldsymbol{0} \\ -\frac{k}{1+k} \boldsymbol{\theta}_2 \end{pmatrix}$$

and

$$R(\hat{\boldsymbol{\theta}}_n^{\mathrm{RR}}(k) : \boldsymbol{I}_p) = \sigma^2 \operatorname{tr} \boldsymbol{N}_1^{-1} + \frac{1}{(1+k)^2} (\sigma^2 \operatorname{tr} \boldsymbol{N}_2^{-1} + k^2 \boldsymbol{\theta}_2^\top \boldsymbol{\theta}_2).$$

Hence, the weighted L_2 risk is obtained as

$$R(\hat{\boldsymbol{\theta}}_n^{\mathrm{RR}}(k); \boldsymbol{N}_1, \boldsymbol{N}_2) = \sigma^2 p_1 + \frac{\sigma^2}{(1+k)^2} (p_2 + k^2 \Delta^2).$$

One may find the optimum value $k = k_o = p_2\Delta^{-2}$, yielding

$$R(\hat{\theta}_n^{\mathrm{RR}}(k_o); N_1, N_2) = \sigma^2\left(p_1 + \frac{p_2\Delta^2}{p_2+\Delta^2}\right). \tag{3.19}$$

3.4 Comparison of Estimators

In this section, we compare various estimators with respect to LSE, in terms of relative weighted L_2-risk efficiency (RWRE).

We recall that for a sparse solution, the L_2 risk of LASSO is $\sigma^2(p_1 + \Delta^2)$, as shown in Eq. (3.14), which is also the "ideal" L_2 risk in an "ideal" diagonal projection scheme. Therefore, we shall use $\sigma^2(p_1 + \Delta^2)$ to compare the L_2-risk function of other estimators.

3.4.1 Comparison of LSE with RLSE

Recall that the RLSE is given by $\hat{\theta}_n = (\tilde{\theta}_{1n}^\top, \mathbf{0}^\top)^\top$. In this case, the RWRE of RLSE vs. LSE is given by

$$\mathrm{RWRE}(\hat{\theta}_n : \tilde{\theta}_n) = \frac{p_1+p_2}{p_1+\Delta^2} = \left(1+\frac{p_2}{p_1}\right)\left(1+\frac{\Delta^2}{p_1}\right)^{-1},$$

which is a decreasing function of Δ^2. So, $0 \le \mathrm{RWRE}(\hat{\theta}_n : \tilde{\theta}_n) \le \left(1+\frac{p_2}{p_1}\right)$.

3.4.2 Comparison of LSE with PTE

Here, it is easy to see that

$$\begin{aligned} R(\hat{\theta}_n^{\mathrm{PT}}(\alpha); N_1, N_2) = {} & p_1 + p_2(1 - H_{p_2+2}(c_\alpha; \Delta^2)) \\ & + \Delta^2[2H_{p_2+2}(c_\alpha; \Delta^2) - H_{p_2+4}(c_\alpha; \Delta^2)]. \end{aligned} \tag{3.20}$$

Then, the RWRE expression for PTE vs. LSE is given by

$$\mathrm{RWRE}(\hat{\theta}_n^{\mathrm{PT}}(\alpha) : \tilde{\theta}_n) = \frac{p_1+p_2}{g(\Delta^2, \alpha)},$$

where

$$g(\Delta^2, \alpha) = p_1 + p_2(1 - H_{p_2+2}(c_\alpha; \Delta^2)) + \Delta^2[2H_{p_2+2}(c_\alpha; \Delta^2) - H_{p_2+4}(c_\alpha; \Delta^2)].$$

Then, the PTE outperforms the LSE for

$$0 \le \Delta^2 \le \frac{p_2 H_{p_2+2}(c_\alpha; \Delta^2)}{2H_{p_2+2}(c_\alpha; \Delta^2) - H_{p_2+4}(c_\alpha; \Delta^2)} = \Delta^2_{\mathrm{PT}}. \tag{3.21}$$

Otherwise, LSE outperforms the PTE in the interval $(\Delta^2_{\mathrm{PT}}, \infty)$. We may mention that $\mathrm{RWRE}(\hat{\theta}_n^{\mathrm{PT}}(\alpha) : \tilde{\theta}_n)$ is a decreasing function of Δ^2. It has a maximum at $\Delta^2 = 0$, decreases crossing the 1-line to a minimum at $\Delta^2 = \Delta^2_{\mathrm{PT}}(\min)$ with a value $M_{\mathrm{PT}}(\alpha)$, and then increases toward 1-line. This means the gains in efficiency of PTE is the highest in the interval given by Eq. (3.21) and loss in efficiency can be noticed outside it.

The $\mathrm{RWRE}(\hat{\theta}_n^{\mathrm{PT}}; \tilde{\theta}_n)$ belongs to the interval

$$M_{\mathrm{PT}}(\alpha) \leq \mathrm{RWRE}(\hat{\theta}_n^{\mathrm{PT}}(\alpha); \tilde{\theta}_n) \leq \left(1+\frac{p_2}{p_1}\right)\left(1+\frac{p_2}{p_1}[1-H_{p_2+2}(c_\alpha;0)]\right)^{-1},$$

where $M_{\mathrm{PT}}(\alpha)$ depends on the size α and is given by

$$M_{\mathrm{PT}}(\alpha) = \left(1+\frac{p_2}{p_1}\right)\left\{1+\frac{p_2}{p_1}[1-H_{p_2+2}(c_\alpha;\Delta^2_{\mathrm{PT}}(\min))]\right.$$
$$\left.+\frac{\Delta^2_{\mathrm{PT}}(\min)}{p_1}[2H_{p_2+2}(c_\alpha;\Delta^2_{\mathrm{PT}}(\min)) - H_{p_2+4}(c_\alpha;\Delta^2_{\mathrm{PT}}(\min))]\right\}^{-1}.$$

The quantity $\Delta^2_{\mathrm{PT}}(\min)$ is the value Δ^2 at which the RWRE value is minimum.

3.4.3 Comparison of LSE with SE and PRSE

Since SE and PRSE need $p_2 \geq 3$ to express their weighted L_2 risk ($\mathrm{WL_2R}$) expressions, we always assume that $p_2 \geq 3$. First, note that

$$R(\hat{\theta}_n^{\mathrm{S}}; \boldsymbol{N}_1, \boldsymbol{N}_2) = p_1 + p_2 - (p_2-2)^2\mathbb{E}[\chi^{-2}_{p_2}(\Delta^2)].$$

As a result, we obtain

$$\mathrm{RWRE}(\hat{\theta}_n^{\mathrm{S}}; \tilde{\theta}_n) = \left(1+\frac{p_2}{p_1}\right)\left(1+\frac{p_2}{p_1}-\frac{(p_2-2)^2}{p_1}\mathbb{E}[\chi^{-2}_{p_2}(\Delta^2)]\right)^{-1}.$$

It is a decreasing function of Δ^2. At $\Delta^2 = 0$, its value is $\left(1+\frac{p_2}{p_1}\right)\left(1+\frac{2}{p_1}\right)^{-1}$; and when $\Delta^2 \to \infty$, its value goes to 1. Hence, for $\Delta^2 \in \mathbb{R}^+$,

$$1 \leq \left(1+\frac{p_2}{p_1}\right)\left(1+\frac{p_2}{p_1}-\frac{(p_2-2)^2}{p_1}\mathbb{E}[\chi^{-2}_{p_2}(\Delta^2)]\right)^{-1} \leq \left(1+\frac{p_2}{p_1}\right)\left(1+\frac{2}{p_1}\right)^{-1}.$$

Hence, the gain in efficiency is the highest when Δ^2 is small and drops toward 1 when Δ^2 is the largest.

Also,

$$\mathrm{RWRE}(\hat{\theta}_n^{\mathrm{S+}}; \tilde{\theta}_n) = \left(1+\frac{p_2}{p_1}\right)\left(1+\frac{p_2}{p_1}-\frac{(p_2-2)^2}{p_1}\mathbb{E}[\chi^{-2}_{p_2}(\Delta^2)]\right.$$
$$-\frac{p_2}{p_1}\mathbb{E}[(1-(p_2-2)\chi^{-2}_{p_2+2}(\Delta^2))^2 I(\chi^2_{p_2+2}(\Delta^2) < (p_2-2))]$$
$$+\frac{\Delta^2}{p_1}\{2\mathbb{E}[(1-(p_2-2)\chi^{-2}_{p_2+2}(\Delta^2))I(\chi^2_{p_2+2}(\Delta^2) < (p_2-2))]$$
$$\left.-\mathbb{E}[(1-(p_2-2)\chi^{-2}_{p_2+4}(\Delta^2))^2 I(\chi^2_{p_2+4}(\Delta^2) < (p_2-2))]\}\right)^{-1}.$$

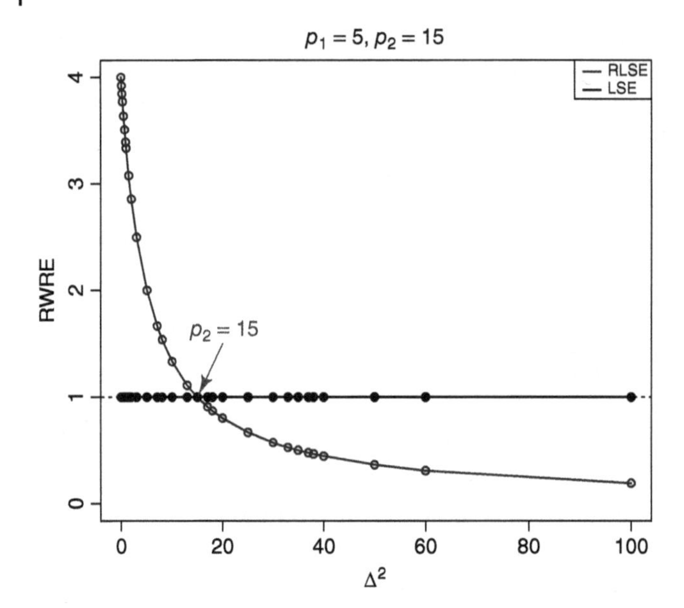

Figure 3.1 RWRE for the restricted estimator.

So that,

$$\mathrm{RWRE}(\hat{\theta}_n^{\mathrm{S+}}; \tilde{\theta}_n) \geq \mathrm{RWRE}(\hat{\theta}_n^{\mathrm{S}}; \tilde{\theta}_n) \geq 1 \qquad \forall \Delta^2 \in \mathbb{R}^+.$$

We also provide a graphical representation (Figures 3.1–3.3) of RWRE of the estimators.

In Section 3.4.4, we show that RRE uniformly dominates all other estimators, although it does not select variables.

3.4.4 Comparison of LSE and RLSE with RRE

First we consider weighted L_2 risk difference of LSE and RRE given by

$$\begin{aligned}\sigma^2(p_1 + p_2) - \sigma^2 p_1 - \sigma^2 \frac{p_2\Delta^2}{p_2 + \Delta^2} &= \sigma^2 p_2 \left(1 - \frac{\Delta^2}{p_2 + \Delta^2}\right) \\ &= \frac{\sigma^2 p_2^2}{p_2 + \Delta^2} > 0, \quad \forall\, \Delta^2 \in \mathbb{R}^+.\end{aligned}$$

Hence, RRE outperforms the LSE uniformly. Similarly, for the RLSE and RRE, the weighted L_2-risk difference is given by

$$\sigma^2(p_1 + \Delta^2) - \left(\sigma^2 p_1 + \frac{\sigma^2 p_2 \Delta^2}{p_2 + \Delta^2}\right) = \frac{\sigma^2 \Delta^4}{p_2 + \Delta^2} > 0.$$

Therefore, RRE performs better than RLSE uniformly.

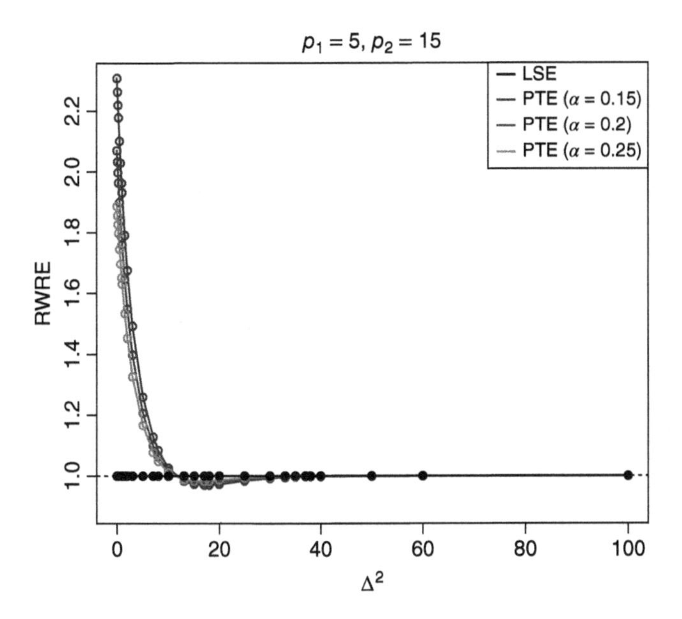

Figure 3.2 RWRE for the preliminary test estimator.

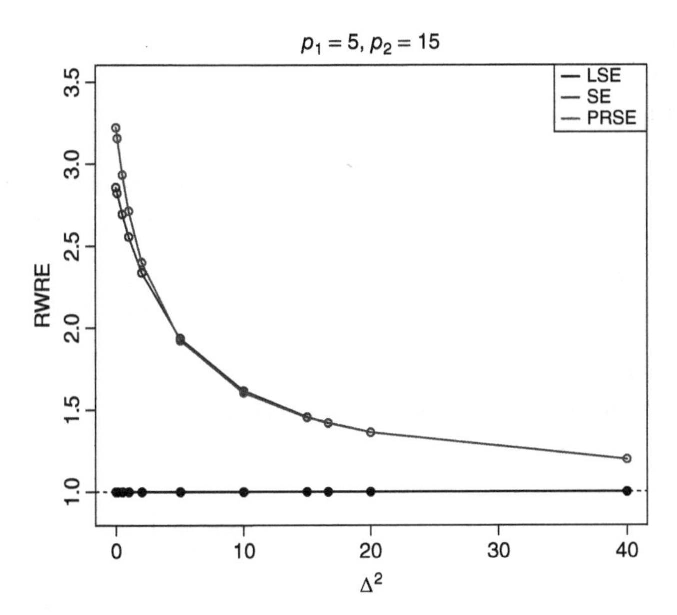

Figure 3.3 RWRE for the Stein-type and its positive-rule estimator.

In addition, the RWRE of RRE vs. LSE equals

$$\text{RWRE}(\hat{\theta}_n^{\text{RR}}(k_o) : \tilde{\theta}_n) = \frac{p_1 + p_2}{p_1 + \frac{p_2\Delta^2}{p_2+\Delta^2}} = \left(1 + \frac{p_2}{p_1}\right)\left(1 + \frac{p_2\Delta^2}{p_1(p_2+\Delta^2)}\right)^{-1},$$

which is a decreasing function of Δ^2 with maximum $\left(1+\frac{p_2}{p_1}\right)$ at $\Delta^2 = 0$ and minimum 1 as $\Delta^2 \to \infty$. So,

$$1 \le \left(1 + \frac{p_2}{p_1}\right)\left(1 + \frac{p_2}{p_1\left(1+\frac{p_2}{\Delta}\right)}\right)^{-1} \le 1 + \frac{p_2}{p_1}; \quad \forall \Delta^2 \in \mathbb{R}^+.$$

3.4.5 Comparison of RRE with PTE, SE, and PRSE

3.4.5.1 Comparison Between $\hat{\theta}_n^{\text{RR}}(k_{\text{opt}})$ and $\hat{\theta}_n^{\text{PT}}(\alpha)$

Here, the weighted L_2-risk difference of $\hat{\theta}_n^{\text{PT}}(\alpha)$ and $\hat{\theta}_n^{\text{RR}}(k_{\text{opt}})$ is given by

$$\begin{aligned} R(\hat{\theta}_n^{\text{PT}}(\alpha); \boldsymbol{N}_1, \boldsymbol{N}_2) &- R(\hat{\theta}_n^{\text{RR}}(k_{\text{opt}}) : \boldsymbol{N}_1, \boldsymbol{N}_2) \\ &= \sigma^2[p_2(1 - H_{p_2+2}(c_\alpha; \Delta^2)), \\ &\quad + \Delta^2\{2H_{p_2+2}(c_\alpha; \Delta^2) - H_{p_2+4}(c_\alpha; \Delta^2)\}] \\ &\quad - \frac{\sigma^2 p_2\Delta^2}{p_2+\Delta^2}. \end{aligned} \tag{3.22}$$

Note that the risk of $\hat{\beta}_{2n}^{\text{PT}}(\alpha)$ is an increasing function of Δ^2 crossing the p_2-line to a maximum and then drops monotonically toward the p_2-line as $\Delta^2 \to \infty$. The value of the risk is $p_2(1 - H_{p_2+2}(\chi^2_{p_2}(\alpha); 0))(< p_2)$ at $\Delta^2 = 0$. On the other hand, $\frac{p_2\Delta^2}{p_2+\Delta^2}$ is an increasing function of Δ^2 below the p_2-line with a minimum value 0 at $\Delta^2 = 0$ and as $\Delta^2 \to \infty$, $\frac{p_2\Delta^2}{p_2+\Delta^2} \to p_2$. Hence, the risk difference in Eq. (3.22) is nonnegative for $\Delta^2 \in R^+$. Thus, the RRE uniformly performs better than PTE.

3.4.5.2 Comparison Between $\hat{\theta}_n^{\text{RR}}(k_{\text{opt}})$ and $\hat{\theta}_n^{\text{S}}$

The weighted L_2-risk difference of $\hat{\theta}_n^{\text{S}}$ and $\hat{\theta}_n^{\text{RR}}(k_{\text{opt}})$ is given by

$$\begin{aligned} R(\hat{\theta}_n^{\text{S}} : \boldsymbol{N}_1, \boldsymbol{N}_2) &- R(\hat{\theta}_n^{\text{RR}}(k_{\text{opt}}) : \boldsymbol{N}_1, \boldsymbol{N}_2) \\ &= \sigma^2(p_1 + p_2 - (p_2-2)^2)\mathbb{E}[\chi_{p_2}^{-2}(\Delta^2)] - \sigma^2\left(p_1 + \frac{p_2\Delta^2}{p_2+\Delta^2}\right) \\ &= \sigma^2\left[p_2 - (p_2-2)^2\mathbb{E}[\chi_{p_2}^{-2}(\Delta^2)] - \frac{p_2\Delta^2}{p_2+\Delta^2}\right]. \end{aligned} \tag{3.23}$$

Note that the first function is increasing in Δ^2 with a value 2 at $\Delta^2 = 0$; and that as $\Delta^2 \to \infty$, it tends to p_2. The second function is also increasing in Δ^2 with a value 0 at $\Delta^2 = 0$ and approaches the value p_2 as $\Delta^2 \to \infty$. Hence, the risk difference is nonnegative for all $\Delta^2 \in R^+$. Consequently, RRE outperforms SE uniformly.

3.4.5.3 Comparison of $\hat{\theta}_n^{\mathrm{RR}}(k_{\mathrm{opt}})$ with $\hat{\theta}_n^{\mathrm{S+}}$

The risk of $\hat{\boldsymbol{\theta}}_n^{\mathrm{S+}}$ is

$$R(\hat{\boldsymbol{\theta}}_n^{\mathrm{S+}}; \boldsymbol{N}_1, \boldsymbol{N}_2) = R(\hat{\boldsymbol{\theta}}_n^{\mathrm{S}}; \boldsymbol{N}_1, \boldsymbol{N}_2) - \mathrm{R}^*, \tag{3.24}$$

where

$$\begin{aligned}\mathrm{R}^* &= \sigma^2 p_2 \mathbb{E}[(1-(p_2-2)\chi_{p_2+2}^{-2}(\Delta^2))^2 I(\chi_{p_2+2}^{-2}(\Delta^2) < p_2 - 2)] \\ &\quad - \sigma^2\Delta^2\{2\mathbb{E}[(1-(p_2-2)\chi_{p_2+2}^{-2}(\Delta^2)) I(\chi_{p_2+2}^{-2}(\Delta^2) < p_2 - 2)] \\ &\quad - \mathbb{E}[(1-(p_2-2)\chi_{p_2+4}^{-2}(\Delta^2))^2 I(\chi_{p_2+4}^{-2}(\Delta^2) < p_2 - 2)]\};\end{aligned} \tag{3.25}$$

and $\mathrm{R}(\hat{\boldsymbol{\theta}}_n^{\mathrm{S}} : \boldsymbol{N}_1, \boldsymbol{N}_2)$ is

$$R(\hat{\boldsymbol{\theta}}_n^{\mathrm{S}} : \boldsymbol{N}_1, \boldsymbol{N}_2) = \sigma^2(p_1 + p_2 - (p_2-2)^2\mathbb{E}[\chi_{p_2}^{-2}(\Delta^2)])$$

The weighted L_2-risk difference of PR and RRE is given by

$$\begin{aligned}&R(\hat{\boldsymbol{\theta}}_n^{\mathrm{S+}} : \boldsymbol{N}_1, \boldsymbol{N}_2) - R(\hat{\boldsymbol{\theta}}_n^{\mathrm{RR}}(k_{\mathrm{opt}}) : \boldsymbol{N}_1, \boldsymbol{N}_2) \\ &\quad = [R(\hat{\boldsymbol{\theta}}_n^{\mathrm{S}} : \boldsymbol{N}_1, \boldsymbol{N}_2) - \mathrm{R}^*] - R(\hat{\boldsymbol{\theta}}_n^{\mathrm{RR}}(k_{\mathrm{opt}}) : \boldsymbol{N}_1, \boldsymbol{N}_2) \geq 0,\end{aligned} \tag{3.26}$$

where

$$R(\hat{\boldsymbol{\theta}}_n^{\mathrm{RR}}(k_{\mathrm{opt}}); \boldsymbol{N}_1, \boldsymbol{N}_2) = \sigma^2\left(p_1 + \frac{p_2\Delta^2}{p_2 + \Delta^2}\right)$$

Consider the $\mathrm{R}(\hat{\boldsymbol{\theta}}_n^{\mathrm{S+}})$. It is a monotonically increasing function of Δ^2. At $\Delta^2 = 0$, its value is $\sigma^2(p_1+2) - \sigma^2 p_2\mathbb{E}[(1-(p_2-2)\chi_{p_2+2}^{-2}(0))^2 I(\chi_{p_2+2}^{-2}(0) < p_2 - 2)] \geq 0$; and as $\Delta^2 \to \infty$, it tends to $\sigma^2(p_1+p_2)$. For $R(\hat{\boldsymbol{\theta}}_n^{\mathrm{RR}}(k_{\mathrm{opt}}); \boldsymbol{N}_1, \boldsymbol{N}_2)$, at $\Delta^2 = 0$, the value is $\sigma^2 p_1$; and as $\Delta^2 \to \infty$, it tends to $\sigma^2(p_1 + p_2)$. Hence, the L_2 risk difference in (3.21) is nonnegative and RRE uniformly outperforms PRSE. Note that the risk difference of $\hat{\boldsymbol{\theta}}_n^{\mathrm{S+}}$ and $\hat{\boldsymbol{\theta}}_n^{\mathrm{RR}}(k_{\mathrm{opt}})$ at $\Delta^2 = 0$ is

$$\begin{aligned}&\sigma^2(p_1+2) - \sigma^2 p_2\mathbb{E}[(1-(p_2-2)\chi_{p_2+2}^{-2}(0))^2 I(\chi_{p_2+2}^{-2}(0) < p_2 - 2)] - \sigma^2 p_1 \\ &\quad = \sigma^2(2 - p_2\mathbb{E}[(1-(p_2-2)\chi_{p_2+2}^{-2}(0))^2 I(\chi_{p_2+2}^{-2}(0) < p_2 - 2)]) \geq 0.\end{aligned} \tag{3.27}$$

Because the expected value in Eq. (3.27) is a decreasing function of DF, and $2 > p_2\mathbb{E}[(1-(p_2-2)\chi_{p_2+2}^{-2}(0))^2 I(\chi_{p_2+2}^{-2}(0) < p_2 - 2)]$. The risk functions of RRE, PTE, SE, and PRSE are plotted in Figures 3.4 and 3.5 for $p_1 = 5, p_2 = 15$ and $p_1 = 7, p_2 = 33$, respectively. These figures are in support of the given comparisons.

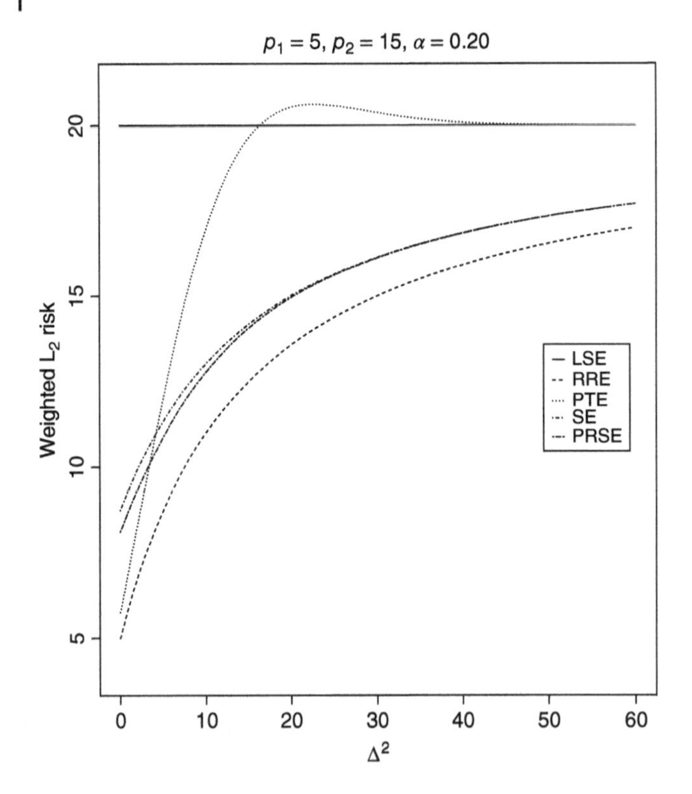

Figure 3.4 Weighted L_2 risk for the ridge, preliminary test, and Stein-type and its positive-rule estimators for $p_1 = 5$, $p_2 = 15$, and $\alpha = 0.20$.

3.4.6 Comparison of LASSO with LSE and RLSE

First, note that if we have for p_1 coefficients, $|\beta_j| > \sigma n_j^{-1/2}$ and also p_2 coefficients are zero in a sparse solution, then the "ideal" weighted L_2 risk is given by $\sigma^2(p_1 + \Delta^2)$. Thereby, we compare all estimators relative to this quantity. Hence, weighted L_2 risk difference between LSE and LASSO is given by

$$\sigma^2(p_1 + p_2) - \sigma^2(p_1 + \Delta^2) = \sigma^2[p_2 - \Delta^2].$$

Hence, if $\Delta^2 \in (0, p_2)$, the LASSO estimator performs better than the LSE; while if $\Delta^2 \in (p_2, \infty)$ the LSE performs better than the LASSO. Consequently, neither LSE nor the LASSO performs better than the other, uniformly.

Next, we compare the RLSE and LASSO. In this case, the weighted L_2-risk difference is given by

$$\sigma^2(p_1 + \Delta^2) - \sigma^2(p_1 + \Delta^2) = 0. \tag{3.28}$$

Hence, LASSO and RLSE are L_2 risk equivalent. And consequently, the LASSO satisfies the oracle properties.

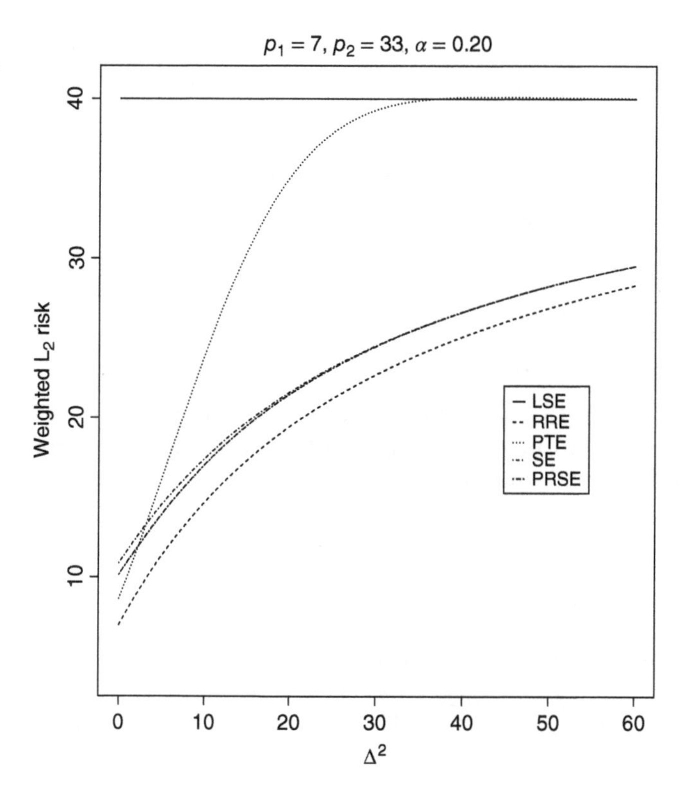

Figure 3.5 RWRE for the ridge, preliminary test, and Stein-type and its positive-rule estimators for $p_1 = 7, p_2 = 33$, and $\alpha = 0.20$.

3.4.7 Comparison of LASSO with PTE, SE, and PRSE

We first consider the PTE vs. LASSO. In this case, the weighted L_2-risk difference is given by

$$\begin{aligned}
&R(\hat{\boldsymbol{\theta}}_n^{\mathrm{PT}}(\alpha) : \boldsymbol{N}_1, \boldsymbol{N}_2) - R(\hat{\boldsymbol{\theta}}_n^{\mathrm{L}}(\lambda) : \boldsymbol{N}_1, \boldsymbol{N}_2) \\
&\quad = \sigma^2[p_2(1 - H_{p_2+2}(c_\alpha; \Delta^2)) - \Delta^2\{1 - 2H_{p_2+2}(c_\alpha; \Delta^2) + H_{p_2+4}(c_\alpha; \Delta^2)\}] \\
&\quad \geq \sigma^2 p_2(1 - H_{p_2+2}(c_\alpha; 0)) \geq 0, \quad \text{if } \Delta^2 = 0.
\end{aligned}$$

Hence, the LASSO outperforms the PTE when $\Delta^2 = 0$. But, when $\Delta^2 \neq 0$, the LASSO outperforms the PTE for

$$0 \leq \Delta^2 \leq \frac{p_2[1 - H_{p_2+2}(c_\alpha; \Delta^2)]}{1 - 2H_{p_2+2}(c_\alpha; \Delta^2) + H_{p_2+4}(c_\alpha; \Delta^2)}.$$

Otherwise, PTE outperforms the LASSO. Hence, neither LASSO nor PTE outmatches the other uniformly.

Next, we consider SE and PRSE vs. LASSO. In these two cases, we have weighted L_2-risk differences given by

$$\begin{aligned}
&R(\hat{\theta}_n^{S} : N_1, N_2) - R(\hat{\theta}_n^{L}(\lambda); N_1, N_2)\\
&\quad = \sigma^2[p_1 + p_2 - (p_2-2)^2\mathbb{E}[\chi^{-2}_{p_2+2}(\Delta^2)] - (p_1 + \Delta^2)]\\
&\quad = \sigma^2[p_2 - (p_2-2)^2\mathbb{E}[\chi^{-2}_{p_2+2}(\Delta^2)] - \Delta^2]
\end{aligned}$$

and from (3.19),

$$\begin{aligned}
&R(\hat{\theta}_n^{S+} : N_1, N_2) - R(\hat{\theta}_n^{L}(\lambda) : N_1, N_2)\\
&\quad = R(\hat{\theta}_n^{S} : N_1, N_2) - R(\hat{\theta}_n^{L}(\lambda) : N_1, N_2) - R^*,
\end{aligned}$$

where R^* is given by (3.20). Therefore, the LASSO outperforms the SE as well as the PRSE in the interval $[0, p_2 - (p_2-2)^2\mathbb{E}[\chi^{-2}_{p_2}(\Delta^2)]]$. Thus, neither SE nor PRSE outperforms the LASSO uniformly.

3.4.8 Comparison of LASSO with RRE

Here, the weighted L_2-risk difference is given by

$$\begin{aligned}
&R_4(\hat{\theta}_n^{L}(\lambda); N_1, N_2) - R_5(\hat{\theta}_n^{RR}(k_o); N_1, N_2)\\
&\quad = \sigma^2\left[(p_1 + \Delta^2) - \left(p_1 + \frac{p_2\Delta^2}{p_2 + \Delta^2}\right)\right]\\
&\quad = \frac{\sigma^2\Delta^2}{p_2 + \Delta^2} \geq 0.
\end{aligned}$$

Hence, the RRE outperforms the LASSO uniformly.

In Figure 3.6, the comparisons of LASSO with other estimators are shown.

3.5 Application

To illustrate the methodologies in Sections 3.3 and 3.4, we consider the following numerical example in this section. The data were generated from $\mathcal{N}(\theta_i, 1.25)$ distribution, where $\theta_i = (3, 1.5, 2.5, 0, 0, 4, 0, 0, 4.5, 0)^\top$ and $\sigma = 1.25$. The respective sample sizes are $n_i = (10, 15, 12, 20, 10, 15, 12, 20, 10, 16)^\top$. The generated data and some summary statistics are presented in Table 3.1.

First, we compute the LASSO estimator and find the following

$$\hat{\theta}_n^{L} = (2.535, 1.120, 2.075, 0.000, 0.000, 3.620, 0.000, 0.000, 4.035, 0.000)^\top$$

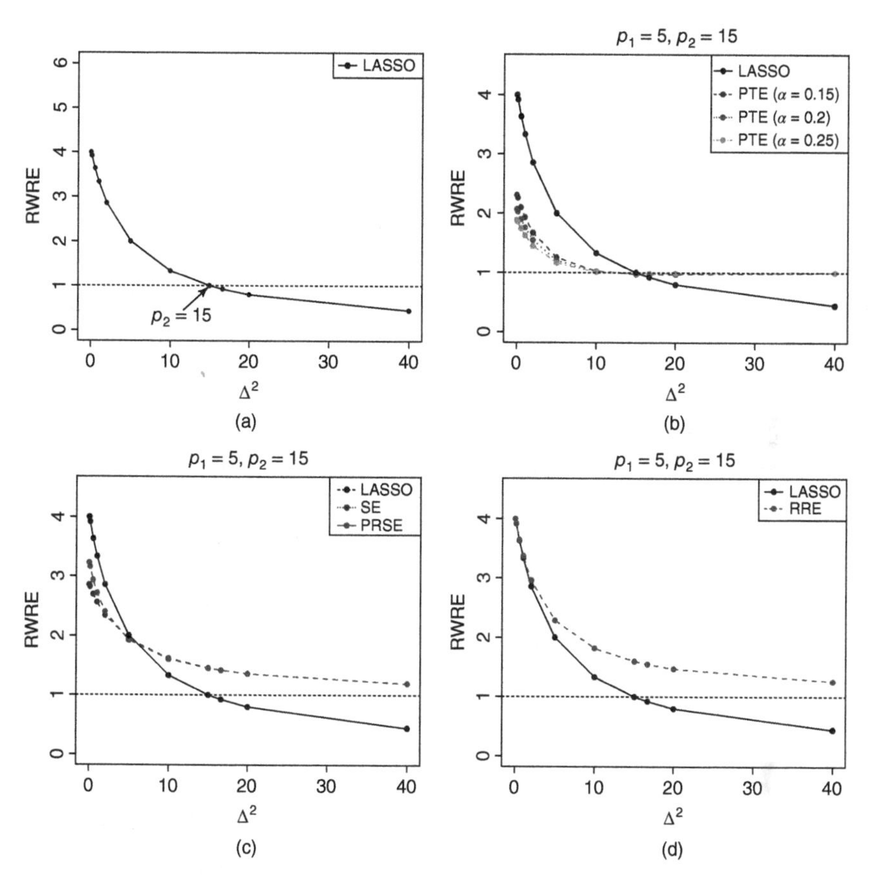

Figure 3.6 RWRE for LASSO, ridge, restricted, preliminary test, and Stein-type and its positive-rule estimators.

Since LASSO kills $\theta_4, \theta_5, \theta_7, \theta_8$, and θ_{10}, we will test the following hypothesis:

$$\mathcal{H}_o : \boldsymbol{\theta}_2 = \mathbf{0}, \quad \text{vs.} \quad \mathcal{H}_A : \boldsymbol{\theta}_2 \neq \mathbf{0},$$

where $\boldsymbol{\theta}_2 = (\theta_4, \theta_5, \theta_7, \theta_8, \theta_{10})^\top$.

Note that the estimate of the overall mean is $\hat{\theta}_0 = 1.250$. The computed value of the test statistic for known $\sigma = 1.25$ is 6.561, which is less than the critical value of $\chi^2_{0.95,5} = 9.488$; hence, the null hypothesis will not be rejected. The estimated values of different estimators for θ are given in Table 3.2.

Table 3.1 One-way ANOVA table.

	y_1	y_2	y_3	y_4	y_5	y_6	y_7	y_8	y_9	y_{10}
1	3.96	2.75	2.15	−0.77	0.58	5.13	0.05	2.37	1.98	−0.81
2	2.55	2.87	3.17	−0.51	−2.10	1.96	−1.05	1.47	3.35	0.95
3	0.94	1.36	1.44	−0.46	−1.44	2.12	0.92	0.43	7.12	−0.17
4	4.25	2.42	2.63	−0.19	0.07	3.38	−0.74	−2.21	3.84	−1.40
5	2.49	2.39	3.28	0.98	−1.60	3.99	1.09	1.40	5.12	−0.27
6	2.68	−0.16	6.68	−2.80	0.85	4.10	−1.85	0.98	3.81	0.81
7	3.68	1.10	0.97	0.72	−0.61	5.16	0.79	−0.37	5.23	−0.34
8	1.80	3.42	0.75	1.26	−2.19	3.66	0.57	0.16	5.31	1.00
9	2.02	1.74	2.51	1.10	−0.13	3.51	−0.44	0.66	3.89	0.36
10	0.46	3.41	0.18	−1.89	−2.28	5.99	−0.62	−0.55	4.06	0.63
11		2.26	3.38	1.19		3.50	−0.51	−0.55		0.45
12		3.44	4.78	−0.06		2.44	−2.08	−0.60		1.31
13		1.44		−1.65		5.94		0.27		−0.69
14		−0.15		−1.18		2.20		0.00		1.29
15		1.45		−0.65		3.27		−1.73		−0.86
16				1.14				−1.12		1.55
17				1.46				−2.78		
18				−0.01				0.57		
19				−0.24				−0.01		
20				0.65				0.71		
$\bar{y}_i$	2.48	1.98	2.66	−0.09	−0.88	3.76	−0.32	−0.45	4.37	0.23
s_i	1.24	1.16	1.81	1.18	1.19	1.32	1.04	1.26	1.39	0.90

Table 3.2 Estimated values of different estimators.

	LSE	RLSE	PTE	LASSO	SE	PRSE
1	2.489	2.489	2.489	2.535	2.489	2.489
2	1.986	1.986	1.986	1.120	1.986	1.986
3	2.663	2.663	2.663	2.075	2.663	2.663
4	3.760	3.760	3.760	0.000	3.760	3.760
5	4.375	4.375	4.375	0.000	4.375	4.375
6	−0.096	0.000	0.000	3.620	−0.067	−0.067
7	−0.886	0.000	0.000	0.000	−0.616	−0.616
8	−0.323	0.000	0.000	0.000	−0.225	−0.225
9	−0.045	0.000	0.000	4.035	−0.031	−0.031
10	0.239	2.489	2.489	0.000	0.166	0.166

For the computation of the estimators, we used formulas in Section 3.2 and prepared Table 3.1. For the comparison of weighted risks among the estimators, see Tables 3.3–3.10.

3.6 Efficiency in Terms of Unweighted L_2 Risk

In the previous sections, we have made all comparisons among the estimators in terms of weighted risk functions. In this section, we provide the L_2-risk efficiency of the estimators in terms of the unweighted (weight = $\boldsymbol{I}_p$) risk expressions.

The unweighted relative efficiency of LASSO:

$$\text{REff}(\hat{\theta}_n^{\text{L}} : \tilde{\theta}_n) = \left(1 + \frac{\text{tr}(\boldsymbol{N}_2^{-1})}{\text{tr}(\boldsymbol{N}_1^{-1})}\right)\left(1 + \frac{\Delta^{*2}}{\text{tr}(\boldsymbol{N}_1^{-1})}\right)^{-1}, \tag{3.29}$$

where $\Delta^{*2} = \Delta^2 \text{Ch}_{\max}(\boldsymbol{N}_2^{-1})$ or $\Delta^{*2} = \Delta^2 \text{Ch}_{\min}(\boldsymbol{N}_1^{-1})$. Note that $\hat{\theta}_n^{\text{L}} = \hat{\theta}_n^{\text{RE}}$.

The unweighted relative efficiency of PTE:

$$\begin{aligned}\text{REff}(\hat{\theta}_n^{\text{PT}}(\alpha) : \tilde{\theta}_n) = &\left(1 + \frac{\text{tr}(\boldsymbol{N}_2^{-1})}{\text{tr}(\boldsymbol{N}_1^{-1})}\right)\\ &\times\left\{1 + \frac{\text{tr}(\boldsymbol{N}_2^{-1})}{\text{tr}(\boldsymbol{N}_1^{-1})}(1 - H_{p_2+2}(c_\alpha; \Delta^2))\right.\\ &\left.+\frac{\Delta^{*2}}{\text{tr}(\boldsymbol{N}_1^{-1})}\{2H_{p_2+2}(c_\alpha; \Delta^2) - H_{p_2+4}(c_\alpha; \Delta^2)\}\right\}^{-1}\end{aligned} \tag{3.30}$$

The unweighted relative efficiency of SE:

$$\begin{aligned}\text{REff}(\hat{\theta}_n^{\text{S}} : \tilde{\theta}_n) = &\left(1 + \frac{\text{tr}(\boldsymbol{N}_2^{-1})}{\text{tr}(\boldsymbol{N}_1^{-1})}\right)\left\{1 + \frac{\text{tr}(\boldsymbol{N}_2^{-1})}{\text{tr}(\boldsymbol{N}_1^{-1})}(1 - (p_2 - 2)A)\right.\\ &\left.+(p_2^2 - 4)\frac{\Delta^{*2}}{\text{tr}(\boldsymbol{N}_1^{-1})}\mathbb{E}[\chi_{p_2+4}^{-4}(\Delta^2)]\right\}^{-1},\end{aligned} \tag{3.31}$$

where

$$A = 2\mathbb{E}[\chi_{p_2+2}^{-2}(\Delta^2)] - (p_2 - 2)\mathbb{E}[\chi_{p_2+4}^{-4}(\Delta^2)]$$

Table 3.3 RWRE for the estimators.

			PTE					
			α					
Δ^2	LSE	RLSE/ LASSO	0.15	0.2	0.25	SE	PRSE	RRE
				$p_1 = 5, p_2 = 15$				
0	1	4.00	2.30	2.07	1.89	2.86	3.22	4.00
0.1	1	3.92	2.26	2.03	1.85	2.82	3.16	3.92
0.5	1	3.64	2.10	1.89	1.74	2.69	2.93	3.64
1	1	3.33	1.93	1.76	1.63	2.56	2.71	3.36
2	1	2.86	1.67	1.55	1.45	2.33	2.40	2.96
3	1	2.50	1.49	1.40	1.33	2.17	2.19	2.67
5	1	2.00	1.26	1.21	1.17	1.94	1.92	2.26
7	1	1.67	1.13	1.10	1.08	1.78	1.77	2.04
10	1	1.33	1.02	1.02	1.01	1.62	1.60	1.81
15	1	1.00	0.97	0.97	0.98	1.46	1.45	1.60
20	1	0.80	0.97	0.98	0.98	1.36	1.36	1.47
30	1	0.57	0.99	0.99	0.99	1.25	1.25	1.33
50	1	0.36	0.99	0.99	1.00	1.16	1.16	1.21
100	1	0.19	1.00	1.00	1.00	1.05	1.05	1.11
				$p_1 = 7, p_2 = 33$				
0	1	5.71	2.86	2.50	2.23	4.44	4.92	5.71
0.1	1	5.63	2.82	2.46	2.20	4.40	4.84	5.63
0.5	1	5.33	2.66	2.34	2.10	4.23	4.57	5.34
1	1	5.00	2.49	2.20	1.98	4.03	4.28	5.02
2	1	4.44	2.21	1.97	1.80	3.71	3.84	4.50
3	1	4.00	1.99	1.79	1.65	3.45	3.51	4.10
5	1	3.33	1.67	1.53	1.43	3.05	3.05	3.53
7	1	2.86	1.46	1.36	1.29	2.76	2.74	3.13
10	1	2.35	1.26	1.20	1.16	2.46	2.44	2.72
15	1	1.82	1.09	1.07	1.05	2.13	2.11	2.31
20	1	1.48	1.02	1.02	1.01	1.92	1.91	2.06
30	1	1.08	0.99	0.99	0.99	1.67	1.67	1.76
33	1	1.00	0.99	0.99	0.99	1.62	1.62	1.70
50	1	0.70	0.99	0.99	0.99	1.43	1.43	1.49
100	1	0.37	1.00	1.00	1.00	1.12	1.12	1.25

Table 3.4 RWRE of the estimators for $p = 10$ and different Δ^2 values for varying p_1.

	$\Delta^2 = 0$				$\Delta^2 = 1$			
Estimators	$p_1 = 2$	$p_1 = 3$	$p_1 = 5$	$p_1 = 7$	$p_1 = 2$	$p_1 = 3$	$p_1 = 5$	$p_1 = 7$
LSE	1.00	1.00	1.00	1.00	1.00	1.00	1.00	1.00
RLSE/LASSO	5.00	3.33	2.00	1.43	3.33	2.50	1.67	1.25
PTE ($\alpha = 0.15$)	2.34	1.98	1.51	1.23	1.75	1.55	1.27	1.09
PTE ($\alpha = 0.2$)	2.06	1.80	1.43	1.19	1.60	1.45	1.22	1.07
PTE ($\alpha = 0.25$)	1.86	1.66	1.36	1.16	1.49	1.37	1.18	1.06
SE	2.50	2.00	1.43	1.11	2.14	1.77	1.33	1.08
PRSE	3.03	2.31	1.56	1.16	2.31	1.88	1.38	1.10
RRE	5.00	3.33	2.00	1.43	3.46	2.58	1.71	1.29
	$\Delta^2 = 5$				$\Delta^2 = 10$			
LSE	1.00	1.00	1.00	1.00	1.00	1.00	1.00	1.00
RLSE/LASSO	1.43	1.25	1.00	0.83	0.83	0.77	0.67	0.59
PTE ($\alpha = 0.15$)	1.05	1.01	0.95	0.92	0.92	0.92	0.92	0.94
PTE ($\alpha = 0.2$)	1.03	1.00	0.95	0.93	0.94	0.93	0.94	0.95
PTE ($\alpha = 0.25$)	1.02	0.99	0.96	0.94	0.95	0.95	0.95	0.97
SE	1.55	1.38	1.15	1.03	1.33	1.22	1.09	1.01
PRSE	1.53	1.37	1.15	1.03	1.32	1.22	1.08	1.01
RRE	1.97	1.69	1.33	1.13	1.55	1.40	1.20	1.07
	$\Delta^2 = 20$				$\Delta^2 = 60$			
LSE	1.00	1.00	1.00	1.00	1.00	1.00	1.00	1.00
RLSE/LASSO	0.45	0.43	0.40	0.37	0.16	0.16	0.15	0.15
PTE ($\alpha = 0.15$)	0.97	0.97	0.98	0.99	1.00	1.00	1.00	1.00
PTE ($\alpha = 0.2$)	0.98	0.98	0.99	0.99	1.00	1.00	1.00	1.00
PTE ($\alpha = 0.25$)	0.98	0.99	0.99	1.00	1.00	1.00	1.00	1.00
SE	1.17	1.12	1.04	1.00	1.06	1.04	1.01	1.00
PRSE	1.17	1.12	1.04	1.00	1.05	1.04	1.01	1.00
RRE	1.30	1.22	1.11	1.04	1.10	1.08	1.04	1.01

Table 3.5 RWRE of the estimators for $p = 20$ and different Δ^2 values for varying p_1.

	$\Delta^2 = 0$				$\Delta^2 = 1$			
Estimators	$p_1 = 2$	$p_1 = 3$	$p_1 = 5$	$p_1 = 7$	$p_1 = 2$	$p_1 = 3$	$p_1 = 5$	$p_1 = 7$
LSE	1.00	1.00	1.00	1.00	1.00	1.00	1.00	1.00
RLSE/LASSO	10.00	6.67	4.00	2.85	6.67	5.00	3.33	2.50
PTE ($\alpha = 0.15$)	3.20	2.84	2.31	1.95	2.50	2.27	1.93	1.68
PTE ($\alpha = 0.2$)	2.70	2.45	2.07	1.80	2.17	2.01	1.76	1.56
PTE ($\alpha = 0.25$)	2.35	2.17	1.89	1.67	1.94	1.82	1.63	1.47
SE	5.00	4.00	2.86	2.22	4.13	3.42	2.56	2.04
PRSE	6.28	4.77	3.22	2.43	4.58	3.72	2.71	2.13
RRE	10.00	6.67	4.00	2.86	6.78	5.07	3.37	2.52
	$\Delta^2 = 5$				$\Delta^2 = 10$			
LSE	1.00	1.00	1.00	1.00	1.00	1.00	1.00	1.00
RLSE/LASSO	2.86	2.50	2.00	1.67	1.67	1.54	1.33	1.18
PTE ($\alpha = 0.15$)	1.42	1.36	1.25	1.17	1.08	1.06	1.02	0.99
PTE ($\alpha = 0.2$)	1.33	1.29	1.20	1.14	1.06	1.04	1.02	0.99
PTE ($\alpha = 0.25$)	1.27	1.23	1.17	1.11	1.04	1.03	1.01	0.99
SE	2.65	2.36	1.94	1.65	2.03	1.87	1.62	1.43
PRSE	2.63	2.34	1.92	1.64	2.01	1.85	1.60	1.42
RRE	3.38	2.91	2.28	1.88	2.37	2.15	1.82	1.58
	$\Delta^2 = 20$				$\Delta^2 = 60$			
LSE	1.00	1.00	1.00	1.00	1.00	1.00	1.00	1.00
RLSE/LASSO	0.91	0.87	0.80	0.74	0.32	0.32	0.31	0.30
PTE ($\alpha = 0.15$)	0.97	0.97	0.97	0.97	1.00	1.00	1.00	1.00
PTE ($\alpha = 0.2$)	0.98	0.98	0.98	0.98	1.00	1.00	1.00	1.00
PTE ($\alpha = 0.25$)	0.99	0.98	0.98	0.99	1.00	1.00	1.00	1.00
SE	1.58	1.51	1.36	1.26	1.21	1.18	1.13	1.09
PRSE	1.58	1.50	1.36	1.25	1.21	1.18	1.13	1.09
RRE	1.74	1.64	1.47	1.34	1.26	1.23	1.18	1.13

Table 3.6 RWRE of the estimators for $p = 40$ and different Δ^2 values for varying p_1.

	$\Delta^2 = 0$				$\Delta^2 = 1$			
Estimators	$p_1 = 2$	$p_1 = 3$	$p_1 = 5$	$p_1 = 7$	$p_1 = 2$	$p_1 = 3$	$p_1 = 5$	$p_1 = 7$
LSE	1.00	1.00	1.00	1.00	1.00	1.00	1.00	1.00
RLSE/LASSO	20.00	13.33	8.00	5.71	13.33	10.00	6.67	5.00
PTE ($\alpha = 0.15$)	4.05	3.74	3.24	2.86	3.32	3.12	2.77	2.49
PTE ($\alpha = 0.2$)	3.29	3.09	2.76	2.50	2.77	2.64	2.40	2.20
PTE ($\alpha = 0.25$)	2.78	2.65	2.42	2.23	2.40	2.30	2.13	1.98
SE	10.00	8.00	5.71	4.44	8.12	6.75	5.05	4.03
PRSE	12.80	9.69	6.52	4.92	9.25	7.51	5.45	4.28
RRE	20.00	13.33	8.00	5.71	13.45	10.07	6.70	5.02
	$\Delta^2 = 5$				$\Delta^2 = 10$			
LSE	1.00	1.00	1.00	1.00	1.00	1.00	1.00	1.00
RLSE/LASSO	5.71	5.00	4.00	3.33	3.33	3.08	2.67	2.35
PTE ($\alpha = 0.15$)	1.9641	1.8968	1.7758	1.6701	1.3792	1.3530	1.3044	1.2602
PTE ($\alpha = 0.2$)	1.75	1.70	1.61	1.53	1.29	1.27	1.24	1.20
PTE ($\alpha = 0.25$)	1.60	1.56	1.50	1.44	1.23	1.22	1.19	1.16
SE	4.87	4.35	3.59	3.05	3.46	3.20	2.78	2.46
PRSE	4.88	4.36	3.59	3.05	3.42	3.16	2.75	2.44
RRE	6.23	5.40	4.27	3.53	4.03	3.68	3.13	2.72
	$\Delta^2 = 20$				$\Delta^2 = 60$			
LSE	1.00	1.00	1.00	1.00	1.00	1.00	1.00	1.00
RLSE/LASSO	1.82	1.74	1.60	1.48	0.64	0.63	0.61	0.60
PTE ($\alpha = 0.15$)	1.05	1.05	1.03	1.02	0.99	0.99	0.99	0.99
PTE ($\alpha = 0.2$)	1.04	1.03	1.02	1.02	0.99	0.99	0.99	0.99
PTE ($\alpha = 0.25$)	1.03	1.02	1.02	1.01	0.99	0.99	1.00	1.00
SE	2.41	2.2946	2.09	1.92	1.52	1.48	1.42	1.36
PRSE	2.41	2.29	2.08	1.91	1.52	1.48	1.42	1.36
RRE	2.65	2.50	2.26	2.06	1.58	1.54	1.47	1.41

Table 3.7 RWRE of the estimators for $p = 60$ and different Δ^2 values for varying p_1.

	$\Delta^2 = 0$				$\Delta^2 = 1$			
Estimators	$p_1 = 2$	$p_1 = 3$	$p_1 = 5$	$p_1 = 7$	$p_1 = 2$	$p_1 = 3$	$p_1 = 5$	$p_1 = 7$
LSE	1.00	1.00	1.00	1.00	1.00	1.00	1.00	1.00
RLSE/LASSO	30.00	20.00	12.00	8.57	20.00	15.00	10.00	7.50
PTE ($\alpha = 0.15$)	4.49	4.23	3.79	3.43	3.80	3.62	3.29	3.02
PTE ($\alpha = 0.2$)	3.58	3.42	3.14	2.91	3.10	2.99	2.78	2.59
PTE ($\alpha = 0.25$)	2.99	2.89	2.70	2.54	2.64	2.56	2.42	2.29
SE	15.00	12.00	8.57	6.67	12.12	10.09	7.55	6.03
PRSE	19.35	14.63	9.83	7.40	13.99	11.34	8.22	6.45
RRE	30.00	20.00	12.00	8.57	20.11	15.06	10.03	7.52
	$\Delta^2 = 5$				$\Delta^2 = 10$			
LSE	1.00	1.00	1.00	1.00	1.00	1.00	1.00	1.00
RLSE/LASSO	8.57	7.50	6.00	5.00	5.00	4.61	4.0000	3.53
PTE ($\alpha = 0.15$)	2.35	2.28	2.16	2.05	1.63	1.60	1.55	1.50
PTE ($\alpha = 0.2$)	2.04	1.99	1.91	1.83	1.49	1.47	1.43	1.39
PTE ($\alpha = 0.25$)	1.83	1.79	1.73	1.67	1.39	1.37	1.34	1.31
SE	7.10	6.35	5.25	4.47	4.89	4.53	3.94	3.50
PRSE	7.17	6.41	5.28	4.50	4.84	4.48	3.91	3.47
RRE	9.09	7.90	6.26	5.19	5.70	5.21	4.45	3.89
	$\Delta^2 = 20$				$\Delta^2 = 60$			
LSE	1.00	1.00	1.00	1.00	1.00	1.00	1.00	1.00
RLSE/LASSO	2.73	2.61	2.40	2.22	0.97	0.95	0.92	0.89
PTE ($\alpha = 0.15$)	1.15	1.14	1.13	1.11	0.99	0.99	0.99	0.99
PTE ($\alpha = 0.2$)	1.11	1.10	1.09	1.08	0.99	0.99	0.99	0.99
PTE ($\alpha = 0.25$)	1.08	1.08	1.07	1.06	0.99	0.99	0.99	0.99
SE	3.25	3.09	2.82	2.60	1.83	1.79	1.72	1.65
PRSE	3.23	3.08	2.81	2.59	1.83	1.79	1.72	1.65
RRE	3.55	3.37	3.05	2.79	1.90	1.86	1.78	1.71

Table 3.8 RWRE values of estimators for $p_1 = 5$ and different values of p_2 and Δ^2.

	LSE	RLSE/ LASSO	PTE α 0.15	0.2	0.25	SE	PRSE	RRE
p_2				$\Delta^2 = 0$				
5	1.00	2.00	1.76	1.51	1.36	1.43	1.56	2.00
15	1.00	4.00	3.11	2.31	1.89	2.86	3.22	4.00
25	1.00	6.00	4.23	2.84	2.20	4.28	4.87	6.00
35	1.00	8.00	5.18	3.24	2.42	5.71	6.52	8.00
55	1.00	12.00	6.71	3.79	2.70	8.57	9.83	12.00
p_2				$\Delta^2 = 0.5$				
5	1.00	1.82	1.58	1.37	1.26	1.37	1.46	1.83
15	1.00	3.64	2.79	2.10	1.74	2.70	2.93	3.65
25	1.00	5.45	3.81	2.61	2.05	4.03	4.43	5.46
35	1.00	7.27	4.68	2.98	2.26	5.36	5.93	7.28
55	1.00	10.91	6.11	3.52	2.55	8.02	8.94	10.92
p_2				$\Delta^2 = 1$				
5	1.00	1.67	1.43	1.27	1.18	1.33	1.38	1.71
15	1.00	3.33	2.53	1.93	1.63	2.56	2.71	3.37
25	1.00	5.00	3.46	2.41	1.92	3.80	4.08	5.03
35	1.00	6.67	4.27	2.77	2.13	5.05	5.45	6.70
55	1.00	10.00	5.61	3.29	2.42	7.55	8.22	10.03
p_2				$\Delta^2 = 5$				
5	1.0000	1.00	0.93	0.95	0.96	1.15	1.15	1.33
15	1.00	2.00	1.47	1.26	1.17	1.94	1.92	2.28
25	1.00	3.00	1.98	1.54	1.35	2.76	2.75	3.27
35	1.00	4.00	2.44	1.77	1.50	3.59	3.59	4.27
55	1.00	6.00	3.27	2.16	1.73	5.25	5.28	6.26

Table 3.9 RWRE values of estimators for $p_1 = 7$ and different values of p_2 and Δ^2.

p_2				$\Delta^2 = 0$				
5	1.00	1.43	1.33	1.23	1.16	1.11	1.16	1.43
15	1.00	2.86	2.41	1.94	1.67	2.22	2.43	2.86
25	1.00	4.28	3.35	2.46	2.00	3.33	3.67	4.28
35	1.00	5.71	4.17	2.86	2.23	4.44	4.92	5.71
55	1.00	8.57	5.54	3.43	2.53	6.67	7.40	8.57
p_2				$\Delta^2 = 0.5$				
5	1.00	1.33	1.23	1.15	1.10	1.09	1.13	1.35
15	1.00	2.67	2.22	1.80	1.56	2.12	2.27	2.67
25	1.00	4.00	3.08	2.29	1.87	3.17	3.41	4.00
35	1.00	5.33	3.84	2.66	2.10	4.23	4.57	5.34
55	1.00	8.00	5.13	3.21	2.40	6.33	6.89	8.00
p_2				$\Delta^2 = 1$				
5	1.00	1.25	1.15	1.09	1.06	1.08	1.10	1.29
15	1.00	2.50	2.05	1.68	1.47	2.04	2.13	2.52
25	1.00	3.75	2.85	2.13	1.77	3.03	3.20	3.77
35	1.00	5.00	3.56	2.49	1.98	4.03	4.28	5.01
55	1.00	7.50	4.77	3.02	2.29	6.03	6.45	7.52
p_2				$\Delta^2 = 5$				
5	1.00	0.83	0.87	0.92	0.94	1.03	1.03	1.13
15	1.00	1.67	1.32	1.17	1.11	1.65	1.64	1.88
25	1.00	2.50	1.78	1.44	1.29	2.34	2.34	2.70
35	1.00	3.33	2.20	1.67	1.44	3.05	3.05	3.53
55	1.00	5.00	2.98	2.05	1.67	4.47	4.50	5.19

Table 3.10 RWRE values of estimators for $p_2 = 5$ and different values of p_1 and Δ^2.

			PTE					
			α					
	LSE	RLSE/ LASSO	0.15	0.2	0.25	SE	PRSE	RRE
5	1.00	2.00	1.76	1.51	1.36	1.43	1.56	2.00
15	1.00	1.33	1.27	1.20	1.15	1.18	1.22	1.33
25	1.00	1.20	1.17	1.127	1.10	1.11	1.14	1.20
35	1.00	1.14	1.12	1.09	1.07	1.08	1.10	1.14
55	1.00	1.09	1.08	1.06	1.04	1.05	1.06	1.09
p_1				$\Delta^2 = 0.5$				
5	1.00	1.82	1.58	1.37	1.26	1.34	1.46	1.83
15	1.00	1.29	1.22	1.16	1.11	1.16	1.19	1.29
25	1.00	1.18	1.14	1.10	1.07	1.10	1.12	1.18
35	1.00	1.13	1.10	1.07	1.05	1.07	1.08	1.13
55	1.00	1.08	1.06	1.05	1.03	1.05	1.05	1.08
p_1				$\Delta^2 = 1$				
5	1.00	1.67	1.43	1.27	1.18	1.33	1.38	1.71
15	1.00	1.25	1.18	1.12	1.08	1.14	1.16	1.26
25	1.00	1.15	1.11	1.08	1.05	1.09	1.10	1.16
35	1.00	1.11	1.08	1.06	1.04	1.07	1.07	1.12
55	1.00	1.07	1.05	1.04	1.03	1.04	1.05	1.07
p_1				$\Delta^2 = 5$				
5	1.00	1.00	0.93	0.95	0.96	1.15	1.15	1.33
15	1.00	1.00	0.97	0.97	0.98	1.07	1.07	1.14
25	1.00	1.00	0.98	0.98	0.98	1.05	1.04	1.09
35	1.00	1.00	0.98	0.99	0.99	1.03	1.03	1.07
55	1.00	1.00	0.99	0.99	0.99	1.02	1.02	1.04

The unweighted relative efficiency of PRSE:

$$\begin{aligned}
\text{REff}(\hat{\theta}_n^{\text{S}+} : \tilde{\theta}_n) = & \left(1 + \frac{\text{tr}(\boldsymbol{N}_2^{-1})}{\text{tr}(\boldsymbol{N}_1^{-1})}\right) \\
& \left\{1 + \frac{\text{tr}(\boldsymbol{N}_2^{-1})}{\text{tr}(\boldsymbol{N}_1^{-1})}(1 - (p_2 - 2)A) + (p_2^2 - 4)\frac{\Delta^{*2}}{\text{tr}(\boldsymbol{N}_1^{-1})}\mathbb{E}\left[\chi_{p_2+4}^{-4}(\Delta^2)\right]\right. \\
& -\frac{\text{tr}(\boldsymbol{N}_2^{-1})}{\text{tr}(\boldsymbol{N}_1^{-1})}\mathbb{E}\left[(1 - (p_2 - 2)\chi_{p_2+2}^{-2}(\Delta^2))^2 I(\chi_{p_2+2}^2(\Delta^2) < (p_2 - 2))\right] \\
& +\frac{\Delta^{*2}}{\text{tr}(\boldsymbol{N}_1^{-1})}\left[2\mathbb{E}\left[(1 - (p_2 - 2)\chi_{p_2+2}^{-2}(\Delta^2))I(\chi_{p_2+2}^2(\Delta^2) < (p_2 - 2))\right]\right. \\
& \left.\left. -\mathbb{E}[(1 - (p_2 - 2)\chi_{p_2+4}^{-2}(\Delta^2))^2 I(\chi_{p_2+4}^2(\Delta^2) < (p_2 - 2))]\right]\right\}^{-1}. \qquad (3.32)
\end{aligned}$$

The unweighted relative efficiency of RRE:

$$\text{REff}(\hat{\theta}_n^{\text{RR}}(k) : \tilde{\theta}_n) = \left(1 + \frac{\text{tr}(\boldsymbol{N}_2^{-1})}{\text{tr}(\boldsymbol{N}_1^{-1})}\right)\left\{1 + \frac{\text{tr}(\boldsymbol{N}_2^{-1})}{\text{tr}(\boldsymbol{N}_1^{-1})}\frac{\Delta^{*2}}{(\text{tr}(\boldsymbol{N}_2^{-1}) + \Delta^{*2})}\right\}^{-1}. \qquad (3.33)$$

3.7 Summary and Concluding Remarks

In this section, we discuss the contents of Tables 3.3–3.10 presented as confirmatory evidence of the theoretical findings of the estimators.

First, we note that we have two classes of estimators, namely, the traditional PTE and SE and the penalty estimators. The RLSE plays an important role due to the fact that LASSO belongs to the class of restricted estimators. We have the following conclusions from our study:

(i) Since the inception of the RRE by Hoerl and Kennard (1970), there have been articles comparing RRE with PTE and SE. From this study, we conclude that the RRE dominates the LSE, PTE, and SE uniformly. The ridge estimator dominates the LASSO estimator uniformly for Δ^2 greater than 0. They are L_2 risk equivalent at $\Delta^2 = 0$; and at this point, LASSO dominates all other estimators. The ridge estimator does not select variables, but the LASSO estimator does. See Table 3.3 and graphs thereof in Figure 3.7.
(ii) The RLSE and LASSO are L_2 risk equivalent. Hence, LASSO satisfies “oracle properties.”
(iii) Under the family of “diagonal linear projection,” the “ideal” L_2 risk of LASSO and subset rule (HTE) are the same and do not depend on the thresholding parameter (κ) for the sparse condition; see Donoho and Johnstone (1994).

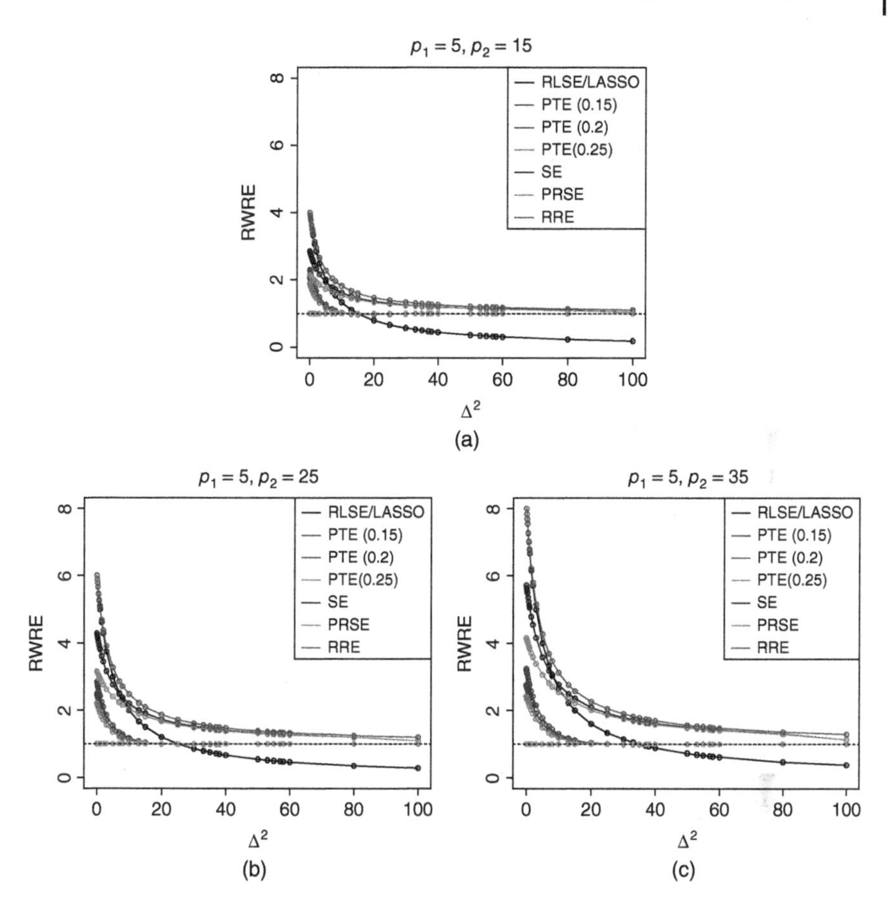

Figure 3.7 RWRE of estimates of a function of Δ^2 for $p_1 = 5$ and different p_2.

(iv) The RWRE of estimators compared to the LSE depends upon the size of p_1, p_2, and divergence parameter, Δ^2. LASSO/RLSE and ridge outperform all the estimators when Δ^2 is 0.

(v) The LASSO satisfies the "oracle properties" and it dominates LSE, PTE, SE, and PRSE in the subinterval of $[0, p_1)$. In this case, with a small number of active parameters, the LASSO and HTE perform best followed by ridge, as pointed out by Tibshirani (1996).

(vi) If p_1 is fixed and p_2 increases, the RWRE of all estimators increases (see Table 3.3).

(vii) If p_2 is fixed and p_1 increases, the RWRE of all estimators decreases. Then, for a given p_2 small and p_1 large, the LASSO, PTE, SE, and PRSE are competitive (see Table 3.6).

(viii) The ridge estimator outperforms the LSE, PTE, SE, and PRSE uniformly. The ridge dominates LASSO and RLSE uniformly for $\Delta^2 > 0$; and at $\Delta^2 = 0$, they are L_2 risk equivalent where Δ^2 is the divergence parameter.

(ix) The PRSE always outperforms SE (see Tables 3.6–3.11).

(x) We illustrated the findings of the paper by a numeral application in Section 3.5

3A. Appendix

We now apply Hansen's method for ANOVA when it is suspected that $\boldsymbol{\theta}$ may be $\mathbf{0}$. In this case, the estimators are

LSE of $\boldsymbol{\theta}$ is $\tilde{\boldsymbol{\theta}}_n$;
RLSE $(\hat{\boldsymbol{\theta}}_n)$ and LASSO $(\hat{\boldsymbol{\theta}}_n^{\mathrm{L}})$ as $\mathbf{0}$ vector;
PTE of $\boldsymbol{\theta}$, $\hat{\boldsymbol{\theta}}_n^{\mathrm{PT}}(\alpha) = \tilde{\boldsymbol{\theta}}_n I(\mathcal{L}_n \geq c_\alpha)$;
SE of $\boldsymbol{\theta}$, $\hat{\boldsymbol{\theta}}_n^{\mathrm{S}} = \tilde{\boldsymbol{\theta}}_n(1 - (p-2)\mathcal{L}_n^{-1})$;
PRSE of $\boldsymbol{\theta}$, $\hat{\boldsymbol{\theta}}_n^{\mathrm{S+}} = \tilde{\boldsymbol{\theta}}_n(1 - (p-2)\mathcal{L}_n^{-1})^{-1}$;
ridge estimator of θ, $\hat{\boldsymbol{\theta}}_n^{\mathrm{RR}}(k) = \frac{1}{1+k}\tilde{\boldsymbol{\theta}}_n$.

Accordingly, the L_2 risks are

$$R(\tilde{\boldsymbol{\theta}}_n, \boldsymbol{N}) = \sigma^2 p,$$

$$R(\hat{\boldsymbol{\theta}}_n, \boldsymbol{N}) = \sigma^2\Delta^{*2}, \quad \Delta^{*2} = \frac{1}{\sigma^2}\boldsymbol{\theta}^\top \boldsymbol{N}\boldsymbol{\theta}$$

$$R(\hat{\boldsymbol{\theta}}^{\mathrm{L}}, \boldsymbol{N}) = \sigma^2\Delta^{*2},$$

$$R(\hat{\boldsymbol{\theta}}_n^{\mathrm{PT}}, \boldsymbol{N}) = \sigma^2 p(1 - H_{p+2}(c_\alpha, \Delta^{*2})) + \sigma^2\Delta^2\{2H_{p+2}(c_\alpha, \Delta^{*2}) - H_{p+4}(c_\alpha, \Delta^{*2})\}$$

$$R(\hat{\boldsymbol{\theta}}_n^{\mathrm{S}}, \boldsymbol{N}) = \sigma^2(p - (p-2)^2\mathbb{E}[\chi_{p_2+2}^{-2}(\Delta^{*2})])$$

$$\begin{aligned} R(\hat{\boldsymbol{\theta}}_n^{\mathrm{S+}}, \boldsymbol{N}) = R(\hat{\boldsymbol{\theta}}_n^{\mathrm{S}}, \boldsymbol{N}) & -\sigma^2 pE[(1-(p-2)\chi_{p_2+2}^{-2}(\Delta^{*2}))^2 I(\chi_{p_2+2}^2(\Delta^{*2}) \leq p-2)] \\ & +\sigma^2\Delta^{*2}\left\{2\mathbb{E}[(1-(p-2)\chi_{p_2+2}^{-2}(\Delta^{*2}))I(\chi_{p_2+2}^2(\Delta^{*2}) \leq p-2)]\right. \\ & \left. -\mathbb{E}[(1-(p-2)\chi_{p_2+4}^{-2}(\Delta^{*2}))^2 I(\chi_{p_2+4}^2(\Delta^{*2}) \leq p-2)]\right\} \end{aligned}$$

$$R(\hat{\boldsymbol{\theta}}_n^{\mathrm{RR}}, \boldsymbol{N}) = \frac{p\Delta^{*2}}{p + \Delta^{*2}}$$

The efficiency table is given in Table 3A.1.

Table 3.11 RWRE values of estimators for $p_2 = 7$ and different values of p_1 and Δ^2.

	LSE	RLSE/ LASSO	PTE α 0.15	PTE α 0.2	PTE α 0.25	SE	PRSE	RRE
p_1				$\Delta^2 = 0$				
3	1.00	3.33	2.60	1.98	1.66	2.00	2.31	3.33
13	1.00	1.54	1.44	1.33	1.24	1.33	1.40	1.54
23	1.00	1.30	1.26	1.20	1.15	1.20	1.23	1.30
33	1.00	1.21	1.18	1.14	1.11	1.14	1.16	1.21
53	1.00	1.13	1.11	1.09	1.07	1.09	1.10	1.13
p_1				$\Delta^2 = 0.5$				
3	1.00	2.86	2.21	1.73	1.49	1.87	2.06	2.88
13	1.00	1.48	1.38	1.27	1.20	1.30	1.35	1.48
23	1.00	1.28	1.22	1.16	1.12	1.18	1.20	1.28
33	1.00	1.19	1.16	1.12	1.09	1.13	1.15	1.19
53	1.00	1.12	1.10	1.07	1.06	1.08	1.09	1.12
p_1				$\Delta^2 = 1$				
3	1.00	2.50	1.93	1.55	1.37	1.77	1.88	2.58
13	1.00	1.43	1.32	1.22	1.16	1.28	1.31	1.44
23	1.00	1.25	1.19	1.13	1.10	1.17	1.18	1.26
33	1.00	1.18	1.14	1.10	1.07	1.12	1.13	1.18
53	1.00	1.11	1.09	1.06	1.05	1.08	1.08	1.11
p_1				$\Delta^2 = 5$				
3	1.00	1.25	1.04	1.01	0.99	1.38	1.372	1.69
13	1.00	1.11	1.02	1.00	0.99	1.16	1.15	1.26
23	1.00	1.07	1.01	1.00	0.99	1.10	1.10	1.16
33	1.00	1.05	1.01	1.00	0.99	1.07	1.07	1.11
53	1.00	1.03	1.01	1.00	0.99	1.05	1.05	1.07

Note that the LASSO solution always puts some (say, p_2) parameters equal to 0 and others (p_1) as $\tilde{\theta}_{jn}$. Then, the oracle solution of the risk is $\sigma^2(p_1 + \Delta^2)$, where $\Delta^{*2} > \Delta^2$ so that efficiency is $p/(p_1 + \Delta^2)$. This happens under our assumptions and not under Hansen's assumption. We get Δ^{*2} corresponding to the infeasible estimator, $\mathbf{0}$. Our assumptions are the right ones for the study of LASSO and related estimators.

Table 3A.1 Sample efficiency table of estimators under Hansen's method.

p	Δ^2	LSE	RLSE/ LASSO	PTE ($\alpha = 0.15$)	PTE ($\alpha = 0.2$)	PTE ($\alpha = 0.25$)	SE	PRSE	RRE
10	0	1.00	∞	2.68	3.38	4.03	5.00	7.03	∞
	0.1	1.00	100.00	2.84	3.54	4.18	4.81	6.56	101.00
	0.5	1.00	20.00	3.46	4.15	4.78	4.19	5.23	21.00
	1	1.00	10.00	4.20	4.87	5.47	3.65	4.25	11.00
	2	1.00	5.00	5.56	6.17	6.69	2.97	3.21	6.00
	5	1.00	2.00	8.59	8.90	9.13	2.09	2.12	3.00
	10	1.00	1.00	10.63	10.52	10.43	1.61	1.61	2.00
	20	1.00	0.50	10.43	10.29	10.20	1.32	1.32	1.50
	50	1.00	0.20	10.00	10.00	10.00	1.13	1.13	1.20
	100	1.00	0.10	10.00	10.00	10.00	1.04	1.04	1.10
20	0	1.00	∞	4.62	5.91	7.13	10.00	15.04	∞
	0.1	1.00	200.00	4.80	6.10	7.32	9.57	13.90	201.00
	0.5	1.00	40.00	5.53	6.84	8.07	8.20	10.73	41.00
	1	1.00	20.00	6.44	7.76	8.97	6.99	8.46	21.00
	2	1.00	10.00	8.19	9.50	10.68	5.48	6.09	11.00
	5	1.00	4.00	12.80	13.91	14.82	3.54	3.63	5.00
	10	1.00	2.00	17.80	18.31	18.68	2.46	2.47	3.00
	20	1.00	1.00	20.45	20.34	20.26	1.78	1.78	2.00
	50	1.00	0.40	20.01	20.00	20.00	1.32	1.32	1.40
	100	1.00	0.20	20.00	20.00	20.00	1.09	1.09	1.20
30	0	1.00	∞	6.45	8.31	10.08	15.00	23.41	∞
	0.1	1.00	300.00	6.66	8.52	10.30	14.33	21.55	301.00
	0.5	1.00	60.00	7.46	9.36	11.15	12.20	16.45	61.00
	1	1.00	30.00	8.47	10.40	12.20	10.33	12.80	31.00
	2	1.00	15.00	10.46	12.43	14.23	7.99	9.04	16.00
	5	1.00	6.00	16.06	17.94	19.54	4.97	5.15	7.00
	10	1.00	3.00	23.29	24.58	25.57	3.30	3.32	4.00
	20	1.00	1.50	29.49	29.66	29.77	2.24	2.24	2.50
	50	1.00	0.60	30.02	30.01	30.01	1.52	1.52	1.60
	100	1.00	0.30	30.00	30.00	30.00	1.14	1.14	1.30

(*Continued*)

Table 3A.1 (Continued)

p	Δ^2	LSE	RLSE/ LASSO	PTE ($\alpha = 0.15$)	PTE ($\alpha = 0.2$)	PTE ($\alpha = 0.25$)	SE	PRSE	RRE
	0	1.00	∞	8.23	10.65	12.97	20.00	31.99	∞
	0.1	1.00	400.00	8.45	10.88	13.20	19.10	29.39	401.00
	0.5	1.00	80.00	9.32	11.79	14.14	16.20	22.28	81.00
	1	1.00	40.00	10.41	12.93	15.31	13.66	17.23	41.00
40	2	1.00	20.00	12.59	15.18	17.58	10.49	12.03	21.00
	5	1.00	8.00	18.91	21.52	23.80	6.41	6.68	9.00
	10	1.00	4.00	27.86	29.96	31.64	4.14	4.18	5.00
	20	1.00	2.00	37.57	38.23	38.68	2.70	2.70	3.00
	50	1.00	0.80	40.04	40.03	40.02	1.71	1.71	1.80
	100	1.00	0.40	40.00	40.00	40.00	1.18	1.18	1.40

Problems

3.1 Show that the test statistic for testing $\mathcal{H}_o : \boldsymbol{\theta}_2 = \mathbf{0}$ vs. $\mathcal{H}_A : \boldsymbol{\theta}_2 \neq \mathbf{0}$ for unknown σ is

$$\mathcal{L}_n = \frac{\tilde{\boldsymbol{\theta}}_{2n} \boldsymbol{N}_2 \tilde{\boldsymbol{\theta}}_{2n}}{p_2 s_n^2}$$

and also show that $\mathcal{L}_n$ has a noncentral F with appropriate DF and noncentrality parameter Δ^2.

3.2 Determine the bias vector of estimators, $\hat{\boldsymbol{\theta}}_n^{\mathrm{HT}}(\kappa)$ and $\hat{\boldsymbol{\theta}}_n^{\mathrm{L}}(\lambda)$ in Eqs. (3.4) and (3.9) respectively.

3.3 Proof that the lower bond of the risk of $\hat{\boldsymbol{\theta}}_n^{\mathrm{HT}}(\kappa)$ is $\sigma^2 \operatorname{tr}(\boldsymbol{N}^{-1}) + \boldsymbol{\theta}^\top \boldsymbol{\theta}$.

3.4 Prove under the regularity conditions that the weighted L_2-risk bounds are given by

$$R(\hat{\boldsymbol{\theta}}_n^{\mathrm{HT}}(\kappa) : \boldsymbol{N}_1, \boldsymbol{N}_2) \leq \begin{cases} \text{(i)} \quad \sigma^2(1+\kappa^2)(p_1+p_2) & \kappa > 1, \\ \text{(ii)} \quad \sigma^2(p_1+p_2) + \boldsymbol{\theta}_1^\top \boldsymbol{N}_1 \boldsymbol{\theta}_1 & \\ \qquad + \boldsymbol{\theta}_2^\top \boldsymbol{N}_2 \boldsymbol{\theta}_2 & \forall \boldsymbol{\theta} \in \mathbb{R}^p \\ \text{(iii)} \quad \sigma^2 \rho_{\mathrm{HT}}(\kappa, 0)(p_1+p_2) & \\ \qquad + 1.2\{\boldsymbol{\theta}_1^\top \boldsymbol{N}_1 \boldsymbol{\theta}_1 + \boldsymbol{\theta}_2^\top \boldsymbol{N}_2 \boldsymbol{\theta}_2\} & 0 < \boldsymbol{\theta} < k\mathbf{1}_p^\top. \end{cases}$$

3.5 Show that the bias expression of LASSO estimator is given by

$$b(\theta_n^{\mathrm{L}}(\lambda)) = \left(\sigma n_j^{-\frac{1}{2}}[\lambda(2\Phi(\Delta_j) - 1); j = 1, \dots, p_1; \Delta_{p_1+1}, \dots, \Delta_p\right)^{\mathrm{T}}.$$

3.6 Show that the bias and MSE of $\hat{\theta}_n^{\mathrm{RR}}(k)$ are, respectively,

$$b(\hat{\theta}_n^{\mathrm{RR}}(k)) = \begin{pmatrix} \mathbf{0} \\ -\frac{k}{1+k}\theta_2 \end{pmatrix}$$

and

$$R(\hat{\theta}_n^{\mathrm{RR}}(k) : \boldsymbol{I}_p) = \sigma^2 \,\mathrm{tr}\boldsymbol{N}_1^{-1} + \frac{1}{(1+k)^2}(\sigma^2 \,\mathrm{tr}(\boldsymbol{N}_2^{-1}) + k^2\theta_2^{\mathrm{T}}\theta_2).$$

3.7 Prove under usual notation that RRE uniformly dominates both LSE and PTE.

3.8 Verify that RRE uniformly dominates both Stein-type and its positive-rule estimators.

3.9 Prove under usual notation that RRE uniformly dominates LASSO.

3.10 Show that the modified LASSO outperforms the SE as well as the PRSE in the interval

$$0 \le \Delta^2 \le p_2 - (p_2 - 2)^2\mathbb{E}[\chi_{p_2}^{-2}(\Delta^2)].$$

4

Seemingly Unrelated Simple Linear Models

In this chapter, we consider an important class of models, namely, the seemingly unrelated simple linear models, belonging to the class of linear hypothesis, useful in the analysis of bioassay data, shelf-life determination of pharmaceutical products, and profitability analysis of factory products in terms of costs and outputs, among other applications. In this model, as in the analysis of variance (ANOVA) model, p independent bivariate samples $\{(x_{\gamma_1}, y_{\gamma_1}), (x_{\gamma_2}, y_{\gamma_2}), \ldots, (x_{\gamma_p}, y_{\gamma_p}) | \gamma = 1, 2, \ldots, p\}$ are considered such that $y_{\gamma_j} \sim \mathcal{N}(\theta_\gamma + \beta_\gamma x_{\gamma j}, \sigma^2)$ for each pair (γ, j) with fixed $x_{\gamma j}$.

The parameters $\boldsymbol{\theta} = (\theta_1, \ldots, \theta_p)^\top$ and $\boldsymbol{\beta} = (\beta_1, \ldots, \beta_p)^\top$ are the intercept and slope vectors of the p-lines, respectively, and σ^2 is the common known variance. In this model, it is common to test the parallelism hypotheses $\mathcal{H}_o : \beta_1 = \cdots = \beta_p = \beta_0$ against the alternative hypothesis, $\mathcal{H}_A : \boldsymbol{\beta} \neq \beta_0 \mathbf{1}_p$. Instead, in many applications, one may suspect some of the elements of the $\boldsymbol{\beta}$-vector may not be significantly different from $\mathbf{0}$, i.e. $\boldsymbol{\beta}$-vector may be sparse; in other words, we partition $\boldsymbol{\beta} = (\boldsymbol{\beta}_1^\top, \boldsymbol{\beta}_2^\top)^\top$ and our suspects, $\boldsymbol{\beta}_2 = \mathbf{0}$. Then, the test statistics $\mathcal{L}_n$ tests the null hypothesis $\mathcal{H}_o : \boldsymbol{\beta}_2 = \mathbf{0}$ vs. $\mathcal{H}_A : \boldsymbol{\beta}_2 \neq \mathbf{0}$. Besides this, the main objective of this chapter is to study some penalty estimators and the preliminary test estimator (PTE) and Stein-type estimator (SE) of θ and β when one suspects that $\boldsymbol{\beta}_2$ may be $\mathbf{0}x$ and compare their properties based on L_2-risk function. For more literature and research on seemingly unrelated linear regression or other models, we refer the readers to Baltagi (1980), Foschi et al. (2003), Kontoghiorghes (2000, 2004), andKontoghiorghes and Clarke (1995), among others.

4.1 Model, Estimation, and Test of Hypothesis

Consider the seemingly unrelated simple linear models

$$y_\gamma = \theta_\gamma \mathbf{1}_{n\gamma} + \boldsymbol{\beta}_\gamma \boldsymbol{x}_\gamma + \epsilon_\gamma, \quad \gamma = 1, \ldots p, \tag{4.1}$$

Theory of Ridge Regression Estimation with Applications, First Edition.
A.K. Md. Ehsanes Saleh, Mohammad Arashi, and B.M. Golam Kibria.

where $\boldsymbol{y}_\gamma = (y_{\gamma_1}, \dots, y_{\gamma_{n_\gamma}})^\top$, $\mathbf{1}_{n\gamma} = (1, \dots, 1)^\top$ an n-tuple of 1s, $\boldsymbol{x}_\gamma = (x_{\gamma_1}, \dots, x_{\gamma_{n_\gamma}})^\top$, and $\epsilon_\gamma \sim \mathcal{N}(0, \sigma^2 \boldsymbol{I}_{n_\gamma})$, $\boldsymbol{I}_{n_\gamma}$ is the n_γ dimensional identity matrices so that $n = n_1 + \cdots + n_p$.

4.1.1 LSE of θ and β

It is easy to see from Saleh (2006, Chapter 6) that the least squares estimator (LSE) of $\boldsymbol{\theta}$ and $\boldsymbol{\beta}$ are

$$\tilde{\boldsymbol{\theta}}_n = \overline{\boldsymbol{y}} - \boldsymbol{T}_n \tilde{\boldsymbol{\beta}}_n \tag{4.2}$$

and

$$\tilde{\boldsymbol{\beta}}_n = (\tilde{\beta}_{1n_1}, \dots, \tilde{\beta}_{pn_p})^\top$$

respectively, where

$$\begin{aligned}
\overline{\boldsymbol{y}} &= (\overline{y}_{1n_1}, \dots, \overline{y}_{pn_p})^\top \\
\boldsymbol{T}_n &= \mathrm{Diag}(\overline{x}_{1n_1}, \dots, \overline{x}_{pn_p}) \\
\tilde{\beta}_{n_\gamma} &= \frac{1}{Q_\gamma}\left(\boldsymbol{x}_\gamma^\top \boldsymbol{y}_\gamma - \frac{1}{n_\gamma}(\mathbf{1}_{n_\gamma}^\top \boldsymbol{x}_\gamma)(\mathbf{1}_{n_\gamma}^\top \boldsymbol{y}_\gamma)\right), \quad Q_\gamma = \boldsymbol{x}_\gamma^\top \boldsymbol{x}_\gamma - \frac{1}{n_\gamma}(\mathbf{1}_{n_\gamma}^\top \boldsymbol{x}_\gamma)^2 \\
\overline{x}_{\gamma_n} &= n_\gamma^{-1}(\mathbf{1}_{n_\gamma}^\top \boldsymbol{x}_\gamma) \\
\overline{y}_{\gamma_n} &= n_\gamma^{-1}(\mathbf{1}_{n_\gamma}^\top \boldsymbol{y}_\gamma) \\
s_e^2 &= (n - 2p)^{-1} \sum_{\gamma=1}^{p} ||\boldsymbol{y}_\gamma - \tilde{\theta}_\gamma \mathbf{1}_{n_\gamma} - \tilde{\beta}_{n_\gamma} x_\gamma||^2.
\end{aligned}$$

4.1.2 Penalty Estimation of β and θ

Following Donoho and Johnstone (1994), Tibshirani (1996) and Saleh et al. (2017), we define the least absolute shrinkage and selection operator (LASSO) estimator of β as

$$\begin{aligned}
&\hat{\boldsymbol{\beta}}_n^{\mathrm{LASSO}}(\lambda) \\
&= \left(\left(\tilde{\beta}_{\gamma,n_\gamma} - \lambda \frac{\sigma}{\sqrt{Q_\gamma}} \mathrm{sgn}(\tilde{\beta}_{\gamma,n_\gamma})\right) I\left(|\tilde{\beta}_{\gamma,n_\gamma}| > \lambda \frac{\sigma}{\sqrt{Q_\gamma}}\right) | \gamma = 1, \dots p\right)^\top \\
&= \left(\frac{\sigma}{\sqrt{Q_\gamma}}(Z_\gamma - \lambda \mathrm{sgn}(Z_\gamma)) I(|Z_\gamma| > \lambda) | \gamma = 1, \dots, p\right)^\top,
\end{aligned} \tag{4.3}$$

where $Z_\gamma = \frac{\sqrt{Q}\tilde{\beta}_{n,\gamma}}{\sigma} \sim \mathcal{N}(\Delta_\gamma, 1)$ with $\Delta_\gamma = \frac{\sqrt{Q_\gamma}\beta_{n,\gamma}}{\sigma}$.

$$\begin{aligned}
\hat{\boldsymbol{\beta}}^{\text{LASSO}}_{\gamma,n_\gamma}(\lambda) &= \frac{\sigma}{\sqrt{Q_\gamma}}(Z_\gamma - \lambda)\ \text{sgn}(Z_\gamma) I(|Z_\gamma| > \lambda) \\
&= \frac{\sigma}{\sqrt{Q_\gamma}}(Z_\gamma - \lambda \text{sgn}(Z_\gamma)) \quad \text{if} \quad |Z_\gamma| > \lambda \\
&= 0 \quad \text{otherwise.}
\end{aligned} \tag{4.4}$$

Thus, we write

$$\hat{\boldsymbol{\beta}}^{\text{LASSO}}_{n}(\lambda) = \begin{pmatrix} \hat{\boldsymbol{\beta}}^{\text{LASSO}}_{1n}(\lambda) \\ \mathbf{0} \end{pmatrix}, \tag{4.5}$$

where $\hat{\boldsymbol{\beta}}^{\text{LASSO}}_{1n}(\lambda) = (\hat{\boldsymbol{\beta}}^{\text{LASSO}}_{1n_1}(\lambda), \dots, \hat{\boldsymbol{\beta}}^{\text{LASSO}}_{1n_p}(\lambda))^\top$.

On the other hand, the ridge estimator of $\boldsymbol{\beta}$ may be defined as

$$\hat{\boldsymbol{\beta}}^{\text{RR}}_{n}(k) = \begin{pmatrix} \tilde{\boldsymbol{\beta}}_{1n} \\ \frac{1}{1+k}\tilde{\boldsymbol{\beta}}_{2n} \end{pmatrix}, \tag{4.6}$$

where the LSE of β is $\begin{pmatrix} \tilde{\boldsymbol{\beta}}_{1n} \\ \tilde{\boldsymbol{\beta}}_{2n} \end{pmatrix}$ and restricted least squares estimator (RLSE) is $\begin{pmatrix} \tilde{\boldsymbol{\beta}}_{1n} \\ \mathbf{0} \end{pmatrix}$.

Consequently, the estimator of $\boldsymbol{\theta}$ is given by

$$\hat{\boldsymbol{\theta}}^{\text{LASSO}}_{n}(\lambda) = \bar{\boldsymbol{y}} - \boldsymbol{T}_n \hat{\boldsymbol{\beta}}^{\text{LASSO}}_{1n}(\lambda) = \begin{pmatrix} \bar{\boldsymbol{y}}_{1n} - \boldsymbol{T}_{n(1)}\hat{\boldsymbol{\beta}}^{\text{LASSO}}_{1n}(\lambda) \\ \bar{\boldsymbol{y}}_{2n} \end{pmatrix}, \tag{4.7}$$

where $\bar{\boldsymbol{y}}_{1n} = (\bar{\boldsymbol{y}}_{1n_1}, \bar{\boldsymbol{y}}_{1n_2}, \dots, \bar{\boldsymbol{y}}_{1n_{p_1}})$ and $\bar{\boldsymbol{y}}_{2n} = (\bar{\boldsymbol{y}}_{p_1+1,n_{p_1+1}}, \bar{\boldsymbol{y}}_{p_1+2,n_{p_1+2}}, \dots, \bar{\boldsymbol{y}}_{p,n_p})$.

Similarly, the ridge estimator of θ is given by

$$\hat{\boldsymbol{\theta}}^{\text{RR}}_{n}(k) = \begin{pmatrix} \bar{\boldsymbol{y}}_{1n} - \boldsymbol{T}_{n(1)}\tilde{\boldsymbol{\beta}}_{1n} \\ \bar{\boldsymbol{y}}_{2n} - \boldsymbol{T}_{n(2)}\tilde{\boldsymbol{\beta}}^{\text{R}}_{2n}(k) \end{pmatrix}. \tag{4.8}$$

4.1.3 PTE and Stein-Type Estimators of β and θ

For the test of $\boldsymbol{\beta}_2 = \mathbf{0}$, where $\boldsymbol{\beta}_2 = (\boldsymbol{\beta}_{\gamma,n_\gamma} | \gamma = p_1 + 1, \dots, p)^\top$, we use the following test statistic:

$$\mathcal{L}_n = \frac{\tilde{\boldsymbol{\beta}}^\top_{2n}\boldsymbol{Q}_2\tilde{\boldsymbol{\beta}}_{2n}}{\sigma^2} = \frac{\sum_{\gamma=1}^{p} Q_\gamma \tilde{\beta}_\gamma n_\gamma}{\sigma^2}, \tag{4.9}$$

where $\boldsymbol{Q}_2 = \text{diag}(Q_{p1+1}, \dots, Q_p)$ and the distribution of $\mathcal{L}_n$ follows a noncentral χ^2 distribution with p_2 degrees of freedom (DF) and noncentrality parameter $\Delta^2 = \frac{\boldsymbol{\beta}_{2n}^\top \boldsymbol{Q}_2 \boldsymbol{\beta}_{2n}}{\sigma^2}$. Then we can define the PTE, SE, and PRSE (positive-rule Stein-type estimator)of $\boldsymbol{\beta}$ as

$$\hat{\boldsymbol{\beta}}_n^{\text{PT}}(\alpha) = \begin{pmatrix} \tilde{\boldsymbol{\beta}}_{1n} \\ \tilde{\boldsymbol{\beta}}_{2n} I(\mathcal{L}_n > \chi^2_{p_2}) \end{pmatrix}$$

$$\hat{\boldsymbol{\beta}}_n^{\text{S}} = \begin{pmatrix} \tilde{\boldsymbol{\beta}}_{1n} \\ \tilde{\boldsymbol{\beta}}_{2n}(1 - (p_2 - 2)\mathcal{L}_n^{-1}) \end{pmatrix}$$

$$\hat{\boldsymbol{\beta}}_n^{\text{S+}} = \begin{pmatrix} \tilde{\boldsymbol{\beta}}_{1n} \\ \tilde{\boldsymbol{\beta}}_{2n}(1 - (p_2 - 2)\mathcal{L}_n^{-1} I(\mathcal{L}_n > (p_2 - 2))) \end{pmatrix}, \tag{4.10}$$

respectively.

For the PTE, SE and PRSE of θ are

$$\hat{\boldsymbol{\theta}}_n^{\text{PT}}(\alpha) = \begin{pmatrix} \bar{y}_{1n} - \boldsymbol{T}_{n(1)} \tilde{\boldsymbol{\beta}}_{1n} \\ \bar{y}_{2n} - \boldsymbol{T}_{n(2)} \hat{\boldsymbol{\beta}}_{2n}^{\text{PT}}(\alpha) \end{pmatrix}$$

$$\hat{\boldsymbol{\theta}}_n^{\text{S}} = \begin{pmatrix} \bar{y}_{1n} - \boldsymbol{T}_{n(1)} \tilde{\boldsymbol{\beta}}_{1n} \\ \bar{y}_{2n} - \boldsymbol{T}_{n(2)} \hat{\boldsymbol{\beta}}_{2n}^{\text{S}} \end{pmatrix}$$

$$\hat{\boldsymbol{\theta}}_n^{\text{S+}} = \begin{pmatrix} \bar{y}_{1n} - \boldsymbol{T}_{n(1)} \tilde{\boldsymbol{\beta}}_{1n} \\ \bar{y}_{2n} - \boldsymbol{T}_{n(2)} \hat{\boldsymbol{\beta}}_{2n}^{\text{S+}} \end{pmatrix} \tag{4.11}$$

4.2 Bias and MSE Expressions of the Estimators

In this section, we present the expressions of bias and mean squared error (MSE) for all the estimators of $\boldsymbol{\beta}$ and $\boldsymbol{\theta}$ as follows.

Theorem 4.1 *Under the assumption of normal distribution of the errors in the model, we have*

$$\begin{pmatrix} \tilde{\boldsymbol{\theta}}_n - \boldsymbol{\theta} \\ \tilde{\boldsymbol{\beta}}_n - \boldsymbol{\beta} \end{pmatrix} \sim \mathcal{N}_{2p}\left(\begin{pmatrix} \mathbf{0} \\ \mathbf{0} \end{pmatrix}; \sigma^2 \begin{pmatrix} \boldsymbol{D}_{11} & -\boldsymbol{T}_n \boldsymbol{D}_{22} \\ -\boldsymbol{T}_n \boldsymbol{D}_{22} & \boldsymbol{D}_{22} \end{pmatrix} \right), \tag{4.12}$$

where

$$\begin{aligned} \boldsymbol{D}_{11} &= \boldsymbol{N}^{-1} + \boldsymbol{T}_n \boldsymbol{D}_{22} \boldsymbol{T}_n \\ \boldsymbol{N} &= \text{Diag}(n_1, \dots, n_p) \\ \boldsymbol{D}_{22}^{-1} &= \text{Diag}(Q_1, \dots, Q_p). \end{aligned} \tag{4.13}$$

Theorem 4.2 *The bias expressions of the estimators of $\boldsymbol{\beta}$ and $\boldsymbol{\theta}$ are given by*

$$
\begin{aligned}
b(\tilde{\boldsymbol{\beta}}_n) &= \mathbf{0} \\
b(\tilde{\theta}_n) &= \mathbf{0} \\
b(\hat{\boldsymbol{\beta}}_n) &= (\mathbf{0}^\top, -\boldsymbol{\beta}_2^\top)^\top \\
b(\hat{\boldsymbol{\theta}}) &= (\mathbf{0}^\top, \boldsymbol{T}_n\boldsymbol{\beta}_2^\top)^\top \\
b(\hat{\boldsymbol{\beta}}_n^{\mathrm{PT}}(\alpha)) &= (\mathbf{0}^\top, -\boldsymbol{\beta}_2^\top H_{p_2+2}(c_\alpha; \Delta^2)^\top)^\top \\
b(\hat{\theta}^{\mathrm{PT}}(\alpha)) &= (\mathbf{0}^\top, \boldsymbol{T}_{n(2)}\boldsymbol{\beta}_2^\top H_{p_2+2}(c_\alpha; \Delta^2)^\top)^\top,
\end{aligned} \tag{4.14}
$$

where $H_\nu(.; \Delta^2)$ is the cumulative distributional function (c.d.f.) of a noncentral χ^2 distribution with ν DF and noncentrality parameter $\Delta^2 = \frac{1}{\sigma^2}\boldsymbol{\beta}_2^\top[\boldsymbol{D}_{22}^{[2]}]^{-1}\boldsymbol{\beta}_2$, where $[\boldsymbol{D}_{22}^{[2]}]^{-1} = \mathrm{Diag}(Q_{p_1+1}, \dots, Q_p)$, $\boldsymbol{\beta}_2 = (\boldsymbol{\beta}_{p_1+1}, \dots, \beta_p)^\top$, and $c_\alpha = \chi^2_{p_2}(\alpha)$ is the upper α-level critical value from a central χ^2 distribution with p_2 DF

Similarly we have

$$
\begin{aligned}
b(\hat{\boldsymbol{\beta}}_n^{\mathrm{S}}) &= (\mathbf{0}^\top, -(p_2-2)\boldsymbol{\beta}_2^\top \mathbb{E}[\chi^2_{p_2+2}(\Delta^2)])^\top \\
&= (\mathbf{0}^\top, (b(\hat{\boldsymbol{\beta}}_{2n}^{\mathrm{S}}))^\top)^\top \\
b(\hat{\theta}_n^{\mathrm{S}}) &= (\mathbf{0}^\top, -(p_2-2)(\boldsymbol{T}_{n(2)}\boldsymbol{\beta}_2)^\top \mathbb{E}[\chi^2_{p_2+2}(\Delta^2)])^\top \\
&= (\mathbf{0}^\top, (b(\boldsymbol{T}_{n(2)}\hat{\boldsymbol{\beta}}_{2n}^{\mathrm{S}}))^\top)^\top,
\end{aligned} \tag{4.15}
$$

where $\mathbb{E}[\chi_\nu^{-2}(\Delta^2)] = \int_0^\infty x^{-1}\mathrm{d}H_\nu(x; \Delta^2)$.

$$
b(\hat{\boldsymbol{\beta}}_n^{\mathrm{S+}}) = \begin{pmatrix} \mathbf{0} \\ b(\hat{\boldsymbol{\beta}}_{2n}^{\mathrm{S}}) - \boldsymbol{\beta}_2\mathbb{E}[(1-(p_2-2)\chi^{-2}_{p_2+2}(\Delta^2))I(\chi^2_{p_2+2}(\Delta^2) < p_2-2)] \end{pmatrix},
$$

$$
b(\hat{\boldsymbol{\theta}}_n^{\mathrm{S+}}) = \begin{pmatrix} \mathbf{0} \\ b(\boldsymbol{T}_{n(2)}\hat{\boldsymbol{\beta}}_{2n}^{\mathrm{S}}) \\ +(\boldsymbol{T}_{n(2)}\boldsymbol{\beta}_2)^\top\mathbb{E}[(1-(p_2-2)\chi^{-2}_{p_2+2}(\Delta^2))I(\chi^2_{p_2+2}(\Delta^2) < p_2-2)] \end{pmatrix} \tag{4.16}
$$

$$
\begin{aligned}
b(\hat{\boldsymbol{\beta}}_n^{\mathrm{L}}) &= \frac{\sigma}{\sqrt{Q_\gamma}}(b(\hat{\boldsymbol{\beta}}_{\gamma,n_\gamma}^{\mathrm{LASSO}})|\gamma = 1, \dots, p)^\top \\
b(\hat{\theta}_n^{\mathrm{L}}) &= -b(\hat{\boldsymbol{\beta}}_n^{\mathrm{L}})\bar{\boldsymbol{x}}
\end{aligned} \tag{4.17}
$$

$$
\begin{aligned}
b(\hat{\boldsymbol{\beta}}_n^{\mathrm{RR}}(k)) &= \begin{pmatrix} \mathbf{0} \\ -\frac{k}{k+1}\boldsymbol{\beta}_2 \end{pmatrix} \\
b(\hat{\theta}_n^{\mathrm{RR}}(k)) &= -\frac{k}{k+1}\boldsymbol{T}_{n(2)}\boldsymbol{\beta}_2.
\end{aligned} \tag{4.18}
$$

Next, we have the following theorem for the L_2 risk of the estimators.

Under the assumption of Theorem 4.1, we have following L_2-risk expressions for the estimators defined in Sections 4.1.1–4.1.3 using the formula

$$R(\boldsymbol{\beta}_n^*; \boldsymbol{W}_1, \boldsymbol{W}_2) = \mathbb{E}[(\boldsymbol{\beta}_{1n}^* - \boldsymbol{\beta}_1)^\top \boldsymbol{W}_1 (\boldsymbol{\beta}_{1n}^* - \boldsymbol{\beta}_1)] + \mathbb{E}[(\boldsymbol{\beta}_{2n}^* - \boldsymbol{\beta}_2)^\top \boldsymbol{W}_2 (\boldsymbol{\beta}_{2n}^* - \boldsymbol{\beta}_2)],$$

where $\boldsymbol{W}_1$ and $\boldsymbol{W}_2$ are the weight matrices.

The L_2 risk of LSE is

$$\begin{aligned} R(\tilde{\boldsymbol{\beta}}_n, \boldsymbol{W}_1, \boldsymbol{W}_2) &= \text{tr}[\boldsymbol{W}_1 \mathbb{E}(\tilde{\boldsymbol{\beta}}_{1n} - \boldsymbol{\beta}_1)(\tilde{\boldsymbol{\beta}}_{1n} - \boldsymbol{\beta}_1)^\top] \\ &\quad + \text{tr}[\boldsymbol{W}_2 \mathbb{E}(\tilde{\boldsymbol{\beta}}_{2n} - \boldsymbol{\beta}_2)(\tilde{\boldsymbol{\beta}}_{2n} - \boldsymbol{\beta}_2)^\top] \\ &= \sigma^2 \{\text{tr}(\boldsymbol{W}_1 [\boldsymbol{D}_{22}^{[1]}]) + \text{tr}(\boldsymbol{W}_2 [\boldsymbol{D}_{22}^{[2]}])\} \\ &= \sigma^2 (p_1 + p_2) \end{aligned}$$

when $\boldsymbol{W}_j = [\boldsymbol{D}_{22}^{[j]}]^{-1}. j = 1, 2$ and

$$\begin{aligned} R(\tilde{\boldsymbol{\theta}}_n, \boldsymbol{W}_1^*, \boldsymbol{W}_2^*) &= \text{tr}[\boldsymbol{W}_1^* \mathbb{E}(\tilde{\boldsymbol{\theta}}_{1n} - \boldsymbol{\theta}_1)(\tilde{\boldsymbol{\theta}}_{1n} - \boldsymbol{\theta}_1)^\top] \\ &\quad + \text{tr}[\boldsymbol{W}_2^* \mathbb{E}(\tilde{\boldsymbol{\beta}}_{2n} - \boldsymbol{\beta}_2)(\tilde{\boldsymbol{\beta}}_{2n} - \boldsymbol{\beta}_2)^\top] \\ &= \sigma^2 \{\text{tr}(\boldsymbol{W}_1^* [\boldsymbol{D}_{11}^{[1]}]) + \text{tr}(\boldsymbol{W}_2^* [\boldsymbol{D}_{11}^{[2]}])\} \\ &= \sigma^2 (p_1 + p_2) \end{aligned} \tag{4.19}$$

when $\boldsymbol{W}_j^* = [\boldsymbol{D}_{11}^{[j]}]^{-1}, j = 1, 2.$

The L_2 risk of RLSE is

$$\begin{aligned} R(\hat{\boldsymbol{\beta}}_n, \boldsymbol{W}_1, \boldsymbol{W}_2) &= \sigma^2 \{\text{tr}(\boldsymbol{W}_1 [\boldsymbol{D}_{22}^{[1]}]) + \text{tr}(\boldsymbol{W}_2 [\boldsymbol{\beta}_2 \boldsymbol{\beta}_2^\top])\} \\ &= \sigma^2 (p_1 + \Delta^2) \end{aligned}$$

when $\boldsymbol{W}_1 = (\boldsymbol{D}_{22}^{[1]})^{-1}$ and $\boldsymbol{W}_2 = (\boldsymbol{D}_{22}^{[2]})^{-1}$ and

$$\begin{aligned} R(\hat{\boldsymbol{\theta}}_n, \boldsymbol{W}_1^*, \boldsymbol{W}_2^*) &= \sigma^2 \{\text{tr}(\boldsymbol{W}_1^* [\boldsymbol{D}_{11}^{[1]}]) + \text{tr}(\boldsymbol{W}_2^* (\boldsymbol{\beta}_2 \boldsymbol{\beta}_2^\top))\} \\ &= \sigma^2 (p_1 + \Delta^2) \end{aligned} \tag{4.20}$$

when $\boldsymbol{W}_1^* = (\boldsymbol{D}_{11}^{[1]})^{-1}$ and $\boldsymbol{W}_2^* = (\boldsymbol{D}_{11}^{[2]})^{-1}$.

The L_2 risk of PTE is

$$\begin{aligned} R(\hat{\boldsymbol{\beta}}_n^{\text{PT}}(\alpha), \boldsymbol{W}_1, \boldsymbol{W}_2) &= \sigma^2 \{\text{tr}(\boldsymbol{W}_1 [\boldsymbol{D}_{22}^{[1]}]) - \text{tr}(\boldsymbol{W}_2 [\boldsymbol{D}_{22}^{[2]}])(1 - H_{p_2+2}(c_\alpha; \Delta^2))\} \\ &\quad + \text{tr}(\boldsymbol{W}_2 [\boldsymbol{\beta}_2 \boldsymbol{\beta}_2^\top])\{2H_{p_2+2}(c_\alpha; \Delta^2) - H_{p_4+2}(c_\alpha; \Delta^2)\} \\ &= \sigma^2 [p_1 + p_2 (1 - H_{p_2+2}(c_\alpha; \Delta^2)) \\ &\quad + \Delta^2 \{2H_{p_2+2}(c_\alpha; \Delta^2) - H_{p_4+2}(c_\alpha; \Delta^2)\}, \end{aligned}$$

where $\boldsymbol{W}_j = \boldsymbol{D}_{22}^{[j]}, j = 1, 2$ and

$$\begin{aligned} R(\hat{\boldsymbol{\theta}}_n^{\text{PT}}(\alpha), \boldsymbol{W}_1^*, \boldsymbol{W}_2^*) &= \sigma^2 \{\text{tr}(\boldsymbol{W}_1^* [\boldsymbol{D}_{11}^{[1]}]) \\ &\quad + \text{tr}(\boldsymbol{W}_2^* (\boldsymbol{T}_{n(2)} \boldsymbol{D}_{22}^{[2]} T_{n(2)}))(1 - H_{p_2+2}(c_\alpha; \Delta^2)) \\ &\quad + \text{tr}(\boldsymbol{W}_2^* (\boldsymbol{T}_{n(2)} \boldsymbol{\beta}_2 \boldsymbol{\beta}_2^\top \boldsymbol{T}_{n(2)})) \\ &\qquad \times \{2H_{p_2+2}(c_\alpha; \Delta^2) - H_{p_4+2}(c_\alpha; \Delta^2)\}\} \\ &= \sigma^2 [p_1 + p_2 (1 - H_{p_2+2}(c_\alpha; \Delta^2))] \\ &\quad + \Delta^2 \{2H_{p_2+2}(c_\alpha; \Delta^2) - H_{p_4+2}(c_\alpha; \Delta^2)\} \end{aligned} \tag{4.21}$$

when $\boldsymbol{W}_1^* = (\boldsymbol{D}_{11}^{[1]})^{-1}$, $\boldsymbol{W}_2^* = (\boldsymbol{T}_{n(2)}\boldsymbol{D}_{22}^{[2]}\boldsymbol{T}_{n(2)})^{-1}$, $c_\alpha = \chi^2_{p_2}(\alpha)$ and $\Delta^2 = \frac{\boldsymbol{\beta}_2^\top \boldsymbol{D}_{22}^{[2]}\boldsymbol{\beta}_2}{\sigma^2}$.

The L_2 risk of PRSE is

$$\begin{aligned} R(\hat{\boldsymbol{\beta}}_n^{S+}, \boldsymbol{W}_1, \boldsymbol{W}_2) &= R(\hat{\boldsymbol{\beta}}_n^{S}, \boldsymbol{W}_1, \boldsymbol{W}_2) \\ &\quad -\sigma^2 p_2 \mathbb{E}[(1-(p_2-2)\chi^{-2}_{p_2+2}(\Delta^2))^2 I(\chi^{-2}_{p_2+2}(\Delta^2) < p_2-2)] \\ &\quad +\Delta^2\{2\mathbb{E}[(1-(p_2-2)\chi^{-2}_{p_2+2}(\Delta^2))^2 I(\chi^{-2}_{p_2+2}(\Delta^2) < p_2-2)] \\ &\quad -\mathbb{E}[(1-(p_2-2)\chi^{-2}_{p_2+4}(\Delta^2))^2 I(\chi^{-2}_{p_2+4}(\Delta^2) < p_2-2)]\} \\ &= R(\hat{\boldsymbol{\theta}}_n^{S+}, \boldsymbol{W}_1^*, \boldsymbol{W}_2^*), \end{aligned} \tag{4.22}$$

where $\boldsymbol{W}_1^* = (\boldsymbol{D}_{11}^{[1]})^{-1}$, and $\boldsymbol{W}_2^* = (\boldsymbol{T}_{n(2)}\boldsymbol{D}_{22}^{[2]}\boldsymbol{T}_{n(2)})^{-1}$.

The LASSO L_2 risk expression for $\hat{\beta}_n^{L}$ is

$$R(\hat{\beta}_n^{L}) = \sigma^2 \left\{ \sum_{\alpha=1}^{p_1} \frac{\overline{x}_\alpha^{2}}{Q_\alpha} \rho_{ST}(\lambda, \Delta_\alpha) + \sum_{\alpha=p_1+1}^{P} \frac{\overline{x}_\alpha^{2}}{Q_\alpha} \Delta_\alpha^2 \right\}. \tag{4.23}$$

The LASSO L_2 risk expression for $\hat{\theta}_n^{L}$ is

$$R(\hat{\beta}_n^{L}) = \sigma^2 \left\{ \sum_{\alpha=1}^{p_1} \left(\frac{1}{n_\alpha} + \frac{\overline{x}_\alpha^{2}}{Q_\alpha} \rho_{ST}(\lambda, \Delta_\alpha) \right) + \sum_{\alpha=p_1+1}^{P} \frac{\overline{x}_\alpha^{2}}{Q_\alpha} \Delta_\alpha^2 \right\}, \tag{4.24}$$

where

$$\begin{aligned} \rho_{ST}(\lambda, \Delta_\alpha) &= 1 + \lambda^2 + (1 - \Delta^2 - \lambda^2)\{\Phi(\lambda - \Delta_\alpha) - \Phi(-\lambda - \Delta_\alpha)\} \\ &\quad -(\lambda - \Delta_\alpha)\phi(\lambda + \Delta_\alpha) - (\lambda + \Delta_\alpha)\phi(\lambda - \Delta_\alpha). \end{aligned} \tag{4.25}$$

The corresponding lower bound of the unweighted risk functions of β and θ are, respectively,

$$\begin{aligned} R(\hat{\beta}_n^{L}) &= \sigma^2 \operatorname{tr}(\boldsymbol{D}_{22}^{[1]}) + \Delta^{*2}, \quad \Delta^{*2} = \Delta^2 \mathrm{Ch}_{\min}(\boldsymbol{T}_{n(2)}^2 \boldsymbol{D}_{22}{}^{[2]})^{-1} \\ R(\hat{\theta}_n^{L}) &= \sigma^2 \operatorname{tr}(\boldsymbol{D}_{11}^{[1]}) + \Delta^{*2}, \quad \Delta^{*2} = \Delta^2 \mathrm{Ch}_{\min}(\boldsymbol{T}_{n(2)}^2 \boldsymbol{D}_{22}{}^{[2]})^{-1}. \end{aligned} \tag{4.26}$$

We will consider the lower bound of L_2 risk of LASSO to compare with the L_2 risk of other estimators. Consequently, the lower bound of the weighted L_2 risk is given by

$$\mathrm{R}(\hat{\boldsymbol{\theta}}_n^{\mathrm{LASSO}}(\lambda) : \boldsymbol{W}_1, \boldsymbol{W}_2) = \sigma^2(p_1 + \Delta^2), \tag{4.27}$$

which is same as the L_2 risk of the ridge regression estimator (RRE).

The L_2 risk of RRE is

$$R(\hat{\boldsymbol{\beta}}_n^{\mathrm{RR}}(k) : \boldsymbol{W}_1, \boldsymbol{W}_2) = \sigma^2 \left(p_1 + \frac{p_2}{p_2 + \Delta^2} \Delta^2 \right), \quad k_{\mathrm{opt}} = \frac{p_2}{\Delta^2}$$

when $\boldsymbol{W}_1 = [\boldsymbol{D}_{22}^{[1]}]^{-1}$ and $\boldsymbol{W}_2 = [\boldsymbol{D}_{22}^{[2]}]^{-1}$ and

$$\mathrm{R}(\hat{\theta}_n^{\mathrm{RR}}(k); \boldsymbol{W}_1^*, \boldsymbol{W}_2^*) = \sigma^2 \left(p_1 + \frac{p_2}{p_2 + \Delta^2} \Delta^2 \right), \quad k_{\mathrm{opt}} = \frac{p_2}{\Delta^2} \tag{4.28}$$

when $\boldsymbol{W}_1^* = [\boldsymbol{D}_{11}^{[1]}]^{-1}$, $\boldsymbol{W}_2^* = [\boldsymbol{T}_{n(2)}\boldsymbol{D}_{22}^{[2]}\boldsymbol{T}_{n(2)}]^{-1}$.

4.3 Comparison of Estimators

In this section, we compare various estimators with respect to the LSE, in terms of relative weighted L_2-risk efficiency (RWRE).

4.3.1 Comparison of LSE with RLSE

Recall that the RLSE is given by $\hat{\boldsymbol{\theta}}_n = (\tilde{\boldsymbol{\theta}}_{1n}^\top, \mathbf{0}^\top)^\top$. In this case, the RWRE of RLSE vs. LSE is given by

$$\text{RWRE}(\hat{\boldsymbol{\theta}}_n; \tilde{\boldsymbol{\theta}}_n) = \frac{p_1 + p_2}{p_1 + \Delta^2} = \left(1 + \frac{p_2}{p_1}\right)\left(1 + \frac{\Delta^2}{p_1}\right)^{-1},$$

which is a decreasing function of Δ^2. So, $0 \le \text{RWRE}(\hat{\boldsymbol{\theta}}_n : \tilde{\boldsymbol{\theta}}_n) \le \left(1 + \frac{p_2}{p_1}\right)$.

4.3.2 Comparison of LSE with PTE

The RWRE expression for PTE vs. LSE is given by

$$\text{RWRE}(\hat{\boldsymbol{\theta}}_n^{\text{PT}}(\alpha) : \tilde{\boldsymbol{\theta}}_n) = \frac{p_1 + p_2}{g(\Delta^2, \alpha)},$$

where

$$g(\Delta^2, \alpha) = p_1 + p_2(1 - H_{p_2+2}(c_\alpha; \Delta^2)) + \Delta^2[2H_{p_2+2}(c_\alpha; \Delta^2) - H_{p_2+4}(c_\alpha; \Delta^2)].$$

Then, the PTE outperforms the LSE for

$$0 \le \Delta^2 \le \frac{p_2 H_{p_2+2}(c_\alpha; \Delta^2)}{2H_{p_2+2}(c_\alpha; \Delta^2) - H_{p_2+4}(c_\alpha; \Delta^2)} = \Delta^2_{\text{PT}}. \tag{4.29}$$

Otherwise, LSE outperforms the PTE in the interval $(\Delta^2_{\text{PT}}, \infty)$.

We may mention that $\text{RWRE}(\hat{\boldsymbol{\theta}}_n^{\text{PT}}(\alpha) : \tilde{\boldsymbol{\theta}}_n)$ is a decreasing function of Δ^2 with a maximum at $\Delta^2 = 0$, then decreases crossing the 1-line to a minimum at $\Delta^2 = \Delta^2_{\text{PT}}(\min)$ with a value $M_{\text{PT}}(\alpha)$, and then increases toward the 1-line. This means the gains in efficiency of PTE is the highest in the interval given by Eq. (4.24) and loss in efficiency can be noticed outside it.

The $\text{RWRE}(\hat{\boldsymbol{\theta}}_n^{\text{PT}}(\alpha); \tilde{\boldsymbol{\theta}}_n)$ belongs to the interval

$$M_{\text{PT}}(\alpha) \le \text{RWRE}(\hat{\boldsymbol{\theta}}_n^{\text{PT}}(\alpha); \tilde{\boldsymbol{\theta}}_n) \le \left(1 + \frac{p_2}{p_1}\right)\left(1 + \frac{p_2}{p_1}[1 - H_{p_2+2}(c_\alpha; 0)]\right)^{-1},$$

where $M_{\text{PT}}(\alpha)$ depends on the size α and given by

$$M_{\text{PT}}(\alpha) = \left(1 + \frac{p_2}{p_1}\right)\left\{1 + \frac{p_2}{p_1}[1 - H_{p_2+2}(c_\alpha; \Delta^2_{\text{PT}}(\min))] + \frac{\Delta^2_{\text{PT}}(\min)}{p_1}[2H_{p_2+2}(c_\alpha; \Delta^2_{\text{PT}}(\min)) - H_{p_2+4}(c_\alpha; \Delta^2_{\text{PT}}(\min))]\right\}^{-1}.$$

The quantity $\Delta^2_{\text{PT}}(\min)$ is the value Δ^2 at which the RWRE value is minimum.

4.3.3 Comparison of LSE with SE and PRSE

We obtain the RWRE as follows:

$$\mathrm{RWRE}(\hat{\theta}_n^{\mathrm{S}}; \tilde{\theta}_n) = \left(1+\frac{p_2}{p_1}\right)\left(1+\frac{p_2}{p_1}-\frac{(p_2-2)^2}{p_1}\mathbb{E}[\chi_{p_2}^{-2}(\Delta^2)]\right)^{-1}.$$

It is a decreasing function of Δ^2. At $\Delta^2 = 0$, its value is $\left(1+\frac{p_2}{p_1}\right)\left(1+\frac{2}{p_1}\right)^{-1}$; and when $\Delta^2 \to \infty$, its value goes to 1. Hence, for $\Delta^2 \in \mathbb{R}^+$,

$$\begin{aligned}1 &\le \left(1+\frac{p_2}{p_1}\right)\left(1+\frac{p_2}{p_1}-\frac{(p_2-2)^2}{p_1}\mathbb{E}[\chi_{p_2}^{-2}(\Delta^2)]\right)^{-1}\\ &\le \left(1+\frac{p_2}{p_1}\right)\left(1+\frac{2}{p_1}\right)^{-1}.\end{aligned}$$

Hence, the gains in efficiency is the highest when Δ^2 is small and drops toward 1 when Δ^2 is the largest. Also,

$$\begin{aligned}\mathrm{RWRE}(\hat{\theta}_n^{\mathrm{S+}}; \tilde{\theta}_n) = \left(1+\frac{p_2}{p_1}\right)\Bigg(1+\frac{p_2}{p_1}&-\frac{(p_2-2)^2}{p_1}\mathbb{E}[\chi_{p_2}^{-2}(\Delta^2)]\\ &-\frac{p_2}{p_1}\mathbb{E}[(1-(p_2-2)\chi_{p_2+2}^{-2}(\Delta^2))^2 I(\chi_{p_2+2}^{2}(\Delta^2)<(p_2-2))]\\ &+\frac{\Delta^2}{p_1}\{2\mathbb{E}[(1-(p_2-2)\chi_{p_2+2}^{-2}(\Delta^2))I(\chi_{p_2+2}^{2}(\Delta^2)<(p_2-2))]\\ &-\mathbb{E}[(1-(p_2-2)\chi_{p_2+4}^{-2}(\Delta^2))^2 I(\chi_{p_2+4}^{2}(\Delta^2)<(p_2-2))]\}\Bigg)^{-1}.\end{aligned}$$

So that,

$$\mathrm{RWRE}(\hat{\theta}_n^{\mathrm{S+}}; \tilde{\theta}_n) \ge \mathrm{RWRE}(\hat{\theta}_n^{\mathrm{S}}; \tilde{\theta}_n) \ge 1 \qquad \forall \Delta^2 \in \mathbb{R}^+.$$

We also provide a graphical representation (Figure 4.1) of RWRE of the estimators.

In the next three subsections, we show that the RRE uniformly dominates all other estimators, although it does not select variables.

4.3.4 Comparison of LSE and RLSE with RRE

First, we consider weighted L_2-risk difference of LSE and RRE given by

$$\begin{aligned}\sigma^2(p_1+p_2) - \sigma^2 p_1 - \sigma^2\frac{p_2\Delta^2}{p_2+\Delta^2} &= \sigma^2 p_2\left(1-\frac{\Delta^2}{p_2+\Delta^2}\right)\\ &= \frac{\sigma^2 p_2^2}{p_2+\Delta^2} > 0,\ \forall\, \Delta^2 \in \mathbb{R}^+.\end{aligned}$$

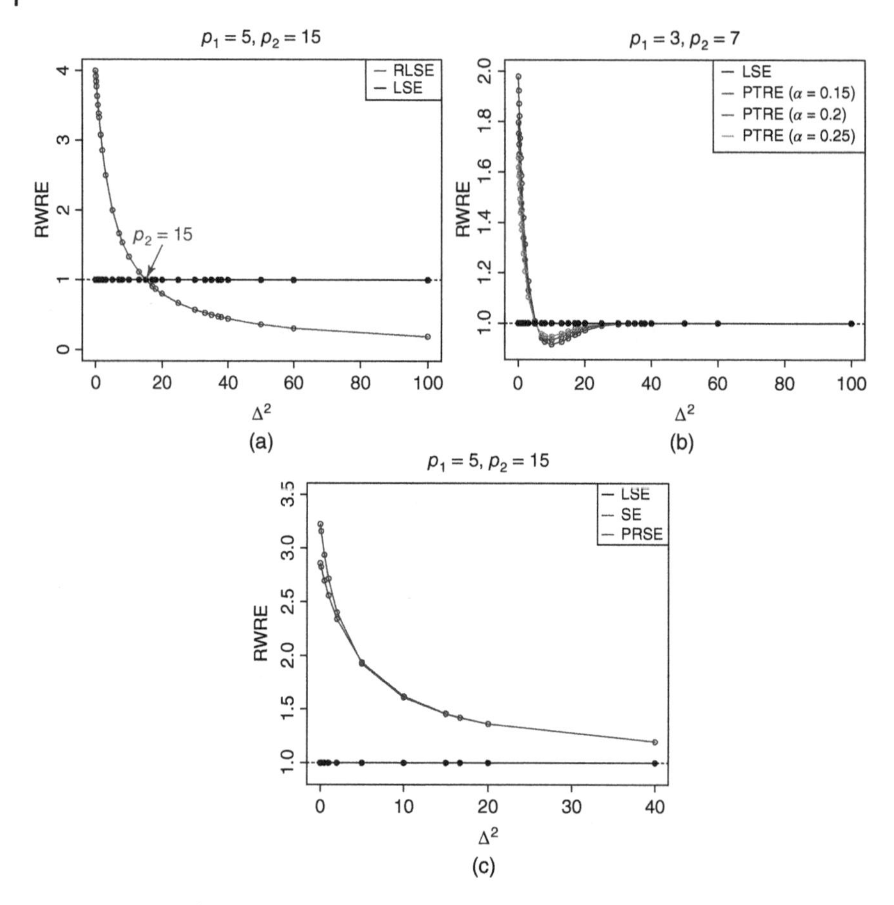

Figure 4.1 RWRE for the restricted, preliminary test, and Stein-type and its positive-rule estimators.

Hence, RRE outperforms the LSE uniformly. Similarly, for the RLSE and RRE, the weighted L_2-risk difference is given by

$$\sigma^2(p_1 + \Delta^2) - \left(\sigma^2 p_1 + \frac{\sigma^2 p_2 \Delta^2}{p_2 + \Delta^2}\right) = \frac{\sigma^2 \Delta^4}{p_2 + \Delta^2} > 0.$$

Therefore, RRE performs better than RLSE uniformly.

In addition, the RWRE of RRE vs. LSE equals

$$\text{RWRE}(\hat{\theta}_n^{\text{RR}}(k_o):\tilde{\theta}_n) = \frac{p_1 + p_2}{p_1 + \frac{p_2\Delta^2}{p_2+\Delta^2}} = \left(1 + \frac{p_2}{p_1}\right)\left(1 + \frac{p_2\Delta^2}{p_1(p_2 + \Delta^2)}\right)^{-1},$$

which is a decreasing function of Δ^2 with maximum $\left(1+\frac{p_2}{p_1}\right)$ at $\Delta^2 = 0$ and minimum 1 as $\Delta^2 \to \infty$. So,

$$1 \leq \left(1+\frac{p_2}{p_1}\right)\left(1+\frac{p_2}{p_1\left(1+\frac{p_2}{\Delta}\right)}\right)^{-1} \leq 1+\frac{p_2}{p_1}; \quad \forall \Delta^2 \in \mathbb{R}^+.$$

4.3.5 Comparison of RRE with PTE, SE, and PRSE

4.3.5.1 Comparison Between $\hat{\theta}_n^{\mathrm{RR}}(k_{\mathrm{opt}})$ and $\hat{\theta}_n^{\mathrm{PT}}$

Here, the weighted L_2-risk difference of $\hat{\theta}_n^{\mathrm{PT}}$ and $\hat{\theta}_n^{\mathrm{RR}}(k_{\mathrm{opt}})$ is given by

$$\begin{aligned}
&\mathrm{R}(\hat{\theta}_n^{\mathrm{PT}}(\alpha): \boldsymbol{W}_1, \boldsymbol{W}_2) - \mathrm{R}(\hat{\theta}_n^{\mathrm{RR}}(k_{\mathrm{opt}}): \boldsymbol{W}_1, \boldsymbol{W}_2) \\
&\quad = \sigma^2[p_2(1 - H_{p_2+2}(c_\alpha; \Delta^2)) + \Delta^2\{2H_{p_2+2}(c_\alpha; \Delta^2) - H_{p_2+4}(c_\alpha; \Delta^2)\}] \\
&\quad\quad - \frac{\sigma^2 p_2 \Delta^2}{p_2 + \Delta^2}.
\end{aligned} \tag{4.30}$$

Note that the risk of $\hat{\boldsymbol{\beta}}_{2n}^{\mathrm{PT}}(\alpha)$ is an increasing function of Δ^2 crossing the p_2-line to a maximum and then drops monotonically toward the p_2-line as $\Delta^2 \to \infty$. The value of the risk is $p_2(1 - H_{p_2+2}(\chi^2_{p_2}(\alpha); 0))(< p_2)$ at $\Delta^2 = 0$. On the other hand, $\frac{p_2\Delta^2}{p_2+\Delta^2}$ is an increasing function of Δ^2 below the p_2-line with a minimum value 0 at $\Delta^2 = 0$ and as $\Delta^2 \to \infty$, $\frac{p_2\Delta^2}{p_2+\Delta^2} \to p_2$. Hence, the risk difference in Eq. (4.30) is nonnegative for $\Delta^2 \in \mathbb{R}^+$. Thus, the RRE uniformly performs better than the PTE.

4.3.5.2 Comparison Between $\hat{\theta}_n^{\mathrm{RR}}(k_{\mathrm{opt}})$ and $\hat{\theta}_n^{\mathrm{S}}$

The weighted L_2-risk difference of $\hat{\theta}_n^{\mathrm{S}}$ and $\hat{\theta}_n^{\mathrm{RR}}(k_{\mathrm{opt}})$ is given by

$$\begin{aligned}
&R(\hat{\theta}_n^{\mathrm{S}}: \boldsymbol{W}_1, \boldsymbol{W}_2) - R(\hat{\theta}_n^{\mathrm{RR}}(k_{\mathrm{opt}}): \boldsymbol{W}_1, \boldsymbol{W}_2) \\
&\quad = \sigma^2(p_1 + p_2 - (p_2-2)^2\mathbb{E}[\chi_{p_2}^{-2}(\Delta^2)]) - \sigma^2\left(p_1 + \frac{p_2\Delta^2}{p_2+\Delta^2}\right) \\
&\quad = \sigma^2\left[p_2 - (p_2-2)^2\mathbb{E}[\chi_{p_2}^{-2}(\Delta^2)] - \frac{p_2\Delta^2}{p_2+\Delta^2}\right].
\end{aligned} \tag{4.31}$$

Note that the first function is increasing in Δ^2 with a value 2 at $\Delta^2 = 0$; and as $\Delta^2 \to \infty$, it tends to p_2. The second function is also increasing in Δ^2 with a value 0 at $\Delta^2 = 0$ and approaches the value p_2 as $\Delta^2 \to \infty$. Hence, the risk difference is nonnegative for all $\Delta^2 \in \mathbb{R}^+$. Consequently, RRE outperforms SE uniformly.

4.3.5.3 Comparison of $\hat{\theta}_n^{\text{RR}}(k_{\text{opt}})$ with $\hat{\theta}_n^{\text{S+}}$

The risk of $\hat{\boldsymbol{\theta}}_n^{\text{S+}}$ is

$$R(\hat{\boldsymbol{\theta}}_n^{\text{S+}} : \boldsymbol{W}_1, \boldsymbol{W}_2) = R(\hat{\boldsymbol{\theta}}_n^{\text{S}} : \boldsymbol{W}_1, \boldsymbol{W}_2) - R^*, \tag{4.32}$$

where

$$\begin{aligned} R^* &= \sigma^2 p_2 \mathbb{E}[(1 - (p_2 - 2)\chi_{p_2+2}^{-2}(\Delta^2))^2 I(\chi_{p_2+2}^{-2}(\Delta^2) < p_2 - 2)] \\ &\quad -\sigma^2 \Delta^2 \{2\mathbb{E}[(1 - (p_2 - 2)\chi_{p_2+2}^{-2}(\Delta^2)) I(\chi_{p_2+2}^{-2}(\Delta^2) < p_2 - 2)] \\ &\quad -\mathbb{E}[(1 - (p_2 - 2)\chi_{p_2+4}^{-2}(\Delta^2))^2 I(\chi_{p_2+4}^{-2}(\Delta^2) < p_2 - 2)]\}, \end{aligned} \tag{4.33}$$

and $R(\hat{\boldsymbol{\theta}}_n^{\text{S}} : \boldsymbol{W}_1, \boldsymbol{W}_2)$ is

$$R(\hat{\boldsymbol{\theta}}_n^{\text{S}} : \boldsymbol{W}_1, \boldsymbol{W}_2) = \sigma^2 (p_1 + p_2 - (p_2 - 2)^2 \mathbb{E}[\chi_{p_2}^{-2}(\Delta^2)]).$$

The weighted L_2-risk difference of PR and RRE is given by

$$\begin{aligned} &R(\hat{\boldsymbol{\theta}}_n^{\text{S+}} : \boldsymbol{W}_1, \boldsymbol{W}_2) - R(\hat{\boldsymbol{\theta}}_n^{\text{RR}}(k_{\text{opt}}) : \boldsymbol{W}_1, \boldsymbol{W}_2) \\ &\quad = [R(\hat{\boldsymbol{\theta}}_n^{\text{S}} : \boldsymbol{W}_1, \boldsymbol{W}_2) - R^*] - R(\hat{\boldsymbol{\theta}}_n^{\text{RR}}(k_{\text{opt}}) : \boldsymbol{W}_1, \boldsymbol{W}_2) \geq 0, \end{aligned} \tag{4.34}$$

where

$$R(\hat{\boldsymbol{\theta}}_n^{\text{RR}}(k_{\text{opt}}) : \boldsymbol{W}_1, \boldsymbol{W}_2) = \sigma^2 \left(p_1 + \frac{p_2 \Delta^2}{p_2 + \Delta^2} \right).$$

Consider the $R(\hat{\boldsymbol{\theta}}_n^{\text{S+}})$. It is a monotonically increasing function of Δ^2. At $\Delta^2 = 0$, its value is $\sigma^2(p_1 + 2) - \sigma^2 p_2 \mathbb{E}[(1 - (p_2 - 2)\chi_{p_2+2}^{-2}(0))^2 I(\chi_{p_2+2}^{-2}(0) < p_2 - 2)] \geq 0$; and as $\Delta^2 \to \infty$, it tends to $\sigma^2(p_1 + p_2)$. For $R(\hat{\boldsymbol{\theta}}_n^{\text{RR}}(k_{\text{opt}}) : \boldsymbol{W}_1, \boldsymbol{W}_2)$, at $\Delta^2 = 0$, the value is $\sigma^2 p_1$; and as $\Delta^2 \to \infty$, it tends to $\sigma^2(p_1 + p_2)$. Hence, the L_2-risk difference in (4.31) is nonnegative and RRE uniformly outperforms PRSE.

Note that the risk difference of $\hat{\boldsymbol{\theta}}_n^{\text{S+}}$ and $\hat{\boldsymbol{\theta}}_n^{\text{RR}}(k_{\text{opt}})$ at $\Delta^2 = 0$ is

$$\begin{aligned} &\sigma^2(p_1 + 2) - \sigma^2 p_2 \mathbb{E}[(1 - (p_2 - 2)\chi_{p_2+2}^{-2}(0))^2 I(\chi_{p_2+2}^{-2}(0) < p_2 - 2)] - \sigma^2 p_1 \\ &\quad = \sigma^2(2 - p_2 \mathbb{E}[(1 - (p_2 - 2)\chi_{p_2+2}^{-2}(0))^2 I(\chi_{p_2+2}^{-2}(0) < p_2 - 2)]) \geq 0, \end{aligned} \tag{4.35}$$

because the expected value in Eq. (4.35) is a decreasing function of DF, and $2 > p_2 \mathbb{E}[(1 - (p_2 - 2)\chi_{p_2+2}^{-2}(0))^2 I(\chi_{p_2+2}^{-2}(0) < p_2 - 2)]$. The risk functions of RRE, PT, SE, and PRSE are plotted in Figures 4.2 and 4.3 for $p_1 = 5, p_2 = 15$, $p_1 = 7, and\, p_2 = 33$, respectively. These figures are in support of the given comparisons.

4.3.6 Comparison of LASSO with RRE

Here, the weighted L_2-risk difference is given by

$$\begin{aligned} &\mathrm{R}(\hat{\boldsymbol{\theta}}_n^{\text{LASSO}}(\lambda); \boldsymbol{W}_1, \boldsymbol{W}_2) - \mathrm{R}(\hat{\boldsymbol{\theta}}_n^{\text{RR}}(k_o); \boldsymbol{W}_1, \boldsymbol{W}_2) \\ &\quad = \sigma^2 \left[(p_1 + \Delta^2) - \left(p_1 + \frac{p_2 \Delta^2}{p_2 + \Delta^2} \right) \right] \\ &\quad = \frac{\sigma^2 \Delta^2}{p_2 + \Delta^2} \geq 0. \end{aligned}$$

Hence, the RRE outperforms the LASSO uniformly.

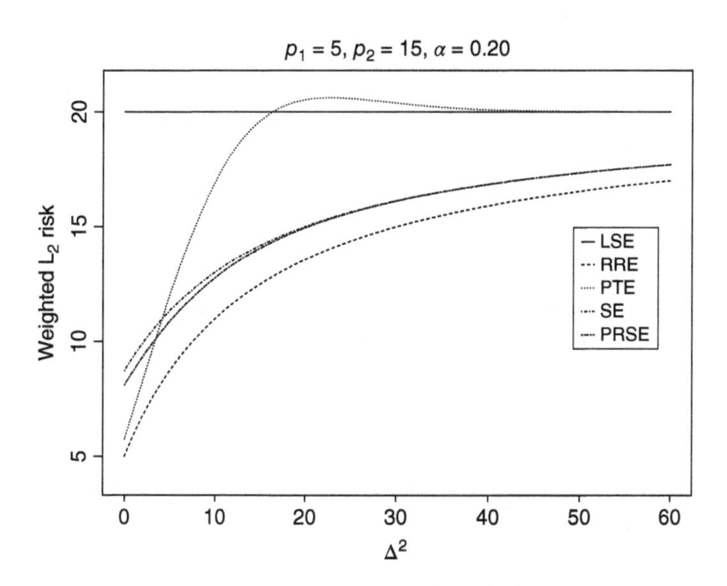

Figure 4.2 Weighted L_2 risk for the ridge, preliminary test, and Stein-type and its positive-rule estimators for $p_1 = 5$, $p_2 = 15$, and $\alpha = 0.20$.

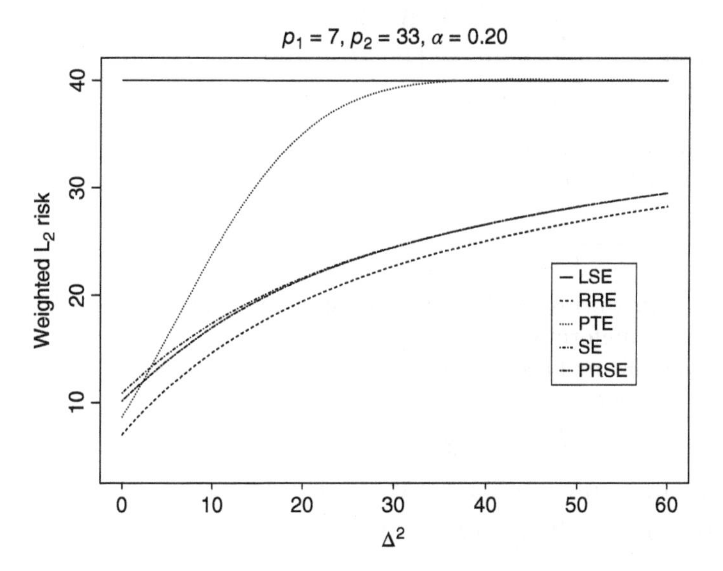

Figure 4.3 Weighted L_2 risk for the ridge, preliminary test, and Stein-type and its positive-rule estimators for $p_1 = 7$, $p_2 = 33$, and $\alpha = 0.20$.

4.3.7 Comparison of LASSO with LSE and RLSE

First, note that if we have for p_1 coefficients, $|\beta_j| > \sigma n_j^{-1/2}$ and also p_2 coefficients are zero in a sparse solution, then the "ideal" weighted L_2-risk is given by $\sigma^2(p_1 + \Delta^2)$. Thereby, we compare all estimators relative to this quantity. Hence, the weighted L_2-risk difference between LSE and LASSO is given by

$$\sigma^2(p_1 + p_2) - \sigma^2(p_1 + \Delta^2) = \sigma^2[p_2 - \Delta^2].$$

Hence, if $\Delta^2 \in (0, p_2)$, the LASSO performs better than the LSE, while if $\Delta^2 \in (p_2, \infty)$ the LSE performs better than the LASSO. Consequently, neither LSE nor the LASSO performs better than the other, uniformly.

Next, we compare the RLSE and LASSO. In this case, the weighted L_2-risk difference is given by

$$\sigma^2(p_1 + \Delta^2) - \sigma^2(p_1 + \Delta^2) = 0.$$

Hence, LASSO and RLSE are L_2 risk equivalent. And consequently, the LASSO satisfies the oracle properties.

4.3.8 Comparison of LASSO with PTE, SE, and PRSE

We first consider the PTE vs. LASSO. In this case, the weighted L_2-risk difference is given by

$$\begin{aligned}
&R(\hat{\boldsymbol{\theta}}_n^{\mathrm{PT}}(\alpha): \boldsymbol{W}_1, \boldsymbol{W}_2) - R(\hat{\boldsymbol{\theta}}_n^{\mathrm{LASSO}}(\lambda): \boldsymbol{W}_1, \boldsymbol{W}_2) \\
&\quad = \sigma^2[p_2(1 - H_{p_2+2}(c_\alpha; \Delta^2)) - \Delta^2\{1 - 2H_{p_2+2}(c_\alpha; \Delta^2) + H_{p_2+4}(c_\alpha; \Delta^2)\}] \\
&\quad \geq \sigma^2 p_2(1 - H_{p_2+2}(c_\alpha; 0)) \geq 0, \quad \text{if } \Delta^2 = 0.
\end{aligned}$$

Hence, the LASSO outperforms the PTE when $\Delta^2 = 0$. But when $\Delta^2 \neq 0$, the LASSO outperforms the PTE for

$$0 \leq \Delta^2 \leq \frac{p_2[1 - H_{p_2+2}(c_\alpha; \Delta^2)]}{1 - 2H_{p_2+2}(c_\alpha; \Delta^2) + H_{p_2+4}(c_\alpha; \Delta^2)}.$$

Otherwise, PTE outperforms the LASSO. Hence, neither LASSO nor PTE outmatches the other uniformly.

Next, we consider SE and PRSE vs. the LASSO. In these two cases, we have weighted L_2-risk differences given by

$$\begin{aligned}
&R(\hat{\boldsymbol{\theta}}_n^{\mathrm{S}}: \boldsymbol{W}_1, \boldsymbol{W}_2) - R(\hat{\boldsymbol{\theta}}_n^{\mathrm{LASSO}}(\lambda); \boldsymbol{W}_1, \boldsymbol{W}_2) \\
&\quad = \sigma^2[p_1 + p_2 - (p_2 - 2)^2\mathbb{E}[\chi_{p_2+2}^{-2}(\Delta^2)] - (p_1 + \Delta^2)] \\
&\quad = \sigma^2[p_2 - (p_2 - 2)^2\mathbb{E}[\chi_{p_2+2}^{-2}(\Delta^2)] - \Delta^2]
\end{aligned}$$

and

$$\begin{aligned}
&R(\hat{\boldsymbol{\theta}}_n^{\mathrm{S+}}: \boldsymbol{W}_1, \boldsymbol{W}_2) - R(\hat{\boldsymbol{\theta}}_n^{\mathrm{LASSO}}(\lambda): \boldsymbol{W}_1, \boldsymbol{W}_2) \\
&\quad = R(\hat{\boldsymbol{\theta}}_n^{\mathrm{S}}: \boldsymbol{W}_1, \boldsymbol{W}_2) - R(\hat{\boldsymbol{\theta}}_n^{\mathrm{L}}(\lambda): \boldsymbol{W}_1, \boldsymbol{W}_2) - R^*.
\end{aligned}$$

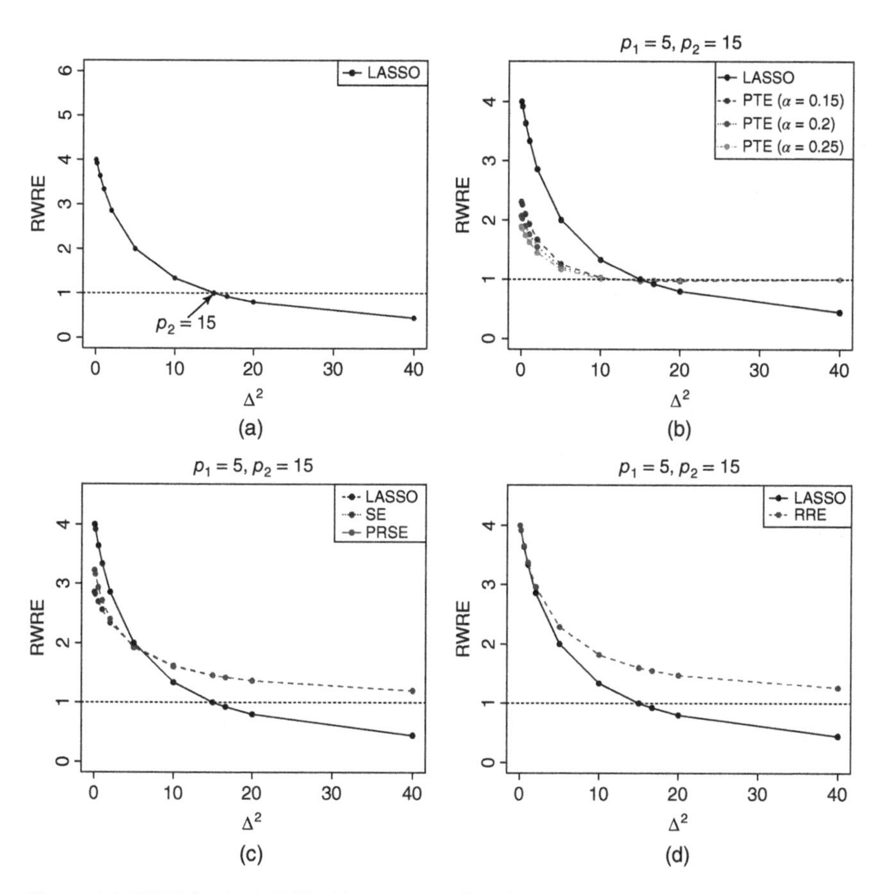

Figure 4.4 RWRE for the LASSO, ridge, restricted, preliminary test, and Stein-type and its positive-rule estimators.

Therefore, the LASSO outperforms the SE as well as the PRSE in the interval $[0, p_2 - (p_2 - 2)^2\mathbb{E}[\chi_{p_2}^{-2}(\Delta^2)]]$. Thus, neither SE nor the PRSE outperform the LASSO uniformly.

In Figure 4.4, the comparisons of LASSO with other estimators are shown.

4.4 Efficiency in Terms of Unweighted L_2 Risk

In the previous sections, we have made all comparisons among the estimators in terms of weighted risk functions. In this section, we provide the L_2-risk efficiency of the estimators in terms of the unweighted (weight $= I_p$) risk expressions for both $\boldsymbol{\beta}$ and $\boldsymbol{\theta}$.

4.4.1 Efficiency for β

The unweighted relative efficiency of LASSO:

$$\text{REff}(\hat{\boldsymbol{\beta}}_n^{\text{L}} : \tilde{\boldsymbol{\beta}}_n) = \left(1 + \frac{\text{tr}(\boldsymbol{D}_{22}^{[2]})}{\text{tr}(\boldsymbol{D}_{22}^{[1]})}\right)\left(1 + \frac{\Delta^{*2}}{\text{tr}(\boldsymbol{D}_{22}^{[1]})}\right)^{-1}. \tag{4.36}$$

Note that the unweighted risk of LASSO and RLSE is the same. The unweighted relative efficiency of PTE:

$$\begin{aligned}\text{REff}(\hat{\boldsymbol{\beta}}_n^{\text{PT}} : \tilde{\boldsymbol{\beta}}_n) = \left(1 + \frac{\text{tr}(\boldsymbol{D}_{22}^{[2]})}{\text{tr}(\boldsymbol{D}_{22}^{[1]})}\right)&\left\{1 + \frac{\text{tr}(\boldsymbol{D}_{22}^{[2]})}{\text{tr}(\boldsymbol{D}_{22}^{[1]})}(1 - H_{p_2+2}(c_\alpha; \Delta^2))\right.\\ &\left.+ \frac{\Delta^{*2}}{\text{tr}(\boldsymbol{D}_{22}^{[1]})}\{2H_{p_2+2}(c_\alpha; \Delta^2) - H_{p_2+4}(c_\alpha; \Delta^2)\}\right\}^{-1}.\end{aligned} \tag{4.37}$$

The unweighted relative efficiency of SE:

$$\begin{aligned}\text{REff}(\hat{\boldsymbol{\beta}}_n^{\text{S}} : \tilde{\boldsymbol{\beta}}_n) = \left(1 + \frac{\text{tr}(\boldsymbol{D}_{22}^{[2]})}{\text{tr}(\boldsymbol{D}_{22}^{[1]})}\right)&\left\{1 + \frac{\text{tr}(\boldsymbol{D}_{22}^{[2]})}{\text{tr}(\boldsymbol{D}_{22}^{[1]})}(1 - (p_2 - 2)A)\right.\\ &\left.+ (p_2^2 - 4)\frac{\Delta^{*2}}{\text{tr}(\boldsymbol{D}_{22}^{[1]})}\mathbb{E}[\chi_{p_2+4}^{-4}(\Delta^2)]\right\}^{-1},\end{aligned} \tag{4.38}$$

where

$$A = 2\mathbb{E}[\chi_{p_2+2}^{-2}(\Delta^2)] - (p_2 - 2)\mathbb{E}[\chi_{p_2+4}^{-4}(\Delta^2)]$$

The unweighted relative efficiency of PRSE:

$$\begin{aligned}&\text{REff}(\hat{\boldsymbol{\beta}}_n^{\text{S+}} : \tilde{\boldsymbol{\beta}}_n)\\ &= \left(1 + \frac{\text{tr}(\boldsymbol{D}_{22}^{[2]})}{\text{tr}(\boldsymbol{D}_{22}^{[1]})}\right)\\ &\quad \left\{1 + \frac{\text{tr}(\boldsymbol{D}_{22}^{[2]})}{\text{tr}(\boldsymbol{D}_{22}^{[1]})}(1 - (p_2 - 2)A) + (p_2^2 - 4)\frac{\Delta^{*2}}{\text{tr}(\boldsymbol{D}_{22}^{[1]})}E[\chi_{p_2+4}^{-4}(\Delta^2)]\right.\\ &\quad - \frac{\text{tr}(\boldsymbol{D}_{22}^{[2]})}{\text{tr}(\boldsymbol{D}_{22}^{[1]})}\mathbb{E}[(1 - (p_2 - 2)\chi_{p_2+2}^{-2}(\Delta^2))^2 I(\chi_{p_2+2}^2(\Delta^2) < (p_2 - 2))]\\ &\quad + \frac{\Delta^{*2}}{\text{tr}(\boldsymbol{D}_{22}^{[1]})}2\mathbb{E}[(1 - (p_2 - 2)\chi_{p_2+2}^{-2}(\Delta^2))I(\chi_{p_2+2}^2(\Delta^2) < (p_2 - 2))]\\ &\quad \left.- \mathbb{E}[(1 - (p_2 - 2)\chi_{p_2+4}^{-2}(\Delta^2))^2 I(\chi_{p_2+4}^2(\Delta^2) < (p_2 - 2))]\right\}^{-1}.\end{aligned} \tag{4.39}$$

The unweighted relative efficiency of RRE:

$$\text{REff}(\hat{\boldsymbol{\beta}}_n^{\text{RR}} : \tilde{\boldsymbol{\beta}}_n) = \left(1 + \frac{\text{tr}(\boldsymbol{D}_{22}^{[2]})}{\text{tr}(\boldsymbol{D}_{22}^{[1]})}\right) \left\{1 + \frac{\text{tr}(\boldsymbol{D}_{22}^{[2]})}{\text{tr}(\boldsymbol{D}_{22}^{[1]})} \frac{\Delta^{*2}}{(\text{tr}(\boldsymbol{D}_{22}^{[2]}) + \Delta^{*2})}\right\}^{-1}. \tag{4.40}$$

4.4.2 Efficiency for θ

The unweighted relative efficiency of LASSO:

$$\text{REff}(\hat{\boldsymbol{\theta}}_n^{\text{L}} : \tilde{\boldsymbol{\theta}}_n) = \left(1 + \frac{\text{tr}(\boldsymbol{D}_{11}^{[2]})}{\text{tr}(\boldsymbol{D}_{11}^{[1]})}\right) \left(1 + \frac{\Delta^{*2}}{\text{tr}(\boldsymbol{D}_{11}^{[1]})}\right)^{-1}. \tag{4.41}$$

Note that the unweighted risk of LASSO and RSLE is the same.

The unweighted relative efficiency of PTE:

$$\text{REff}(\hat{\theta}_n^{\text{PT}} : \tilde{\theta}_n) = \left(1 + \frac{\text{tr}(\boldsymbol{D}_{11}^{[2]})}{\text{tr}(\boldsymbol{D}_{11}^{[1]})}\right) \left\{1 + \frac{\text{tr}(\boldsymbol{D}_{11}^{[2]})}{\text{tr}(\boldsymbol{D}_{11}^{[1]})}(1 - H_{p_2+2}(c_\alpha; \Delta^2)) \right.$$
$$\left. + \frac{\Delta^{*2}}{\text{tr}(\boldsymbol{D}_{11}^{[1]})}\{2H_{p_2+2}(c_\alpha; \Delta^2) - H_{p_2+4}(c_\alpha; \Delta^2)\}\right\}^{-1}. \tag{4.42}$$

The unweighted relative efficiency of SE:

$$\text{REff}(\hat{\boldsymbol{\theta}}_n^{\text{S}} : \tilde{\boldsymbol{\theta}}_n) = \left(1 + \frac{\text{tr}(\boldsymbol{D}_{11}^{[2]})}{\text{tr}(\boldsymbol{D}_{11}^{[1]})}\right) \left\{1 + \frac{\text{tr}(\boldsymbol{D}_{11}^{[2]})}{\text{tr}(\boldsymbol{D}_{11}^{[1]})}(1 - (p_2 - 2)A) \right.$$
$$\left. + (p_2^2 - 4)\frac{\Delta^{*2}}{\text{tr}(\boldsymbol{D}_{11}^{[1]})}\mathbb{E}[\chi_{p_2+4}^{-4}(\Delta^2)]\right\}^{-1}, \tag{4.43}$$

where

$$A = 2\mathbb{E}[\chi_{p_2+2}^{-2}(\boldsymbol{\Delta}^2)] - (p_2 - 2)\mathbb{E}[\chi_{p_2+4}^{-4}(\Delta^2)]$$

The unweighted relative efficiency of PRSE:

$$\text{REff}(\hat{\boldsymbol{\theta}}_n^{\text{S+}} : \tilde{\boldsymbol{\theta}}_n)$$
$$= \left(1 + \frac{\text{tr}(\boldsymbol{D}_{11}^{[2]})}{\text{tr}(\boldsymbol{D}_{11}^{[1]})}\right)$$
$$\left\{1 + \frac{\text{tr}(\boldsymbol{D}_{11}^{[2]})}{\text{tr}(\boldsymbol{D}_{11}^{[1]})}(1 - (p_2 - 2)A) + (p_2^2 - 4)\frac{\Delta^{*2}}{\text{tr}(\boldsymbol{D}_{11}^{[1]})}\mathbb{E}[\chi_{p_2+4}^{-4}(\Delta^2)] \right.$$
$$- \frac{\text{tr}(\boldsymbol{D}_{11}^{[2]})}{\text{tr}(\boldsymbol{D}_{11}^{[1]})}\mathbb{E}[(1 - (p_2 - 2)\chi_{p_2+2}^{-2}(\Delta^2))^2 I(\chi_{p_2+2}^2(\Delta^2) < (p_2 - 2))]$$
$$+ \frac{\Delta^{*2}}{\text{tr}(\boldsymbol{D}_{11}^{[1]})}[2\mathbb{E}[(1 - (p_2 - 2)\chi_{p_2+2}^{-2}(\Delta^2))I(\chi_{p_2+2}^2(\Delta^2) < (p_2 - 2))]$$
$$\left. - \mathbb{E}[(1 - (p_2 - 2)\chi_{p_2+4}^{-2}(\Delta^2))^2 I(\chi_{p_2+4}^2(\Delta^2) < (p_2 - 2))]\right\}^{-1}. \tag{4.44}$$

The unweighted relative efficiency of RRE:

$$\text{REff}(\hat{\theta}_n^{\text{RR}} : \tilde{\theta}_n) = \left(1 + \frac{\text{tr}(\boldsymbol{D}_{11}^{[2]})}{\text{tr}(\boldsymbol{D}_{11}^{[1]})}\right) \left\{1 + \frac{\text{tr}(\boldsymbol{D}_{11}^{[2]})}{\text{tr}(\boldsymbol{D}_{11}^{[1]})} \frac{\Delta^{*2}}{(\text{tr}(\boldsymbol{D}_{11}^{[2]})) + \Delta^{*2})}\right\}^{-1}. \tag{4.45}$$

4.5 Summary and Concluding Remarks

In this section, we discuss the contents of Tables 4.1–4.9 presented as confirmatory evidence of the theoretical findings of the estimators. First, we note that we have two classes of estimators, namely, the traditional PTE and SE and the penalty estimators. The restricted LSE plays an important role due to the fact that LASSO belongs to the class of restricted estimators.

We have the following conclusions from our study.

(i) Since the inception of the RRE by Hoerl and Kennard (1970), there have been articles comparing RRE with PTE and the SE. From this study, we conclude that the RRE dominates the LSE, PTE, and the SE uniformly. The PRE dominates the LASSO estimator uniformly for Δ^2 greater than 0. They are L_2 risk equivalent at $\Delta^2 = 0$ and at this point LASSO dominates all other estimators. The ridge estimator does not select variables but the LASSO estimator does. See Table 4.1 and graphs in Figure 4.5.

(ii) The RLSE and LASSO are L_2 risk equivalent. Hence, LASSO satisfies "oracle properties."

(iii) Under the family of "diagonal linear projection," the "ideal" L_2 risk of LASSO and subset rule (hard threshold estimator, HTE) are same and do not depend on the thresholding parameter (κ) under sparse condition. SeeDonoho and Johnstone (1994).

(iv) The RWRE of estimators compared to the LSE depends upon the size of p_1, p_2, and the divergence parameter, Δ^2. LASSO/RLSE and RRE outperform all the estimators when Δ^2 is 0.

(v) The LASSO satisfies the "oracle properties" and it dominates LSE, PTE, SE, and PRSE in the subinterval of $[0, p_1)$. In this case, with a small number of active parameters, the LASSO and HTE perform best followed by RRE as pointed out by Tibshirani (1996).

(vi) If p_1 is fixed and p_2 increases, the RWRE of all estimators increases; see Tables 4.6 and 4.7.

(vii) If p_2 is fixed and p_1 increases, the RWRE of all estimators decreases. Then, for a given p_2 small and p_1 large, the LASSO, PTE, SE, and PRSE are competitive. See Tables 4.8 and 4.9.

(viii) The PRE outperforms the LSE, PTE, SE, and PRSE uniformly. The PRE dominates LASSO and RLSE uniformly for $\Delta^2 > 0$; and at $\Delta^2 = 0$, they are L_2 risk equivalent where Δ^2 is the divergence parameter.

(ix) The PRSE always outperforms SE; see Tables 4.1–4.9.

Table 4.1 RWRE for the estimators.

			PTE					
			α					
Δ^2	LSE	RLSE/ LASSO	0.15	0.2	0.25	SE	PRSE	RRE
				$p_1 = 5, p_2 = 15$				
0	1	4.00	2.30	2.07	1.89	2.86	3.22	4.00
0.1	1	3.92	2.26	2.03	1.85	2.82	3.16	3.92
0.5	1	3.64	2.10	1.89	1.74	2.69	2.93	3.64
1	1	3.33	1.93	1.76	1.63	2.56	2.71	3.36
2	1	2.86	1.67	1.55	1.45	2.33	2.40	2.96
3	1	2.50	1.49	1.40	1.33	2.17	2.19	2.67
5	1	2.00	1.26	1.21	1.17	1.94	1.92	2.26
7	1	1.67	1.13	1.10	1.08	1.78	1.77	2.04
10	1	1.33	1.02	1.02	1.01	1.62	1.60	1.81
15	1	1.00	0.97	0.97	0.98	1.46	1.45	1.60
20	1	0.80	0.97	0.98	0.98	1.36	1.36	1.47
30	1	0.57	0.99	0.99	0.99	1.25	1.25	1.33
50	1	0.36	0.99	0.99	1.00	1.16	1.16	1.21
100	1	0.19	1.00	1.00	1.00	1.05	1.05	1.11
				$p_1 = 7, p_2 = 33$				
0	1	5.71	2.86	2.50	2.23	4.44	4.92	5.71
0.1	1	5.63	2.82	2.46	2.20	4.40	4.84	5.63
0.5	1	5.33	2.66	2.34	2.10	4.23	4.57	5.34
1	1	5.00	2.49	2.20	1.98	4.03	4.28	5.02
2	1	4.44	2.21	1.97	1.80	3.71	3.84	4.50
3	1	4.00	1.99	1.79	1.65	3.45	3.51	4.10
5	1	3.33	1.67	1.53	1.43	3.05	3.05	3.53
7	1	2.86	1.46	1.36	1.29	2.76	2.74	3.13
10	1	2.35	1.26	1.20	1.16	2.46	2.44	2.72
15	1	1.82	1.09	1.07	1.05	2.13	2.11	2.31
20	1	1.48	1.02	1.02	1.01	1.92	1.91	2.06
30	1	1.08	0.99	0.99	0.99	1.67	1.67	1.76
33	1	1.00	0.99	0.99	0.99	1.62	1.62	1.70
50	1	0.70	0.99	0.99	0.99	1.43	1.43	1.49
100	1	0.37	1.00	1.00	1.00	1.12	1.12	1.25

Table 4.2 RWRE of the estimators for $p = 10$ and different Δ^2-value for varying p_1.

	$\Delta^2 = 0$				$\Delta^2 = 1$			
Estimators	$p_1 = 2$	$p_1 = 3$	$p_1 = 5$	$p_1 = 7$	$p_1 = 2$	$p_1 = 3$	$p_1 = 5$	$p_1 = 7$
LSE	1.00	1.00	1.00	1.00	1.00	1.00	1.00	1.00
RLSE/LASSO	5.00	3.33	2.00	1.43	3.33	2.50	1.67	1.25
PTE ($\alpha = 0.15$)	2.34	1.98	1.51	1.23	1.75	1.55	1.27	1.09
PTE ($\alpha = 0.2$)	2.06	1.80	1.43	1.19	1.60	1.45	1.22	1.07
PTE ($\alpha = 0.25$)	1.86	1.66	1.36	1.16	1.49	1.37	1.18	1.06
SE	2.50	2.00	1.43	1.11	2.14	1.77	1.33	1.08
PRSE	3.03	2.31	1.56	1.16	2.31	1.88	1.38	1.10
RRE	5.00	3.33	2.00	1.43	3.46	2.58	1.71	1.29
	$\Delta^2 = 5$				$\Delta^2 = 10$			
LSE	1.00	1.00	1.00	1.00	1.00	1.00	1.00	1.00
RLSE/LASSO	1.43	1.25	1.00	0.83	0.83	0.77	0.67	0.59
PTE ($\alpha = 0.15$)	1.05	1.01	0.95	0.92	0.92	0.92	0.92	0.94
PTE ($\alpha = 0.2$)	1.03	1.00	0.95	0.93	0.94	0.93	0.94	0.95
PTE ($\alpha = 0.25$)	1.02	0.99	0.96	0.94	0.95	0.95	0.95	0.97
SE	1.55	1.38	1.15	1.03	1.33	1.22	1.09	1.01
PRSE	1.53	1.37	1.15	1.03	1.32	1.22	1.08	1.01
RRE	1.97	1.69	1.33	1.13	1.55	1.40	1.20	1.07
	$\Delta^2 = 20$				$\Delta^2 = 60$			
LSE	1.00	1.00	1.00	1.00	1.00	1.00	1.00	1.00
RLSE/LASSO	0.45	0.43	0.40	0.37	0.16	0.16	0.15	0.15
PTE ($\alpha = 0.15$)	0.97	0.97	0.98	0.99	1.00	1.00	1.00	1.00
PTE ($\alpha = 0.2$)	0.98	0.98	0.99	0.99	1.00	1.00	1.00	1.00
PTE ($\alpha = 0.25$)	0.98	0.99	0.99	1.00	1.00	1.00	1.00	1.00
SE	1.17	1.12	1.04	1.00	1.06	1.04	1.01	1.00
PRSE	1.17	1.12	1.04	1.00	1.05	1.04	1.01	1.00
RRE	1.30	1.22	1.11	1.04	1.10	1.08	1.04	1.01

Table 4.3 RWRE of the estimators for $p = 20$ and different Δ^2 values for varying p_1.

	$\Delta^2 = 0$				$\Delta^2 = 1$			
Estimators	$p_1 = 2$	$p_1 = 3$	$p_1 = 5$	$p_1 = 7$	$p_1 = 2$	$p_1 = 3$	$p_1 = 5$	$p_1 = 7$
LSE	1.00	1.00	1.00	1.00	1.00	1.00	1.00	1.00
RLSE/LASSO	10.00	6.67	4.00	2.85	6.67	5.00	3.33	2.50
PTE ($\alpha = 0.15$)	3.20	2.84	2.31	1.95	2.50	2.27	1.93	1.68
PTE ($\alpha = 0.2$)	2.70	2.45	2.07	1.80	2.17	2.01	1.76	1.56
PTE ($\alpha = 0.25$)	2.35	2.17	1.89	1.67	1.94	1.82	1.63	1.47
SE	5.00	4.00	2.86	2.22	4.13	3.42	2.56	2.04
PRSE	6.28	4.77	3.22	2.43	4.58	3.72	2.71	2.13
RRE	10.00	6.67	4.00	2.86	6.78	5.07	3.37	2.52
	$\Delta^2 = 5$				$\Delta^2 = 10$			
LSE	1.00	1.00	1.00	1.00	1.00	1.00	1.00	1.00
RLSE/LASSO	2.86	2.50	2.00	1.67	1.67	1.54	1.33	1.18
PTE ($\alpha = 0.15$)	1.42	1.36	1.25	1.17	1.08	1.06	1.02	0.99
PTE ($\alpha = 0.2$)	1.33	1.29	1.20	1.14	1.06	1.04	1.02	0.99
PTE ($\alpha = 0.25$)	1.27	1.23	1.17	1.11	1.04	1.03	1.01	0.99
SE	2.65	2.36	1.94	1.65	2.03	1.87	1.62	1.43
PRSE	2.63	2.34	1.92	1.64	2.01	1.85	1.60	1.42
RRE	3.38	2.91	2.28	1.88	2.37	2.15	1.82	1.58
	$\Delta^2 = 20$				$\Delta^2 = 60$			
LSE	1.00	1.00	1.00	1.00	1.00	1.00	1.00	1.00
RLSE/LASSO	0.91	0.87	0.80	0.74	0.32	0.32	0.31	0.30
PTE ($\alpha = 0.15$)	0.97	0.97	0.97	0.97	1.00	1.00	1.00	1.00
PTE ($\alpha = 0.2$)	0.98	0.98	0.98	0.98	1.00	1.00	1.00	1.00
PTE ($\alpha = 0.25$)	0.99	0.98	0.98	0.99	1.00	1.00	1.00	1.00
SE	1.58	1.51	1.36	1.26	1.21	1.18	1.13	1.09
PRSE	1.58	1.50	1.36	1.25	1.21	1.18	1.13	1.09
RRE	1.74	1.64	1.47	1.34	1.26	1.23	1.18	1.13

Table 4.4 RWRE of the estimators for $p = 40$ and different Δ^2 values for varying p_1.

	$\Delta^2 = 0$				$\Delta^2 = 1$			
Estimators	$p_1 = 2$	$p_1 = 3$	$p_1 = 5$	$p_1 = 7$	$p_1 = 2$	$p_1 = 3$	$p_1 = 5$	$p_1 = 7$
LSE	1.00	1.00	1.00	1.00	1.00	1.00	1.00	1.00
RLSE/LASSO	20.00	13.33	8.00	5.71	13.33	10.00	6.67	5.00
PTE ($\alpha = 0.15$)	4.05	3.74	3.24	2.86	3.32	3.12	2.77	2.49
PTE ($\alpha = 0.2$)	3.29	3.09	2.76	2.50	2.77	2.64	2.40	2.20
PTE ($\alpha = 0.25$)	2.78	2.65	2.42	2.23	2.40	2.30	2.13	1.98
SE	10.00	8.00	5.71	4.44	8.12	6.75	5.05	4.03
PRSE	12.80	9.69	6.52	4.92	9.25	7.51	5.45	4.28
RRE	20.00	13.33	8.00	5.71	13.45	10.07	6.70	5.02
	$\Delta^2 = 5$				$\Delta^2 = 10$			
LSE	1.00	1.00	1.00	1.00	1.00	1.00	1.00	1.00
RLSE/LASSO	5.71	5.00	4.00	3.33	3.33	3.08	2.67	2.35
PTE ($\alpha = 0.15$)	1.9641	1.8968	1.7758	1.6701	1.3792	1.3530	1.3044	1.2602
PTE ($\alpha = 0.2$)	1.75	1.70	1.61	1.53	1.29	1.27	1.24	1.20
PTE ($\alpha = 0.25$)	1.60	1.56	1.50	1.44	1.23	1.22	1.19	1.16
SE	4.87	4.35	3.59	3.05	3.46	3.20	2.78	2.46
PRSE	4.88	4.36	3.59	3.05	3.42	3.16	2.75	2.44
RRE	6.23	5.40	4.27	3.53	4.03	3.68	3.13	2.72
	$\Delta^2 = 20$				$\Delta^2 = 60$			
LSE	1.00	1.00	1.00	1.00	1.00	1.00	1.00	1.00
RLSE/LASSO	1.82	1.74	1.60	1.48	0.64	0.63	0.61	0.60
PTE ($\alpha = 0.15$)	1.05	1.05	1.03	1.02	0.99	0.99	0.99	0.99
PTE ($\alpha = 0.2$)	1.04	1.03	1.02	1.02	0.99	0.99	0.99	0.99
PTE ($\alpha = 0.25$)	1.03	1.02	1.02	1.01	0.99	0.99	1.00	1.00
SE	2.41	2.2946	2.09	1.92	1.52	1.48	1.42	1.36
PRSE	2.41	2.29	2.08	1.91	1.52	1.48	1.42	1.36
RRE	2.65	2.50	2.26	2.06	1.58	1.54	1.47	1.41

Table 4.5 RWRE of the estimators for $p = 60$ and different Δ^2 values for varying p_1.

	$\Delta^2 = 0$				$\Delta^2 = 1$			
Estimators	$p_1 = 2$	$p_1 = 3$	$p_1 = 5$	$p_1 = 7$	$p_1 = 2$	$p_1 = 3$	$p_1 = 5$	$p_1 = 7$
LSE	1.00	1.00	1.00	1.00	1.00	1.00	1.00	1.00
RLSE/LASSO	30.00	20.00	12.00	8.57	20.00	15.00	10.00	7.50
PTE ($\alpha = 0.15$)	4.49	4.23	3.79	3.43	3.80	3.62	3.29	3.02
PTE ($\alpha = 0.2$)	3.58	3.42	3.14	2.91	3.10	2.99	2.78	2.59
PTE ($\alpha = 0.25$)	2.99	2.89	2.70	2.54	2.64	2.56	2.42	2.29
SE	15.00	12.00	8.57	6.67	12.12	10.09	7.55	6.03
PRSE	19.35	14.63	9.83	7.40	13.99	11.34	8.22	6.45
RRE	30.00	20.00	12.00	8.57	20.11	15.06	10.03	7.52
	$\Delta^2 = 5$				$\Delta^2 = 10$			
LSE	1.00	1.00	1.00	1.00	1.00	1.00	1.00	1.00
RLSE/LASSO	8.57	7.50	6.00	5.00	5.00	4.61	4.0000	3.53
PTE ($\alpha = 0.15$)	2.35	2.28	2.16	2.05	1.63	1.60	1.55	1.50
PTE ($\alpha = 0.2$)	2.04	1.99	1.91	1.83	1.49	1.47	1.43	1.39
PTE ($\alpha = 0.25$)	1.83	1.79	1.73	1.67	1.39	1.37	1.34	1.31
SE	7.10	6.35	5.25	4.47	4.89	4.53	3.94	3.50
PRSE	7.17	6.41	5.28	4.50	4.84	4.48	3.91	3.47
RRE	9.09	7.90	6.26	5.19	5.70	5.21	4.45	3.89
	$\Delta^2 = 20$				$\Delta^2 = 60$			
LSE	1.00	1.00	1.00	1.00	1.00	1.00	1.00	1.00
RLSE/LASSO	2.73	2.61	2.40	2.22	0.97	0.95	0.92	0.89
PTE ($\alpha = 0.15$)	1.15	1.14	1.13	1.11	0.99	0.99	0.99	0.99
PTE ($\alpha = 0.2$)	1.11	1.10	1.09	1.08	0.99	0.99	0.99	0.99
PTE ($\alpha = 0.25$)	1.08	1.08	1.07	1.06	0.99	0.99	0.99	0.99
SE	3.25	3.09	2.82	2.60	1.83	1.79	1.72	1.65
PRSE	3.23	3.08	2.81	2.59	1.83	1.79	1.72	1.65
RRE	3.55	3.37	3.05	2.79	1.90	1.86	1.78	1.71

Table 4.6 RWRE values of estimators for $p_1 = 5$ and different values of p_2 and Δ^2.

	LSE	RLSE/LASSO	PTE α 0.15	PTE α 0.2	PTE α 0.25	SE	PRSE	RRE
p_2				$\Delta^2 = 0$				
5	1.00	2.00	1.76	1.51	1.36	1.43	1.56	2.00
15	1.00	4.00	3.11	2.31	1.89	2.86	3.22	4.00
25	1.00	6.00	4.23	2.84	2.20	4.28	4.87	6.00
35	1.00	8.00	5.18	3.24	2.42	5.71	6.52	8.00
55	1.00	12.00	6.71	3.79	2.70	8.57	9.83	12.00
p_2				$\Delta^2 = 0.5$				
5	1.00	1.82	1.58	1.37	1.26	1.37	1.46	1.83
15	1.00	3.64	2.79	2.10	1.74	2.70	2.93	3.65
25	1.00	5.45	3.81	2.61	2.05	4.03	4.43	5.46
35	1.00	7.27	4.68	2.98	2.26	5.36	5.93	7.28
55	1.00	10.91	6.11	3.52	2.55	8.02	8.94	10.92
p_2				$\Delta^2 = 1$				
5	1.00	1.67	1.43	1.27	1.18	1.33	1.38	1.71
15	1.00	3.33	2.53	1.93	1.63	2.56	2.71	3.37
25	1.00	5.00	3.46	2.41	1.92	3.80	4.08	5.03
35	1.00	6.67	4.27	2.77	2.13	5.05	5.45	6.70
55	1.00	10.00	5.61	3.29	2.42	7.55	8.22	10.03
p_2				$\Delta^2 = 5$				
5	1.0000	1.00	0.93	0.95	0.96	1.15	1.15	1.33
15	1.00	2.00	1.47	1.26	1.17	1.94	1.92	2.28
25	1.00	3.00	1.98	1.54	1.35	2.76	2.75	3.27
35	1.00	4.00	2.44	1.77	1.50	3.59	3.59	4.27
55	1.00	6.00	3.27	2.16	1.73	5.25	5.28	6.26

Table 4.7 RWRE values of estimators for $p_1 = 7$ and different values of p_2 and Δ^2.

p_2				$\Delta^2 = 0$				
5	1.00	1.43	1.33	1.23	1.16	1.11	1.16	1.43
15	1.00	2.86	2.41	1.94	1.67	2.22	2.43	2.86
25	1.00	4.28	3.35	2.46	2.00	3.33	3.67	4.28
35	1.00	5.71	4.17	2.86	2.23	4.44	4.92	5.71
55	1.00	8.57	5.54	3.43	2.53	6.67	7.40	8.57
p_2				$\Delta^2 = 0.5$				
5	1.00	1.33	1.23	1.15	1.10	1.09	1.13	1.35
15	1.00	2.67	2.22	1.80	1.56	2.12	2.27	2.67
25	1.00	4.00	3.08	2.29	1.87	3.17	3.41	4.00
35	1.00	5.33	3.84	2.66	2.10	4.23	4.57	5.34
55	1.00	8.00	5.13	3.21	2.40	6.33	6.89	8.00
p_2				$\Delta^2 = 1$				
5	1.00	1.25	1.15	1.09	1.06	1.08	1.10	1.29
15	1.00	2.50	2.05	1.68	1.47	2.04	2.13	2.52
25	1.00	3.75	2.85	2.13	1.77	3.03	3.20	3.77
35	1.00	5.00	3.56	2.49	1.98	4.03	4.28	5.01
55	1.00	7.50	4.77	3.02	2.29	6.03	6.45	7.52
p_2				$\Delta^2 = 5$				
5	1.00	0.83	0.87	0.92	0.94	1.03	1.03	1.13
15	1.00	1.67	1.32	1.17	1.11	1.65	1.64	1.88
25	1.00	2.50	1.78	1.44	1.29	2.34	2.34	2.70
35	1.00	3.33	2.20	1.67	1.44	3.05	3.05	3.53
55	1.00	5.00	2.98	2.05	1.67	4.47	4.50	5.19

Table 4.8 RWRE values of estimators for $p_2 = 5$ and different values of p_1 and Δ^2.

			PTE					
		RLSE/		α				
	LSE	LASSO	0.15	0.2	0.25	SE	PRSE	RRE
p_1				$\Delta^2 = 0$				
5	1.00	2.00	1.76	1.51	1.36	1.43	1.56	2.00
15	1.00	1.33	1.27	1.20	1.15	1.18	1.22	1.33
25	1.00	1.20	1.17	1.127	1.10	1.11	1.14	1.20
35	1.00	1.14	1.12	1.09	1.07	1.08	1.10	1.14
55	1.00	1.09	1.08	1.06	1.04	1.05	1.06	1.09
p_1				$\Delta^2 = 0.5$				
5	1.00	1.82	1.58	1.37	1.26	1.34	1.46	1.83
15	1.00	1.29	1.22	1.16	1.11	1.16	1.19	1.29
25	1.00	1.18	1.14	1.10	1.07	1.10	1.12	1.18
35	1.00	1.13	1.10	1.07	1.05	1.07	1.08	1.13
55	1.00	1.08	1.06	1.05	1.03	1.05	1.05	1.08
p_1				$\Delta^2 = 1$				
5	1.00	1.67	1.43	1.27	1.18	1.33	1.38	1.71
15	1.00	1.25	1.18	1.12	1.08	1.14	1.16	1.26
25	1.00	1.15	1.11	1.08	1.05	1.09	1.10	1.16
35	1.00	1.11	1.08	1.06	1.04	1.07	1.07	1.12
55	1.00	1.07	1.05	1.04	1.03	1.04	1.05	1.07
p_1				$\Delta^2 = 5$				
5	1.00	1.00	0.93	0.95	0.96	1.15	1.15	1.33
15	1.00	1.00	0.97	0.97	0.98	1.07	1.07	1.14
25	1.00	1.00	0.98	0.98	0.98	1.05	1.04	1.09
35	1.00	1.00	0.98	0.99	0.99	1.03	1.03	1.07
55	1.00	1.00	0.99	0.99	0.99	1.02	1.02	1.04

Table 4.9 RWRE values of estimators for $p_2 = 7$ and different values of p_1 and Δ^2.

			PTE					
			α					
	LSE	RLSE/ LASSO	0.15	0.2	0.25	SE	PRSE	RRE
p_1				$\Delta^2 = 0$				
3	1.00	3.33	2.60	1.98	1.66	2.00	2.31	3.33
13	1.00	1.54	1.44	1.33	1.24	1.33	1.40	1.54
23	1.00	1.30	1.26	1.20	1.15	1.20	1.23	1.30
33	1.00	1.21	1.18	1.14	1.11	1.14	1.16	1.21
53	1.00	1.13	1.11	1.09	1.07	1.09	1.10	1.13
p_1				$\Delta^2 = 0.5$				
3	1.00	2.86	2.21	1.73	1.49	1.87	2.06	2.88
13	1.00	1.48	1.38	1.27	1.20	1.30	1.35	1.48
23	1.00	1.28	1.22	1.16	1.12	1.18	1.20	1.28
33	1.00	1.19	1.16	1.12	1.09	1.13	1.15	1.19
53	1.00	1.12	1.10	1.07	1.06	1.08	1.09	1.12
p_1				$\Delta^2 = 1$				
3	1.00	2.50	1.93	1.55	1.37	1.77	1.88	2.58
13	1.00	1.43	1.32	1.22	1.16	1.28	1.31	1.44
23	1.00	1.25	1.19	1.13	1.10	1.17	1.18	1.26
33	1.00	1.18	1.14	1.10	1.07	1.12	1.13	1.18
53	1.00	1.11	1.09	1.06	1.05	1.08	1.08	1.11
p_1				$\Delta^2 = 5$				
3	1.00	1.25	1.04	1.01	0.99	1.38	1.372	1.69
13	1.00	1.11	1.02	1.00	0.99	1.16	1.15	1.26
23	1.00	1.07	1.01	1.00	0.99	1.10	1.10	1.16
33	1.00	1.05	1.01	1.00	0.99	1.07	1.07	1.11
53	1.00	1.03	1.01	1.00	0.99	1.05	1.05	1.07

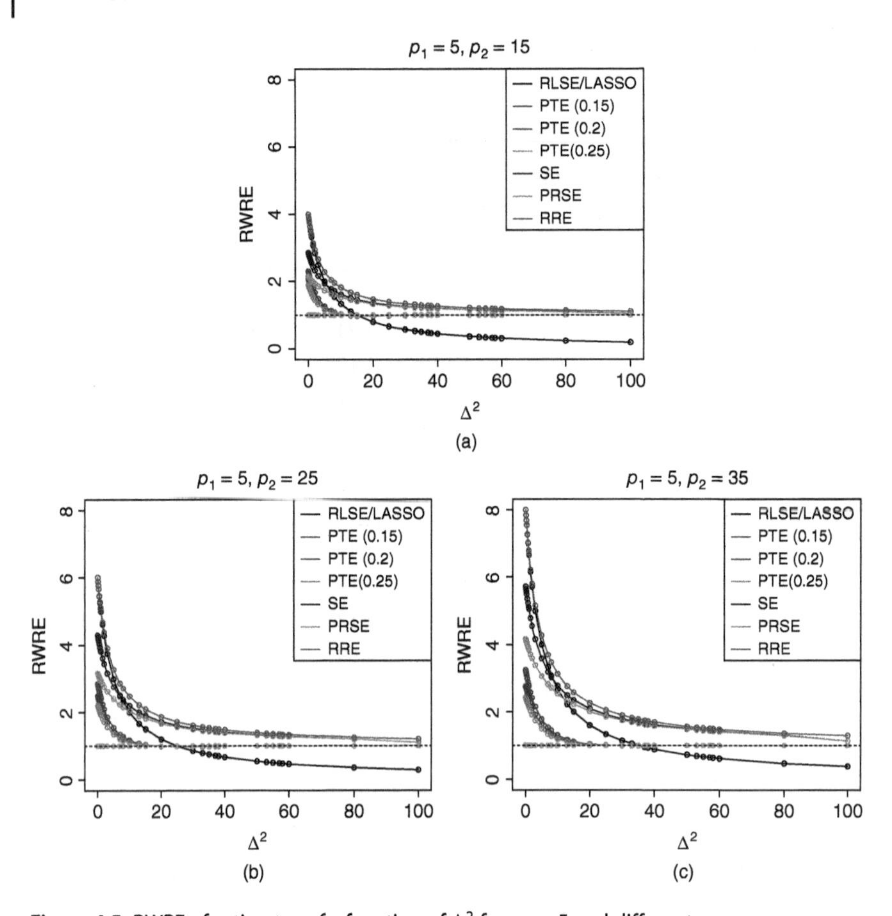

Figure 4.5 RWRE of estimates of a function of Δ^2 for $p_1 = 5$ and different p_2.

Problems

4.1 Show that the test statistic for testing $\mathcal{H}_o : \boldsymbol{\beta}_2 = \mathbf{0}$ vs. $\mathcal{H}_A : \boldsymbol{\beta}_2 \neq \mathbf{0}$ for unknown σ is

$$\mathcal{L}_n = \frac{\tilde{\boldsymbol{\theta}}_{2n}^\top \mathbf{Q}_2 \tilde{\boldsymbol{\theta}}_{2n}}{\sigma^2}$$

and also show that $\mathcal{L}_n$ has a noncentral χ^2 with appropriate DF and noncentrality parameter Δ^2.

4.2 Determine the bias vector of estimators, $\hat{\theta}_n^{\mathrm{S}}$ and $\hat{\theta}_n^{L}$ in Eqs. (4.15) and (4.17), respectively.

4.3 Show that the bias and MSE of $\hat{\boldsymbol{\theta}}_n^{\rm RR}(k)$ are, respectively,

$$b(\hat{\boldsymbol{\theta}}_n^{\rm RR}(k)) = \begin{pmatrix} \mathbf{0} \\ -\frac{k}{1+k}\boldsymbol{T}_{n(2)}\boldsymbol{\beta}_2 \end{pmatrix} \tag{4.46}$$

and

$$\mathrm{R}(\hat{\boldsymbol{\theta}}_n^{\rm RR}(k): \boldsymbol{W}_1^*, \boldsymbol{W}_2^*) = \sigma^2 \left(p_1 + \frac{p_2}{p_2 + \Delta^2}\right), \quad k = p_2\Delta^{-2} \tag{4.47}$$

4.4 Prove under usual notation that the RRE uniformly dominates both LSE and PTEs.

4.5 Verify that RRE uniformly dominates both Stein-type and its positive-rule estimators.

4.6 Prove under usual notation that the RRE uniformly dominates LASSO.

4.7 Show that the modified LASSO outperforms the SE as well as the PRSE in the interval

$$0 \le \Delta^2 \le p_2 - (p_2 - 2)^2 \mathbb{E}[\chi_{p_2}^{-2}(\Delta^2)].$$

5

Multiple Linear Regression Models

5.1 Introduction

Traditionally, we use least squares estimators (LSEs) for a linear model which provide *minimum variance unbiased estimators.* However, data analysts point out two deficiencies of LSEs, namely, the *prediction accuracy* and the *interpretation.* To overcome these concerns, Tibshirani (1996) proposed the least absolute shrinkage and selection operator (LASSO). It defines a continuous shrinking operation that can produce coefficients that are exactly zero and is competitive with *subset selection* and *ridge regression* estimators (RREs), retaining the good properties of both the estimators. The LASSO simultaneously estimates and selects the coefficients of a given linear model.

However, the preliminary test estimator (PTE) and the Stein-type estimator only shrink toward the target value and do not select coefficients for appropriate prediction and interpretation.

LASSO is related to the estimators, such as nonnegative garrote by Breiman (1996), smoothly clipped absolute derivation (SCAD) by Fan and Li (2001), elastic net by Zou and Hastie (2005), adaptive LASSO by Zou (2006), hard threshold LASSO by Belloni and Chernozhukov (2013), and many other versions. A general form of an extension of LASSO-type estimation called *the bridge estimation,* by Frank and Friedman (1993), is worth pursuing.

This chapter is devoted to the comparative study of the finite sample performance of the primary penalty estimators, namely, LASSO and the RREs. They are compared to the LSE, restricted least squares estimator (RLSE), PTE, SE, and positive-rule Stein-type estimator (PRSE) in the context of the multiple linear regression model. The question of comparison between the RRE (first discovery of penalty estimator) and the Stein-type estimator is well known and is established by Draper and Nostrand (1979), among others. So far, the literature is full of simulated results without any theoretical backups, and definite conclusions are not available whether the design matrix is orthogonal or nonorthogonal. In this chapter, as in the analysis of variance (ANOVA) model, we try to

Theory of Ridge Regression Estimation with Applications, First Edition.
A.K. Md. Ehsanes Saleh, Mohammad Arashi, and B.M. Golam Kibria.

cover some detailed theoretical derivations/comparisons of these estimators in the well-known multiple linear model.

5.2 Linear Model and the Estimators

Consider the multiple linear model,

$$Y = X\beta + \epsilon, \tag{5.1}$$

where X is the design matrix such that $C_n = X^\top X$, $\beta = (\beta_1, \dots, \beta_p)^\top$, and $Y = (Y_1, \dots, Y_n)^\top$ is the response vector. Also, $\epsilon = (\epsilon_1, \dots, \epsilon_n)^\top$ is the n-vector of errors such that $\mathbb{E}(\epsilon\epsilon^\top) = \sigma^2 I_n$, σ^2 is the known variance of any ϵ_i $(i = 1, \dots, n)$.

It is well known that the LSE of β, say, $\tilde{\beta}_n = C_n^{-1} X^\top Y$, has the distribution

$$(\tilde{\beta}_n - \beta) \sim \mathcal{N}_p(\mathbf{0}, \sigma^2 C^{-1}), \quad C^{-1} = (C^{ij}),\ i,j = 1, 2, \dots, p. \tag{5.2}$$

We designate $\tilde{\beta}_n$ as the LSE of β.

In many situations, a sparse model is desired such as high-dimensional settings. Under the sparsity assumption, we partition the coefficient vector and the design matrix as

$$\beta = \underset{p_1 \times 1}{(\beta_1^\top,} \ \underset{p_2 \times 1}{\beta_2^\top)^\top}, \qquad X = \underset{n \times p_1}{(X_1,} \ \underset{n \times p_2}{X_2)}, \tag{5.3}$$

where $p = p_1 + p_2$.

Hence, (5.1) may also be written as

$$Y = X_1\beta_1 + X_2\beta_2 + \epsilon, \tag{5.4}$$

where β_1 may stand for the main effects and β_2 for the interaction which may be insignificant, although one is interested in the estimation and selection of the main effects. Thus, the problem of estimating β is reduced to the estimation of β_1 when β_2 is suspected to be equal to $\mathbf{0}$. Under this setup, the LSE of β is

$$\tilde{\beta}_n = \begin{pmatrix} \tilde{\beta}_{1n} \\ \tilde{\beta}_{2n} \end{pmatrix}; \tag{5.5}$$

and if $\beta = (\beta_1^\top, \mathbf{0}^\top)^\top$, it is

$$\hat{\beta}_n = \begin{pmatrix} \hat{\beta}_{1n} \\ \mathbf{0} \end{pmatrix}, \tag{5.6}$$

where $\hat{\beta}_{1n} = (X_1^\top X_1)^{-1} X_1^\top Y$.

Note that the marginal distribution of $\tilde{\beta}_{1n}$ is $\mathcal{N}_{p_1}(\beta_1, \sigma^2 C_{11\cdot 2}^{-1})$ and that of $\tilde{\beta}_{2n}$ is $\mathcal{N}_{p_2}(\beta_2, \sigma^2 C_{22\cdot 1}^{-1})$. Hence, the weighted L_2 risk of $\tilde{\beta}_n = (\tilde{\beta}_{1n}^\top, \tilde{\beta}_{2n}^\top)^\top$ is given by

$$R(\tilde{\beta}_n; C_{11\cdot 2}, C_{22\cdot 1}) = \sigma^2(p_1 + p_2). \tag{5.7}$$

Similarly, the weighted L_2 risk of $\hat{\boldsymbol{\beta}}_n = (\hat{\boldsymbol{\beta}}_{1n}^{\top}, \mathbf{0}^{\top})^{\top}$ is given by

$$R(\hat{\boldsymbol{\beta}}_n; \boldsymbol{C}_{11\cdot 2}, \boldsymbol{C}_{22\cdot 1}) = \sigma^2\left(\text{tr}(\boldsymbol{C}_{11}^{-1}\boldsymbol{C}_{11\cdot 2}) + \frac{1}{\sigma^2}\boldsymbol{\beta}_2^{\top}\boldsymbol{C}_{22\cdot 1}\boldsymbol{\beta}_2\right)$$
$$= \sigma^2(\text{tr}(\boldsymbol{C}_{11}^{-1}\boldsymbol{C}_{11\cdot 2}) + \Delta^2), \quad \Delta^2 = \frac{1}{\sigma^2}\boldsymbol{\beta}_2^{\top}\boldsymbol{C}_{22\cdot 1}\boldsymbol{\beta}_2, \quad (5.8)$$

since the covariance matrix of $(\hat{\boldsymbol{\beta}}_{1n}^{\top}, \mathbf{0}^{\top})^{\top}$ is $(\sigma^2\boldsymbol{C}_{11}^{-1}, -\boldsymbol{\beta}_2\boldsymbol{\beta}_2^{\top})^{\top}$ and computation of the risk function of RLSE with $\boldsymbol{W}_1 = \boldsymbol{C}_{11\cdot 2}$ and $\boldsymbol{W}_2 = \boldsymbol{C}_{22\cdot 1}$ yields the result (5.7).

Our focus in this chapter is on the comparative study of the performance properties of three penalty estimators compared to the PTE and the Stein-type estimator. We refer to Saleh (2006) for the comparative study of PTE and the Stein-type estimator, when the design matrix is nonorthogonal. We extend the study to include the penalty estimators, which has not been theoretically done yet, except for simulation studies.

5.2.1 Penalty Estimators

Motivated by the idea that only a few regression coefficients contribute to the signal, we consider threshold rules that retain only observed data that exceed a multiple of the noise level. Accordingly, we consider the *subset selection* rule of Donoho and Johnstone (1994) known as the *hard threshold* rule, as given by

$$\hat{\boldsymbol{\beta}}_n^{\text{HT}}(\kappa) = (\tilde{\beta}_{jn} I(|\tilde{\beta}_{jn}| > \kappa\sigma\sqrt{C^{jj}}) | j = 1, \ldots, p)^{\top},$$
$$= (\sigma\sqrt{C^{jj}} Z_j I(|Z_j| > \kappa) | j = 1, \ldots, p)^{\top}, \quad (5.9)$$

where $\tilde{\beta}_{jn}$ is the jth element of $\tilde{\boldsymbol{\beta}}_n$, $I(A)$ is an indicator function of the set A, and marginally

$$Z_j = \frac{\tilde{\beta}_{jn}}{\sigma\sqrt{C^{jj}}} \sim \mathcal{N}(\Delta_j, 1), \quad j = 1, \ldots, p, \quad (5.10)$$

where $\Delta_j = \beta_j / \sigma\sqrt{C^{jj}}$.

Here, Z_j is the test statistic for testing the null hypothesis $\mathcal{H}_o : \beta_j = 0$ vs. $\mathcal{H}_A : \beta_j \neq 0$. The quantity κ is called the threshold parameter. The components of $\hat{\boldsymbol{\beta}}_n^{\text{HT}}(\kappa)$ are kept as $\tilde{\beta}_{jn}$ if they are significant and zero, otherwise. It is apparent that each component of $\hat{\boldsymbol{\beta}}_n^{\text{HT}}(\kappa)$ is a PTE of the predictor concerned. The components of $\hat{\boldsymbol{\beta}}_n^{\text{HT}}(\kappa)$ are PTEs and discrete variables and lose some optimality properties. Hence, one may define a continuous version of (5.9) based on marginal distribution of $\tilde{\beta}_{jn}$ ($j = 1, \ldots, p$).

In accordance with the principle of the PTE approach (see Saleh 2006), we define the Stein-type estimator as the continuous version of PTE based on the marginal distribution of $\tilde{\beta}_{jn} \sim \mathcal{N}(\beta_j, \sigma^2 C^{jj}), j = 1, 2, \ldots, p$ given by

$$\begin{aligned}\hat{\boldsymbol{\beta}}_n^{\mathrm{S}}(\kappa) &= \left(\tilde{\beta}_{jn} - \kappa\sigma\sqrt{C^{jj}}\frac{\tilde{\beta}_{jn}}{|\tilde{\beta}_{jn}|} \middle| j = 1, 2, \ldots, p \right)^{\top} \\ &= (\sigma\sqrt{C^{jj}}\mathrm{sgn}(Z_j)(|Z_j| - \kappa)|j = 1, 2, \ldots, p)^{\top} \\ &= (\hat{\beta}_{1n}^{\mathrm{S}}(\kappa), \ldots, \hat{\beta}_{pn}^{\mathrm{S}}(\kappa))^{\top}. \end{aligned} \tag{5.11}$$

See Saleh (2006, p. 83) for details.

In order to develop LASSO for our case, we propose the following modified least absolute shrinkage and selection operator (MLASSO) given by

$$\hat{\boldsymbol{\beta}}_n^{\mathrm{MLASSO}}(\kappa) = (\hat{\beta}_{1n}^{\mathrm{MLASSO}}(\kappa), \ldots, \hat{\beta}_{pn}^{\mathrm{MLASSO}}(\kappa)) \tag{5.12}$$

where for $j = 1, \ldots, p$,

$$\hat{\beta}_{jn}^{\mathrm{MLASSO}}(\kappa) = \sigma\sqrt{C^{jj}}\ \mathrm{sgn}(Z_j)(|Z_j| - \kappa)^{+}. \tag{5.13}$$

The estimator $\hat{\boldsymbol{\beta}}_n^{\mathrm{MLASSO}}(\kappa)$ defines a continuous shrinkage operation that produces a sparse solution.

The formula (5.13) is obtained as follows:
Differentiating $(\boldsymbol{Y} - \boldsymbol{X\beta})^{\top}(\boldsymbol{Y} - \boldsymbol{X\beta}) + 2\kappa\sigma \sum_{j=1}^{p} (\boldsymbol{C}^{jj})^{-\frac{1}{2}}|\beta_j|$ where $\kappa > 0$, we obtain the following equation

$$-2\tilde{\boldsymbol{\beta}}_n + 2\hat{\boldsymbol{\beta}}_n^{\mathrm{MLASSO}}(\kappa) + 2\kappa\sigma\mathrm{Diag}\left([\boldsymbol{C}^{11}]^{\frac{1}{2}}, \ldots, [\boldsymbol{C}^{pp}]^{\frac{1}{2}}\right)\mathrm{sgn}\left(\hat{\boldsymbol{\beta}}_n^{\mathrm{MLASSO}}(\kappa)\right) = 0, \tag{5.14}$$

where $\mathrm{sgn}(\boldsymbol{\beta}) = (\mathrm{sgn}(\beta_1), \ldots, \mathrm{sgn}(\beta_p))^{\top}$ and $\boldsymbol{C}^{jj}$ is the jth diagonal element of $\boldsymbol{C}^{-1}$. Now, the jth marginal component of (5.13) is given by $\hat{\boldsymbol{\beta}}_{jn}^{\mathrm{MLASSO}}(\kappa)$,

$$-\tilde{\boldsymbol{\beta}}_{jn} + \hat{\boldsymbol{\beta}}_{jn}^{\mathrm{MLASSO}}(\kappa) + \kappa\sigma\sqrt{\boldsymbol{C}^{jj}}\mathrm{sgn}(\hat{\boldsymbol{\beta}}_{jn}^{\mathrm{MLASSO}}(\kappa)) = 0;\ j = 1, 2, \ldots, p. \tag{5.15}$$

Now, we have two cases:

(i) $\mathrm{sgn}(\hat{\boldsymbol{\beta}}_{jn}^{\mathrm{MLASSO}}(\kappa)) = +1$, then (5.14) reduces to

$$-Z_j + \left[\frac{\hat{\boldsymbol{\beta}}_{jn}^{\mathrm{MLASSO}}(\kappa)}{\sigma\sqrt{\boldsymbol{C}^{jj}}}\right] + \kappa = 0, \quad j = 1, 2, \ldots, p, \tag{5.16}$$

where $Z_j = \frac{\tilde{\beta}_{jn}}{\sigma\sqrt{\boldsymbol{C}^{jj}}}$. Hence,

$$0 < \hat{\boldsymbol{\beta}}_{jn}^{\mathrm{MLASSO}}(\kappa) = \sigma\sqrt{\boldsymbol{C}^{jj}}(Z_j - \kappa) = \sigma\sqrt{\boldsymbol{C}^{jj}}(|Z_j| - \kappa), \tag{5.17}$$

with, clearly, $Z_j > 0$ and $|Z_j| > \kappa$.

(ii) $\text{sgn}(\hat{\boldsymbol{\beta}}_{jn}^{\text{MLASSO}}(\kappa)) = -1$, then we have

$$-Z_j + \left[\frac{\hat{\boldsymbol{\beta}}_{jn}^{\text{MLASSO}}(\kappa)}{\sigma\sqrt{C^{jj}}}\right] - \kappa = 0; \quad j = 1, 2, \ldots, p. \tag{5.18}$$

Hence,

$$0 > \frac{\hat{\boldsymbol{\beta}}_{jn}^{\text{MLASSO}}(\kappa)}{\sigma\sqrt{C^{jj}}} = Z_j + \kappa = -|Z_j| + \kappa$$

$$\hat{\boldsymbol{\beta}}_{jn}^{\text{MLASSO}}(\kappa) = -(|Z_j| - \kappa) = -\sigma\sqrt{C^{jj}}(|Z_j| - \kappa) \tag{5.19}$$

with, clearly, $Z_j < 0$ and $|Z_j| > \kappa$.

(iii) For $\hat{\boldsymbol{\beta}}_{jn}^{\text{MLASSO}}(\kappa) = 0$, we have $-Z_j + \kappa\gamma = 0$ for some $\gamma \in [-1, 1]$. Hence, we obtain $Z_j = \kappa\gamma$, which implies $|Z_j| \le \kappa$.

Combining (5.17), (5.19), and (iii), we obtain (5.13).

Finally, we consider the *RREs* of $(\boldsymbol{\beta}_1^\top, \boldsymbol{\beta}_2^\top)^\top$. They are obtained using marginal distributions of $\tilde{\beta}_{jn} \sim \mathcal{N}(\beta_j, \sigma^2 C^{jj}), j = 1, \ldots, p$, as

$$\hat{\boldsymbol{\beta}}^{\text{RR}}(k) = \begin{pmatrix} \tilde{\boldsymbol{\beta}}_{1n} \\ \frac{1}{1+\kappa}\tilde{\boldsymbol{\beta}}_{2n} \end{pmatrix}, \tag{5.20}$$

to accommodate the sparsity condition; see Tibshirani (1996) on the summary of properties discussed earlier.

In the next section, we define the traditional shrinkage estimators.

5.2.2 Shrinkage Estimators

We recall that the unrestricted estimator of $(\boldsymbol{\beta}_1^\top, \boldsymbol{\beta}_2^\top)^\top$ is given by $(\tilde{\boldsymbol{\beta}}_{1n}^\top, \tilde{\boldsymbol{\beta}}_{2n}^\top)^\top$. Using marginal distributions, we have

$$\tilde{\boldsymbol{\beta}}_{1n} \sim \mathcal{N}_{p_1}(\boldsymbol{\beta}_1, \sigma^2 \boldsymbol{C}_{11\cdot 2}^{-1})$$
$$\tilde{\boldsymbol{\beta}}_{2n} \sim \mathcal{N}_{p_2}(\boldsymbol{\beta}_2, \sigma^2 \boldsymbol{C}_{22\cdot 1}^{-1}).$$

The restricted parameter may be denoted by $\boldsymbol{\beta}_{\text{R}}^\top = (\boldsymbol{\beta}_1^\top, \boldsymbol{0}^\top)$. Thus, the restricted estimator of $\boldsymbol{\beta}$ is $\hat{\boldsymbol{\beta}}_n = (\hat{\boldsymbol{\beta}}_{1n}^\top, \boldsymbol{0}^\top)^\top$; see (5.5). Next, we consider the PTE of $\boldsymbol{\beta}$. For this, we first define the test statistic for testing the sparsity hypothesis $\mathcal{H}_o : \boldsymbol{\beta}_2 = \boldsymbol{0}$ vs. $\mathcal{H}_A : \boldsymbol{\beta}_2 \ne \boldsymbol{0}$ as

$$\mathcal{L}_n = \frac{\tilde{\boldsymbol{\beta}}_{2n}^\top \boldsymbol{C}_{22\cdot 1} \tilde{\boldsymbol{\beta}}_{2n}}{\sigma^2}.$$

Indeed, $\mathcal{L}_n = \chi^2_{p_2}$ (chi-square with p_2 degrees of freedom (DF)).

Thus, define the PTE of $(\boldsymbol{\beta}_1^\top, \boldsymbol{\beta}_2^\top)^\top$ with an upper α-level of significance as

$$\hat{\boldsymbol{\beta}}_n^{\mathrm{PT}} = \begin{pmatrix} \tilde{\boldsymbol{\beta}}_{1n} \\ \hat{\boldsymbol{\beta}}_{2n}^{\mathrm{PT}}(\alpha) \end{pmatrix}, \tag{5.21}$$

where α stands for the level of significance of the test using $\mathcal{L}_n$,

$$\hat{\boldsymbol{\beta}}_{2n}^{\mathrm{PT}}(\alpha) = \tilde{\boldsymbol{\beta}}_{2n} I(\mathcal{L}_n > c_\alpha).$$

In a similar manner, we define the James–Stein estimator given by

$$\hat{\boldsymbol{\beta}}_n^{\mathrm{S}} = \begin{pmatrix} \tilde{\boldsymbol{\beta}}_{1n} \\ \hat{\boldsymbol{\beta}}_{2n}^{\mathrm{S}} \end{pmatrix},$$

where

$$\hat{\boldsymbol{\beta}}_{2n}^{\mathrm{S}} = \tilde{\boldsymbol{\beta}}_{2n}(1 - d\mathcal{L}_n^{-1}), \qquad d = p_2 - 2.$$

The estimator $\hat{\boldsymbol{\beta}}_{1n}^{\mathrm{S}}$ is not a convex combination of $\tilde{\boldsymbol{\beta}}_{1n}$ and $\hat{\boldsymbol{\beta}}_{1n}$ and may change the sign opposite to the unrestricted estimator, due to the presence of the term $(1 - d\mathcal{L}_n^{-1})$. This is the situation for $\hat{\boldsymbol{\beta}}_{2n}^{\mathrm{S}}$ as well. To avoid this anomaly, we define the PRSE, $\hat{\boldsymbol{\beta}}_n^{\mathrm{S+}}$ as

$$\hat{\boldsymbol{\beta}}_n^{\mathrm{S+}} = \begin{pmatrix} \tilde{\boldsymbol{\beta}}_{1n} \\ \hat{\boldsymbol{\beta}}_{2n}^{\mathrm{S+}} \end{pmatrix},$$

where

$$\hat{\boldsymbol{\beta}}_{2n}^{\mathrm{S+}} = \tilde{\boldsymbol{\beta}}_{2n}(1 - d\mathcal{L}_n^{-1}) I(\mathcal{L}_n > d), \quad d = p_2 - 2.$$

5.3 Bias and Weighted L_2 Risks of Estimators

First, we consider the bias and L_2 risk expressions of the penalty estimators.

5.3.1 Hard Threshold Estimator

Using the results of Donoho and Johnstone (1994), we write the bias and L_2 risk of the hard threshold estimator (HTE), under nonorthogonal design matrices.

The bias and L_2 risk expressions of $\hat{\boldsymbol{\beta}}_n^{\mathrm{HT}}(\kappa)$ are given by

$$b\left(\hat{\boldsymbol{\beta}}_n^{\mathrm{HT}}(\kappa)\right) = \left(-\sigma\sqrt{C^{jj}}\Delta_j H_3(\kappa^2; \Delta_j^2)\middle| j = 1, \ldots, p\right)^\top, \tag{5.22}$$

where $H_\nu(\kappa^2; \Delta_j^2)$ is the cumulative distribution function (c.d.f.) of a noncentral chi-square distribution with 3 DF and noncentrality parameter $\Delta_j^2/2$

$(j = 1, \ldots, p)$ and the mean square error of $\hat{\boldsymbol{\beta}}_n^{\mathrm{HT}}(\kappa)$ is given by

$$
\begin{aligned}
R(\hat{\boldsymbol{\beta}}_n^{\mathrm{HT}}(\kappa)) &= \sum_{j=1}^{p} \mathbb{E}[\tilde{\beta}_{jn} I(|\tilde{\beta}_{jn}| > \kappa\sigma\sqrt{C^{jj}}) - \beta_j]^2 \\
&= \sigma^2 \sum_{j=1}^{p} C^{jj} \left\{ (1 - H_3(\kappa^2; \Delta_j^2)) \right. \\
&\quad \left. + \Delta_j^2 (2H_3(\kappa^2; \Delta_j^2) - H_5(\kappa^2; \Delta_j^2)) \right\}. \qquad (5.23)
\end{aligned}
$$

Since

$$
[\tilde{\beta}_{jn} I(|\tilde{\beta}_{jn}| > \kappa\sigma\sqrt{C^{jj}}) - \beta_j]^2 \le (\tilde{\beta}_{jn} - \beta_j)^2 + \beta_j^2, \quad j = 1, \ldots, p.
$$

Hence,

$$
R(\hat{\boldsymbol{\beta}}_n^{\mathrm{HT}}(\kappa)) \le \sigma^2 \operatorname{tr}(\boldsymbol{C}^{-1}) + \boldsymbol{\beta}^\top \boldsymbol{\beta} \quad (\text{free of } \kappa) \quad \forall \boldsymbol{\beta} \in \mathbb{R}^p. \qquad (5.24)
$$

Thus, we have the revised form of Lemma 1 of Donoho and Johnstone (1994).

Lemma 5.1 *We have*

$$
R(\hat{\boldsymbol{\beta}}_n^{\mathrm{HT}}(\kappa)) \le \begin{cases} \sigma^2(1+\kappa^2)\operatorname{tr} \boldsymbol{C}^{-1} & \forall\, \boldsymbol{\beta} \in \mathbb{R}^p,\ \kappa > 1 \\ \sigma^2 \operatorname{tr} \boldsymbol{C}^{-1} + \boldsymbol{\beta}^\top \boldsymbol{\beta} & \forall\, \boldsymbol{\beta} \in \mathbb{R}^p \\ \rho_{\mathrm{HT}}(\kappa; 0)\sigma^2 \operatorname{tr} \boldsymbol{C}^{-1} + 1.2\, \boldsymbol{\beta}^\top \boldsymbol{\beta} & \boldsymbol{0} < \boldsymbol{\beta} < \kappa \boldsymbol{1}_p^\top, \end{cases} \qquad (5.25)
$$

where $\rho_{\mathrm{HT}}(\kappa, 0) = 2[1 - \Phi(\kappa) + \kappa\Phi(-\kappa)]$.

The upper bound of $\sigma^2 \operatorname{tr} \boldsymbol{C}^{-1} + \boldsymbol{\beta}^\top \boldsymbol{\beta}$ in Lemma 5.1 is independent of κ. We may obtain the upper bound of the weighted L_2 risk of $\hat{\boldsymbol{\beta}}_n^{\mathrm{HT}}(\kappa)$ as given here by

$$
R\left(\hat{\boldsymbol{\beta}}_n^{\mathrm{HT}}(\kappa); \boldsymbol{C}_{11\cdot2}, \boldsymbol{C}_{22\cdot1}\right) \le \begin{cases} \sigma^2(1+\kappa^2)(p_1+p_2) & \kappa > 1 \\ (\sigma^2 p_1 + \boldsymbol{\beta}_1^\top \boldsymbol{C}_{11\cdot2}\boldsymbol{\beta}_1) & \forall\, \boldsymbol{\beta}_1 \in \mathbb{R}^{p_1} \\ +(\sigma^2 p_2 + \boldsymbol{\beta}_2^\top \boldsymbol{C}_{22\cdot1}\boldsymbol{\beta}_2) & \forall\, \boldsymbol{\beta}_2 \in \mathbb{R}^{p_2} \\ \rho_{\mathrm{HT}}(\kappa; 0)\sigma^2 p_1 + 1.2\boldsymbol{\beta}_1^\top \boldsymbol{C}_{11\cdot2}\boldsymbol{\beta}_1 & \\ +\rho_{\mathrm{HT}}(\kappa; 0)\sigma^2 p_2 + 1.2\boldsymbol{\beta}_2^\top \boldsymbol{C}_{22\cdot1}\boldsymbol{\beta}_2 & \boldsymbol{0} < \boldsymbol{\beta} < \kappa \boldsymbol{1}_p^\top. \end{cases} \qquad (5.26)
$$

If we have the sparse solution with p_1 nonzero coefficients, $|\beta_j| > \sigma\sqrt{C^{jj}}$, $(j = 1, \ldots, p_1)$ and p_2 zero coefficients

$$
\hat{\boldsymbol{\beta}}_n^{\mathrm{HT}}(\kappa) = \begin{pmatrix} \tilde{\boldsymbol{\beta}}_{1n} \\ \boldsymbol{0} \end{pmatrix}. \qquad (5.27)
$$

Thus, the upper bound of a weighted L_2 risk using Lemma 5.1 and (5.26), is given by

$$R(\hat{\boldsymbol{\beta}}_n^{\mathrm{HT}}(\kappa); \boldsymbol{C}_{11\cdot2}, \boldsymbol{C}_{22\cdot1}) \le \sigma^2(p_1 + \Delta^2), \quad \Delta^2 = \frac{1}{\sigma^2}\boldsymbol{\beta}_2^\top \boldsymbol{C}_{22\cdot1}\boldsymbol{\beta}_2, \tag{5.28}$$

which is independent of κ.

5.3.2 Modified LASSO

In this section, we provide expressions of bias and mean square errors and weighted L_2 risk. The bias expression for the modified LASSO is given by

$$\begin{aligned} & b(\hat{\boldsymbol{\beta}}_n^{\mathrm{MLASSO}}(\kappa)) \\ & = (\sigma\sqrt{C^{jj}}\left[\kappa\ (2\Phi(\Delta_j) - 1)(1 - H_3(\kappa; \Delta_j)) + \Delta_j H_3(\kappa^2; \Delta_j)\right] \Big| j = 1, \ldots, p)^\top . \end{aligned} \tag{5.29}$$

The L_2 risk of the modified LASSO is given by

$$R(\hat{\boldsymbol{\beta}}_n^{\mathrm{MLASSO}}(\kappa)) = \sigma^2 \sum_{j=1}^{p} C^{jj} \rho_{\mathrm{ST}}(\kappa, \Delta_j), \tag{5.30}$$

where

$$\begin{aligned} \rho_{\mathrm{ST}}(\kappa, \Delta_j) = {} & 1 + \kappa^2 + (\Delta_j^2 - \kappa^2 - 1)\{\Phi(\kappa - \Delta_j) - \Phi(-\kappa - \Delta_j)\} \\ & - \{(\kappa - \Delta_j)\varphi(\kappa + \Delta_j) + (\kappa + \Delta_j)\varphi(\kappa - \Delta_j)\} \end{aligned} \tag{5.31}$$

and $\rho_{\mathrm{ST}}(\kappa, 0) = (1 + \kappa^2)(2\Phi(\kappa) - 1) - 2\kappa\phi(\kappa)$.

Further, Donoho and Johnstone (1994) gives us the revised Lemma 5.2.

Lemma 5.2 *Under the assumption of this section,*

$$R(\hat{\boldsymbol{\beta}}_n^{\mathrm{MLASSO}}(\kappa)) \le \begin{cases} \sigma^2(1 + \kappa^2)\mathrm{tr}\ \boldsymbol{C}^{-1} & \forall\, \boldsymbol{\beta} \in \mathbb{R}^p,\ \kappa > 1 \\ \sigma^2 \mathrm{tr}\ \boldsymbol{C}^{-1} + \boldsymbol{\beta}^\top \boldsymbol{\beta} & \forall\, \boldsymbol{\beta} \in \mathbb{R}^p \\ \sigma^2 \rho_{\mathrm{ST}}(\kappa, 0)\mathrm{tr}\ \boldsymbol{C}^{-1} + \boldsymbol{\beta}^\top \boldsymbol{\beta} & \beta_j^2 \in \mathbb{R}^+, j = 1, \ldots, p. \end{cases} \tag{5.32}$$

The second upper bound in Lemma 5.2 is free of κ. If we have a sparse solution with p_1 nonzero and p_2 zero coefficients such as

$$\hat{\boldsymbol{\beta}}_n^{\mathrm{MLASSO}}(\kappa) = \begin{pmatrix} \tilde{\boldsymbol{\beta}}_{1n} \\ \mathbf{0} \end{pmatrix}, \tag{5.33}$$

then the weighted L_2-risk bound is given by

$$R(\hat{\boldsymbol{\beta}}_n^{\mathrm{MLASSO}}(\kappa); \boldsymbol{C}_{11\cdot2}, \boldsymbol{C}_{22\cdot1}) \le \sigma^2(p_1 + \Delta^2), \quad \Delta^2 = \frac{1}{\sigma^2}\boldsymbol{\beta}_2^\top \boldsymbol{C}_{22\cdot1}\boldsymbol{\beta}_2, \tag{5.34}$$

which is independent of κ.

5.3.3 Multivariate Normal Decision Theory and Oracles for Diagonal Linear Projection

Consider the following problem in multivariate normal decision theory. We are given the LSE of $\boldsymbol{\beta}$, namely, $\tilde{\boldsymbol{\beta}}_n = (\tilde{\beta}_{1n}, \ldots, \tilde{\beta}_{pn})^\top$ according to

$$\tilde{\beta}_{jn} = \beta_j + \sigma\sqrt{C^{jj}}Z_j, \quad Z_j \sim \mathcal{N}(0,1), \tag{5.35}$$

where $\sigma^2 C^{jj}$ is the marginal variance of $\tilde{\beta}_{jn}$, $j = 1, \ldots, p$, and noise level and $\{\beta_j\}_{j=1}^p$ are the object of interest.

We consider a family of diagonal linear projections,

$$\boldsymbol{T}_{\mathrm{DP}}(\hat{\boldsymbol{\beta}}_n^{\mathrm{MLASSO}}, \boldsymbol{\delta}) = (\delta_1\hat{\boldsymbol{\beta}}_{1n}^{\mathrm{MLASSO}}, \ldots, \delta_2\hat{\boldsymbol{\beta}}_{pn}^{\mathrm{MLASSO}})^\top, \quad \delta_j \in \{0,1\}. \tag{5.36}$$

Such estimators *keep* or *kill* coordinate. The ideal diagonal coefficients, in this case, are $I(|\beta_j| > \sigma\sqrt{C^{jj}})$. These coefficients estimate those β_j's which are larger than the noise level $\sigma\sqrt{C^{jj}}$, yielding the lower bound on the risk as

$$R(\boldsymbol{T}_{\mathrm{DP}}) = \sum_{j=1}^{p} \min(\beta_j^2, \sigma^2 \boldsymbol{C}^{(jj)}). \tag{5.37}$$

As a special case of (5.36), we obtain

$$R(\boldsymbol{T}_{\mathrm{DP}}) = \begin{cases} \sigma^2 \mathrm{tr}(\boldsymbol{C}^{-1}) & \text{if all } |\beta_j| \geq \sigma\sqrt{C^{jj}}, j = 1, \ldots, p \\ \boldsymbol{\beta}^\top\boldsymbol{\beta} & \text{if all } |\beta_j| < \sigma\sqrt{C^{jj}}, j = 1, \ldots, p. \end{cases} \tag{5.38}$$

In general, the risk $R(\boldsymbol{T}_{\mathrm{DP}})$ cannot be attained for all $\boldsymbol{\beta}$ by any estimator, linear or nonlinear. However, for the sparse case, if p_1 is the number of nonzero coefficients, $|\beta_j| > \sigma\sqrt{C^{jj}}$; $(j = 1, 2, \ldots, p_1)$ and p_2 is the number of zero coefficients, then (5.38) reduces to the lower bound given by

$$R(\boldsymbol{T}_{\mathrm{DP}}) = \sigma^2 \mathrm{tr}\ \boldsymbol{C}_{11\cdot2}^{-1} + \boldsymbol{\beta}_2^\top\boldsymbol{\beta}_2. \tag{5.39}$$

Consequently, the weighted L_2-risk lower bound is given by (5.39) as

$$R(\boldsymbol{T}_{\mathrm{DP}}; \boldsymbol{C}_{11\cdot2}, \boldsymbol{C}_{22\cdot1}) = \sigma^2(p_1 + \Delta^2), \quad \Delta^2 = \frac{1}{\sigma^2}\boldsymbol{\beta}^2 C_{22\cdot1}\boldsymbol{\beta}_2. \tag{5.40}$$

As we mentioned earlier, ideal risk cannot be attained, in general, by any estimator, linear or non-linear. However, in the case of MLASSO and HTE, we revise Theorems 1–4 of Donoho and Johnstone (1994) as follows.

Theorem 5.1 *Assume (5.35). The MLASSO estimator defined by (5.12) with $\kappa^* = \sqrt{2\ln(p)}$ satisfies*

$$R(\hat{\boldsymbol{\beta}}_n^{\mathrm{MLASSO}}(\kappa^*)) \leq (2\ln(p) + 1)\left\{\sigma^2 + \sum_{j=1}^{p} \min(\beta_j^2, \sigma^2 C^{jj})\right\}; \quad \forall \boldsymbol{\beta} \in \mathbb{R}^p.$$

The inequality says that we can mimic the performance of an oracle plus one extra parameter, σ^2, to within a factor of essentially $2\ln(p)$.

However, it is natural and more revealing to look for *optimal thresholds,* κ_p^*, which yield the smallest possible constant Λ_p^* in place of $(2\ln(p)+1)$ among soft threshold estimators. We state this in the following theorem.

Theorem 5.2 *Assume (5.35). The minimax threshold κ_p^* defined by the minimax quantities*

$$\Lambda_p^* = \inf_{\kappa} \sup_{\beta} \frac{\sigma^2 C^{jj} \rho_{\mathrm{ST}}(\kappa, \Delta_j)}{\frac{1}{\mathrm{tr}(C^{-1})} + \min(\Delta_j, 1)} \tag{5.41}$$

$$\kappa_p^* = \text{The largest } \kappa \text{ attains } \Lambda_p^* \tag{5.42}$$

and satisfies the equation

$$(p+1)\rho_{\mathrm{ST}}(\kappa, 0) = \rho_{\mathrm{ST}}(\kappa, \infty) \tag{5.43}$$

yields the estimator

$$\hat{\boldsymbol{\beta}}_n^{\mathrm{MLASSO}}(\kappa_p^*) = (\hat{\beta}_{1n}^{\mathrm{MLASSO}}(\kappa_p^*), \ldots, \hat{\beta}_{pn}^{\mathrm{MLASSO}}(\kappa_p^*))^{\top}, \tag{5.44}$$

which is given by

$$R(\hat{\boldsymbol{\beta}}_n^{\mathrm{MLASSO}}(\kappa_p^*)) \leq \Lambda_p^* \left\{ \sigma^2 + \sum_{j=1}^{p} \min(\beta_j^2, \sigma^2 C^{jj}) \right\}, \quad \forall \boldsymbol{\beta} \in \mathbb{R}^p. \tag{5.45}$$

The coefficients defined in Λ_p^* satisfy $\Lambda_p^* \leq (2\ln(p)+1)$ and $\kappa_p^* \leq \sqrt{2\ln(p)}$. Asymptotically, as $p \to \infty$,

$$\Lambda_n^* \approx 2\ln(p), \qquad \kappa_p^* \approx (2\ln(p))^{\frac{1}{2}}.$$

Theorem 5.3 *The following results hold under the same assumption as in Theorems 5.1 and 5.2,*

$$\inf \sup_{\beta \in \mathbb{R}^p} \frac{R(\hat{\boldsymbol{\beta}}_n^{\mathrm{MLASSO}}(\kappa_p^*))}{\sigma^2 + \sum_{j=1}^{p} \min(\beta_j^2, \sigma^2 C^{jj})} \approx 2\ln(p) \quad \text{as} \quad p \to \infty.$$

Finally, we deal with the theorem related to the HTE (subset selection rule).

Theorem 5.4 *With $\{\coprod_n\}$, a thresholding sequence sufficiently close to $\sqrt{2\ln(p)}$, the HTE satisfies form $Q_p \approx 2\ln(p_n)$ in inequality*

$$R(\hat{\boldsymbol{\beta}}_n^{\mathrm{HT}}(\kappa_p^*)) \leq Q_p \left\{ \sigma^2 + \sum_{j=1}^{p} \min(\beta_j^2, \sigma^2 C^{jj}) \right\}, \quad \forall \boldsymbol{\beta} \in \mathbb{R}^p,$$

where Q_p is the pth component of the $\{\coprod_n\}$.

Here, sufficiently close to $\sqrt{2\ln(p)}$ means $(1-\gamma)\ln(\ln(p)) \le \coprod_p - 2\ln(p) \le o(\ln(p))$ for some $\gamma > 0$.

5.3.4 Ridge Regression Estimator

We have defined RRE as $\hat{\boldsymbol{\beta}}_n(k) = (\tilde{\boldsymbol{\beta}}_{1n}^\top, \tilde{\boldsymbol{\beta}}_{2n}^\top/(1+k))^\top$ in Eq. (5.20). The bias and L_2 risk are then given by

$$b(\hat{\boldsymbol{\beta}}_n^{RR}(k)) = \begin{pmatrix} \mathbf{0} \\ -\frac{k}{1+k}\boldsymbol{\beta}_2 \end{pmatrix},$$

$$R(\hat{\boldsymbol{\beta}}_n^{RR}(k)) = \sigma^2 \mathrm{tr}(\boldsymbol{C}_{11\cdot2}^{-1}) + \frac{1}{(1+k)^2}[\sigma^2 \mathrm{tr}\ \boldsymbol{C}_{22\cdot1}^{-1} + k^2\boldsymbol{\beta}_2^\top\boldsymbol{\beta}_2]. \tag{5.46}$$

The weighted L_2 risk is then given by

$$R(\hat{\boldsymbol{\beta}}_n^{RR}(k); \boldsymbol{C}_{11\cdot2}, \boldsymbol{C}_{22\cdot1}) = \sigma^2 p_1 + \frac{\sigma^2}{(1+\kappa)^2}[p_2 + \kappa^2\Delta^2], \quad \Delta^2 = \frac{1}{\sigma^2}\boldsymbol{\beta}_2^\top \boldsymbol{C}_{22\cdot1}\boldsymbol{\beta}_2. \tag{5.47}$$

The optimum value of κ is obtained as $\kappa^* = p_2\Delta^{-2}$; so that

$$R(\hat{\boldsymbol{\beta}}_n(p_2\Delta^{-2}); \boldsymbol{C}_{11\cdot2}, \boldsymbol{C}_{22\cdot1}) = \sigma^2 p_1 + \frac{\sigma^2 p_2 \Delta^2}{(p_2 + \Delta^2)}. \tag{5.48}$$

5.3.5 Shrinkage Estimators

We know from Section 5.2.2, the LSE of $\boldsymbol{\beta}$ is $(\tilde{\boldsymbol{\beta}}_{1n}^\top, \tilde{\boldsymbol{\beta}}_{2n}^\top)^\top$ with bias $(\mathbf{0}^\top, \mathbf{0}^\top)^\top$ and weighted L_2 risk given by (5.7), while the restricted estimator of $(\boldsymbol{\beta}_1^\top, \mathbf{0}^\top)^\top$ is $(\hat{\boldsymbol{\beta}}_{1n}^\top, \mathbf{0}^\top)^\top$. Then, the bias is equal to $(\mathbf{0}^\top, -\boldsymbol{\beta}_2^\top)^\top$ and the weighted L_2 risk is given by (5.8).

Next, we consider the PTE of $\boldsymbol{\beta} = (\boldsymbol{\beta}_1^\top, \boldsymbol{\beta}_2^\top)$ given by (5.20). Then, the bias and weighted L_2 risk are given by

$$b(\hat{\boldsymbol{\beta}}_n^{PT}) = \begin{pmatrix} \mathbf{0} \\ -\boldsymbol{\beta}_2 H_{p_2+2}(c_\alpha; \Delta^2) \end{pmatrix}$$

$$\begin{aligned} R(\hat{\boldsymbol{\beta}}_n^{PT}; \boldsymbol{C}_{11\cdot2}, \boldsymbol{C}_{22\cdot1}) &= \sigma^2 p_1 + \sigma^2 p_2[1 - H_{p_2+2}(c_\alpha; \Delta^2)] \\ &\quad + \Delta^2[2H_{p_2+2}(c_\alpha; \Delta^2) - H_{p_2+4}(c_\alpha; \Delta^2)]. \end{aligned} \tag{5.49}$$

For the Stein estimator, we have

$$b(\hat{\boldsymbol{\beta}}_n^{S}) = \begin{pmatrix} \mathbf{0} \\ -(p_2-2)\boldsymbol{\beta}_2 \mathbb{E}[\chi_{p_2+2}^{-2}(\Delta^2)] \end{pmatrix},$$

$$R(\hat{\boldsymbol{\beta}}_n^{S}; \boldsymbol{C}_{11\cdot2}, \boldsymbol{C}_{22\cdot1}) = \sigma^2 p_1 + \sigma^2 p_2 - \sigma^2(p_2-2)^2\mathbb{E}[\chi_{p_2}^{-2}(\Delta^2)]. \tag{5.50}$$

Similarly, the bias and weighted L_2 risk of the PRSE are given by

$$b(\hat{\boldsymbol{\beta}}_n^{S+}) = \begin{pmatrix} \mathbf{0} \\ -(p_2-2)\boldsymbol{\beta}_2\left\{H_{p_2+2}(c_\alpha;\Delta^2)\right. \\ \left. +\mathbb{E}[\chi^{-2}_{p_2+2}(\Delta^2)I(\chi^2_{p_2+2}(\Delta^2)<p_2-2)]\right\} \end{pmatrix},$$

$$\begin{aligned} R(\hat{\boldsymbol{\beta}}_n^{S+};\boldsymbol{C}_{11\cdot2},\boldsymbol{C}_{22\cdot1}) &= R(\hat{\boldsymbol{\beta}}_n^{S};\boldsymbol{C}_{11\cdot2},\boldsymbol{C}_{22\cdot1}) \\ &\quad -\sigma^2 p_2\mathbb{E}[(1-(p_2-2)\chi^{-2}_{p_2+2}(\Delta^2))^2 I(\chi^{-2}_{p_2+2}(\Delta^2)<p_2-2)] \\ &\quad +\Delta^2\left\{2\mathbb{E}[(1-(p_2-2)\chi^{-2}_{p_2+2}(\Delta^2))I(\chi^{-2}_{p_2+2}(\Delta^2)<p_2-2)]\right. \\ &\quad \left.+\mathbb{E}[(1-(p_2-2)\chi^{-2}_{p_2+4}(\Delta^2))^2 I(\chi^{-2}_{p_2+4}(\Delta^2)<p_2-2)]\right\}. \end{aligned} \tag{5.51}$$

5.4 Comparison of Estimators

In this section, we compare various estimators with respect to the LSE, in terms of relative weighted L_2-risk efficiency (RWRE).

5.4.1 Comparison of LSE with RLSE

In this case, the RWRE of RLSE vs. LSE is given by

$$\begin{aligned} \mathrm{RWRE}(\hat{\boldsymbol{\beta}}_n:\tilde{\boldsymbol{\beta}}_n) &= \frac{p_1+p_2}{\mathrm{tr}(\boldsymbol{C}_{11}^{-1}\boldsymbol{C}_{11\cdot2})+\Delta^2} \\ &= \left(1+\frac{p_2}{p_1}\right)\left(1-\frac{\mathrm{tr}(\boldsymbol{M}_0)}{p_1}+\frac{\Delta^2}{p_1}\right)^{-1}; \\ \boldsymbol{M}_0 &= \boldsymbol{C}_{11}^{-1}\boldsymbol{C}_{12}\boldsymbol{C}_{22}^{-1}\boldsymbol{C}_{21}, \end{aligned} \tag{5.52}$$

which is a decreasing function of Δ^2. So,

$$0 \le \mathrm{RWRE}(\hat{\boldsymbol{\beta}}_n:\tilde{\boldsymbol{\beta}}_n) \le \left(1+\frac{p_2}{p_1}\right)\left(1-\frac{\mathrm{tr}\,\boldsymbol{M}_0}{p_1}\right)^{-1}.$$

In order to compute $\mathrm{tr}(\boldsymbol{M}_0)$, we need to find $\boldsymbol{C}_{11}$, $\boldsymbol{C}_{22}$, and $\boldsymbol{C}_{12}$. These are obtained by generating explanatory variables by the following equation based on McDonald and Galarneau (1975),

$$x_{ij} = \sqrt{1-\rho^2}z_{ij}+\rho z_{ip}, \quad i=1,\ldots,n; j=1,\ldots,p,. \tag{5.53}$$

where z_{ij} are independent $\mathcal{N}(0,1)$ pseudo-random numbers and ρ^2 is the correlation between any two explanatory variables. In this study, we take $\rho^2 = 0.1, 0.2, 0.8$, and 0.9 which shows variables are lightly collinear and severely collinear. In our case, we chose $n = 100$ and various (p_1, p_2). The resulting output is then used to compute $\mathrm{tr}(\boldsymbol{M}_0)$.

5.4.2 Comparison of LSE with PTE

Here, the RWRE expression for PTE vs. LSE is given by

$$\mathrm{RWRE}(\hat{\boldsymbol{\beta}}_n^{\mathrm{PT}}(\alpha):\tilde{\boldsymbol{\beta}}_n) = \frac{p_1+p_2}{g(\Delta^2,\alpha)}, \tag{5.54}$$

where

$$g(\Delta^2,\alpha) = p_1 + p_2(1-H_{p_2+1}(c_\alpha;\Delta)) + \Delta^2[2H_{p_2+2}(c_\alpha;\Delta^2) - H_{p_2+4}(c_\alpha;\Delta^2)].$$

Then, the PTE outperforms the LSE for

$$0 \le \Delta^2 \le \frac{p_2 H_{p_2+2}(c_\alpha;\Delta^2)}{2H_{p_2+2}(c_\alpha;\Delta^2) - H_{p_2+4}(c_\alpha;\Delta^2)} = \Delta^2_{\mathrm{PT}}. \tag{5.55}$$

Otherwise, LSE outperforms the PTE in the interval $(\Delta^2_{\mathrm{PT}}(\alpha),\infty)$. We may mention that $\mathrm{RWRE}(\hat{\boldsymbol{\beta}}_n^{\mathrm{PT}}(\alpha):\tilde{\boldsymbol{\beta}}_n)$ is a decreasing function of Δ^2 with a maximum at $\Delta^2=0$, then decreases crossing the 1-line to a minimum at $\Delta^2=\Delta^2_{\mathrm{PT}}(\min)$ with a value $M_{\mathrm{PT}}(\alpha)$, and then increases toward the 1-line.

The $\mathrm{RWRE}(\hat{\boldsymbol{\beta}}_n^{\mathrm{PT}};\tilde{\boldsymbol{\beta}}_n)$ belongs to the interval

$$M_{\mathrm{PT}}(\alpha) \le \mathrm{RWRE}(\hat{\boldsymbol{\beta}}_n^{\mathrm{PT}};\tilde{\boldsymbol{\beta}}_n) \le \left(1+\frac{p_2}{p_1}\right)\left(1+\frac{p_2}{p_1}[1-H_{p_2+2}(c_\alpha;0)]\right)^{-1},$$

where $M_{\mathrm{PT}}(\alpha)$ depends on the size of α and given by

$$\begin{aligned} M_{\mathrm{PT}}(\alpha) = \left(1+\frac{p_2}{p_1}\right)\Bigg\{&1+\frac{p_2}{p_1}[1-H_{p_2+2}(c_\alpha;\Delta^2_{\mathrm{PT}}(\min))] \\ &+\frac{\Delta^2_{\mathrm{PT}}(\min)}{p_1}[2H_{p_2+2}(c_\alpha;\Delta^2_{\mathrm{PT}}(\min)) - H_{p_2+4}(c_\alpha;\Delta^2_{\mathrm{PT}}(\min))]\Bigg\}^{-1}. \end{aligned}$$

The quantity $\Delta^2_{\mathrm{PT}}(\min)$ is the value Δ^2 at which the RWRE value is minimum.

5.4.3 Comparison of LSE with SE and PRSE

Since SE and PRSE need $p_2 \ge 3$ to express their weighted L_2-risk expressions, we assume always $p_2 \ge 3$. We have

$$\mathrm{RWRE}(\hat{\boldsymbol{\beta}}_n^{\mathrm{S}};\tilde{\boldsymbol{\beta}}_n) = \left(1+\frac{p_2}{p_1}\right)\left(1+\frac{p_2}{p_1}-\frac{(p_2-2)^2}{p_1}\mathbb{E}[\chi^{-2}_{p_2}(\Delta^2)]\right)^{-1}. \tag{5.56}$$

It is a decreasing function of Δ^2. At $\Delta^2=0$, its value is $\left(1+\frac{p_2}{p_1}\right)\left(1+\frac{2}{p_1}\right)^{-1}$; and when $\Delta^2\to\infty$, its value goes to 1. Hence, for $\Delta^2\in\mathbb{R}^+$,

$$1 \le \left(1+\frac{p_2}{p_1}\right)\left(1+\frac{p_2}{p_1}-\frac{(p_2-2)^2}{p_1}\mathbb{E}[\chi^{-2}_{p_2}(\Delta^2)]\right)^{-1} \le \left(1+\frac{p_2}{p_1}\right)\left(1+\frac{2}{p_1}\right)^{-1}.$$

Also,

$$\begin{aligned}\text{RWRE}(\hat{\boldsymbol{\beta}}_n^{\text{S+}};\tilde{\boldsymbol{\beta}}_n) = \left(1+\frac{p_2}{p_1}\right)\Bigg(1+\frac{p_2}{p_1}-\frac{(p_2-2)^2}{p_1}\mathbb{E}[\chi_{p_2}^{-2}(\Delta^2)] \\ -\frac{p_2}{p_1}\mathbb{E}[(1-(p_2-2)\chi_{p_2+2}^{-2}(\Delta^2))^2 I(\chi_{p_2+2}^2(\Delta^2)<(p_2-2))] \\ +\frac{\Delta^2}{p_1}\Big\{2\mathbb{E}[(1-(p_2-2)\chi_{p_2+2}^{-2}(\Delta^2))I(\chi_{p_2+2}^2(\Delta^2)<(p_2-2))] \\ -\mathbb{E}[(1-(p_2-2)\chi_{p_2+4}^{-2}(\Delta^2))^2 I(\chi_{p_2+4}^2(\Delta^2)<(p_2-2))]\Big\}\Bigg)^{-1}.\end{aligned} \tag{5.57}$$

So that,

$$\text{RWRE}(\hat{\boldsymbol{\beta}}_n^{\text{S+}};\tilde{\boldsymbol{\beta}}_n) \geq \text{RWRE}(\hat{\boldsymbol{\beta}}_n^{\text{S}};\tilde{\boldsymbol{\beta}}_n) \geq 1 \qquad \forall \Delta^2 \in \mathbb{R}^+.$$

5.4.4 Comparison of LSE and RLSE with RRE

First, we consider the weighted L_2-risk difference of LSE and RRE given by

$$\begin{aligned}\sigma^2(p_1+p_2)-\sigma^2 p_1-\sigma^2\frac{p_2\Delta^2}{p_2+\Delta^2} &= \sigma^2 p_2\left(1-\frac{\Delta^2}{p_2+\Delta^2}\right) \\ &= \frac{\sigma^2 p_2^2}{p_2+\Delta^2} > 0,\ \forall\ \Delta^2 \in \mathbb{R}^+.\end{aligned} \tag{5.58}$$

Hence, RRE outperforms the LSE uniformly. Similarly, for the RLSE and RRE, the weighted L_2-risk difference is given by

$$\begin{aligned}&\sigma^2(\text{tr}\ \boldsymbol{C}_{11}^{-1}\boldsymbol{C}_{11\cdot 2}+\Delta^2)-\sigma^2\left(p_1+\frac{p_2\Delta^2}{p_2+\Delta^2}\right) \\ &= \sigma^2\left\{[\text{tr}\ \boldsymbol{C}_{11}^{-1}\boldsymbol{C}_{11\cdot 2}-p_1]+\frac{\Delta^4}{p_2+\Delta^2}\right\} \\ &= \sigma^2\left(\frac{\Delta^4}{p_2+\Delta^2}-\text{tr}(\boldsymbol{M}_0)\right).\end{aligned} \tag{5.59}$$

If $\Delta^2 = 0$, then (5.59) is negative. Hence, RLSE outperforms RRE at this point. Solving the equation

$$\frac{\Delta^4}{p_2+\Delta^2} = \text{tr}(\boldsymbol{M}_0). \tag{5.60}$$

For Δ^2, we get

$$\Delta_0^2 = \frac{1}{2}\text{tr}(\boldsymbol{M}_0)\left\{1 + \sqrt{1 + \frac{4p_2}{\text{tr}(\boldsymbol{M}_0)}}\right\}. \tag{5.61}$$

If $0 \le \Delta^2 \le \Delta_0^2$, then RLSE performs better than the RRE; and if $\Delta^2 \in (\Delta_0^2, \infty)$, RRE performs better than RLSE. Thus, neither RLSE nor RRE outperforms the other uniformly.

In addition, the RWRE of RRE vs. LSE equals

$$\text{RWRE}(\hat{\boldsymbol{\beta}}_n(\kappa^*):\tilde{\boldsymbol{\beta}}_n) = \frac{p_1 + p_2}{p_1 + \frac{p_2\Delta^2}{p_2+\Delta^2}} = \left(1 + \frac{p_2}{p_1}\right)\left(1 + \frac{p_2\Delta^2}{p_1(p_2 + \Delta^2)}\right)^{-1}, \tag{5.62}$$

which is a decreasing function of Δ^2 with maximum $\left(1 + \frac{p_2}{p_1}\right)$ at $\Delta^2 = 0$ and minimum 1 as $\Delta^2 \to \infty$. So,

$$1 \le \left(1 + \frac{p_2}{p_1}\right)\left(1 + \frac{p_2}{p_1\left(1 + \frac{p_2}{\Delta}\right)}\right)^{-1} \le 1 + \frac{p_2}{p_1}; \quad \forall \Delta^2 \in \mathbb{R}^+.$$

5.4.5 Comparison of RRE with PTE, SE, and PRSE

5.4.5.1 Comparison Between $\hat{\theta}_n^{\text{RR}}(k_{\text{opt}})$ and $\hat{\theta}_n^{\text{PT}}(\alpha)$

Here, the weighted L_2-risk difference of $\hat{\boldsymbol{\theta}}_n^{\text{PT}}(\alpha)$ and $\hat{\boldsymbol{\theta}}_n^{\text{RR}}(k_{\text{opt}})$ is given by

$$\begin{aligned} R(\hat{\boldsymbol{\theta}}_n^{\text{PT}}(\alpha):\boldsymbol{C}_{11\cdot2},\boldsymbol{C}_{22\cdot1}) &- R(\hat{\boldsymbol{\theta}}_n^{\text{RR}}(k_{\text{opt}}):\boldsymbol{N}_1,\boldsymbol{N}_2) \\ &= \sigma^2[p_2(1 - H_{p_2+2}(c_\alpha;\Delta^2)) + \Delta^2\{2H_{p_2+2}(c_\alpha;\Delta^2) - H_{p_2+4}(c_\alpha;\Delta^2)\}] \\ &\quad - \frac{\sigma^2 p_2\Delta^2}{p_2 + \Delta^2}. \end{aligned} \tag{5.63}$$

Note that the risk of $\hat{\boldsymbol{\beta}}_{2n}^{\text{PT}}(\alpha)$ is an increasing function of Δ^2 crossing the p_2-line to a maximum then drops monotonically toward p_2-line as $\Delta^2 \to \infty$. The value of the risk is $p_2(1 - H_{p_2+2}(\chi_{p_2}^2(\alpha);0))(< p_2)$ at $\Delta^2 = 0$. On the other hand, $\frac{p_2\Delta^2}{p_2+\Delta^2}$ is an increasing function of Δ^2 below the p_2-line with a minimum value 0 at $\Delta^2 = 0$ and as $\Delta^2 \to \infty$, $\frac{p_2\Delta^2}{p_2+\Delta^2} \to p_2$. Hence, the risk difference in Eq. (5.63) is nonnegative for $\Delta^2 \in \mathbb{R}^+$. Thus, the RRE uniformly performs better than PTE.

5.4.5.2 Comparison Between $\hat{\theta}_n^{RR}(k_{opt})$ and $\hat{\theta}_n^{S}$

The weighted L_2-risk difference of $\hat{\theta}_n^{S}$ and $\hat{\theta}_n^{RR}(k_{opt})$ is given by

$$
\begin{aligned}
&R(\hat{\theta}_n^{S}:C_{11\cdot2}, C_{22\cdot1}) - R(\hat{\theta}_n^{RR}(k_{opt}):C_{11\cdot2}, C_{22\cdot1}) \\
&\quad = \sigma^2(p_1 + p_2 - (p_2-2)^2\mathbb{E}[\chi_{p_2}^{-2}(\Delta^2)]) - \sigma^2\left(p_1 + \frac{p_2\Delta^2}{p_2+\Delta^2}\right) \\
&\quad = \sigma^2\left[p_2 - (p_2-2)^2\mathbb{E}[\chi_{p_2}^{-2}(\Delta^2)] - \frac{p_2\Delta^2}{p_2+\Delta^2}\right].
\end{aligned} \tag{5.64}
$$

Note that the first function is increasing in Δ^2 with a value 2 at $\Delta^2 = 0$ and as $\Delta^2 \to \infty$, it tends to p_2. The second function is also increasing in Δ^2 with a value 0 at $\Delta^2 = 0$ and approaches the value p_2 as $\Delta^2 \to \infty$. Hence, the risk difference is nonnegative for all $\Delta^2 \in \mathbb{R}^+$. Consequently, RRE outperforms SE uniformly.

5.4.5.3 Comparison of $\hat{\theta}_n^{RR}(k_{opt})$ with $\hat{\theta}_n^{S+}$

The risk of $\hat{\theta}_n^{S+}$ is

$$
R(\hat{\theta}_n^{S+}; C_{11\cdot2}, C_{22\cdot1}) = R(\hat{\theta}_n^{S}; C_{11\cdot2}, C_{22\cdot1}) - R^*, \tag{5.65}
$$

where

$$
\begin{aligned}
R^* &= \sigma^2 p_2\mathbb{E}[(1-(p_2-2)\chi_{p_2+2}^{-2}(\Delta^2))^2 I(\chi_{p_2+2}^{-2}(\Delta^2) < p_2-2)] \\
&\quad -\sigma^2\Delta^2\left\{2\mathbb{E}[(1-(p_2-2)\chi_{p_2+2}^{-2}(\Delta^2))I(\chi_{p_2+2}^{-2}(\Delta^2) < p_2-2)]\right. \\
&\quad \left.-\mathbb{E}[(1-(p_2-2)\chi_{p_2+4}^{-2}(\Delta^2))^2 I(\chi_{p_2+4}^{-2}(\Delta^2) < p_2-2)]\right\},
\end{aligned} \tag{5.66}
$$

and $R(\hat{\theta}_n^{S}:C_{11\cdot2}, C_{22\cdot1})$ is

$$
R(\hat{\theta}_n^{S}:C_{11\cdot2}, C_{22\cdot1}) = \sigma^2(p_1 + p_2 - (p_2-2)^2\mathbb{E}[\chi_{p_2}^{-2}(\Delta^2)]).
$$

The weighted L_2-risk difference of PRSE and RRE is given by

$$
\begin{aligned}
&R(\hat{\theta}_n^{S+}:C_{11\cdot2}, C_{22\cdot1}) - R(\hat{\theta}_n^{RR}(k_{opt}):C_{11\cdot2}, C_{22\cdot1}) \\
&\quad = [R(\hat{\theta}_n^{S}:C_{11\cdot2}, C_{22\cdot1}) - R^*] - R(\hat{\theta}_n^{RR}(k_{opt}):C_{11\cdot2}, C_{22\cdot1}) \geq 0,
\end{aligned} \tag{5.67}
$$

where

$$
R(\hat{\theta}_n^{RR}(k_{opt}):C_{11\cdot2}, C_{22\cdot1}2) = \sigma^2\left(p_1 + \frac{p_2\Delta^2}{p_2+\Delta^2}\right).
$$

Consider the $R(\hat{\theta}_n^{S+})$. It is a monotonically increasing function of Δ^2. At $\Delta^2 = 0$, its value is

$$
\sigma^2(p_1+2) - \sigma^2 p_2\mathbb{E}[(1-(p_2-2)\chi_{p_2+2}^{-2}(0))^2 I(\chi_{p_2+2}^{-2}(0) < p_2-2)] \geq 0
$$

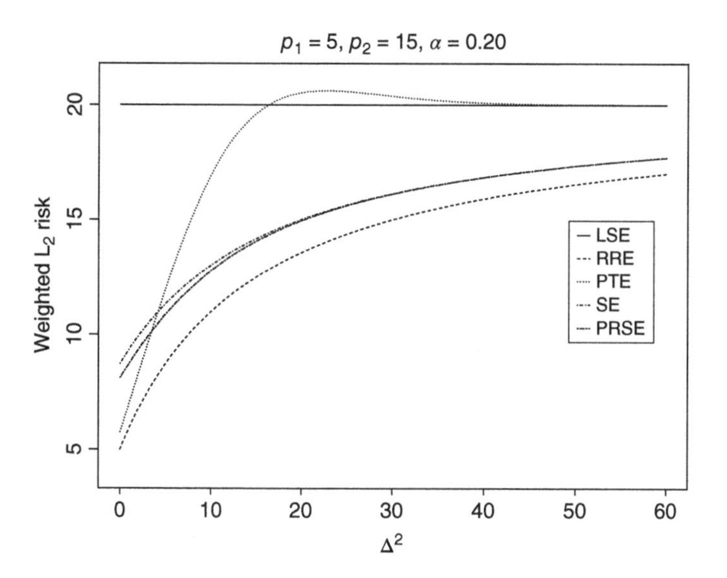

Figure 5.1 Weighted L_2 risk for the ridge, preliminary test, and Stein-type and its positive-rule estimators for $p_1 = 5, p_2 = 15$, and $\alpha = 0.20$.

and as $\Delta^2 \to \infty$, it tends to $\sigma^2(p_1 + p_2)$. For $R(\hat{\theta}_n^{\rm RR}(k_{\rm opt}) : \boldsymbol{C}_{11\cdot 2}, \boldsymbol{C}_{22\cdot 1}2)$, at $\Delta^2 = 0$, the value is $\sigma^2 p_1$; and as $\Delta^2 \to \infty$, it tends to $\sigma^2(p_1 + p_2)$. Hence, the L_2 risk difference in (5.67) is nonnegative and RRE uniformly outperforms PRSE.

Note that the risk difference of $\hat{\theta}_n^{\rm S+}$ and $\hat{\theta}_n^{\rm RR}(k_{\rm opt})$ at $\Delta^2 = 0$ is

$$\begin{aligned} &\sigma^2(p_1 + 2) - \sigma^2 p_2 \mathbb{E}[(1 - (p_2 - 2)\chi_{p_2+2}^{-2}(0))^2 I(\chi_{p_2+2}^{-2}(0) < p_2 - 2)] - \sigma^2 p_1 \\ &\quad = \sigma^2(2 - p_2 \mathbb{E}[(1 - (p_2 - 2)\chi_{p_2+2}^{-2}(0))^2 I(\chi_{p_2+2}^{-2}(0) < p_2 - 2)]) \geq 0 \end{aligned} \tag{5.68}$$

because the expected value in Eq. (5.68) is a decreasing function of DF, and $2 > p_2 \mathbb{E}[(1 - (p_2 - 2)\chi_{p_2+2}^{-2}(0))^2 I(\chi_{p_2+2}^{-2}(0) < p_2 - 2)]$. The risk functions of RRE, PT, SE, and PRSE are plotted in Figures 5.1 and 5.2 for $p_1 = 5, p_2 = 15$ and $p_1 = 7, p_2 = 33$, respectively. These figures are in support of the given comparisons.

5.4.6 Comparison of MLASSO with LSE and RLSE

First, note that if p_1 coefficients $|\beta_j| > \sigma\sqrt{C^{jj}}$ and p_2 coefficients are zero in a sparse solution, the lower bound of the weighted L_2 risk is given by $\sigma^2(p_1 + \Delta^2)$. Thereby, we compare all estimators relative to this quantity. Hence, the weighted L_2-risk difference between LSE and MLASSO is given by

$$\sigma^2(p_1 + p_2) - \sigma^2(p_1 + \Delta^2 - \text{tr}(\boldsymbol{M}_0)) = \sigma^2[(p_2 + \text{tr}(\boldsymbol{M}_0)) - \Delta^2]. \tag{5.69}$$

Hence, if $\Delta^2 \in (0, p_2 + \text{tr}(\boldsymbol{M}_0))$, the MLASSO performs better than the LSE; while if $\Delta^2 \in (p_2 + \text{tr}(\boldsymbol{M}_0), \infty)$ the LSE performs better than the MLASSO.

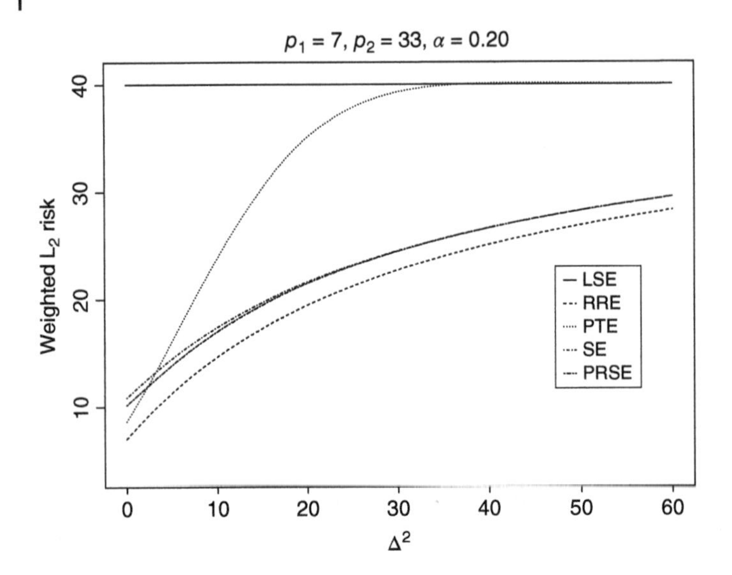

Figure 5.2 Weighted L_2 risk for the ridge, preliminary test, and Stein-type and its positive-rule estimators for $p_1 = 7$, $p_2 = 33$, and $\alpha = 0.20$.

Consequently, neither LSE nor the MLASSO performs better than the other uniformly.

Next, we compare the RLSE and MLASSO. In this case, the weighted L_2-risk difference is given by

$$\sigma^2(p_1 + \Delta^2 - \text{tr}(\boldsymbol{M}_0)) - \sigma^2(p_1 + \Delta^2) = -\sigma^2(\text{tr}(\boldsymbol{M}_0)) < 0. \tag{5.70}$$

Hence, the RLSE uniformly performs better than the MLASSO.

If $\text{tr}(\boldsymbol{M}_0) = 0$, MLASSO and RLSE are L_2-risk equivalent. If the LSE estimators are independent, then $\text{tr}(\boldsymbol{M}_0) = 0$. Hence, MLASSO satisfies the oracle properties.

5.4.7 Comparison of MLASSO with PTE, SE, and PRSE

We first consider the PTE vs. MLASSO. In this case, the weighted L_2-risk difference is given by

$$\begin{aligned}
&R(\hat{\boldsymbol{\beta}}_n^{\text{PT}}; \boldsymbol{C}_{11\cdot2}, \boldsymbol{C}_{22\cdot1}) - R(\hat{\boldsymbol{\beta}}_n^{\text{MLASSO}}(\kappa^*); \boldsymbol{C}_{11\cdot2}, \boldsymbol{C}_{22\cdot1}) \\
&\quad = \sigma^2 \left[p_2(1 - H_{p_2+2}(c_\alpha; \Delta^2)) - \Delta^2\{1 - 2H_{p_2+2}(c_\alpha; \Delta^2) + H_{p_2+4}(c_\alpha; \Delta^2)\}\right] \\
&\quad \geq \sigma^2 p_2(1 - H_{p_2+2}(c_\alpha; 0)) \geq 0, \quad \text{if } \Delta^2 = 0.
\end{aligned} \tag{5.71}$$

Hence, the MLASSO outperforms the PTE when $\Delta^2 = 0$. When $\Delta^2 \neq 0$, the MLASSO outperforms the PTE for

$$0 \leq \Delta^2 \leq \frac{p_2[1 - H_{p_2+2}(c_\alpha; \Delta^2)]}{1 - 2H_{p_2+2}(c_\alpha; \Delta^2) + H_{p_2+4}(c_\alpha; \Delta^2)}. \tag{5.72}$$

Otherwise, PTE outperforms the MLASSO. Hence, neither outperforms the other uniformly.

Next, we consider SE and PRSE vs. the MLASSO. In these two cases, we have weighted L_2-risk differences given by

$$\begin{aligned}
&R(\hat{\boldsymbol{\beta}}_n^{\mathrm{S}}; \mathbf{C}_{11\cdot2}, \mathbf{C}_{22\cdot1}) - R(\hat{\boldsymbol{\beta}}_n^{\mathrm{MLASSO}}(\kappa^*); \mathbf{C}_{11\cdot2}, \mathbf{C}_{22\cdot1}) \\
&\quad = \sigma^2[p_1 + p_2 - (p_2 - 2)^2\mathbb{E}[\chi_{p_2+2}^{-2}(\Delta^2)] - (p_1 + \Delta^2)] \\
&\quad = \sigma^2[p_2 - (p_2 - 2)^2\mathbb{E}[\chi_{p_2+2}^{-2}(\Delta^2)] - \Delta^2]
\end{aligned} \tag{5.73}$$

and from (5.65)

$$\begin{aligned}
&R(\hat{\boldsymbol{\beta}}_n^{\mathrm{S+}}; \mathbf{C}_{11\cdot2}, \mathbf{C}_{22\cdot1}) - R(\hat{\boldsymbol{\beta}}_n^{\mathrm{MLASSO}}(\kappa^*); \mathbf{C}_{11\cdot2}, \mathbf{C}_{22\cdot1}) \\
&\quad = R(\hat{\boldsymbol{\beta}}_n^{\mathrm{S}}; \mathbf{C}_{11\cdot2}, \mathbf{C}_{22\cdot1}) - R(\hat{\boldsymbol{\beta}}_n^{\mathrm{MLASSO}}(\kappa^*); \mathbf{C}_{11\cdot2}, \mathbf{C}_{22\cdot1}) - R^*,
\end{aligned} \tag{5.74}$$

where R^* is given by (5.66). Hence, the MLASSO outperforms the SE as well as the PRSE in the interval

$$0 \leq \Delta^2 \leq p_2 - (p_2 - 2)^2\mathbb{E}[\chi_{p_2}^{-2}(\Delta^2)]. \tag{5.75}$$

Thus, neither SE nor the PRSE outperforms the MLASSO uniformly.

5.4.8 Comparison of MLASSO with RRE

Here, the weighted L_2-risk difference is given by

$$\begin{aligned}
&R(\hat{\boldsymbol{\beta}}_n^{\mathrm{MLASSO}}(\kappa^*); \mathbf{C}_{11\cdot2}, \mathbf{C}_{22\cdot1}) - R(\hat{\boldsymbol{\beta}}_n^{\mathrm{RR}}(k_{\mathrm{opt}}); \mathbf{C}_{11\cdot2}, \mathbf{C}_{22\cdot1}) \\
&\quad = \sigma^2\left[(p_1 + \Delta^2) - \left(p_1 + \frac{p_2\Delta^2}{p_2 + \Delta^2}\right)\right] = \frac{\sigma^2\Delta^2}{p_2 + \Delta^2} \geq 0.
\end{aligned} \tag{5.76}$$

Hence, the RRE outperforms the MLASSO uniformly.

5.5 Efficiency in Terms of Unweighted L_2 Risk

In the previous sections, we have made all comparisons among the estimators in terms of weighted risk functions. In this section, we provide the L_2-risk efficiency of the estimators in terms of the unweighted (weight $= \boldsymbol{I}_p$) risk expressions.

The unweighted relative efficiency of the MLASSO:

$$\text{REff}(\hat{\boldsymbol{\beta}}_n^{\text{MLASSO}} : \tilde{\boldsymbol{\beta}}_n) = \left(1 + \frac{\text{tr}(\boldsymbol{C}_{22\cdot1}^{-1})}{\text{tr}(\boldsymbol{C}_{11\cdot2}^{-1})}\right)\left(1 + \frac{\Delta^{*2}}{\text{tr}(\boldsymbol{C}_{11\cdot2}^{-1})}\right)^{-1}, \tag{5.77}$$

where $\Delta^{*2} = \Delta^2 \text{Ch}_{\max}(\boldsymbol{C}_{22\cdot1}^{-1})$ or $\Delta^{*2} = \Delta^2 \text{Ch}_{\min}(\boldsymbol{C}_{11\cdot2}^{-1})$.

The unweighted relative efficiency of the ridge estimator:

$$\text{REff}(\hat{\boldsymbol{\beta}}_n : \tilde{\boldsymbol{\beta}}_n) = \left(1 + \frac{\text{tr}(\boldsymbol{C}_{22\cdot1}^{-1})}{\text{tr}(\boldsymbol{C}_{11\cdot2}^{-1})}\right)\left(1 - \frac{1}{\text{tr}(\boldsymbol{C}_{11\cdot2}^{-1})}\{\boldsymbol{M}_0 - \Delta^{*2}\}\right)^{-1}. \tag{5.78}$$

The unweighted relative efficiency of PTE:

$$\text{REff}(\hat{\boldsymbol{\beta}}_n^{\text{PT}} : \tilde{\boldsymbol{\beta}}_n) = \left(1 + \frac{\text{tr}(\boldsymbol{C}_{22\cdot1}^{-1})}{\text{tr}(\boldsymbol{C}_{11\cdot2}^{-1})}\right)\left\{1 + \frac{\text{tr}(\boldsymbol{C}_{22\cdot1}^{-1})}{\text{tr}(\boldsymbol{C}_{11\cdot2}^{-1})}(1 - H_{p_2+2}(c_\alpha; \Delta^2)) \right.$$
$$\left. + \frac{\Delta^{*2}}{\text{tr}(\boldsymbol{C}_{11\cdot2}^{-1})}\{2H_{p_2+2}(c_\alpha; \Delta^2) - H_{p_2+4}(c_\alpha; \Delta^2)\}\right\}^{-1}. \tag{5.79}$$

The unweighted relative efficiency of SE:

$$\text{REff}(\hat{\boldsymbol{\beta}}_n^{\text{S}} : \tilde{\boldsymbol{\beta}}_n) = \left(1 + \frac{\text{tr}(\boldsymbol{C}_{22\cdot1}^{-1})}{\text{tr}(\boldsymbol{C}_{11\cdot2}^{-1})}\right)\left\{1 + \frac{\text{tr}(\boldsymbol{C}_{22\cdot1}^{-1})}{\text{tr}(\boldsymbol{C}_{11\cdot2}^{-1})}(1 - (p_2 - 2)A) \right.$$
$$\left. + (p_2^2 - 4)\frac{\Delta^{*2}}{\text{tr}(\boldsymbol{C}_{11\cdot2}^{-1})}\mathbb{E}[\chi_{p_2+4}^{-4}(\Delta^2)]\right\}^{-1}, \tag{5.80}$$

where

$$A = 2\mathbb{E}[\chi_{p_2+2}^{-2}(\Delta^2)] - (p_2 - 2)\mathbb{E}[\chi_{p_2+4}^{-4}(\Delta^2)].$$

The unweighted relative efficiency of PRSE:

$$\text{REff}(\hat{\boldsymbol{\beta}}_n^{\text{S+}} : \tilde{\boldsymbol{\beta}}_n) = \left(1 + \frac{\text{tr}(\boldsymbol{C}_{22\cdot1}^{-1})}{\text{tr}(\boldsymbol{C}_{11\cdot2}^{-1})}\right)$$
$$\left\{1 + \frac{\text{tr}(\boldsymbol{C}_{22\cdot1}^{-1})}{\text{tr}(\boldsymbol{C}_{11\cdot2}^{-1})}(1 - (p_2 - 2)A) + (p_2^2 - 4)\frac{\Delta^{*2}}{\text{tr}(\boldsymbol{C}_{11\cdot2}^{-1})}\mathbb{E}[\chi_{p_2+4}^{-4}(\Delta^2)] \right.$$
$$- \frac{\text{tr}(\boldsymbol{C}_{22\cdot1}^{-1})}{\text{tr}(\boldsymbol{C}_{11\cdot2}^{-1})}\mathbb{E}[(1 - (p_2 - 2)\chi_{p_2+2}^{-2}(\Delta^2))^2 I(\chi_{p_2+2}^2(\Delta^2) < (p_2 - 2))]$$
$$+ \frac{\Delta^{*2}}{\text{tr}(\boldsymbol{C}_{11\cdot2}^{-1})}[2\mathbb{E}[(1 - (p_2 - 2)\chi_{p_2+2}^{-2}(\Delta^2))I(\chi_{p_2+2}^2(\Delta^2) < (p_2 - 2))]$$
$$\left. - \mathbb{E}\left[(1 - (p_2 - 2)\chi_{p_2+4}^{-2}(\Delta^2))^2 I(\chi_{p_2+4}^2(\Delta^2) < (p_2 - 2))\right]\right\}^{-1}. \tag{5.81}$$

The unweighted relative efficiency of RRE:

$$\mathrm{REff}(\hat{\boldsymbol{\beta}}_n^{\mathrm{RR}} : \tilde{\boldsymbol{\beta}}_n) = \left(1 + \frac{\mathrm{tr}(\boldsymbol{C}_{22\cdot1}^{-1})}{\mathrm{tr}(\boldsymbol{C}_{11\cdot2}^{-1})}\right) \left\{1 + \frac{\mathrm{tr}(\boldsymbol{C}_{22\cdot1}^{-1})}{\mathrm{tr}(\boldsymbol{C}_{11\cdot2}^{-1})} \frac{\Delta^{*2}}{\mathrm{tr}(\boldsymbol{C}_{22\cdot1}^{-1} + \Delta^{*2})}\right\}^{-1}. \tag{5.82}$$

5.6 Summary and Concluding Remarks

In this section, we discuss the contents of Tables 5.1–5.10 presented as confirmatory evidence of the theoretical findings of the estimators.

First, we note that we have two classes of estimators, namely, the traditional PTE and the Stein-type estimator and the penalty estimators. The RLSE plays an important role due to the fact that LASSO belongs to the class of restricted estimators. We have the following conclusion from our study.

(i) Since the inception of the RRE by Hoerl and Kennard (1970), there have been articles comparing the ridge estimator with PTE and the Stein-type estimator. We have now definitive conclusion that the RRE dominates the LSE and PTE and the Stein-type estimator uniformly (see Table 5.1). The ridge estimator dominates the MLASSO estimator uniformly for $\Delta^2 > 0$, while they are L_2-risk equivalent at $\Delta^2 = 0$. The ridge estimator does not select variables but the MLASSO estimator does.
(ii) The RLSE and MLASSO are competitive, although MLASSO lags behind RLSE uniformly. Both estimators outperform the LSE, PTE, SE, and PRSE in a subinterval of $[0, p_2]$ (see Table 5.1).
(iii) The lower bound of L_2 risk of HTE and MLASSO is the same and independent of the threshold parameter (κ). But the upper bound of L_2 risk is dependent on κ.
(iv) Maximum of RWRE occurs at $\Delta^2 = 0$, which indicates that the LSE underperforms all estimators for any value of (p_1, p_2). Clearly, RLSE outperforms all estimators for any (p_1, p_2) at $\Delta^2 = 0$. However, as Δ^2 deviates from 0, the PTE and the Stein-type estimator outperform LSE, RLSE, and MLASSO (see Table 5.1).
(v) If p_1 is fixed and p_2 increases, the relative RWRE of all estimators increases (see Table 5.4).
(vi) If p_2 is fixed and p_1 increases, the RWRE of all estimators decreases. Then, for p_2 small and p_2 large, the MLASSO, PTE, SE, and PRSE are competitive (see Tables 5.9 and 5.10).
(vii) The PRSE always outperforms SE (see Tables 5.1–5.10).

Now, we describe Table 5.1. This table presents the RWRE of the seven estimators for $p_1 = 5$, $p_2 = 15$ and $p_1 = 7$, $p_2 = 33$ against Δ^2-values. Using a sample of size $n = 100$, the $\boldsymbol{X}$ matrix is produced. We use the model given by

Table 5.1 Relative weighted L_2-risk efficiency for the estimators for $p_1 = 5, p_2 = 15$.

			RLSE					PTE					
	Δ^2	LSE	$\rho^2 = 0.1$	$\rho^2 = 0.2$	$\rho^2 = 0.8$	$\rho^2 = 0.9$	MLASSO	$\alpha = 0.15$	$\alpha = 0.2$	$\alpha = 0.25$	SE	PRSE	RRE
	0	1	4.91	5.13	5.73	5.78	4.00	2.30	2.06	1.88	2.85	3.22	4.00
	0.1	1	4.79	5.00	5.57	5.62	3.92	2.26	2.03	1.85	2.82	3.15	3.92
	0.5	1	4.37	4.54	5.01	5.05	3.63	2.10	1.89	1.74	2.69	2.93	3.64
	1	1	3.94	4.08	4.45	4.48	3.33	1.93	1.76	1.62	2.55	2.71	3.36
	2	1	3.29	3.39	3.64	3.66	2.85	1.67	1.54	1.45	2.33	2.40	2.95
	3	1	2.82	2.89	3.08	3.09	2.50	1.49	1.39	1.32	2.17	2.19	2.66
$\Delta_0^2(\rho^2 = 0.1)$	4.20	1	2.41	2.46	2.59	2.60	2.17	1.33	1.26	1.21	2.01	2.01	2.41
$\Delta_0^2(\rho^2 = 0.2)$	4.64	1	2.29	2.34	2.45	2.46	2.07	1.29	1.23	1.18	1.97	1.96	2.34
	5	1	2.20	2.24	2.35	2.36	2.00	1.25	1.20	1.16	1.93	1.92	2.28
$\Delta_0^2(\rho^2 = 0.8)$	5.57	1	2.07	2.11	2.20	2.21	1.89	1.21	1.16	1.13	1.88	1.86	2.20
$\Delta_0^2(\rho^2 = 0.9)$	5.64	1	2.05	2.09	2.19	2.19	1.88	1.20	1.16	1.13	1.87	1.86	2.19
	7	1	1.80	1.83	1.90	1.91	1.66	1.12	1.09	1.07	1.77	1.76	2.04
	10	1	1.42	1.43	1.48	1.48	1.33	1.02	1.01	1.01	1.61	1.60	1.81
p_2	15	1	1.04	1.05	1.08	1.08	1.00	0.97	0.97	0.98	1.45	1.45	1.60
$p_2 + \mathrm{tr}(\boldsymbol{M}_0)(\rho^2 = 0.1)$	15.92	1	1.00	1.00	1.03	1.03	0.95	0.97	0.97	0.98	1.43	1.43	1.57
$p_2 + \mathrm{tr}(\boldsymbol{M}_0)(\rho^2 = 0.2)$	16.10	1	0.99	1.00	1.02	1.02	0.94	0.97	0.97	0.98	1.43	1.42	1.56
$p_2 + \mathrm{tr}(\boldsymbol{M}_0)(\rho^2 = 0.8)$	16.51	1	0.97	0.97	1.00	1.00	0.92	0.97	0.97	0.98	1.42	1.42	1.52
$p_2 + \mathrm{tr}(\boldsymbol{M}_0)(\rho^2 = 0.9)$	16.54	1	0.97	0.97	0.99	1.00	0.92	0.97	0.97	0.98	1.42	1.42	1.55
	20	1	0.83	0.83	0.85	0.85	0.80	0.97	0.98	0.98	1.36	1.36	1.47
	30	1	0.58	0.59	0.59	0.59	0.57	0.99	0.99	0.99	1.25	1.25	1.33
	50	1	0.36	0.37	0.37	0.37	0.36	0.99	0.99	1.00	1.15	1.15	1.20
	100	1	0.19	0.19	0.19	0.19	0.19	1.00	1.00	1.00	1.04	1.04	1.10

Table 5.2 Relative weighted L_2-risk efficiency for the estimators for $p_1 = 7, p_2 = 33$.

			RLSE					PTE					
	Δ^2	LSE	$\rho^2 = 0.1$	$\rho^2 = 0.2$	$\rho^2 = 0.8$	$\rho^2 = 0.9$	MLASSO	$\alpha = 0.15$	$\alpha = 0.2$	$\alpha = 0.25$	SE	PRSE	RRE
	0	1.00	8.93	9.23	9.80	9.81	5.71	2.85	2.49	2.22	4.44	4.91	5.71
	0.1	1.00	8.74	9.02	9.56	9.58	5.63	2.81	2.46	2.20	4.39	4.84	5.63
	0.5	1.00	8.03	8.27	8.72	8.74	5.33	2.66	2.33	2.09	4.22	4.57	5.33
	1	1.00	7.30	7.49	7.86	7.88	5.00	2.49	2.20	1.98	4.03	4.28	5.01
	2	1.00	6.17	6.31	6.57	6.58	4.44	2.20	1.97	1.79	3.71	3.83	4.50
	3	1.00	5.34	5.45	5.64	5.65	4.00	1.98	1.79	1.65	3.44	3.50	4.10
	5	1.00	4.21	4.28	4.40	4.40	3.33	1.67	1.53	1.43	3.05	3.05	3.52
	7	1.00	3.48	3.52	3.60	3.61	2.85	1.45	1.36	1.29	2.76	2.74	3.13
	10	1.00	2.76	2.78	2.83	2.84	2.35	1.26	1.20	1.16	2.45	2.43	2.72
$\Delta_0^2(\rho^2 = 0.1)$	10.46	1.00	2.67	2.70	2.74	2.75	2.29	1.23	1.18	1.14	2.41	2.39	2.67
$\Delta_0^2(\rho^2 = 0.2)$	10.79	1.00	2.61	2.64	2.68	2.68	2.24	1.22	1.17	1.13	2.39	2.37	2.64
$\Delta_0^2(\rho^2 = 0.8)$	11.36	1.00	2.52	2.54	2.58	2.58	2.17	1.19	1.15	1.12	2.34	2.33	2.58
$\Delta_0^2(\rho^2 = 0.9)$	11.38	1.00	2.52	2.54	2.58	2.58	2.17	1.19	1.15	1.12	2.34	2.32	2.58
	15	1.00	2.05	2.06	2.09	2.09	1.81	1.09	1.06	1.05	2.12	2.11	2.31
	20	1.00	1.63	1.64	1.66	1.66	1.48	1.02	1.01	1.01	1.91	1.91	2.05
	30	1.00	1.16	1.16	1.17	1.17	1.08	0.99	0.99	0.99	1.66	1.66	1.76
p_2	33	1.00	1.06	1.07	1.07	1.07	1.00	0.99	0.99	0.99	1.61	1.61	1.70
$p_2 + \text{tr}(\boldsymbol{M}_0)(\rho^2 = 0.1)$	35.52	1.00	1.00	1.00	1.01	1.01	0.94	0.99	0.99	0.99	1.58	1.57	1.65
$p_2 + \text{tr}(\boldsymbol{M}_0)(\rho^2 = 0.2)$	35.66	1.00	0.99	1.00	1.00	1.00	0.93	0.99	0.99	0.99	1.57	1.57	1.65
$p_2 + \text{tr}(\boldsymbol{M}_0)(\rho^2 = 0.8)$	35.91	1.00	0.99	0.99	1.00	1.02	0.93	0.99	0.99	0.99	1.57	1.57	1.65
$p_2 + \text{tr}(\boldsymbol{M}_0)(\rho^2 = 0.9)$	35.92	1.00	0.99	0.99	0.99	1.00	0.93	0.99	0.99	0.99	1.57	1.57	1.65
	50	1.00	0.73	0.73	0.73	0.73	0.70	0.99	0.99	0.99	1.43	1.43	1.48
	100	1.00	0.38	0.38	0.38	0.38	0.37	1.00	1.00	1.00	1.12	1.12	1.25

Table 5.3 Relative weighted L_2-risk efficiency of the estimators for $p = 10$ and different Δ^2 values for varying p_1.

	$\Delta^2 = 0$				$\Delta^2 = 1$			
Estimators	$p_1 = 2$	$p_1 = 3$	$p_1 = 5$	$p_1 = 7$	$p_1 = 2$	$p_1 = 3$	$p_1 = 5$	$p_1 = 7$
LSE	1.00	1.00	1.00	1.00	1.00	1.00	1.00	1.00
RLSE ($\rho^2 = 0.1$)	5.68	3.71	2.15	1.49	3.62	2.70	1.77	1.30
RLSE ($\rho^2 = 0.2$)	6.11	3.93	2.23	1.52	3.79	2.82	1.82	1.32
RLSE ($\rho^2 = 0.8$)	9.45	5.05	2.54	1.67	4.85	3.35	2.03	1.43
RLSE ($\rho^2 = 0.9$)	10.09	5.18	2.58	1.68	5.02	3.41	2.05	1.44
MLASSO	5.00	3.33	2.00	1.42	3.33	2.50	1.66	1.25
PTE ($\alpha = 0.15$)	2.34	1.97	1.51	1.22	1.75	1.55	1.27	1.08
PTE ($\alpha = 0.2$)	2.06	1.79	1.42	1.19	1.60	1.44	1.22	1.06
PTE ($\alpha = 0.25$)	1.86	1.65	1.36	1.16	1.49	1.37	1.18	1.05
SE	2.50	2.00	1.42	1.11	2.13	1.77	1.32	1.07
PRSE	3.03	2.31	1.56	1.16	2.31	1.88	1.38	1.10
RRE	5.00	3.33	2.00	1.42	3.46	2.58	1.71	1.29
	$\Delta^2 = 5$				$\Delta^2 = 10$			
LSE	1.00	1.00	1.00	1.00	1.00	1.00	1.00	1.00
RLSE ($\rho^2 = 0.1$)	1.47	1.30	1.03	0.85	0.85	0.78	0.68	0.59
RLSE ($\rho^2 = 0.2$)	1.50	1.32	1.05	0.86	0.85	0.79	0.69	0.60
RLSE ($\rho^2 = 0.8$)	1.65	1.43	1.12	0.91	0.90	0.83	0.71	0.62
RLSE ($\rho^2 = 0.9$)	1.66	1.44	1.12	0.91	0.90	0.83	0.72	0.62
MLASSO	1.42	1.25	1.00	0.83	0.83	0.76	0.66	0.58
PTE ($\alpha = 0.15$)	1.05	1.00	0.94	0.91	0.92	0.91	0.91	0.93
PTE ($\alpha = 0.2$)	1.03	1.00	0.95	0.93	0.93	0.93	0.93	0.95
PTE ($\alpha = 0.25$)	1.02	0.99	0.95	0.94	0.94	0.94	0.95	0.96
SE	1.55	1.38	1.15	1.02	1.32	1.22	1.08	1.01
PRSE	1.53	1.37	1.15	1.02	1.31	1.21	1.08	1.01
RRE	1.96	1.69	1.33	1.12	1.55	1.40	1.20	1.07
	$\Delta^2 = 20$				$\Delta^2 = 60$			
LSE	1.00	1.00	1.00	1.00	1.00	1.00	1.00	1.00
RLSE ($\rho^2 = 0.1$)	0.45	0.44	0.40	0.37	0.16	0.15	0.15	0.14
RLSE ($\rho^2 = 0.2$)	0.46	0.44	0.40	0.37	0.16	0.15	0.15	0.15
RLSE ($\rho^2 = 0.8$)	0.47	0.45	0.41	0.38	0.16	0.16	0.15	0.15
RLSE ($\rho^2 = 0.9$)	0.47	0.45	0.41	0.38	0.16	0.16	0.15	0.15
MLASSO	0.45	0.43	0.40	0.37	0.16	0.15	0.15	0.14
PTE ($\alpha = 0.15$)	0.96	0.97	0.98	0.99	1.00	1.00	1.00	1.00
PTE ($\alpha = 0.2$)	0.97	0.98	0.98	0.99	1.00	1.00	1.00	1.00
PTE ($\alpha = 0.25$)	0.98	0.98	0.99	0.99	1.00	1.00	1.00	1.00
SE	1.17	1.12	1.04	1.00	1.05	1.04	1.01	1.00
PRSE	1.17	1.11	1.04	1.00	1.05	1.04	1.01	1.00
RRE	1.29	1.22	1.11	1.04	1.10	1.07	1.04	1.01

Table 5.4 Relative weighted L_2-risk efficiency of the estimators for $p = 20$ and different Δ^2 values for varying p_1.

	$\Delta^2 = 0$				$\Delta^2 = 1$			
Estimators	$p_1 = 2$	$p_1 = 3$	$p_1 = 5$	$p_1 = 7$	$p_1 = 2$	$p_1 = 3$	$p_1 = 5$	$p_1 = 7$
LSE	1.00	1.00	1.00	1.00	1.00	1.00	1.00	1.00
RLSE ($\rho^2 = 0.1$)	12.98	8.50	4.92	3.39	7.86	5.96	3.95	2.90
RLSE ($\rho^2 = 0.2$)	14.12	9.05	5.14	3.54	8.27	6.22	4.09	2.98
RLSE ($\rho^2 = 0.8$)	21.56	11.44	5.74	3.76	10.47	7.36	4.59	3.25
RLSE ($\rho^2 = 0.9$)	22.85	11.73	5.79	3.78	10.65	7.39	4.49	3.18
MLASSO	10.00	6.66	4.00	2.85	6.66	5.00	3.33	2.50
PTE ($\alpha = 0.15$)	3.20	2.83	2.30	1.94	2.49	2.27	1.93	1.67
PTE ($\alpha = 0.2$)	2.69	2.44	2.06	1.79	2.17	2.01	1.76	1.56
PTE ($\alpha = 0.25$)	2.34	2.16	1.88	1.66	1.94	1.82	1.62	1.47
SE	5.00	4.00	2.85	2.22	4.12	3.42	2.55	2.04
PRSE	6.27	4.77	3.22	2.43	4.57	3.72	2.71	2.13
RRE	10.00	6.66	4.00	2.85	6.78	5.07	3.36	2.52
	$\Delta^2 = 5$				$\Delta^2 = 10$			
LSE	1.00	1.00	1.00	1.00	1.00	1.00	1.00	1.00
RLSE ($\rho^2 = 0.1$)	3.05	2.71	2.20	1.83	1.73	1.61	1.42	1.25
RLSE ($\rho^2 = 0.2$)	3.11	2.77	2.25	1.87	1.75	1.63	1.44	1.27
RLSE ($\rho^2 = 0.8$)	3.37	2.96	2.35	1.93	1.82	1.70	1.48	1.30
RLSE ($\rho^2 = 0.9$)	3.40	2.98	2.36	1.94	1.83	1.70	1.48	1.30
MLASSO	2.85	2.50	2.00	1.66	1.66	1.53	1.33	1.17
PTE ($\alpha = 0.15$)	1.42	1.36	1.25	1.17	1.07	1.05	1.02	0.99
PTE ($\alpha = 0.2$)	1.33	1.28	1.20	1.13	1.05	1.04	1.01	0.99
PTE ($\alpha = 0.25$)	1.26	1.23	1.16	1.11	1.04	1.03	1.01	0.99
SE	2.65	2.35	1.93	1.64	2.02	1.86	1.61	1.43
PRSE	2.63	2.34	1.92	1.63	2.00	1.85	1.60	1.42
RRE	3.38	2.91	2.28	1.88	2.37	2.15	1.81	1.58
	$\Delta^2 = 20$				$\Delta^2 = 60$			
LSE	1.00	1.00	1.00	1.00	1.00	1.00	1.00	1.00
RLSE ($\rho^2 = 0.1$)	0.92	0.87	0.83	0.77	0.32	0.32	0.31	0.30
RLSE ($\rho^2 = 0.2$)	0.93	0.90	0.83	0.77	0.32	0.32	0.31	0.30
RLSE ($\rho^2 = 0.8$)	0.95	0.91	0.85	0.79	0.32	0.32	0.31	0.30
RLSE ($\rho^2 = 0.9$)	0.95	0.92	0.85	0.79	0.32	0.32	0.31	0.30
MLASSO	0.90	0.86	0.80	0.74	0.32	0.31	0.30	0.29
PTE ($\alpha = 0.15$)	0.97	0.97	0.97	0.97	1.00	1.00	1.00	1.00
PTE ($\alpha = 0.2$)	0.98	0.98	0.98	0.98	1.00	1.00	1.00	1.00
PTE ($\alpha = 0.25$)	0.98	0.98	0.98	0.98	1.00	1.00	1.00	1.00
SE	1.57	1.49	1.36	1.25	1.20	1.18	1.13	1.09
PRSE	1.57	1.49	1.36	1.25	1.20	1.18	1.13	1.09
RRE	1.74	1.64	1.47	1.34	1.26	1.23	1.17	1.13

Table 5.5 Relative weighted L_2-risk efficiency of the estimators for $p = 40, 60$ and different Δ^2 values for varying p_1.

	$\Delta^2 = 0$				$\Delta^2 = 1$			
Estimators	$p_1 = 2$	$p_1 = 3$	$p_1 = 5$	$p_1 = 7$	$p_1 = 2$	$p_1 = 3$	$p_1 = 5$	$p_1 = 7$
LSE	1.00	1.00	1.00	1.00	1.00	1.00	1.00	1.00
RLSE ($\rho^2 = 0.1$)	34.97	22.73	13.02	8.95	18.62	14.47	9.82	7.31
RLSE ($\rho^2 = 0.2$)	38.17	24.29	13.64	9.24	19.49	15.09	10.16	7.50
RLSE ($\rho^2 = 0.8$)	57.97	30.39	15.05	9.80	23.60	17.24	10.93	7.87
RLSE ($\rho^2 = 0.9$)	61.39	30.94	15.20	9.83	24.15	17.42	11.01	7.89
MLASSO	20.00	13.33	8.00	5.71	13.33	10.00	6.66	5.00
PTE ($\alpha = 0.15$)	4.04	3.73	3.23	2.85	3.32	3.11	2.76	2.49
PTE ($\alpha = 0.2$)	3.28	3.08	2.76	2.49	2.77	2.63	2.39	2.20
PTE ($\alpha = 0.25$)	2.77	2.64	2.41	2.22	2.39	2.30	2.13	1.98
SE	10.00	8.00	5.71	4.44	8.12	6.75	5.05	4.03
PRSE	12.80	9.69	6.52	4.91	9.24	7.50	5.45	4.28
RRE	20.00	13.33	8.00	5.71	13.44	10.06	6.69	5.01
	$\Delta^2 = 5$				$\Delta^2 = 10$			
LSE	1.00	1.00	1.00	1.00	1.00	1.00	1.00	1.00
RLSE ($\rho^2 = 0.1$)	6.50	5.91	4.95	4.22	3.58	3.39	3.05	2.76
RLSE ($\rho^2 = 0.2$)	6.58	6.00	5.04	4.30	3.61	3.43	3.09	2.79
RLSE ($\rho^2 = 0.8$)	7.02	6.32	5.22	4.40	3.73	3.53	3.15	2.83
RLSE ($\rho^2 = 0.9$)	7.06	6.35	5.23	4.40	3.75	3.54	3.16	2.84
MLASSO	5.71	5.00	4.00	3.33	3.33	3.07	2.66	2.35
PTE ($\alpha = 0.15$)	1.96	1.89	1.77	1.67	1.37	1.35	1.30	1.26
PTE ($\alpha = 0.2$)	1.75	1.70	1.61	1.53	1.29	1.27	1.23	1.20
PTE ($\alpha = 0.25$)	1.60	1.56	1.49	1.43	1.23	1.21	1.18	1.16
SE	4.87	4.35	3.58	3.05	3.45	3.19	2.77	2.45
PRSE	4.88	4.35	3.58	3.05	3.41	3.16	2.75	2.43
RRE	6.23	5.40	4.26	3.52	4.03	3.67	3.13	2.72
	$\Delta^2 = 20$				$\Delta^2 = 60$			
LSE	1.00	1.00	1.00	1.00	1.00	1.00	1.00	1.00
RLSE ($\rho^2 = 0.1$)	1.89	1.83	1.73	1.63	0.65	0.64	0.63	0.62
RLSE ($\rho^2 = 0.2$)	1.93	1.87	1.76	1.66	0.65	0.64	0.63	0.62
RLSE ($\rho^2 = 0.8$)	1.92	1.86	1.76	1.66	0.65	0.65	0.63	0.62
RLSE ($\rho^2 = 0.9$)	1.93	1.87	1.76	1.66	0.65	0.65	0.63	0.62
MLASSO	1.81	1.73	1.60	1.48	0.64	0.63	0.61	0.59
PTE ($\alpha = 0.15$)	1.05	1.04	1.03	1.02	0.99	0.99	0.99	0.99
PTE ($\alpha = 0.2$)	1.03	1.03	1.02	1.01	0.99	0.99	0.99	0.99
PTE ($\alpha = 0.25$)	1.02	1.02	1.01	1.01	0.99	0.99	1.00	1.00
SE	2.41	2.29	2.08	1.91	1.51	1.48	1.42	1.36
PRSE	2.40	2.28	2.08	1.91	1.51	1.48	1.42	1.36
RRE	2.64	2.50	2.25	2.05	1.58	1.54	1.47	1.41

Table 5.6 Relative weighted L_2-risk efficiency of the estimators for $p = 60$ and different Δ^2 values for varying p_1.

	$\Delta^2 = 0$				$\Delta^2 = 1$			
Estimators	$p_1 = 2$	$p_1 = 3$	$p_1 = 5$	$p_1 = 7$	$p_1 = 2$	$p_1 = 3$	$p_1 = 5$	$p_1 = 7$
LSE	1.00	1.00	1.00	1.00	1.00	1.00	1.00	1.00
RLSE ($\rho^2 = 0.1$)	78.67	50.45	28.36	19.22	33.91	27.33	19.24	14.553
RLSE ($\rho^2 = 0.2$)	85.74	53.89	29.66	19.83	35.15	28.31	19.82	14.89
RLSE ($\rho^2 = 0.8$)	130.12	67.07	32.66	20.97	40.85	31.57	21.12	15.52
RLSE ($\rho^2 = 0.9$)	138.26	68.81	33.13	21.00	41.62	31.95	21.31	15.58
MLASSO	30.00	20.00	12.00	8.57	20.00	15.00	10.00	7.50
PTE ($\alpha = 0.15$)	4.49	4.22	3.78	3.42	3.80	3.61	3.29	3.02
PTE ($\alpha = 0.2$)	3.58	3.42	3.14	2.90	3.10	2.98	2.77	2.59
PTE ($\alpha = 0.25$)	2.99	2.88	2.70	2.53	2.64	2.56	2.41	2.28
SE	15.00	12.00	8.57	6.66	12.12	10.08	7.55	6.03
PRSE	19.35	14.63	9.83	7.40	13.98	11.33	8.22	6.45
RRE	30.00	20.00	12.00	8.57	20.11	15.06	10.02	7.51
	$\Delta^2 = 5$				$\Delta^2 = 10$			
LSE	1.00	1.00	1.00	1.00	1.00	1.00	1.00	1.00
RLSE ($\rho^2 = 0.1$)	10.39	9.67	8.42	7.38	5.56	5.35	4.94	4.57
RLSE ($\rho^2 = 0.2$)	10.50	9.79	8.53	7.47	5.60	5.39	4.98	4.60
RLSE ($\rho^2 = 0.8$)	10.96	10.16	8.76	7.62	5.72	5.50	5.06	4.66
RLSE ($\rho^2 = 0.9$)	11.02	10.20	8.79	7.63	5.74	5.58	5.07	4.66
MLASSO	8.57	7.50	6.00	5.00	5.00	4.61	4.00	3.52
PTE ($\alpha = 0.15$)	2.35	2.28	2.16	2.04	1.63	1.60	1.54	1.50
PTE ($\alpha = 0.2$)	2.04	1.99	1.90	1.82	1.48	1.46	1.42	1.39
PTE ($\alpha = 0.25$)	1.82	1.79	1.72	1.67	1.38	1.37	1.34	1.31
SE	7.09	6.35	5.25	4.47	4.88	4.53	3.94	3.50
PRSE	7.16	6.40	5.28	4.49	4.83	4.48	3.91	3.47
RRE	9.08	7.89	6.26	5.18	5.69	5.21	4.45	3.89
	$\Delta^2 = 20$				$\Delta^2 = 60$			
LSE	1.00	1.00	1.00	1.00	1.00	1.00	1.00	1.00
RLSE ($\rho^2 = 0.1$)	2.88	2.83	2.71	2.59	0.98	0.98	0.96	0.95
RLSE ($\rho^2 = 0.2$)	2.89	2.84	2.72	2.60	0.98	0.98	0.96	0.95
RLSE ($\rho^2 = 0.8$)	2.93	2.87	2.74	2.62	0.99	0.98	0.97	0.95
RLSE ($\rho^2 = 0.9$)	2.93	2.87	2.74	2.62	0.99	0.98	0.97	0.95
MLASSO	2.72	2.60	2.40	2.22	0.96	0.95	0.92	0.89
PTE ($\alpha = 0.15$)	1.15	1.14	1.12	1.11	0.99	0.99	0.99	0.99
PTE ($\alpha = 0.2$)	1.11	1.10	1.09	1.08	0.99	0.99	0.99	0.99
PTE ($\alpha = 0.25$)	1.08	1.07	1.07	1.06	0.99	0.99	0.99	0.99
SE	3.25	3.09	2.82	2.60	1.83	1.79	1.77	1.65
PRSE	3.23	3.08	2.81	2.55	1.83	1.79	1.71	1.65
RRE	3.55	3.36	3.05	2.78	1.90	1.86	1.78	1.70

Table 5.7 Relative weighted L_2-risk efficiency values of estimators for $p_1 = 5$ and different values of p_2 and Δ^2.

		RLSE			PTE					
		ρ^2			α					
	LSE	0.8	0.9	MLASSO	0.15	0.2	0.25	SE	PRSE	RRE
p_2					$\Delta^2 = 0$					
5	1.00	2.00	2.02	1.71	1.56	1.39	1.28	1.33	1.42	1.71
15	1.00	4.14	4.15	3.14	2.62	2.06	1.74	2.44	2.68	3.14
25	1.00	6.81	6.84	4.57	3.52	2.55	2.04	3.55	3.92	4.57
35	1.00	10.28	10.32	6.00	4.32	2.92	2.26	4.66	5.16	6.00
55	1.00	21.72	21.75	8.85	5.66	3.47	2.56	6.88	7.65	8.85
p_2					$\Delta^2 = 0.5$					
5	1.00	1.85	1.86	1.60	1.43	1.29	1.20	1.29	1.35	1.60
15	1.00	3.78	3.79	2.93	2.40	1.90	1.63	2.33	2.49	2.93
25	1.00	6.15	6.18	4.26	3.24	2.36	1.92	3.38	3.64	4.27
35	1.00	9.16	9.19	5.60	3.98	2.72	2.13	4.43	4.80	5.60
55	1.00	18.48	18.49	8.26	5.24	3.25	2.42	6.54	7.11	8.27
p_2					$\Delta^2 = 1$					
5	1.00	1.71	1.73	1.50	1.33	1.21	1.14	1.26	1.29	1.53
15	1.00	3.48	3.49	2.75	2.22	1.78	1.54	2.24	2.34	2.77
25	1.00	5.61	5.63	4.00	3.00	2.21	1.81	3.23	3.42	4.01
35	1.00	8.26	8.28	5.25	3.69	2.55	2.02	4.23	4.50	5.26
55	1.00	16.07	16.09	7.75	4.88	3.06	2.31	6.23	6.67	7.76
p_2					$\Delta^2 = 5$					
5	1.00	1.09	1.09	1.00	0.94	0.95	0.96	1.12	1.12	1.26
15	1.00	2.13	2.13	1.83	1.41	1.23	1.14	1.78	1.77	2.04
25	1.00	3.29	3.30	2.66	1.86	1.48	1.31	2.48	2.47	2.86
35	1.00	4.62	4.63	3.50	2.28	1.71	1.46	3.19	3.19	3.69
55	1.00	7.88	7.89	5.16	3.04	2.08	1.69	4.61	4.64	5.35

Table 5.8 Relative weighted L_2-risk efficiency values of estimators for $p_1 = 7$ and different values of p_2 and Δ^2.

		RLSE			PTE					
		ρ^2			α					
	LSE	0.8	0.9	MLASSO	0.15	0.2	0.25	SE	PRSE	RRE
p_2					$\Delta^2 = 0$					
3	1.00	1.67	1.68	1.42	1.33	1.22	1.16	1.11	1.16	1.42
13	1.00	3.76	3.78	2.85	2.41	1.94	1.66	2.22	2.43	2.85
23	1.00	6.38	6.42	4.28	3.34	2.46	1.99	3.33	3.67	4.28
33	1.00	9.79	9.84	5.71	4.16	2.85	2.22	4.44	4.91	5.71
53	1.00	21.01	21.05	8.57	5.54	3.42	2.53	6.66	7.40	8.57
p_2					$\Delta^2 = 0.5$					
3	1.00	1.54	1.55	1.33	1.23	1.14	1.10	1.09	1.12	1.34
13	1.00	3.43	3.45	2.66	2.22	1.80	1.56	2.12	2.26	2.67
23	1.00	5.77	5.79	4.00	3.08	2.28	1.87	3.17	3.41	4.00
33	1.00	8.72	8.76	5.33	3.84	2.66	2.09	4.22	4.57	5.33
53	1.00	17.87	17.90	8.00	5.13	3.21	2.40	6.33	6.88	8.00
p_2					$\Delta^2 = 1$					
3	1.00	1.43	1.44	1.25	1.14	1.08	1.05	1.07	1.10	1.29
13	1.00	3.16	3.18	2.50	2.05	1.67	1.47	2.04	2.13	2.52
23	1.00	5.26	5.28	3.75	2.85	2.13	1.76	3.03	3.20	3.76
33	1.00	7.86	7.90	5.00	3.56	2.49	1.98	4.03	4.28	5.01
53	1.00	15.55	15.57	7.50	4.77	3.02	2.28	6.03	6.45	7.51
p_2					$\Delta^2 = 5$					
3	1.00	0.91	0.91	0.83	0.86	0.91	0.94	1.02	1.02	1.12
13	1.00	1.93	1.94	1.66	1.31	1.17	1.11	1.64	1.63	1.88
23	1.00	3.09	3.10	2.50	1.77	1.43	1.28	2.34	2.33	2.70
33	1.00	4.40	4.41	3.33	2.20	1.67	1.43	3.05	3.05	3.52
53	1.00	7.63	7.63	5.00	2.97	2.04	1.67	4.47	4.49	5.18

Table 5.9 Relative weighted L_2-risk efficiency values of estimators for $p_2 = 5$ and different values of p_1 and Δ^2.

		RLSE			PTE					
		ρ^2			α					
	LSE	0.8	0.9	MLASSO	0.15	0.2	0.25	SE	PRSE	RRE
p_1					$\Delta^2 = 0$					
5	1.00	2.55	2.57	2.00	1.76	1.51	1.36	1.42	1.56	2.00
15	1.00	1.48	1.48	1.33	1.27	1.20	1.15	1.17	1.21	1.33
25	1.00	1.30	1.30	1.20	1.16	1.12	1.09	1.11	1.13	1.20
35	1.00	1.22	1.22	1.14	1.12	1.09	1.07	1.08	1.09	1.14
55	1.00	1.15	1.15	1.09	1.07	1.05	1.04	1.05	1.06	1.09
p_1					$\Delta^2 = 0.5$					
5	1.00	2.26	2.28	1.81	1.57	1.37	1.26	1.37	1.43	1.83
15	1.00	1.42	1.43	1.29	1.22	1.15	1.11	1.15	1.18	1.29
25	1.00	1.27	1.27	1.17	1.13	1.10	1.07	1.09	1.11	1.17
35	1.00	1.20	1.20	1.12	1.10	1.07	1.05	1.07	1.08	1.12
55	1.00	1.14	1.14	1.08	1.06	1.04	1.03	1.04	1.05	1.08
p_1					$\Delta^2 = 1$					
5	1.00	2.03	2.04	1.66	1.43	1.27	1.18	1.32	1.38	1.71
15	1.00	1.38	1.38	1.25	1.17	1.11	1.08	1.14	1.16	1.26
25	1.00	1.24	1.24	1.15	1.11	1.07	1.05	1.09	1.10	1.16
35	1.00	1.18	1.18	1.11	1.08	1.05	1.04	1.06	1.07	1.11
55	1.00	1.13	1.13	1.07	1.05	1.03	1.02	1.04	1.04	1.07
p_1					$\Delta^2 = 5$					
5	1.00	1.12	1.12	1.00	0.93	0.94	0.95	1.15	1.15	1.33
15	1.00	1.08	1.08	1.00	0.96	0.97	0.97	1.07	1.07	1.14
25	1.00	1.06	1.06	1.00	0.97	0.98	0.98	1.04	1.04	1.09
35	1.00	1.06	1.06	1.00	0.98	0.98	0.98	1.03	1.03	1.06
55	1.00	1.05	1.05	1.00	0.98	0.99	0.99	1.02	1.02	1.04

Table 5.10 Relative weighted L_2-risk efficiency values of estimators for $p_2 = 7$ and different values of p_1 and Δ^2.

	LSE	RLSE $\rho^2 = 0.8$	RLSE $\rho^2 = 0.9$	MLASSO	PTE $\alpha = 0.15$	PTE $\alpha = 0.2$	PTE $\alpha = 0.25$	SE	PRSE	RRE
p_1					$\Delta^2 = 0$					
3	1.00	5.05	5.18	3.33	2.60	1.97	1.65	2.00	2.31	3.33
13	1.00	1.76	1.77	1.53	1.44	1.32	1.24	1.33	1.39	1.53
23	1.00	1.44	1.45	1.30	1.25	1.19	1.15	1.20	1.23	1.30
33	1.00	1.32	1.32	1.21	1.18	1.14	1.11	1.14	1.16	1.21
53	1.00	1.22	1.22	1.13	1.11	1.08	1.07	1.09	1.10	1.13
p_1					$\Delta^2 = 0.5$					
3	1.00	4.03	4.12	2.85	2.21	1.73	1.49	1.87	2.06	2.88
13	1.00	1.69	1.69	1.48	1.37	1.26	1.19	1.30	1.34	1.48
23	1.00	1.41	1.41	1.27	1.22	1.16	1.12	1.18	1.20	1.27
33	1.00	1.30	1.30	1.19	1.15	1.11	1.08	1.13	1.14	1.19
53	1.00	1.21	1.21	1.12	1.10	1.07	1.05	1.08	1.09	1.12
p_1					$\Delta^2 = 1$					
3	1.00	3.35	3.41	2.50	1.92	1.55	1.37	1.77	1.88	2.58
13	1.00	1.62	1.62	1.42	1.31	1.21	1.15	1.27	1.30	1.44
23	1.00	1.38	1.38	1.25	1.19	1.13	1.09	1.17	1.18	1.25
33	1.00	1.28	1.28	1.17	1.13	1.09	1.07	1.12	1.13	1.18
53	1.00	1.20	1.20	1.11	1.08	1.06	1.04	1.07	1.08	1.11
p_1					$\Delta^2 = 5$					
3	1.00	1.43	1.44	1.25	1.04	1.00	0.99	1.38	1.37	1.69
13	1.00	1.22	1.22	1.11	1.02	1.00	0.99	1.16	1.15	1.25
23	1.00	1.16	1.16	1.07	1.01	1.00	0.99	1.10	1.09	1.15
33	1.00	1.14	1.14	1.05	1.01	1.00	0.99	1.07	1.07	1.11
53	1.00	1.11	1.11	1.03	1.00	1.00	0.99	1.04	1.04	1.07

Eq. (5.54) for chosen values $\rho^2 = 0.1, 0.2$ and $0.8, 0.9$. Therefore, REff values of RLSE has four entries – two for low correlation and two for high correlation. Some Δ^2-values are given as p_2 and $p_2 + \text{tr}(\boldsymbol{M}_0)$ for chosen ρ^2-values. Now, one may use the table for the performance characteristics of each estimator compared to any other.

Tables 5.2–5.6 give the RWRE values of estimators for $p_1 = 2, 3, 5$, and 7 for $p = 10, 20, 40$, and 60.

Tables 5.7 and 5.8 give the RWRE values of estimators for $p_1 = 5$ and $p_2 = 5, 15, 25, 35$, and 55, and also, for $p_1 = 7$ and $p_2 = 3, 13, 23, 33$, and 53 to see the effect of p_2 variation on relative weighted L_2-risk efficiency.

Tables 5.9 and 5.10 give the RWRE values of estimators for $p_2 = 5$ and $p_1 = 5, 15, 25, 35$, and 55, and also for $p_2 = 7$ and $p_1 = 3, 13, 23, 33$, and 53 to see the effect of p_1 variation on RWRE.

Problems

5.1 Verify (5.9).

5.2 Show that the mean square error of $\hat{\boldsymbol{\beta}}_n^{\text{HT}}(\kappa)$ is given by

$$R(\hat{\boldsymbol{\beta}}_n^{\text{HT}}(\kappa)) = \sigma^2 \sum_{j=1}^{p} C^{jj} \left\{ (1 - H_3(\kappa^2; \Delta_j^2)) + \Delta_j^2 (2H_3(\kappa^2; \Delta_j^2) - H_5(\kappa^2; \Delta_j^2)) \right\}.$$

5.3 Prove inequality (5.25).

5.4 Prove Theorem 5.1.

5.5 Prove Theorem 5.3.

5.6 Show that the weighted L_2 risk of PTE is

$$R(\hat{\boldsymbol{\beta}}_n^{\text{PT}}; \boldsymbol{C}_{11\cdot 2}, \boldsymbol{C}_{22\cdot 1}) = \sigma^2 p_1 + \sigma^2 p_2 [1 - H_{p_2+2}(c_\alpha; \Delta^2)] + \Delta^2 [2H_{p_2+2}(c_\alpha; \Delta^2) - H_{p_2+4}(c_\alpha; \Delta^2)].$$

5.7 Verify (5.47).

5.8 Show that the risk function of $R(\hat{\boldsymbol{\beta}}_n^{\text{S+}})$ is

$$R(\hat{\boldsymbol{\beta}}_n^{\text{S+}}; \boldsymbol{C}_{11\cdot2}, \boldsymbol{C}_{22\cdot1}) = R(\hat{\boldsymbol{\beta}}_n^{\text{S}}; \boldsymbol{C}_{11\cdot2}, \boldsymbol{C}_{22\cdot1})$$
$$- \sigma^2 p_2 \mathbb{E}[(1 - (p_2 - 2)\chi^{-2}_{p_2+2}(\Delta^2))^2 I(\chi^{-2}_{p_2+2}(\Delta^2) < p_2 - 2)]$$
$$+ \Delta^2 \left\{ 2\mathbb{E}[(1 - (p_2 - 2)\chi^{-2}_{p_2+2}(\Delta^2)) I(\chi^{-2}_{p_2+2}(\Delta^2) < p_2 - 2)] \right.$$
$$\left. + \mathbb{E}[(1 - (p_2 - 2)\chi^{-2}_{p_2+4}(\Delta^2))^2 I(\chi^{-2}_{p_2+4}(\Delta^2) < p_2 - 2)] \right\}.$$

5.9 Show that the MLASSO dominates the PTE when

$$0 \leq \Delta^2 \leq \frac{p_2[1 - H_{p_2+2}(c_\alpha; \Delta^2)]}{1 - 2H_{p_2+2}(c_\alpha; \Delta^2) + H_{p_2+4}(c_\alpha; \Delta^2)}.$$

5.8 Show that the risk function of [illegible] is

[illegible]

5.9 Show that the LASSO estimates the PLS [illegible]

[illegible]

6

Ridge Regression in Theory and Applications

The multiple linear regression model is one of the best known and widely used among the models for statistical data analysis in every field of sciences and engineering as well as in social sciences, economics, and finance. The subject of this chapter is the study of the rigid regression estimator (RRE) for the regression coefficients, its characteristic properties, and comparing its relation with the least absolute shrinkage and selection operator (LASSO). Further, we consider the preliminary test estimator (PTE) and the Stein-type ridge estimator in low dimension and study their dominance properties. We conclude the chapter with the asymptotic distributional theory of the ridge estimators following Knight and Fu (2000).

6.1 Multiple Linear Model Specification

Consider the multiple linear model with coefficient vector, $\boldsymbol{\beta} = (\beta_1, \dots, \beta_p)^\top$ given by

$$\boldsymbol{Y} = \boldsymbol{X}\boldsymbol{\beta} + \boldsymbol{\epsilon}, \tag{6.1}$$

where $\boldsymbol{Y} = (y_1, \dots, y_n)^\top$ is a vector of n responses, $\boldsymbol{X}$ is an $n \times p$ design matrix of rank $p(\leq n)$, and $\boldsymbol{\epsilon}$ is an n-vector of independently and identically distributed (i.i.d.) random variables with distribution $\mathcal{N}_n(\mathbf{0}, \sigma^2\boldsymbol{I}_n)$, with $\boldsymbol{I}_n$, the identity matrix of order n.

6.1.1 Estimation of Regression Parameters

Using the model (6.1) and the error distribution of $\boldsymbol{\epsilon}$ we obtain the maximum likelihood estimator/least squares estimator (MSE/LSE) of $\boldsymbol{\beta}$ by minimizing

$$\min_{\boldsymbol{\beta}\in\mathbb{R}^p}\{(\boldsymbol{Y} - \boldsymbol{X}\boldsymbol{\beta})^\top(\boldsymbol{Y} - \boldsymbol{X}\boldsymbol{\beta})\} \tag{6.2}$$

to get $\tilde{\boldsymbol{\beta}}_n = (\boldsymbol{X}^\top\boldsymbol{X})^{-1}\boldsymbol{X}\boldsymbol{Y} = \boldsymbol{C}^{-1}\boldsymbol{X}^\top\boldsymbol{Y}$, where $\boldsymbol{C} = \boldsymbol{X}^\top\boldsymbol{X} = (C_{ij})$, $i,j = 1, \dots, p$.

Theory of Ridge Regression Estimation with Applications, First Edition.
A.K. Md. Ehsanes Saleh, Mohammad Arashi, and B.M. Golam Kibria.

Sometimes the method is written as

$$Y = X_1\beta_1 + X_2\beta_2 + \epsilon, \tag{6.3}$$

where $\beta = (\beta_1^\top, \beta_2^\top)^\top$, and $X = (X_1, X_2)$ where X_1 and X_2 are $n \times p_1$ and $n \times p_2$ submatrices, respectively, β_1 of dimension p_1 stands for the main effect, and β_2 of dimension p_2 stands for the interactions; and we like estimate $(\beta_1^\top, \beta_2^\top)$. Here, $p_1 + p_2 = p$.

In this case, we may write

$$\begin{aligned}
\begin{pmatrix} \tilde{\beta}_{1n} \\ \tilde{\beta}_{2n} \end{pmatrix} &= \begin{pmatrix} X_1^\top X_1 & X_1^\top X_2 \\ X_2^\top X_1 & X_2^\top X_2 \end{pmatrix}^{-1} \begin{pmatrix} X_1^\top Y \\ X_2^\top Y \end{pmatrix} \\
&= \begin{pmatrix} C_{11} & C_{12} \\ C_{21} & C_{22} \end{pmatrix}^{-1} \begin{pmatrix} X_1^\top Y \\ X_2^\top Y \end{pmatrix} \\
&= \begin{pmatrix} C^{11} & C^{12} \\ C^{21} & C^{22} \end{pmatrix} \begin{pmatrix} X_1^\top Y \\ X_2^\top Y \end{pmatrix} \\
&= \begin{pmatrix} C^{11}X_1^\top Y + C^{12}X_2^\top Y \\ C^{21}X_1^\top Y + C^{22}X_2^\top Y \end{pmatrix},
\end{aligned} \tag{6.4}$$

where $C^{-1} = (C^{ij})$, $i, j = 1, \dots, p$. Hence

$$\tilde{\beta}_{1n} = C^{11}X_1^\top Y + C^{12}X_2^\top Y$$

$$\tilde{\beta}_{2n} = C^{21}X_1^\top Y + C^{22}X_2^\top Y,$$

where

$$\begin{aligned}
C^{11} &= C_{11\cdot 2}^{-1} = C_{11} - C_{12}C_{22}^{-1}C_{21} \\
C^{12} &= -C_{11\cdot 2}^{-1}C_{12}C_{22}^{-1}, \\
C^{21} &= -C_{22}^{-1}C_{21}C_{11\cdot 2}^{-1} \\
C^{22} &= C_{22\cdot 1}^{-1} = C_{22} - C_{21}C_{11}^{-1}C_{12},
\end{aligned}$$

respectively.

It is well known that $\mathbb{E}(\tilde{\beta}_n) = \beta$ and the covariance matrix of $\tilde{\beta}_n$ is given by

$$\mathrm{Cov}(\tilde{\beta}_n) = \sigma^2 \begin{pmatrix} C_{11} & C_{12} \\ C_{21} & C_{22} \end{pmatrix}^{-1} = \sigma^2 \begin{pmatrix} C^{11} & C^{12} \\ C^{21} & C^{22} \end{pmatrix}. \tag{6.5}$$

Further, an unbiased estimator of σ^2 is given by

$$s_n^2 = \frac{1}{n-p}(Y - X\tilde{\beta}_n)^\top (Y - X\tilde{\beta}_n). \tag{6.6}$$

Using normal theory it may be shown that $\left(\tilde{\boldsymbol{\beta}}_n, (n-p)\frac{s_n^2}{\sigma^2}\right)$ are statistically independent and $\tilde{\boldsymbol{\beta}}_n \sim \mathcal{N}_p(\boldsymbol{\beta}, \sigma^2\boldsymbol{C}^{-1})$ and $(n-p)\frac{s_n^2}{\sigma^2}$ follows a central chi-square distribution with $(n-p)$ degrees of freedom (DF).

The L_2 risk of $\boldsymbol{\beta}_n^*$, any estimator of $\boldsymbol{\beta}$, is defined by

$$R(\boldsymbol{\beta}_n^*) = \mathbb{E}\|\boldsymbol{\beta}_n^* - \boldsymbol{\beta}\|^2.$$

Then, the L_2 risk of $\tilde{\boldsymbol{\beta}}_n$ is given by

$$R(\tilde{\boldsymbol{\beta}}_n) = \sigma^2 \operatorname{tr}(\boldsymbol{C}^{-1}). \tag{6.7}$$

If $0 < \lambda_1 \leq \cdots \leq \lambda_p$ be the eigenvalues of $\boldsymbol{C}$, then we may write

$$R(\tilde{\boldsymbol{\beta}}_n) = \sigma^2 \sum_{i=1}^{p} \frac{1}{\lambda_i} > \sigma^2 \lambda_1^{-1}. \tag{6.8}$$

Similarly, one notes that

$$\begin{aligned}\mathbb{E}(\tilde{\boldsymbol{\beta}}_{1n}) &= [\boldsymbol{C}_{11\cdot2}^{-1}\boldsymbol{X}_1^\top - \boldsymbol{C}_{11\cdot2}^{-1}\boldsymbol{C}_{12}\boldsymbol{C}_{22}^{-1}\boldsymbol{X}_2^\top](\boldsymbol{X}_1\boldsymbol{\beta}_1 + \boldsymbol{X}_2\boldsymbol{\beta}_2)\\ &= (\boldsymbol{C}_{11\cdot2}^{-1}\boldsymbol{C}_{11} - \boldsymbol{C}_{11\cdot2}^{-1}\boldsymbol{C}_{12}\boldsymbol{C}_{22}^{-1}\boldsymbol{C}_{21})\boldsymbol{\beta}_1\\ &\quad + (\boldsymbol{C}_{11\cdot2}^{-1}\boldsymbol{C}_{12} - \boldsymbol{C}_{11\cdot2}^{-1}\boldsymbol{C}_{12}\boldsymbol{C}_{22}^{-1}\boldsymbol{C}_{22})\boldsymbol{\beta}_2\\ &= \boldsymbol{\beta}_1.\end{aligned} \tag{6.9}$$

In a similar manner, $\mathbb{E}(\tilde{\boldsymbol{\beta}}_{2n}) = \boldsymbol{\beta}_2$.

Now, we find the $\text{Cov}(\tilde{\boldsymbol{\beta}}_{1n})$ and $\text{Cov}(\tilde{\boldsymbol{\beta}}_{2n})$ as given below

$$\begin{aligned}\text{Cov}(\tilde{\boldsymbol{\beta}}_{1n}) &= \sigma^2[\boldsymbol{C}_{11\cdot2}^{-1}(\boldsymbol{X}_1^\top - \boldsymbol{C}_{12}\boldsymbol{C}_{22}^{-1}\boldsymbol{X}_2^\top)(\boldsymbol{X}_1 - \boldsymbol{X}_2\boldsymbol{C}_{22}^{-1}\boldsymbol{C}_{21})\boldsymbol{C}_{11\cdot2}^{-1}]\\ &= \sigma^2[\boldsymbol{C}_{11\cdot2}^{-1}(\boldsymbol{C}_{11} - \boldsymbol{C}_{12}\boldsymbol{C}_{22}^{-1}\boldsymbol{C}_{21})\boldsymbol{C}_{11\cdot2}^{-1}]\\ &= \sigma^2\boldsymbol{C}_{11\cdot2}^{-1}\boldsymbol{C}_{11\cdot2}\boldsymbol{C}_{11\cdot2}^{-1}\\ &= \sigma^2\boldsymbol{C}_{11\cdot2}^{-1}.\end{aligned} \tag{6.10}$$

Similarly, we obtain

$$\text{Cov}(\tilde{\boldsymbol{\beta}}_{2n}) = \sigma^2\boldsymbol{C}_{22\cdot1}^{-1}, \quad \text{and} \quad \text{Cov}(\tilde{\boldsymbol{\beta}}_{1n}, \tilde{\boldsymbol{\beta}}_{2n}) = -\sigma^2\boldsymbol{C}_{11\cdot2}^{-1}\boldsymbol{C}_{12}\boldsymbol{C}_{22}^{-1}. \tag{6.11}$$

6.1.2 Test of Hypothesis for the Coefficients Vector

Suppose we want to test the null-hypothesis $\mathcal{H}_o : \boldsymbol{\beta} = \boldsymbol{0}$ vs. $\mathcal{H}_A : \boldsymbol{\beta} \neq \boldsymbol{0}$. Then, we use the test statistic (likelihood ratio test),

$$\mathcal{L}_n = \frac{\tilde{\boldsymbol{\beta}}_n^\top \boldsymbol{C}\tilde{\boldsymbol{\beta}}_n}{p s_n^2}, \tag{6.12}$$

which follows a noncentral F-distribution with $(p, n-p)$ DF and noncentrality parameter, Δ^2, defined by

$$\Delta^2 = \frac{\boldsymbol{\beta}^\top \boldsymbol{C} \boldsymbol{\beta}}{\sigma^2}. \tag{6.13}$$

Similarly, if we want to test the subhypothesis $\mathcal{H}_o : (\boldsymbol{\beta}_1^\top, \boldsymbol{\beta}_2^\top) = (\boldsymbol{\beta}_1^\top, \boldsymbol{0}^\top)$ vs. $\mathcal{H}_A : (\boldsymbol{\beta}_1^\top, \boldsymbol{\beta}_2^\top) \neq (\boldsymbol{\beta}_1^\top, \boldsymbol{0}^\top)$, then one may use the test statistic

$$\mathcal{L}_m^* = \frac{\tilde{\boldsymbol{\beta}}_{2n}^\top \boldsymbol{C}_{22\cdot1} \tilde{\boldsymbol{\beta}}_{2n}}{p_2 s_{2n}^2}, \tag{6.14}$$

where $s_{2n}^2 = m^{-1}(\boldsymbol{Y} - \boldsymbol{X}_1^\top \hat{\boldsymbol{\beta}}_{1n})^\top (\boldsymbol{Y} - \boldsymbol{X}_1^\top \hat{\boldsymbol{\beta}}_{1n})$, $m = n - p_1$ and

$$\begin{aligned} \hat{\boldsymbol{\beta}}_{1n} &= (\boldsymbol{X}_1^\top \boldsymbol{X}_1)^{-1} \boldsymbol{X}_1 \boldsymbol{Y} \\ &= \tilde{\boldsymbol{\beta}}_{1n} + \boldsymbol{C}_{12} \boldsymbol{C}_{22}^{-1} \tilde{\boldsymbol{\beta}}_{2n}. \end{aligned} \tag{6.15}$$

Note that $\mathcal{L}_m^*$ follows a noncentral F-distribution with (p_2, m) DF and noncentrality parameter Δ_2^2 given by

$$\Delta_2^2 = \frac{\boldsymbol{\beta}_2^\top \boldsymbol{C}_{22\cdot1} \boldsymbol{\beta}_2}{\sigma^2}. \tag{6.16}$$

Let $F(\alpha)$ be the upper α-level critical value for the test of $\mathcal{H}_o$, then we reject $\mathcal{H}_o$ whenever $\mathcal{L}_n$ or $\mathcal{L}_m^* > F(\alpha)$; otherwise, we accept $\mathcal{H}_o$.

6.2 Ridge Regression Estimators (RREs)

From Section 6.1.1, we may see that

(i) $\mathbb{E}(\tilde{\boldsymbol{\beta}}_n) = \boldsymbol{\beta}$
(ii) cov-matrix of $\tilde{\boldsymbol{\beta}}_n$ is $\sigma^2 \boldsymbol{C}^{-1}$.

Hence, L_2-risk function of $\tilde{\boldsymbol{\beta}}_n$ is σ^2 tr $\boldsymbol{C}^{-1}$. Further, when $\boldsymbol{\epsilon} \sim \mathcal{N}_n(\boldsymbol{0}, \sigma^2 \boldsymbol{I}_n)$, the variance of $(\tilde{\boldsymbol{\beta}}_n - \boldsymbol{\beta})^\top (\tilde{\boldsymbol{\beta}}_n - \boldsymbol{\beta})$ is $2\sigma^4$ tr $\boldsymbol{C}^{-2}$.

Set $0 \leq \lambda_{\min} = \lambda_1 \leq \cdots \leq \lambda_p = \lambda_{\max}$. Then, it is seen that the lower bound of the L_2 risk and the variance are $\sigma^2/\lambda_{\min}$ and $2\sigma^4/\lambda_{\min}$, respectively. Hence, the shape of the parameter space is such that reasonable data collection may result in a $\boldsymbol{C}$ matrix with one or more small eigenvalues. As a result, L_2 distance of $\tilde{\boldsymbol{\beta}}_n$ to $\boldsymbol{\beta}$ will tend to be large. In particular, coefficients tend to become large in absolute value and may even have wrong signs and be unstable. As a result, such difficulties increase as more prediction vectors deviate from orthogonality. As the design matrix, $\boldsymbol{X}$ deviates from orthogonality, $\lambda_{\min}$ becomes smaller and $\tilde{\boldsymbol{\beta}}_n$ will be farther away from $\boldsymbol{\beta}$. The ridge regression method is a remedy to circumvent these problems with the LSE.

For the model (6.1), Hoerl and Kennard (1970) defined the RRE of $\boldsymbol{\beta}$ as

$$\hat{\boldsymbol{\beta}}_n^{\mathrm{RR}}(k) = (\boldsymbol{C} + k\boldsymbol{I}_p)^{-1} \boldsymbol{X}^\top \boldsymbol{Y}, \quad \boldsymbol{C} = \boldsymbol{X}^\top \boldsymbol{X}. \tag{6.17}$$

The basic idea of this type is from Tikhonov (1963) where the tuning parameter, k, is to be determined from the data $\{(\boldsymbol{x}_i, Y_i) | i = 1, \ldots, n\}$. The expression (6.17) may be written as

$$\hat{\boldsymbol{\beta}}_n^{\mathrm{RR}}(k) = (\boldsymbol{I}_p + k\boldsymbol{C}^{-1})^{-1}\tilde{\boldsymbol{\beta}}_n, \quad \tilde{\boldsymbol{\beta}}_n = \boldsymbol{C}^{-1}\boldsymbol{X}^\top\boldsymbol{Y}. \tag{6.18}$$

If $\boldsymbol{C}$ is invertible, whereas if $\boldsymbol{C}$ is ill-conditioned or near singular, then (6.17) is the appropriate expression for the estimator of $\boldsymbol{\beta}$.

The derivation of (6.17) is very simple. Instead of minimizing the LSE objective function, one may minimize the objective function where $\boldsymbol{\beta}^\top\boldsymbol{\beta}$ is the penalty function

$$(\boldsymbol{Y} - \boldsymbol{X}\boldsymbol{\beta})^\top(\boldsymbol{Y} - \boldsymbol{X}\boldsymbol{\beta}) + k\boldsymbol{\beta}^\top\boldsymbol{\beta} \tag{6.19}$$

with respect to (w.r.t.) $\boldsymbol{\beta}$ yielding the normal equations,

$$(\boldsymbol{X}^\top\boldsymbol{X} + k\boldsymbol{I}_p)\boldsymbol{\beta} = \boldsymbol{X}^\top\boldsymbol{Y} \Rightarrow \hat{\boldsymbol{\beta}}_n^{\mathrm{RR}}(k) = (\boldsymbol{C} + k_{\mathrm{opt}}\boldsymbol{I}_p)^{-1}\boldsymbol{X}^\top\boldsymbol{Y}. \tag{6.20}$$

Here, we have equal weight to the components of $\boldsymbol{\beta}$.

6.3 Bias, MSE, and L_2 Risk of Ridge Regression Estimator

In this section, we consider bias, mean squared error (MSE), and L_2-risk expressions of the RRE.

First, we consider the equal weight RRE, $\hat{\boldsymbol{\beta}}_n^{\mathrm{RR}}(k_{\mathrm{opt}}) = (\boldsymbol{C} + k_{\mathrm{opt}}\boldsymbol{I}_p)^{-1}\boldsymbol{X}^\top\boldsymbol{Y}$. The bias, MSE, and L_2 risk are obtained as

$$\begin{aligned} b(\hat{\boldsymbol{\beta}}_n^{\mathrm{RR}}(k)) &= \mathbb{E}[\hat{\boldsymbol{\beta}}_n^{\mathrm{RR}}(k)] - \boldsymbol{\beta} \\ &= (\boldsymbol{C} + k\boldsymbol{I}_p)^{-1}\boldsymbol{C}\boldsymbol{\beta} - \boldsymbol{\beta} \\ &= -k(\boldsymbol{C} + k\boldsymbol{I}_p)^{-1}\boldsymbol{\beta}, \end{aligned} \tag{6.21}$$

$$\begin{aligned} \mathrm{MSE}(\hat{\boldsymbol{\beta}}_n^{\mathrm{RR}}(k)) &= \mathrm{Cov}(\hat{\boldsymbol{\beta}}_n^{\mathrm{RR}}(k)) + [b(\hat{\boldsymbol{\beta}}_n^{\mathrm{RR}}(k))][b(\hat{\boldsymbol{\beta}}_n^{\mathrm{RR}}(k))]^\top \\ &= \sigma^2(\boldsymbol{C} + k\boldsymbol{I}_p)^{-1}\boldsymbol{C}(\boldsymbol{C} + k\boldsymbol{I}_p)^{-1} \\ &\quad + k^2(\boldsymbol{C} + k\boldsymbol{I}_p)^{-1}\boldsymbol{\beta}\boldsymbol{\beta}^\top(\boldsymbol{C} + k\boldsymbol{I}_p)^{-1} \end{aligned} \tag{6.22}$$

$$R(\hat{\boldsymbol{\beta}}_n^{\mathrm{RR}}(k)) = \sigma^2 \operatorname{tr}((\boldsymbol{C} + k\boldsymbol{I}_p)^{-2}\boldsymbol{C}) + k^2\boldsymbol{\beta}^\top(\boldsymbol{C} + k\boldsymbol{I}_p)^{-2}\boldsymbol{\beta}, \tag{6.23}$$

respectively.

One may write

$$\begin{aligned} \hat{\boldsymbol{\beta}}_n^{\mathrm{RR}}(k) &= \boldsymbol{W}\boldsymbol{X}^\top\boldsymbol{Y}, \quad \boldsymbol{W} = (\boldsymbol{C} + k\boldsymbol{I}_p)^{-1} \\ &= \boldsymbol{Z}\tilde{\boldsymbol{\beta}}_n, \quad \boldsymbol{Z} = (\boldsymbol{I}_p + k\boldsymbol{C}^{-1})^{-1} = \boldsymbol{I}_p - k\boldsymbol{W}. \end{aligned} \tag{6.24}$$

Let $\lambda_i(\boldsymbol{W})$ and $\lambda_i(\boldsymbol{Z})$ be the ith eigenvalue of $\boldsymbol{W}$ and $\boldsymbol{Z}$, respectively. Then

$$\begin{aligned}\lambda_i(\boldsymbol{W}) &= (\lambda_i + k)^{-1},\\ \lambda_i(\boldsymbol{Z}) &= \lambda_i(\lambda_i + k)^{-1}\end{aligned} \tag{6.25}$$

$i = 1, \dots, p$.

Note that

(i) $\hat{\boldsymbol{\beta}}_n^{\text{RR}}(k)$, for $k \neq 0$, is shorter than $\tilde{\boldsymbol{\beta}}_n$, that is

$$[\hat{\boldsymbol{\beta}}_n^{\text{RR}}(k)]^\top[\hat{\boldsymbol{\beta}}_n^{\text{RR}}(k)] < \tilde{\boldsymbol{\beta}}_n^\top\tilde{\boldsymbol{\beta}} \tag{6.26}$$

(ii)
$$[\hat{\boldsymbol{\beta}}_n^{\text{RR}}(k)]^\top[\hat{\boldsymbol{\beta}}_n^{\text{RR}}(k)] \leq \xi_{\max}(\boldsymbol{Z})\tilde{\boldsymbol{\beta}}_n^\top\tilde{\boldsymbol{\beta}}, \tag{6.27}$$

where $\xi_{\max}(\boldsymbol{Z}) = \lambda_p(\lambda_p + k)^{-1}$.

For the estimate of $\hat{\boldsymbol{\beta}}_n^{\text{RR}}(k)$, the residual sum of squares (RSS) is

$$\begin{aligned}\text{RSS} &= [\boldsymbol{Y} - \boldsymbol{X}\hat{\boldsymbol{\beta}}_n^{\text{RR}}(k)]^\top[\boldsymbol{Y} - \boldsymbol{X}\hat{\boldsymbol{\beta}}_n^{\text{RR}}(k)]\\ &= \boldsymbol{Y}^\top\boldsymbol{Y} - 2[\hat{\boldsymbol{\beta}}_n^{\text{RR}}(k)]^\top\boldsymbol{X}^\top\boldsymbol{Y} + [\hat{\boldsymbol{\beta}}_n^{\text{RR}}(k)]^\top\boldsymbol{X}^\top\boldsymbol{X}[\hat{\boldsymbol{\beta}}_n^{\text{RR}}(k)].\end{aligned} \tag{6.28}$$

The expression shows that the RSS in (6.28) is equal to the total sum of squares due to $\hat{\boldsymbol{\beta}}_n^{\text{RR}}(k)$ with a modification upon the squared length of $\hat{\boldsymbol{\beta}}_n^{\text{RR}}(k)$.

Now, we explain the properties of L_2 risk of the RSS given in (6.28), which may be written as

$$\begin{aligned}R(\hat{\boldsymbol{\beta}}_n^{\text{RR}}(k)) &= \sigma^2\sum_{i=1}^{p}\frac{\lambda_i}{(\lambda_i + k)^2} + k^2\boldsymbol{\beta}^\top(\boldsymbol{C} + k\boldsymbol{I}_p)^{-2}\boldsymbol{\beta}\\ &= \text{Var}(k) + b(k), \quad \text{say.}\end{aligned} \tag{6.29}$$

Figure 6.1, the graph of the L_2 risk, depicts the two components as a function of k. It shows the relationships between variances, the quadratic bias and k, the tuning parameter.

Total variance decreases as k increases, while quadratic bias increases with k. As indicated by the dotted line graph representing RRE L_2 risk, there are values of k for which $R(\hat{\boldsymbol{\beta}}_n^{\text{RR}}(k))$ is less than or equal to $R(\tilde{\boldsymbol{\beta}}_n)$.

It may be noted that $\text{Var}(k)$ is a monotonic decreasing function of k_{opt}, which $b(k)$ is a monotonic increasing function of k. Now, find the derivatives of these two functions at the origin ($k = 0$) as

$$\lim_{k_{\text{opt}}\to 0^+}\frac{\text{d Var}(k)}{\text{d}k} = -2\sigma^2\sum_{j=1}^{p}\frac{1}{\lambda_j} \tag{6.30}$$

$$\lim_{k\to 0^+}\frac{\text{d}b(k)}{\text{d}k} = 0. \tag{6.31}$$

Hence, $\text{Var}(k)$ has a negative derivative, which tends to $-2p\sigma^2$ as $k \to 0^+$ for an orthogonal $\boldsymbol{X}^\top\boldsymbol{X}$ and approaches ∞ as $\boldsymbol{X}^\top\boldsymbol{X}$ becomes ill-conditioned and

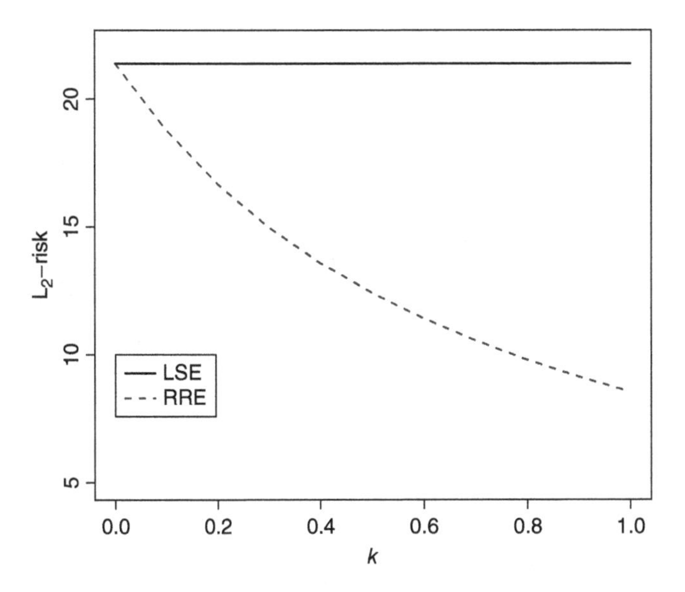

Figure 6.1 Graph of L_2 risk of RRE in case of $p = 10$.

$\lambda_{\min} \to 0$, while as $k \to 0^+$, $b(k)$ is flat and zero at $k = 0$. These properties indicate that if we choose $k > 0$ appropriately, our estimator will inherit a little bias and substantially reduce variances, and thereby the L_2-risk will decrease to improve the prediction.

We now prove the existence of k to validate the commentaries put forward by the following theorems.

Theorem 6.1 *Under the assumptions of this section,*

(i) $\mathrm{Var}(k)$ *is a continuous, monotonically decreasing function of* k, $\frac{\mathrm{dVar}(k)}{\mathrm{d}k}$ *approaches* $-\infty$ *as* $k \to 0^+$ *and* $\lambda_{\min} \to 0$. *This implies that the matrix* $\boldsymbol{X}^\top\boldsymbol{X}$ *approaches singularity.*
(ii) The quadratic bias, $b(k)$, *is a continuous, monotonically increasing function of* k.
(iii) $b(k)$ *approaches* $\boldsymbol{\beta}^\top\boldsymbol{\beta}$ *as an upper limit.*
(iv) As $k \to 0^+$, $\frac{\mathrm{d}b(k)}{\mathrm{d}k} \to 0$.

Proof: Here we provide the proofs of (ii)–(iv).

(ii) Writing $\boldsymbol{X}^\top\boldsymbol{X} = \boldsymbol{P}^\top\boldsymbol{\Lambda}\boldsymbol{P}$, with $\boldsymbol{P}$ orthogonal, we get

$$b(k) = k^2 \sum_{i=1}^{p} \frac{\alpha_i^2}{(\lambda_i + k)^2}, \quad \boldsymbol{\alpha} = \boldsymbol{P}\boldsymbol{\beta}, \tag{6.32}$$

where α_i is the ith element of $\boldsymbol{\alpha}$. Since $\lambda_i > 0$ and $k \geq 0$ for each of the components in (6.32), there are no singularities in the sum. Clearly $b(0) = 0$. Then, $b(k)$ is a continuous function of $k \geq 0$. Thus, one may write

$$b(k) = \sum_{i=1}^{p} \frac{\alpha_i^2}{\left[1 + \frac{\lambda_i}{k}\right]^2}. \tag{6.33}$$

Since for every $i = 1, \ldots, p$, $\lambda_i > 0$, λ_i/k are monotone decreasing in k and each term is monotone increasing, the sum is therefore a monotone increasing function of k.

(iii) The result follows from the fact that

$$\lim_{k\to\infty} b(k) = \sum_{i=1}^{p} \alpha_i^2 = \boldsymbol{\alpha}^\top\boldsymbol{\alpha} = \boldsymbol{\beta}^\top\boldsymbol{P}\boldsymbol{P}^\top\boldsymbol{\beta} = \boldsymbol{\beta}^\top\boldsymbol{\beta}.$$

(iv) Taking derivatives from $b(k)$ w.r.t. k gives

$$\frac{\mathrm{d}b(k)}{\mathrm{d}k} = 2k\sum_{i=1}^{p} \frac{\lambda_i\alpha_i^2}{(\lambda_i + k)^2}.$$

Each component of the sum is a continuous function and the limit of each term is zero as $k \to 0^+$.

□

A direct consequence of the given result is the following existence theorem.

Theorem 6.2 *There always exists a $k > 0$ such that $R(\hat{\boldsymbol{\beta}}_n^{\mathrm{RR}}(k)) \leq R(\tilde{\boldsymbol{\beta}}_n)$.*

Proof: Consider the derivative of the L_2 risk of $\hat{\boldsymbol{\beta}}_n^{\mathrm{RR}}(k)$ w.r.t. k

$$\begin{aligned}\frac{\mathrm{d}R(\hat{\boldsymbol{\beta}}_n^{\mathrm{RR}}(k))}{\mathrm{d}k} &= \frac{\mathrm{dVar}(k)}{\mathrm{d}k} + \frac{\mathrm{d}b(k)}{\mathrm{d}k} \\ &= -2\sigma^2\sum_{i=1}^{p}\frac{\lambda_i}{(\lambda_i+k)^3} + k\sum_{i=1}^{p}\frac{\lambda_i\alpha_i^2}{(\lambda_i+k)^3}\end{aligned} \tag{6.34}$$

where $\mathrm{Var}(0) = \sigma^2\sum_{i=1}^{p}\lambda_i^{-1}$ and $b(0) = 0$. In Theorem 6.1, we showed $\mathrm{Var}(k)$ and $b(k)$ are monotonically decreasing and increasing, respectively. Their first derivatives are always nonpositive and nonnegative, respectively. We have shown that there exists always a k for which the abovementioned derivative is less than zero. This condition holds with a value of k as

$$k < \frac{\sigma^2}{\alpha_{\max}^2}, \quad \alpha_{\max} = \max_{1\leq j\leq p}(\alpha_j)$$

by writing the L_2-risk difference as

$$\sum_{i=1}^{p} \frac{\lambda_i}{(\lambda_i + k)^3}(\alpha_i^2 - \sigma^2) < 0, \text{ whenever } k < \frac{\sigma^2}{\alpha_{\max}^2}. \qquad \square$$

6.4 Determination of the Tuning Parameters

Suppose $\boldsymbol{C}$ is nonsingular, then the LSE of $\boldsymbol{\beta}$ is $\boldsymbol{C}^{-1}\boldsymbol{X}^\top\boldsymbol{Y}$. Let $\hat{\boldsymbol{\beta}}_n^{\mathrm{RR}}(k)$ be the equal weight RRE of $\boldsymbol{\beta}$ deleting the kth data point $(\boldsymbol{x}_k^\top, Y_k)$. If k_{opt} is chosen properly, then the kth component of $\hat{\boldsymbol{\beta}}_n^{\mathrm{RR}}(k)$ predicts Y_k well. The generalized cross-validation (GCV) is defined as the weight average of predicted square errors

$$\mathrm{Var}(k) = \frac{1}{n}\sum_{k=1}^{n}\left([\boldsymbol{x}_k^\top \hat{\boldsymbol{\beta}}_n^{\mathrm{RR}}(k)] - y_k\right)^2 w_k(k), \tag{6.35}$$

where $w_k(k) = (1 - a_{kk}(k)) / \left[1 - \frac{1}{n}\,\mathrm{tr}\boldsymbol{A}(k)\right]$ and $a_{kk}(k)$ is the kth diagonal element of $\boldsymbol{A}(k) = \boldsymbol{X}(\boldsymbol{X}^\top\boldsymbol{X} + k\boldsymbol{I}_p)^{-1}\boldsymbol{X}^\top$. See Golub et al. (1979) for more details.

A computationally efficient version of the GCV function of the RRE is

$$\mathrm{Var}(k) = \frac{1}{n}\frac{\|(\boldsymbol{I}_n - \boldsymbol{A}(k))\boldsymbol{Y}\|^2}{\left[\frac{1}{n}\,\mathrm{tr}(\boldsymbol{I}_n - \boldsymbol{A}(k))\right]^2}. \tag{6.36}$$

The GCV estimator of k is then given by

$$k_{\mathrm{opt}} = \mathrm{argmin}_{k\geq 0}\,\mathrm{Var}(k). \tag{6.37}$$

The GCV theorem (Golub et al. 1979) guarantees the asymptotic efficiency of the GCV estimator under $p < n$ (and also $p > n$) setups.

We later see that the statistical estimator of k_{opt} is given by $ps_n^2/\tilde{\boldsymbol{\beta}}_n^\top\tilde{\boldsymbol{\beta}}_n$, where $s_n^2 = (n-p)^{-1}(\boldsymbol{Y} - \boldsymbol{X}\tilde{\boldsymbol{\beta}}_n)^\top(\boldsymbol{Y} - \boldsymbol{X}\tilde{\boldsymbol{\beta}}_n)$.

6.5 Ridge Trace

Ridge regression has two interpretations of its characteristics. The first is the *ridge trace*. It is a two-dimensional plot of $\tilde{\beta}_{jn}(k_{\mathrm{opt}})$ and the RSS given by

$$\mathrm{RSS} = \left[\boldsymbol{Y} - \boldsymbol{X}\hat{\boldsymbol{\beta}}_n^{\mathrm{RR}}(k_{\mathrm{opt}})\right]^\top \left[\boldsymbol{Y} - \boldsymbol{X}\hat{\boldsymbol{\beta}}_n^{\mathrm{RR}}(k_{\mathrm{opt}})\right],$$

for a number of values of $k_{\mathrm{opt}} \in [0, 1]$. The trace serves to depict the complex relationships that exist between nonorthogonal prediction vectors and the effect of these interrelationships on the estimation of $\boldsymbol{\beta}$. The second is the way to estimate k_{opt} that gives a better estimator of $\boldsymbol{\beta}$ by damping the effect of the lower bound mentioned earlier. Ridge trace is a diagnostic test that gives a readily interpretable picture of the effects of nonorthogonality and may guide to a better point estimate.

Now, we discuss how ridge regression can be used by statisticians with the gasoline mileage data from Montgomery et al. (2012). The data contains the following variables: y = miles/gallon, x_1 = displacement (cubic in.), x_2 = horsepower (ft-lb), x_3 = torque (ft-lb), x_4 = compression ratio, x_5 = rear axle ratio, x_6 = carburetor (barrels), x_7 = number of transmission speeds, x_8 = overall length, $x_9 = E$ = width (in.), x_{10} = weight (lb), x_{11} = type of transmission.

Table 6.1 is in correlation format of $\boldsymbol{X}^\top\boldsymbol{X}$ and $\boldsymbol{X}^\top\boldsymbol{Y}$, where $\boldsymbol{X}$ is centered.

We may find that the eigenvalues of the matrix $\boldsymbol{X}^\top\boldsymbol{X}$ are given by $\lambda_1 = 223.33$, $\lambda_2 = 40.70$, $\lambda_3 = 22.43$, $\lambda_4 = 16.73$, $\lambda_5 = 6.13$, $\lambda_6 - 4.11$, $\lambda_7 = 2.76$, $\lambda_8 = 1.45$, $\lambda_9 = 0.96$, $\lambda_{10} = 0.24$, and $\lambda_{11} = 0.11$.

The condition number is $\kappa = 223.338/0.110 = 2030.345$, which indicates serious multicollinearity. They are nonzero real numbers and $\sum_{j=1}^{11} \lambda_j^{-1} = 15.88$, which is 1.6 times more than 10 in the orthogonal case. Thus, it shows that the expected distance of the BLUE (best linear unbiased estimator), $\tilde{\beta}_n$ from β is $15.88 \times \sigma^2$. Thus, the parameter space of β is 11-dimensional, but most of the variations are due to the largest two eigenvalues.

Table 6.1 Correlation coefficients for gasoline mileage data.

	y	x_1	x_2	x_3	x_4	x_5	x_6	x_7	x_8	x_9	x_{10}	x_{11}
y	1.0	−0.8	−0.7	−0.8	0.4	0.6	−0.4	0.7	−0.7	−0.7	−0.8	−0.7
x_1	−0.8	1.0	0.9	0.9	−0.3	−0.6	0.6	−0.7	0.8	0.7	0.9	0.8
x_2	−0.7	0.9	1.0	0.9	−0.2	−0.5	0.7	−0.6	0.8	0.7	0.8	0.7
x_3	−0.8	0.9	0.9	1.0	−0.3	−0.6	0.6	−0.7	0.8	0.7	0.9	0.8
x_4	0.4	−0.3	−0.2	−0.3	1.0	0.4	0.0	0.5	−0.3	−0.3	−0.3	−0.4
x_5	0.6	−0.6	−0.5	−0.6	0.4	1.0	−0.2	0.8	−0.5	−0.4	−0.5	−0.7
x_6	−0.4	0.6	0.7	0.6	0.0	−0.2	1.0	−0.2	0.4	0.3	0.5	0.3
x_7	0.7	−0.7	−0.6	−0.7	0.5	0.8	−0.2	1.0	−0.6	−0.6	−0.7	−0.8
x_8	−0.7	0.8	0.8	0.8	−0.3	−0.5	0.4	−0.6	1.0	0.8	0.9	0.6
x_9	−0.7	0.7	0.7	0.7	−0.3	−0.4	0.3	−0.6	0.8	1.0	0.8	0.6
x_{10}	−0.8	0.9	0.8	0.9	−0.3	−0.5	0.5	−0.7	0.9	0.8	1.0	0.7
x_{11}	−0.7	0.8	0.7	0.8	−0.4	−0.7	0.3	−0.8	0.6	0.6	0.7	1.0

Variance inflation factor (*VIF*): First, we want to find VIF, which is defined as

$$\mathrm{VIF}_j = \frac{1}{1 - R_j^2},$$

where R_j^2 is the coefficient of determination obtained when $\boldsymbol{x}_j$ is regressed on the remaining $p - 1$ regressors. A VIF_j greater than 5 or 10 indicates that the associated regression coefficients are poor estimates because of multicollinearity.

The VIF of the variables in this data set are as follows: $\mathrm{VIF}_1 = 119.48$, $\mathrm{VIF}_2 = 42.80$, $\mathrm{VIF}_3 = 149.23$, $\mathrm{VIF}_4 = 2.06$, $\mathrm{VIF}_5 = 7.73$, $\mathrm{VIF}_6 = 5.32$, $\mathrm{VIF}_7 = 11.76$, $\mathrm{VIF}_8 = 20.92$, $\mathrm{VIF}_9 = 9.40$, $\mathrm{VIF}_{10} = 85.74$, and $\mathrm{VIF}_{11} = 5.14$. These VIFs certainly indicate severe multicollinearity in the data. It is also evident from the correlation matrix in Table 6.1. This is the most appropriate data set to analyze ridge regression.

On the other hand, one may look at the ridge trace and discover many finer details of each factor and the optimal value of k_{opt}, the tuning parameter of the ridge regression errors. The ridge trace gives a two-dimensional portrayal of the effect of factor correlations and making possible assessments that cannot be made even if all regressions are computed. Figure 6.2 depicts the analysis using the ridge trace.

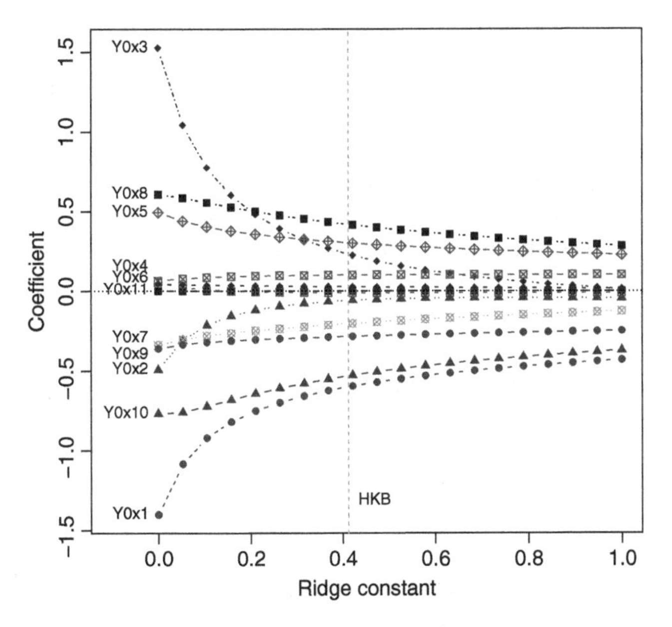

Figure 6.2 Ridge trace for gasoline mileage data.

Notice that the absolute value of each factor tends to the LSE as k_{opt} goes to zero. From Figure 6.2 we can see that reasonable coefficient stability is achieved in the range $4.0 < k_{\text{opt}} < 4.2$. For more comments, see Hoerl and Kennard (1970).

6.6 Degrees of Freedom of RRE

RRE constraints the flexibility of the model by adding a quadratic penalty term to the least squares objective function (see Section 6.2).

When $k_{\text{opt}} = 0$, the RRE equals the LSE. The larger the value of k_{opt}, the greater the penalty for having large coefficients.

In our study, we use L_2-risk expression to assess the overall quality of an estimator. From (6.7) we see that the L_2 risk of the LSE, $\tilde{\boldsymbol{\beta}}_n$ is $\sigma^2 \operatorname{tr} \boldsymbol{C}^{-1}$ and from (6.23) we find the L_2 risk of RRE is

$$\sigma^2((\boldsymbol{C} + k_{\text{opt}}\boldsymbol{I}_p)^{-1}\boldsymbol{C}) + k_{\text{opt}}^2 \boldsymbol{\beta}^\top (\boldsymbol{C} + k_{\text{opt}}\boldsymbol{I}_p)^{-1}\boldsymbol{\beta}.$$

These expressions give us the distance of LSE and RRE from the true value of $\boldsymbol{\beta}$, respectively. LSE is a BLUE of $\boldsymbol{\beta}$. But RRE is a biased estimator of $\boldsymbol{\beta}$. It modifies the LSE by introducing a little bias to improve its performance. The RRE existence in Theorem 6.2 shows that for $0 < k_{\text{opt}} < \sigma^2/\alpha_{\max}^2$, $R(\tilde{\boldsymbol{\beta}}_n) \geq R(\hat{\boldsymbol{\beta}}_n^{\text{RR}}(k_{\text{opt}}))$. Thus, for an appropriate choice of k_{opt}, RRE is closer to $\boldsymbol{\beta}$ than to the LSE. In this sense, RRE is more reliable than the LSE.

On top of the abovementioned property, RRE improves the model performance over the LSE and variable subset selection. RRE also reduces the model DF For example, LSE has p parameters and therefore uses p DF.

To see how RRE reduces the number of parameters, we define linear smoother given by

$$\hat{\boldsymbol{Y}} = \boldsymbol{S}\boldsymbol{Y}, \tag{6.38}$$

where the linear operator $\boldsymbol{S}(n \times n)$ is a smoother matrix and $\hat{\boldsymbol{Y}}$ contains the predicted value of $\boldsymbol{Y}$. The DF of $\boldsymbol{S}$ is given by the trace of the smoother matrix, $\operatorname{tr}(\boldsymbol{S})$. The LSE and RRE are both linear smoother, since the predicted values of either estimators are given by the product

$$\hat{\boldsymbol{Y}} = \boldsymbol{H}_{k_{\text{opt}}}\boldsymbol{Y}, \tag{6.39}$$

where $\boldsymbol{H}_{k_{\text{opt}}}$ is the hat matrix of the RRE. Let $\boldsymbol{H}_{k_{\text{opt}}} = \boldsymbol{H}_{\text{o}}$ be the hat matrix of the LSE, then $\operatorname{tr}(\boldsymbol{H}_{\text{o}}) = \operatorname{tr}(\boldsymbol{X}(\boldsymbol{X}^\top\boldsymbol{X})^{-1}\boldsymbol{X}^\top) = p$ is the DF of LSE.

To find the DF of RRE, let $\boldsymbol{\Lambda} = \operatorname{Diag}(\lambda_1, \ldots, \lambda_p)$, $\lambda_j > 0$, $j = 1, \ldots, p$ be the eigenvalue matrix corresponding to $\boldsymbol{X}^\top\boldsymbol{X}$. We consider the singular value decomposition (SVD) of $\boldsymbol{X} = \boldsymbol{U}\boldsymbol{\Lambda}\boldsymbol{V}^\top$, where $\boldsymbol{U}(n \times p)$ and $\boldsymbol{V}(p \times p)$ are column orthogonal matrices; then

$$\begin{aligned}\mathrm{DF}(k_{\mathrm{opt}}) &= \mathrm{tr}[\boldsymbol{X}(\boldsymbol{X}^\top\boldsymbol{X} + k_{\mathrm{opt}}\boldsymbol{I}_p)^{-1}\boldsymbol{X}^\top] \\ &= \mathrm{tr}[\boldsymbol{U}\boldsymbol{\Lambda}\boldsymbol{V}^\top(\boldsymbol{V}\boldsymbol{\Lambda}^2\boldsymbol{V}^\top + k_{\mathrm{opt}}\boldsymbol{V}\boldsymbol{V}^\top)^{-1}\boldsymbol{V}\boldsymbol{\Lambda}\boldsymbol{U}^\top] \\ &= \mathrm{tr}[\boldsymbol{U}\boldsymbol{\Lambda}\boldsymbol{V}^\top[\boldsymbol{V}(\boldsymbol{\Lambda}^2 + k_{\mathrm{opt}}\boldsymbol{I}_p)\boldsymbol{V}^\top]^{-1}\boldsymbol{V}\boldsymbol{\Lambda}\boldsymbol{U}^\top] \\ &= \mathrm{tr}[\boldsymbol{U}\boldsymbol{\Lambda}(\boldsymbol{\Lambda}^2 + k_{\mathrm{opt}}\boldsymbol{I}_p)^{-1}\boldsymbol{\Lambda}\boldsymbol{U}^\top] \\ &= \mathrm{tr}[\boldsymbol{\Lambda}^2(\boldsymbol{\Lambda}^2 + k_{\mathrm{opt}}\boldsymbol{I}_p)^{-1}] \\ &= \sum_{j=1}^{p} \frac{\lambda_j^2}{\lambda_j^2 + k_{\mathrm{opt}}},\end{aligned} \tag{6.40}$$

where $\frac{\lambda_j^2}{\lambda_j^2+k_{\mathrm{opt}}}$ are called the shrinkage fractions. When k_{opt} is positive, the shrinkage fractions are all less than 1 and, hence, $\mathrm{tr}(\boldsymbol{H}_{k_{\mathrm{opt}}}) < p$. Thus, the effective number of parameters in a ridge regression model is less than the actual number of predictors. The variable subset selection method explicitly drops the variables from the model, while RRE reduces the effects of the unwanted variables without dropping them.

A simple way of finding a value of k_{opt} is to set the equation $\mathrm{DF}(k_{\mathrm{opt}}) = \nu$ and solve for ν. It is easy to see that

$$\mathrm{DF}(k_{\mathrm{opt}}) = \sum_{j=1}^{p} \frac{\lambda_j}{\lambda_j + k_{\mathrm{opt}}} \le \frac{\lambda_1 p}{\lambda_1 + k_{\mathrm{opt}}}. \tag{6.41}$$

Then, solve $\frac{\lambda_1 p}{\lambda_1+k_{\mathrm{opt}}} = \nu$. Hence, an optimum value of k_{opt} falls in the interval

$$\left[0, \frac{\lambda_1(p-\nu)}{\nu}\right]$$

such that $\mathrm{DF}(k_{\mathrm{opt}}) = \nu$.

6.7 Generalized Ridge Regression Estimators

In Sections 6.2 and 6.3, we discussed ordinary RREs and the associated bias and L_2-risk properties. In this RRE, all coefficients are equally weighted; but in reality, the coefficients deserve unequal weight. To achieve this goal, we define the generalized ridge regression estimator (GRRE) of $\boldsymbol{\beta}$ as

$$\hat{\boldsymbol{\beta}}_n^{\mathrm{GRR}}(\boldsymbol{K}) = (\boldsymbol{C} + \boldsymbol{K})^{-1}\boldsymbol{X}^\top\boldsymbol{Y}, \quad \boldsymbol{C} = \boldsymbol{X}^\top\boldsymbol{X}, \tag{6.42}$$

where $\boldsymbol{K} = \mathrm{Diag}(k_1, \ldots, k_p)$, $k_i \ge 0$, $i = 1, \ldots, p$. If $\boldsymbol{K} = k_{\mathrm{opt}}\boldsymbol{I}_p$, we get (6.17). One may derive (6.38) by minimizing the objective function with $\boldsymbol{\beta}^\top\boldsymbol{K}\boldsymbol{\beta}$ as the penalty function

$$(\boldsymbol{Y} - \boldsymbol{X}\boldsymbol{\beta})^\top(\boldsymbol{Y} - \boldsymbol{X}\boldsymbol{\beta}) + \boldsymbol{\beta}^\top\boldsymbol{K}\boldsymbol{\beta} \tag{6.43}$$

w.r.t. $\boldsymbol{\beta}$ to obtain the normal equations

$$(\boldsymbol{X}^\top \boldsymbol{X} + \boldsymbol{K})\boldsymbol{\beta} = \boldsymbol{X}^\top \boldsymbol{Y} \Rightarrow \hat{\boldsymbol{\beta}}_n^{\text{GRR}}(\boldsymbol{K}) = (\boldsymbol{C} + \boldsymbol{K})^{-1}\boldsymbol{X}^\top \boldsymbol{Y}.$$

Here, we have put unequal weight on the components of $\boldsymbol{\beta}$.

The bias, MSE, and L_2-risk expressions of the GRRE given by

$$\begin{aligned} b(\hat{\boldsymbol{\beta}}_n^{\text{GRR}}(\boldsymbol{K})) &= (\boldsymbol{C} + \boldsymbol{K})^{-1}\boldsymbol{C}\boldsymbol{\beta} - \boldsymbol{\beta} \\ &= -(\boldsymbol{C} + \boldsymbol{K})^{-1}\boldsymbol{K}\boldsymbol{\beta}, \end{aligned} \tag{6.44}$$

$$\begin{aligned} \text{MSE}(\hat{\boldsymbol{\beta}}_n^{\text{GRR}}(\boldsymbol{K})) &= \sigma^2(\boldsymbol{C} + \boldsymbol{K})^{-1}\boldsymbol{C}(\boldsymbol{C} + \boldsymbol{K})^{-1} \\ &\quad + (\boldsymbol{C} + \boldsymbol{K})^{-1}\boldsymbol{K}\boldsymbol{\beta}\boldsymbol{\beta}^\top\boldsymbol{K}(\boldsymbol{C} + \boldsymbol{K})^{-1}, \end{aligned} \tag{6.45}$$

$$R(\hat{\boldsymbol{\beta}}_n^{\text{GRR}}(\boldsymbol{K})) = \sigma^2 \operatorname{tr}[(\boldsymbol{C} + \boldsymbol{K})^{-2}\boldsymbol{C}] + \boldsymbol{\beta}^\top\boldsymbol{K}(\boldsymbol{C} + \boldsymbol{K})^{-2}\boldsymbol{K}\boldsymbol{\beta}, \tag{6.46}$$

respectively.

In the next section, we show the application of the GRRE to obtain the adaptive RRE.

6.8 LASSO and Adaptive Ridge Regression Estimators

If the L_2 norm in the penalty function of the LSE objective function

$$(\boldsymbol{Y} - \boldsymbol{X}\boldsymbol{\beta})^\top(\boldsymbol{Y} - \boldsymbol{X}\boldsymbol{\beta}) + k\boldsymbol{\beta}^\top\boldsymbol{\beta}$$

is replaced by the L_1 norm $\mathbf{1}_p^\top||\boldsymbol{\beta}|| = |\beta_1| + \cdots + |\beta_p|$, the resultant estimator is

$$\tilde{\boldsymbol{\beta}}_n^{\text{L}}(\lambda) = \operatorname{argmin}_{\boldsymbol{\beta}}\{||\boldsymbol{Y} - \boldsymbol{X}\boldsymbol{\beta}||^2 + \lambda\mathbf{1}_p^\top||\boldsymbol{\beta}||\}. \tag{6.47}$$

This estimator makes some coefficients zero, making the LASSO estimator different from RRE where most of coefficients become small but nonzero. Thus, LASSO simultaneously selects and estimates at the same time.

The LASSO estimator may be written explicitly as

$$\tilde{\boldsymbol{\beta}}_n^{\text{L}}(\lambda) = (\operatorname{sgn}(\tilde{\beta}_{jn})\left(|\tilde{\beta}_{jn}| - \lambda\sigma\sqrt{C^{jj}}\right)I\left(|\tilde{\beta}_{jn}| > \lambda\sigma\sqrt{C^{jj}}\right)|j = 1, \ldots, p)^\top, \tag{6.48}$$

where C^{jj} is the jth diagonal element of $\boldsymbol{C}^{-1}$ and $\tilde{\boldsymbol{\beta}}_n = (\tilde{\beta}_{1n}, \ldots, \tilde{\beta}_{pn})^\top$.

Now, if we consider the estimation of $\boldsymbol{K}$ in the GRRE, it may be easy to see that if $0 < k_j < \frac{\sigma^2}{\alpha_j}$, $j = 1, \ldots, p$, $\boldsymbol{\alpha} = \boldsymbol{P}\boldsymbol{\beta}$, where $\boldsymbol{P}$ is an orthogonal matrix; see Hoerl and Kennard (1970).

To avoid simultaneous estimation of the weights $(k_1, \ldots, k_p)$, we assume that λ in (6.47) is equal to the harmonic mean of $(k_1, \ldots, k_p)$, i.e.

$$\begin{aligned} k_{\text{HM}} &= \frac{p}{k_1^{-1} + \cdots + k_p^{-1}} \quad \text{or} \quad p(\mathbf{1}_p^\top\boldsymbol{K}^{-1}\mathbf{1}_p)^{-1} = k_{\text{HM}} \\ k_j &> 0, \quad j = 1, \ldots, p. \end{aligned} \tag{6.49}$$

The value of $\lambda = k_{\text{HM}}$ controls the global complexity. This constraint is the link between the p tuning parameters of $\hat{\boldsymbol{\beta}}_n^{\text{GRR}}(\boldsymbol{K})$ and tuning parameter k_{opt} of $\hat{\boldsymbol{\beta}}_n^{\text{RR}}(k_{\text{opt}})$.

The adaptive generalized ridge regression estimator (AGRRE) is obtained by minimizing the following objective function:

$$(\boldsymbol{Y} - \boldsymbol{X}\boldsymbol{\beta})^{\top}(\boldsymbol{Y} - \boldsymbol{X}\boldsymbol{\beta}) + \boldsymbol{\beta}^{\top}\boldsymbol{K}\boldsymbol{\beta} + \gamma[k_{\text{HM}}(\boldsymbol{1}_p^{\top}\boldsymbol{K}^{-1}\boldsymbol{1}_p) - p] + \boldsymbol{\eta}^{\top}\boldsymbol{K}\boldsymbol{1}_p, \tag{6.50}$$

which is the Lagrangian form of the objective function

$$(\boldsymbol{Y} - \boldsymbol{X}\boldsymbol{\beta})^{\top}(\boldsymbol{Y} - \boldsymbol{X}\boldsymbol{\beta}) + \boldsymbol{\beta}^{\top}\boldsymbol{K}\boldsymbol{\beta} \tag{6.51}$$

and γ and $\boldsymbol{\eta}$ are the Lagrangian for (6.50).

Following Avalos et al. (2007), a necessary condition for the optimality is obtained by deriving the Lagrangian w.r.t. $\boldsymbol{K}$, given by

$$k_j = \frac{\sqrt{\gamma}}{|\beta_j|}, \quad j = 1, \ldots, p. \tag{6.52}$$

Putting the expression in the constraint (6.49), we obtain

$$\gamma = \left(\frac{k_{\text{HM}}}{p}\boldsymbol{1}_p^{\top}\|\boldsymbol{\beta}\|\right)^2. \tag{6.53}$$

The optimal $\boldsymbol{K}$ is then obtained from k_{opt} and β_j's so that (6.51) may be written as

$$(\boldsymbol{Y} - \boldsymbol{X}\boldsymbol{\beta})^{\top}(\boldsymbol{Y} - \boldsymbol{X}\boldsymbol{\beta}) + \frac{k_{\text{HM}}}{p}(\boldsymbol{1}_p^{\top}\|\boldsymbol{\beta}\|)^2, \tag{6.54}$$

which is equivalent to minimizing

$$(\boldsymbol{Y} - \boldsymbol{X}\boldsymbol{\beta})^{\top}(\boldsymbol{Y} - \boldsymbol{X}\boldsymbol{\beta}) \quad \text{subject to} \quad \boldsymbol{1}_p^{\top}\|\boldsymbol{\beta}\| \le t \tag{6.55}$$

for some t, which is exactly LASSO problem.

Minimizing (6.55), we obtain p normal equations for $k = 1, \ldots, p$, as

$$\begin{cases} \sum_{i=1}^{n} x_{ik}\left(\sum_{j=1}^{p}(y_i - x_{ij}\tilde{\beta}_{jn})\right) + \frac{k_{\text{HM}}}{p}\,\text{sgn}(\tilde{\beta}_{kn})\sum_{j=1}^{p}|\tilde{\beta}_{jn}| = 0 \\ \qquad\qquad \text{or} \quad \beta_k = 0. \end{cases} \tag{6.56}$$

The solution (6.56) may be obtained following Grandvalet (1998) as given here:Let

$$\gamma_j = \sqrt{\frac{k_j}{k_{\text{HM}}}}\beta_j, \quad \text{and} \quad D_j = \sqrt{\frac{k_{\text{HM}}}{k_j}}, \quad j = 1, \ldots, p. \tag{6.57}$$

Then, the problem (6.51) with constraint (6.52) may be written as

$$\min_{\boldsymbol{D},\boldsymbol{\gamma}}\{(\boldsymbol{Y} - \boldsymbol{X}\boldsymbol{D}\boldsymbol{\gamma})^{\top}(\boldsymbol{Y} - \boldsymbol{X}\boldsymbol{D}\boldsymbol{\gamma}) + k_{\text{HM}}\boldsymbol{\gamma}^{\top}\boldsymbol{\gamma}\}, \tag{6.58}$$

where $\gamma = (\gamma_1, \dots, \gamma_p)^\top$ with $D = \text{Diag}(D_1, \dots, D_p)$, subject to $\mathbf{1}_p^\top D^2 \mathbf{1}_p = p$, $D_j > 0, j = 1, \dots, p$.

After some tedious algebra, as in Avalos et al. (2007), we obtain the RRE of γ as

$$\tilde{\gamma}_n(k_{\text{HM}}) = \left[DX^\top XD + k_{\text{HM}} I_p\right]^{-1} DX^\top Y \tag{6.59}$$

$$= (I_p + k_{\text{HM}} \Sigma_n^{-1})^{-1} \Sigma_n^{-1} DX^\top Y, \quad \Sigma_n = DX^\top XD$$

$$= (I_p + k_{\text{HM}} \Sigma_n^{-1})^{-1} \tilde{\gamma}_n, \quad \tilde{\gamma}_n = \Sigma_n^{-1} DX^\top Y. \tag{6.60}$$

In the next section, we present the algorithm to obtain LASSO solutions.

6.9 Optimization Algorithm

For the computation of LASSO estimators, Tibshirani (1996) used quadratic programming. In this section, we use a fixed point algorithm from the expression (6.51) which is the GRRE. Thus, we solve a sequence of weighted ridge regression problems, as suggested by Knight and Fu (2000). First, we estimate K-matrix based on the estimators of k_{HM} as well as $(k_1, \dots, k_p)$. For the estimator of k_{HM}, we use from Hoerl et al. (1975):

$$\tilde{k}_{\text{HM}} = \frac{p s_n^2}{\tilde{\beta}_n^\top \tilde{\beta}_n}, \quad \text{and} \quad \tilde{k}_{jn} = \frac{s_n^2}{\tilde{\beta}_{jn}^2}, \quad j = 1, \dots, p, \tag{6.61}$$

where

$$s_n^2 = (n-p-1)^{-1} \|Y - \bar{y}\mathbf{1} - X\tilde{\beta}_n\|^2, \quad \bar{y} = \frac{1}{n}\sum_{i=1}^n y_i, \text{ and } \tilde{\beta}_n = (\tilde{\beta}_{1n}, \dots, \tilde{\beta}_{pn})^\top.$$

Hence, the estimate of the diagonal matrix K is given by $\tilde{K}_n = \text{Diag}(\tilde{K}_{1n}, \dots, \tilde{K}_{pn})$, where

$$\tilde{K}_{jn} = \sqrt{\frac{p\tilde{\beta}_{jn}^2}{\tilde{\beta}_n^\top \tilde{\beta}_n}}, \quad j = 1, \dots, p. \tag{6.62}$$

Alternatively, we may use the estimator, $\overline{K}_n = \text{Diag}(\overline{K}_{1n}, \dots, \overline{K}_{pn})$, where

$$\overline{K}_{jn} = \sqrt{\frac{p|\tilde{\beta}_{jn}|}{\sum_{j=1}^p |\tilde{\beta}_{jn}|}}, \quad j = 1, \dots, p. \tag{6.63}$$

The derivation of this estimator is explained here.

We consider the LASSO problem, which is

$$\frac{1}{2}\|X\beta - Y\|^2 + \nu \sum_{j=1}^p |\beta_j|. \tag{6.64}$$

Let $\hat{\boldsymbol{\beta}}_n = (\hat{\beta}_{1n}, \dots, \hat{\beta}_{pn})^\top$ be the LASSO solution. For any j, such that $\hat{\beta}_{jn} \neq 0$, the optimality conditions are

$$\boldsymbol{X}_j^\top(\boldsymbol{X}\hat{\boldsymbol{\beta}}_n - \boldsymbol{Y}) + \nu \operatorname{sgn}(\hat{\beta}_{jn}) = 0, \tag{6.65}$$

where $\boldsymbol{X}_j$ is the jth column of $\boldsymbol{X}$.

Take the equivalent adaptive ridge problem:

$$\frac{1}{2}\|\boldsymbol{X}\boldsymbol{\beta} - \boldsymbol{Y}\|^2 + \sum_{j=1}^{p} k_j \beta_j^2, \tag{6.66}$$

whose solution is $\hat{\boldsymbol{\beta}}_n$ (since the problems are equivalent). For any j, such that $\hat{\beta}_{jn} \neq 0$, the optimality conditions are

$$\boldsymbol{X}_j^\top(\boldsymbol{X}\hat{\boldsymbol{\beta}}_n - \boldsymbol{Y}) + 2k_j\hat{\beta}_{jn} = 0. \tag{6.67}$$

Hence, we have, for all j, such that $\hat{\beta}_{jn} \neq 0$:

$$\nu = 2k_j|\hat{\beta}_{jn}|; \tag{6.68}$$

and since (see (6.49))

$$\frac{1}{p}\sum_{j=1}^{p}\frac{1}{k_j} = \frac{1}{k_{\mathrm{HM}}}, \tag{6.69}$$

we have

$$k_{\mathrm{HM}} = \frac{2pk_j|\beta_j|}{2\sum_{j=1}^{p}|\beta_j|} \quad \text{and} \quad \frac{k_{\mathrm{HM}}}{k_j} = \frac{p|\beta_j|}{\sum|\beta_j|} = \overline{C}_j,\ j = 1, \dots, p. \tag{6.70}$$

To obtain the LASSO estimator, we start the iterative process based on (6.61) as given here:

$$\hat{\boldsymbol{\beta}}_n^{\mathrm{RR}(0)}(\tilde{k}_{\mathrm{HM}}) = \boldsymbol{C}_n^*(\boldsymbol{\Sigma}_n^* + \tilde{k}_{\mathrm{HM}}\boldsymbol{I}_p)^{-1}\boldsymbol{C}^*\boldsymbol{X}^\top\boldsymbol{Y}, \quad \boldsymbol{C}^* = \tilde{\boldsymbol{C}} \tag{6.71}$$

We can define successive estimates by

$$\tilde{\boldsymbol{\beta}}_n^{(t)} = \psi\left(\boldsymbol{\beta}_n^{(t-1)}(\tilde{k}_{\mathrm{HM}})\right)\boldsymbol{X}^\top\boldsymbol{Y}, \tag{6.72}$$

where

$$\psi\left(\boldsymbol{\beta}_n^{(t-1)}(\tilde{k}_{\mathrm{HM}})\right) = \boldsymbol{C}^{*(t-1)}\left[\boldsymbol{\Sigma}_n^{*(t-1)} + \tilde{k}_{\mathrm{HM}}\boldsymbol{I}_n\right]^{-1}\boldsymbol{C}^{*(t-1)}, \tag{6.73}$$

with $\boldsymbol{C}^{*(t-1)}$ is the diagonal matrix with some elements as zero. The expression (6.73) is similar to Knight and Fu (2000).

The sequence $\{\boldsymbol{\beta}_n^{(t-1)}(\tilde{k}_{\mathrm{HM}})\}$ does not necessarily converge to the global minimum, but seems to work well if multiple starting points are used. The resulting LASSO estimator produces p_1 nonzero coefficients and p_2 zero coefficients such that $p_1 + p_2 = p$. This information will be useful to define the PTE and the Stein-type shrinkage estimator to assess the performance characteristics of the LASSO estimators.

An estimator of the MSE matrix is given by

$$\begin{aligned}\widehat{\text{MSE}}(\hat{\boldsymbol{\beta}}_n^{\text{RR}}(\tilde{k}_{\text{HM}})) = s_n^2 \boldsymbol{C}^{*(t)}\left[\boldsymbol{I}_p + \tilde{k}_{\text{HM}}(\boldsymbol{\Sigma}^{*(t)})^{-1}\right]^{-1}\left[\boldsymbol{\Sigma}_n^{*(t)}\right]^{-1} \\ \times\left[\boldsymbol{I}_p + \tilde{k}_{\text{HM}}(\boldsymbol{\Sigma}^{*(t)})^{-1}\right]^{-1}\boldsymbol{C}^{*(t)} \\ + \tilde{k}_{\text{HM}}^2\left[\boldsymbol{\Sigma}_n^{*(t)} + \tilde{k}_{\text{HM}}\boldsymbol{I}_p\right]^{-1}\tilde{\boldsymbol{\beta}}_n^{(1)}\tilde{\boldsymbol{\beta}}_n^{(1)\top}\left[\boldsymbol{\Sigma}^{*(t)} + \tilde{k}_{\text{HM}}\boldsymbol{I}_p\right]^{-1},\end{aligned} \tag{6.74}$$

where some elements of $\boldsymbol{C}^{*(t)}$ are zero.

6.9.1 Prostate Cancer Data

The prostate cancer data attributed to Stamey et al. (1989) was given by Tibshirani (1996) for our analysis. They examined the correlation between the level of the prostate-specific antigen and a number of clinical measures in men who were about to undergo a radical prostatectomy. The factors were considered as follows: log(cancer volume) (lcavol), log(prostate weight) (l weight), age, log(benign prostatic hyperplasia amount) (lbph), seminal vesicle invasion (svi), log(capsular penetration) (lcp), Gleason score (gleason), and precebathe Gleason scores 4 or 5 (pgg45). Tibshirani (1996) fitted a log linear model to log(prostate specific antigen) after first standardizing the predictors. Our data consists of the sample size 97 instead of 95 considered by Tibshirani (1996). The following results (Table 6.2) have been obtained for the prostate example.

Table 6.2 Estimated coefficients (standard errors) for prostate cancer data using LS, LASSO, and ARR estimators.

LSE (s.e.)	LASSO (s.e.)	ARRE (s.e)
2.47 (0.07)	2.47 (0.07)	2.47 (0.07)
0.66 (0.10)	0.55 (0.10)	0.69 (0.00)
0.26 (0.08)	0.17 (0.08)	0.15 (0.00)
−0.15 (0.08)	0.00 (0.08)	0.00 (0.00)
0.14 (0.08)	0.00 (0.08)	0.00 (0.00)
0.31 (0.10)	0.18 (0.10)	0.13 (0.00)
−0.14 (0.12)	0.00 (0.12)	0.00 (0.00)
0.03 (0.11)	0.00 (0.12)	0.00 (0.00)
0.12 (0.12)	0.00 (0.12)	0.00 (0.00)

LS, least specific; ARR, adaptive ridge regression.

6.10 Estimation of Regression Parameters for Low-Dimensional Models

Consider the multiple regression model

$$Y = X\beta + \epsilon, \quad \epsilon \sim \mathcal{N}_n(\mathbf{0}, \sigma^2 I_n) \tag{6.75}$$

with the $n \times p$ design matrix, X, where $n > p$.

6.10.1 BLUE and Ridge Regression Estimators

The LSE of β is the value (or values) of β for which the L_2-norm $\|Y - X\beta\|^2$ is least. In the case where $p < n$ and X has rank p, the $p \times p$ matrix $X^\top X$ is invertible and the LSE of β is unique and given by

$$\tilde{\beta}_n = (X^\top X)^{-1} X^\top Y. \tag{6.76}$$

If X is not of rank p, (6.76) is no longer valid. In particular, (6.76) is not valid when $p > n$.

Now, we consider the model

$$Y = X_1\beta_1 + X_2\beta_2 + \epsilon \tag{6.77}$$

where X_i is $n \times p_i$ $(i = 1, 2)$ matrix and suspect that the sparsity condition $\beta_2 = 0$ may hold. Under this setup, the unrestricted ridge estimator (URE) of β is

$$\begin{aligned}\hat{\beta}_n^{\mathrm{RR}}(k) &= (X^\top X + kI_p)^{-1} X^\top y \\ &= (I_p + k(X^\top X)^{-1})^{-1}\tilde{\beta}_n.\end{aligned} \tag{6.78}$$

Now, we consider the partition of the LSE, $\tilde{\beta}_n = (\tilde{\beta}_{1n}^\top, \tilde{\beta}_{2n}^\top)^\top$, where $\tilde{\beta}_{in}$ is a p_i-vector, $i = 1, 2$ so that $p_1 + p_2 = p$. Note that $\tilde{\beta}_{1n}$ and $\tilde{\beta}_{2n}$ are given by (6.4). We know that the marginal distribution of $\tilde{\beta}_{1n} \sim \mathcal{N}_{p_1}(\beta_1, \sigma^2 C_{11\cdot2}^{-1})$ and $\tilde{\beta}_{2n} \sim \mathcal{N}_{p_2}(\beta_2, \sigma^2 C_{11\cdot2}^{-1})$, respectively. Thus, we may define the corresponding RREs as

$$\hat{\beta}_{1n}^{\mathrm{RR}}(k) = W_1\tilde{\beta}_{1n}; \qquad W_1 = (I_{p_1} + kC_{11\cdot2}^{-1})^{-1} \tag{6.79}$$

$$\hat{\beta}_{2n}^{\mathrm{RR}}(k) = W_2\tilde{\beta}_{2n}; \qquad W_2 = (I_{p_2} + kC_{22\cdot1}^{-1})^{-1}, \tag{6.80}$$

respectively, for β_1 and β_2.

If we consider that $\beta_2 = \mathbf{0}$ holds, then we have the restricted regression parameter $\beta_{\mathrm{R}} = (\beta_1^\top, \mathbf{0}^\top)^\top$, which is estimated by the restricted ridge regression estimator (RRRE)

$$\hat{\beta}_n^{\mathrm{RR}(R)}(k) = \begin{pmatrix} \hat{\beta}_{1n}^{\mathrm{RR}(R)}(k) \\ \mathbf{0} \end{pmatrix},$$

where

$$\hat{\boldsymbol{\beta}}_{1n}^{\mathrm{RR}(R)}(k) = \boldsymbol{W}_0\tilde{\boldsymbol{\beta}}_{1n} \qquad \boldsymbol{W}_0 = (\boldsymbol{I}_{p_1} + k\boldsymbol{C}_{11}^{-1})^{-1}. \tag{6.81}$$

On the other hand, if $\boldsymbol{\beta}_2$ is suspected to be $\mathbf{0}$, then we may test the validity of $\boldsymbol{\beta}_2 = \mathbf{0}$ based on the statistic,

$$\mathcal{L}_n = \frac{1}{\sigma^2}\left[\tilde{\boldsymbol{\beta}}_{2n}^{\top}\boldsymbol{C}_{22\cdot1}\tilde{\boldsymbol{\beta}}_{2n}\right], \tag{6.82}$$

where σ^2 is assumed to be known. Then, $\mathcal{L}_n$ follows a chi-squared distribution with p_2 DF under $\boldsymbol{\beta}_2 = \mathbf{0}$. Let us then consider an α-level critical value $c_\alpha = \chi^2_{p_2}(\alpha)$ from the null distribution of $\mathcal{L}_n$.

Define the PTE as

$$\hat{\boldsymbol{\beta}}_{n}^{\mathrm{RR}(PT)}(k,\alpha) = \begin{pmatrix} \hat{\boldsymbol{\beta}}_{1n}^{\mathrm{RR}}(k) \\ \boldsymbol{W}_2\tilde{\boldsymbol{\beta}}_{2n}I(\mathcal{L}_n > c_\alpha) \end{pmatrix}. \tag{6.83}$$

Similarly, we may define the SE as

$$\hat{\boldsymbol{\beta}}_{n}^{\mathrm{RR(S)}}(k) = \begin{pmatrix} \hat{\boldsymbol{\beta}}_{1n}^{\mathrm{RR}}(k) \\ \boldsymbol{W}_2\tilde{\boldsymbol{\beta}}_{2n}(1 - d\mathcal{L}_n^{-1}) \end{pmatrix}, \quad d = p_2 - 2 \tag{6.84}$$

and the positive-rule Stein-type estimator (PRSE) as

$$\hat{\boldsymbol{\beta}}_{n}^{\mathrm{RR(S+)}}(k) = \begin{pmatrix} \hat{\boldsymbol{\beta}}_{1n}^{\mathrm{RR}}(k) \\ \boldsymbol{W}_2\tilde{\boldsymbol{\beta}}_{2n}(1 - d\mathcal{L}_n^{-1})^{+} \end{pmatrix}. \tag{6.85}$$

6.10.2 Bias and L$_2$-risk Expressions of Estimators

In this section, we present the bias and L$_2$ risk of the estimators as follows:

(1) Unrestricted ridge estimator (URE)

$$\begin{aligned} b(\hat{\boldsymbol{\beta}}_{1n}^{\mathrm{RR}}(k)) &= -k(\boldsymbol{C}_{11\cdot2} + k\boldsymbol{I}_{p_1})^{-1}\boldsymbol{\beta}_1 \\ b(\hat{\boldsymbol{\beta}}_{2n}^{\mathrm{RR}}(k)) &= -k(\boldsymbol{C}_{22\cdot1} + k\boldsymbol{I}_{p_2})^{-1}\boldsymbol{\beta}_2. \end{aligned} \tag{6.86}$$

The L$_2$-risk of $\hat{\boldsymbol{\beta}}_{1n}^{\mathrm{RR}}(k)$ is obtained as follows:

$$\begin{aligned} R(\hat{\boldsymbol{\beta}}_{1n}^{\mathrm{RR}}(k)) &= \sigma^2\ \mathrm{tr}(\boldsymbol{W}_1\boldsymbol{C}_{11\cdot2}^{-1}\boldsymbol{W}_1) \\ &\quad + k^2\boldsymbol{\beta}_1^{\top}(\boldsymbol{C}_{11\cdot2} + k\boldsymbol{I}_{p_1})^{-2}\boldsymbol{\beta}_1, \end{aligned} \tag{6.87}$$

where $\boldsymbol{W}_1 = (\boldsymbol{I}_{p_1} + k\boldsymbol{C}_{11\cdot2}^{-1})^{-1}$. Then the weighted risk of $\hat{\boldsymbol{\beta}}_{1n}^{\mathrm{RR}}(k)$, with weight $\boldsymbol{A}_1 = (\boldsymbol{I}_{p_1} + k\boldsymbol{C}_{11\cdot2}^{-1})\boldsymbol{C}_{11\cdot2}(\boldsymbol{I}_{p_1} + k\boldsymbol{C}_{11\cdot2}^{-1})$ is obtained as

$$R(\hat{\boldsymbol{\beta}}_{1n}^{\mathrm{RR}}(k) : \boldsymbol{A}_1, \boldsymbol{I}_{p_2}) = \sigma^2\left(p_1 + k^2\frac{\boldsymbol{\beta}_1^{\top}\boldsymbol{C}_{11\cdot2}\boldsymbol{\beta}_1}{\sigma^2}\right). \tag{6.88}$$

Similarly, the risk of $\hat{\boldsymbol{\beta}}_{2n}^{\text{RR}}(k)$ is obtained as follows:

$$R(\hat{\boldsymbol{\beta}}_{2n}^{\text{RR}}(k)) = \sigma^2 \operatorname{tr}[\boldsymbol{W}_2^\top \boldsymbol{C}_{22\cdot1}^{-1} \boldsymbol{W}_2] + k^2 \boldsymbol{\beta}_2^\top (\boldsymbol{C}_{22\cdot1} + k\boldsymbol{I}_{p_2})^{-2} \boldsymbol{\beta}_2, \tag{6.89}$$

where $\boldsymbol{W}_2 = (I_{p_2} + kC_{22\cdot1}^{-1})^{-1}$. Then, the weighted risk of $\tilde{\boldsymbol{\beta}}_{2n}^{\text{RR}}(k)$, with weight $\boldsymbol{A}_2 = (\boldsymbol{I}_{p_2} + k\boldsymbol{C}_{22\cdot1}^{-1})\boldsymbol{C}_{22\cdot1}(\boldsymbol{I}_{p_2} + k\boldsymbol{C}_{22\cdot1}^{-1})$ is obtained as

$$\begin{aligned} R(\hat{\boldsymbol{\beta}}_{2n}^{\text{RR}}(k) : \boldsymbol{I}_{p_1}, \boldsymbol{A}_2) &= \sigma^2 \left(p_2 + k^2 \frac{\boldsymbol{\beta}_2^\top \boldsymbol{C}_{22\cdot1} \boldsymbol{\beta}_2}{\sigma^2} \right) \\ &= \sigma^2 (p_2 + k^2 \Delta^2). \end{aligned} \tag{6.90}$$

Finally, the weighted L_2 risk of $\hat{\boldsymbol{\beta}}_n^{\text{RR}}(k)$

$$R(\hat{\boldsymbol{\beta}}_n^{\text{RR}}(k) : \boldsymbol{I}_{p_1}, \boldsymbol{A}_2) = \sigma^2 \left[p + \frac{k^2}{\sigma^2} (\boldsymbol{\beta}_1^\top \boldsymbol{C}_{11\cdot2} \boldsymbol{\beta}_1 + \Delta^2) \right]. \tag{6.91}$$

(2) Restricted ridge estimator (RRRE)

$$b(\hat{\boldsymbol{\beta}}_{1n}^{\text{RR}(R)}(k)) = -k(\boldsymbol{C}_{11} + k\boldsymbol{I}_{p_1})^{-1} \boldsymbol{\beta}_1 + (\boldsymbol{C}_{11} + k\boldsymbol{I}_{p_1})^{-1} \boldsymbol{C}_{12} \boldsymbol{\beta}_2. \tag{6.92}$$

Then, the L_2-risk of $\hat{\boldsymbol{\beta}}_{1n}^{\text{RR}(R)}(k)$ is

$$\begin{aligned} R(\hat{\boldsymbol{\beta}}_{1n}^{\text{RR}(R)}(k)) = \sigma^2 \operatorname{tr}[&(\boldsymbol{C}_{11} + k\boldsymbol{I}_{p_1})^{-1} \boldsymbol{C}_{11} (\boldsymbol{C}_{11} + k\boldsymbol{I}_{p_1})^{-1}] \\ &+ k^2 \boldsymbol{\beta}_1^\top (\boldsymbol{C}_{11} + k\boldsymbol{I}_{p_1})^{-2} \boldsymbol{\beta}_1 \\ &- 2k \boldsymbol{\beta}_1^\top (\boldsymbol{C}_{11} + k\boldsymbol{I}_{p_1})^{-2} \boldsymbol{C}_{12} \boldsymbol{\beta}_2 \\ &+ \boldsymbol{\beta}_2^\top \boldsymbol{C}_{21} (\boldsymbol{C}_{11} + k\boldsymbol{I}_{p_1})^{-2} \boldsymbol{C}_{12} \boldsymbol{\beta}_2. \end{aligned} \tag{6.93}$$

Now consider the following weighted risk function:

$$\begin{aligned} \boldsymbol{D} &= [(\boldsymbol{C}_{11} + k\boldsymbol{I}_{p_1})^{-1} \boldsymbol{C}_{11} (\boldsymbol{C}_{11} + k\boldsymbol{I}_{p_1})^{-1}]^{-1} \\ &= (\boldsymbol{C}_{11} + k\boldsymbol{I}_{p_1}) \boldsymbol{C}_{11}^{-1} (\boldsymbol{C}_{11} + k\boldsymbol{I}_{p_1}). \end{aligned}$$

Thus,

$$\begin{aligned} R(\hat{\boldsymbol{\beta}}_{1n}^{\text{RR}(R)}(k) : \boldsymbol{D}, \boldsymbol{I}_{p_2}) &= \sigma^2 p_1 + k^2 \boldsymbol{\beta}_1^\top \boldsymbol{C}_{11}^{-1} \boldsymbol{\beta}_1 - 2k \boldsymbol{\beta}_1^\top \boldsymbol{C}_{11}^{-1} \boldsymbol{C}_{12} \boldsymbol{\beta}_2 \\ &\quad + \boldsymbol{\beta}_2^\top \boldsymbol{C}_{21} \boldsymbol{C}_{11}^{-1} \boldsymbol{C}_{12} \boldsymbol{\beta}_2. \end{aligned} \tag{6.94}$$

(3) Preliminary test ridge regression estimator (PTRRE)
The bias and weighted risk functions of PTRE are respectively obtained as follows:

$$\begin{aligned} b(\hat{\boldsymbol{\beta}}_{2n}^{\text{RR}(PT)}(k, \alpha)) = &-(\boldsymbol{C}_{22\cdot1} + k\boldsymbol{I}_{p_2})^{-1} \\ &\times (k\boldsymbol{I}_{p_2} + \boldsymbol{C}_{22\cdot1} H_{p_2+2}(c_\alpha; \Delta^2)) \boldsymbol{\beta}_2 \end{aligned}$$

$$
\begin{aligned}
R(\hat{\boldsymbol{\beta}}_{2n}^{\mathrm{RR}(PT)}(k,\alpha)) = {} & \sigma^2 \operatorname{tr}(\boldsymbol{W}_2\boldsymbol{C}_{22\cdot1}^{-1}\boldsymbol{W}_2)(1-H_{p_2+2}(c_\alpha;\Delta^2)) \\
& + (\boldsymbol{\beta}_2^\top \boldsymbol{W}_2^\top \boldsymbol{W}_2\boldsymbol{\beta}_2) \\
& \times \{2H_{p_2+2}(c_\alpha;\Delta^2) - H_{p_2+4}(c_\alpha;\Delta^2)\} \\
& + 2k\boldsymbol{\beta}_2^\top \boldsymbol{W}_2(\boldsymbol{C}_{22\cdot1}+k\boldsymbol{I}_{p_2})^{-1}\boldsymbol{\beta}_2 H_{p_2+2}(c_\alpha;\Delta^2) \\
& + k^2\boldsymbol{\beta}_2^\top(\boldsymbol{C}_{22\cdot1}+k\boldsymbol{I}_{p_2})^{-2}\boldsymbol{\beta}_2. \qquad (6.95)
\end{aligned}
$$

The weighted (weight, $\boldsymbol{M} = (\boldsymbol{W}_2\boldsymbol{C}_{22\cdot1}^{-1}\boldsymbol{W}_2)^{-1}$ is obtained as follows:

$$
\begin{aligned}
R(\hat{\boldsymbol{\beta}}_{2n}^{\mathrm{RR}(PT)}(k,\alpha) : \boldsymbol{I}_{p_1},\boldsymbol{M}) = \sigma^2 \big[& p_2(1-H_{p_2+2}(c_\alpha;\Delta^2)) \\
& + \Delta^2\{2H_{p_2+2}(c_\alpha;\Delta^2) - H_{p_2+4}(c_\alpha;\Delta^2) + k^2\} \\
& + 2\sigma^{-2}k\boldsymbol{\beta}_2^\top\boldsymbol{\beta}_2 H_{p_2+2}(c_\alpha;\Delta^2)\big]. \qquad (6.96)
\end{aligned}
$$

(4) Stein-type ridge regression estimator (SRRE)
The bias and weighted risk functions of SRRE are respectively obtained as follows:

$$
\begin{aligned}
b\left(\hat{\boldsymbol{\beta}}_{2n}^{\mathrm{RR(S)}}(k)\right) = {} & -(\boldsymbol{C}_{22\cdot1}+k\boldsymbol{I}_{p_2})^{-1} \\
& \times (k\boldsymbol{I}_{p_2} + (p_2-2)\boldsymbol{C}_{22\cdot1}\mathbb{E}[\chi_{p_2+2}^{-2}(\Delta^2)])\boldsymbol{\beta}_2 \\
R\left(\hat{\boldsymbol{\beta}}_{2n}^{\mathrm{RR(S)}}(k)\right) = {} & \sigma^2 \operatorname{tr}(\boldsymbol{W}_2\boldsymbol{C}_{22\cdot1}^{-1}\boldsymbol{W}_2) \\
& - \sigma^2(p_2-2)\operatorname{tr}(\boldsymbol{W}_2\boldsymbol{C}_{22\cdot1}^{-1}\boldsymbol{W}_2)^{-1} \\
& \times \{2\mathbb{E}[\chi_{p_2+2}^{-2}(\Delta^2)] - (p_2-2)\mathbb{E}[\chi_{p_2+2}^{-4}(\Delta^2)]\} \\
& + (p_2^2-4)(\boldsymbol{W}_2\boldsymbol{\beta}_2^\top\boldsymbol{\beta}_2\boldsymbol{W}_2)\mathbb{E}[\chi_{p_2+4}^{-4}(\Delta^2)] \\
& + 2k\boldsymbol{\beta}_2^\top\boldsymbol{W}_2(\boldsymbol{C}_{22\cdot1}+k\boldsymbol{I}_{p_2})^{-1}\boldsymbol{\beta}_2\mathbb{E}[\chi_{p_2+2}^{-2}(\Delta^2)] \\
& + k^2\boldsymbol{\beta}_2^\top(\boldsymbol{C}_{22\cdot1}+k\boldsymbol{I}_{p_2})^{-2}\boldsymbol{\beta}_2. \qquad (6.97)
\end{aligned}
$$

The weighted ($\boldsymbol{M} = (\boldsymbol{W}_2\boldsymbol{C}_{22\cdot1}^{-1}\boldsymbol{W}_2)^{-1}$) risk function of the Stein-type estimator is obtained as follows

$$
\begin{aligned}
R\left(\hat{\boldsymbol{\beta}}_{2n}^{\mathrm{RR(S)}}(k) : \boldsymbol{I}_{p_1},\boldsymbol{M}\right) = \sigma^2 \big[& p_2 - \sigma^2(p_2^2-2p_2) \\
& \times \{2\mathbb{E}[\chi_{p_2+2}^{-2}(\Delta^2)] - (p_2-2)\mathbb{E}[\chi_{p_2+2}^{-4}(\Delta^2)]\} \\
& + \Delta^2\{k^2 + (p_2^2-4)\mathbb{E}[\chi_{p_2+4}^{-4}(\Delta^2)]\} \\
& + 2k(p_2-2)\sigma^{-2}\boldsymbol{\beta}_2^\top\boldsymbol{\beta}_2\mathbb{E}[\chi_{p_2+2}^{-2}(\Delta^2)]\big]. \qquad (6.98)
\end{aligned}
$$

(5) Positive-rule Stein-type ridge estimator (PRSRRE)
The bias and weighted ($\boldsymbol{M} = (\boldsymbol{W}_2\boldsymbol{C}_{22\cdot1}^{-1}\boldsymbol{W}_2)^{-1}$) risk functions of PTSRRE are, respectively, obtained as follows:

$$
\begin{aligned}
b(\hat{\boldsymbol{\beta}}_{2n}^{\mathrm{RR(S+)}}(k)) = {} & b(\hat{\boldsymbol{\beta}}_{2n}^{\mathrm{RR(S)}}(k)) \\
& - (\boldsymbol{C}_{22\cdot1}+k\boldsymbol{I}_{p_2})^{-1}(k\boldsymbol{I}_{p_2} + \boldsymbol{C}_{22\cdot1}A_1)\boldsymbol{\beta}_2
\end{aligned}
$$

$$
\begin{aligned}
R(\hat{\boldsymbol{\beta}}_{2n}^{\mathrm{RR(S+)}}(k) : \boldsymbol{I}_{p_1}, \boldsymbol{M}) &= R(\hat{\boldsymbol{\beta}}_{2n}^{\mathrm{RR(S)}}(k) : \boldsymbol{I}_{p_1}, \boldsymbol{M}) \\
&\quad - \sigma^2\{p_2(1-A_2)+\Delta^2\{2A_1-A_3\}2\sigma^{-2}k\boldsymbol{\beta}_2^{\top}\boldsymbol{\beta}_2 A_1\} \\
&= \sigma^2[p_2 - (p_2^2 - 2) \\
&\quad \times \{2\mathbb{E}[\chi_{p_2+2}^{-2}(\Delta^2)] - (p_2 - 2\mathbb{E}[\chi_{p_2+2}^{-4}(\Delta^2)])\} \\
&\quad + \Delta^2\{(p_2^2-4)\mathbb{E}[\chi_{p_2+4}^{-4}(\Delta^2)] + k^2 + (2A_1 - A_3)\} \\
&\quad + 2k\sigma^{-2}\boldsymbol{\beta}_2^{\top}\boldsymbol{\beta}_2(1 + A_1)],
\end{aligned} \tag{6.99}
$$

where

$$
\begin{aligned}
A_1 &= \mathbb{E}[(1 - (p_2 - 2)\chi_{p_2+2}^{-2}(\Delta^2))I(\chi_{p_2+2}^{-2}(\Delta^2) < p - 2)] \\
A_2 &= \mathbb{E}[(1 - (p_2 - 2)\chi_{p_2+4}^{-2}(\Delta^2))^2 I(\chi_{p_2+4}^{-2}(\Delta^2) < p - 2)] \\
A_3 &= \mathbb{E}[(1 - (p_2 - 2)\chi_{p_2+4}^{-2}(\Delta^2))I(\chi_{p_2+4}^{-2}(\Delta^2) < p - 2)].
\end{aligned}
$$

6.10.3 Comparison of the Estimators

Here, we compare the URE, SRE, and PRSRRE using the weighted L_2-risks criterion given in Theorem 6.3.

Theorem 6.3 *Under the assumed condition, we may order the risks of URE, SRE, and PRSRRE as follows:*

$$
R(\hat{\boldsymbol{\beta}}_{2n}^{\mathrm{RR(S+)}}(k)) \leq R(\hat{\boldsymbol{\beta}}_{2n}^{\mathrm{RR(S)}}(k)) \leq R(\hat{\boldsymbol{\beta}}_{2n}^{\mathrm{RR}}(k)), \quad \forall (k, \Delta^2).
$$

Proof: First consider the risk difference of RRE and SRRE.

The weighted L_2-risk difference of the RRE and SRRE is given by

$$
\begin{aligned}
&p_2 + k^2\Delta^2 - p_2 + (p_2^2 - 2p_2) \\
&\quad \times \{2\mathbb{E}[\chi_{p_2+2}^{-2}(\Delta^2)] - (p_2 - 2)\mathbb{E}[\chi_{p_2+2}^{-4}(\Delta^2)]\} \\
&\quad - \Delta^2\{k^2 + (p_2^2 - 4)\mathbb{E}[\chi_{p_2+4}^{-4}(\Delta^2)]\} - 2k\sigma^{-2}\boldsymbol{\beta}_2^{\top}\boldsymbol{\beta}_2\mathbb{E}[\chi_{p_2+2}^{-2}(\Delta^2)] \\
&= (p_2^2 - 2)\{2\mathbb{E}[\chi_{p_2+2}^{-2}(\Delta^2)] - (p_2 - 2)\mathbb{E}[\chi_{p_2+2}^{-4}(\Delta^2)]\} \\
&\quad - \Delta^2(p_2^2 - 4)\mathbb{E}[\chi_{p_2+4}^{-4}(\Delta^2)] - 2k(p_2 - 2)\sigma^{-2}\boldsymbol{\beta}_2^{\top}\boldsymbol{\beta}_2\mathbb{E}[\chi_{p_2+2}^{-2}(\Delta^2)].
\end{aligned} \tag{6.100}
$$

This difference is nonnegative. Thus, SRRE uniformly dominates the RRE.

The weighted L_2-risk difference of PRSRRE and RRE is given by

$$
\begin{aligned}
&- (p_2^2 - 2)\{2\mathbb{E}[\chi_{p_2+2}^{-2}(\Delta^2)] - (p_2 - 2)\mathbb{E}[\chi_{p_2+2}^{-4}(\Delta^2)]\} \\
&+ \Delta^2\{(p_2^2 - 4)\mathbb{E}[\chi_{p_2+4}^{-4}(\Delta^2)] + (2A_1 - A_3)\} \\
&+ 2k(p_2 - 2)\sigma^{-2}\boldsymbol{\beta}_2^{\top}\boldsymbol{\beta}_2(\mathbb{E}[\chi_{p_2+2}^{-2}(\Delta^2)] + A_1).
\end{aligned} \tag{6.101}
$$

This difference is negative and the risk of PRSRRE is uniformly smaller than RRE.

Finally, the weighted L_2-risk difference of the PRSRRE and SRE is given by

$$-\sigma^2\{p_2(1-A_2)+\Delta^2\{2A_1-A_3\}+2\sigma^{-2}k\boldsymbol{\beta}_2^\top\boldsymbol{\beta}_2A_1\}\leq 0. \tag{6.102}$$

That means PRSRRE uniformly dominates SRE. □

6.10.4 Asymptotic Results of RRE

First, we assume that

$$Y_{ni}=\boldsymbol{\beta}_n^\top\boldsymbol{x}_{ni}+\epsilon_{ni},\quad i=1,\dots,n, \tag{6.103}$$

where for each n, $\epsilon_{n1},\dots,\epsilon_{nn}$ are i.i.d. random variables with mean zero and variance σ^2. Also, we assume that $\boldsymbol{x}_{ni}$s satisfy

$$\frac{1}{n}\sum_{i=1}^{n}\boldsymbol{x}_{ni}\boldsymbol{x}_{ni}^\top\rightarrow\boldsymbol{C},\quad \text{as}\quad n\rightarrow\infty, \tag{6.104}$$

for some positive definite matrix $\boldsymbol{C}$ and

$$\frac{1}{n}\max_{1\leq i\leq n}\boldsymbol{x}_{ni}^\top\boldsymbol{x}_{ni}\rightarrow 0,\quad \text{as}\quad n\rightarrow\infty. \tag{6.105}$$

Suppose that

$$\boldsymbol{\beta}_n=\boldsymbol{\beta}+\frac{\boldsymbol{\delta}}{\sqrt{n}},\quad \boldsymbol{\delta}=(\delta_1,\dots,\delta_p)^\top \tag{6.106}$$

and define $\hat{\boldsymbol{\beta}}_n^{\mathrm{RR}}(k_n)$ to minimize

$$\sum_{i=1}^{n}(Y_{ni}-\boldsymbol{\phi}^\top\boldsymbol{x}_{ni})^2+k_n\boldsymbol{\phi}^\top\boldsymbol{\phi}. \tag{6.107}$$

Then, we have the following theorem from Knight and Fu (2000).

Theorem 6.4 *If $\boldsymbol{\beta}=\boldsymbol{0}$ and $\frac{k_n}{n}\rightarrow k_o\geq 0$, then*

$$\sqrt{n}(\hat{\boldsymbol{\beta}}_n(k_n)-\boldsymbol{\beta}_n)\xrightarrow{\mathcal{D}}\arg\min V(\boldsymbol{u}),$$

where

$$V(\boldsymbol{u})=-2\boldsymbol{u}^\top\boldsymbol{Z}+\boldsymbol{u}^\top\boldsymbol{C}\boldsymbol{u}+k_o\sum_{j=1}^{p}(u_j+\delta_j)^2$$

with $\boldsymbol{Z}\sim\mathcal{N}_p(\boldsymbol{0},\sigma^2\boldsymbol{C})$.

This theorem suggests that the advantages of ridge estimators are limited to situations where all coefficients are relatively small.

Proof: Define $V_n(\boldsymbol{u})$ as

$$V_n(\boldsymbol{u}) = \sum_{i=1}^{n}\left[\left(\epsilon_{ni} - \frac{1}{\sqrt{n}}\boldsymbol{u}^\top \boldsymbol{x}_{ni}\right)^2 - \epsilon_{ni}^2\right]$$
$$+ k_n \sum_{j=1}^{p}\left[\left(\beta_j + \frac{1}{\sqrt{n}}u_j\right)^2 - \beta_j^2\right], \tag{6.108}$$

where $\boldsymbol{u} = (u_1, \dots, u_p)^\top$. Now, $V_n(\boldsymbol{u})$ is minimized at $\sqrt{n}(\hat{\boldsymbol{\beta}}_n^{\mathrm{RR}}(k_n) - \boldsymbol{\beta}_n)$. Clearly,

$$\sum_{i=1}^{n}\left[\left(\epsilon_{ni} - \frac{1}{\sqrt{n}}\boldsymbol{u}^\top \boldsymbol{x}_{ni}\right)^2 - \epsilon_{ni}^2\right] \xrightarrow{\mathcal{D}} -2\boldsymbol{u}^\top \boldsymbol{Z} + \boldsymbol{u}^\top \boldsymbol{C}\boldsymbol{u}. \tag{6.109}$$

The result follows from the fact that

$$k_n \sum_{j=1}^{p}\left[\left(\beta_j + \frac{1}{\sqrt{n}}u_j\right)^2 - \beta_j^2\right] \xrightarrow{\mathcal{D}} k_o(\boldsymbol{u} + \boldsymbol{\delta})^\top(\boldsymbol{u} + \boldsymbol{\delta}). \tag{6.110}$$

□

The next theorem gives the asymptotic distribution of $\sqrt{n}(\hat{\boldsymbol{\beta}}_n^{\mathrm{RR}}(k) - \boldsymbol{\beta}_n)$.

Theorem 6.5 *Under* (6.106) *and* $\boldsymbol{\delta} = (\boldsymbol{\delta}_1^\top, \boldsymbol{\delta}_2^\top)^\top$

$$\sqrt{n}(\hat{\boldsymbol{\beta}}_n^{\mathrm{RR}}(k) - \boldsymbol{\beta}_n) \xrightarrow{\mathcal{D}} (\boldsymbol{C} + k_o \boldsymbol{I}_p)^{-1}(\boldsymbol{Z} - k_o \boldsymbol{\delta}),$$

where $\boldsymbol{Z} \sim \mathcal{N}_p(\boldsymbol{0}, \sigma^2 \boldsymbol{C})$.

Proof: Differentiating (6.108) w.r.t. $\boldsymbol{u}$ and solving for $\boldsymbol{u}$ from the equation

$$V'(\boldsymbol{u}) = -2\boldsymbol{Z} + \boldsymbol{C}\boldsymbol{u} + k_o(\boldsymbol{u} + \boldsymbol{\delta}) = \boldsymbol{0},$$

we obtain

$$(\boldsymbol{C} + k_o \boldsymbol{I}_p)\boldsymbol{u} = \boldsymbol{Z} - k_o \boldsymbol{\delta}. \tag{6.111}$$

Hence, $\boldsymbol{u} = (\boldsymbol{C} + k_o \boldsymbol{I}_p)^{-1}(\boldsymbol{Z} - k_o \boldsymbol{\delta})$. □

From Theorem 6.5,

$$\sqrt{n}\left(\hat{\boldsymbol{\beta}}_n^{\mathrm{RR}}(k_o) - \frac{\boldsymbol{\delta}}{\sqrt{n}}\right) \xrightarrow{\mathcal{D}} \mathcal{N}_p(-k_o(\boldsymbol{C} + k_o \boldsymbol{I}_p)^{-1}\boldsymbol{\delta}, \sigma^2 \boldsymbol{\Sigma}),$$

where $\boldsymbol{\Sigma} = (\boldsymbol{C} + k_o \boldsymbol{I}_p)^{-1}\boldsymbol{C}(\boldsymbol{C} + k_o \boldsymbol{I}_p)^{-1}$.

6.11 Summary and Concluding Remarks

In this chapter, we considered the unrestricted estimator and shrinkage estimators, namely, restricted estimator, PTE, Stein-type estimator, PRSE and two penalty estimators, namely, RRE and LASSO for estimating the regression parameters of the linear regression models. We also discussed about the determination of the tuning parameter. A detailed discussion on LASSO and adaptive ridge regression estimators are given. The optimization algorithm for estimating the LASSO estimator is discussed, and prostate cancer data are used to illustrate the optimization algorithm.

Problems

6.1 Show that $X^\top X$ approaches singularity as the minimum eigenvalue tends to 0.

6.2 Display the graph of ridge trace for the Portland cement data in Section 6.5.

6.3 Consider the model, $Y = X\beta + \epsilon$ and suppose that we want to test the null-hypothesis $\mathcal{H}_o : \beta = \mathbf{0}$ (vs.) $\mathcal{H}_A : \beta \neq \mathbf{0}$. Then, show that the likelihood ratio test statistic is

$$\mathcal{L}_n = \frac{\tilde{\beta}_n^\top C \tilde{\beta}_n}{p s_n^2},$$

which follows a noncentral F-distribution with $(p, n-p)$ DF and noncentrality parameter, $\Delta^2 = \frac{\beta^\top C \beta}{\sigma^2}$.

6.4 Similarly, if we want to test the subhypothesis $\mathcal{H}_o : (\beta_1^\top, \beta_2^\top) = (\beta_1^\top, \mathbf{0}^\top)$ vs. $\mathcal{H}_A : (\beta_1^\top, \beta_2^\top) \neq (\beta_1^\top, \mathbf{0}^\top)$, then show that the appropriate test statistic is,

$$\mathcal{L}_m^* = \frac{\tilde{\beta}_{2n}^\top C_{22\cdot1} \tilde{\beta}_{2n}}{p_2 s_{2n}^2},$$

where

$$\hat{\beta}_{1n} = (X_1^\top X_1)^{-1} X_1 Y = \tilde{\beta}_{1n} + C_{12} C_{22}^{-1} \tilde{\beta}_{2n},$$

and $s_{2n}^2 = m^{-1}(Y - X_1^\top \hat{\beta}_{1n})^\top (Y - X_1^\top \hat{\beta}_{1n})$, $m = n - p_1$.

6.5 Show that the risk function for GRRE is

$$R(\hat{\beta}_n^{\mathrm{GRR}}(K)) = \sigma^2 \,\mathrm{tr}[(C+K)^{-2} C] + \beta^\top K (C+K)^{-2} K \beta. \tag{6.112}$$

6.6 Verify (6.60).

6.7 Verify (6.74).

6.8 Consider a real data set, where the design matrix elements are moderate to highly correlated, then find the efficiency of the estimators using unweighted risk functions. Find parallel formulas for the efficiency expressions and compare the results with that of the efficiency using weighted risk function. Are the two results consistent?

7

Partially Linear Regression Models

7.1 Introduction

In this chapter, the problem of ridge estimation is studied in the context of partially linear models (PLMs). In a nutshell, PLMs are smoothed models that include both parametric and nonparametric parts. They allow more flexibility compared to full/nonparametric regression models.

Consider the usual PLM with the form

$$y_i = \boldsymbol{x}_i^\top \boldsymbol{\beta} + f(t_i) + \epsilon_i, \qquad i = 1, \dots, n \tag{7.1}$$

where $\boldsymbol{x}_i^\top = (x_{i1}, \dots, x_{ip})$ is a vector of explanatory variables, $\boldsymbol{\beta} = (\beta_1, \dots, \beta_p)^\top$ is an unknown p-dimensional parameter vector, the t_i's are known and nonrandom in some bounded domain $D \in \mathbb{R}$, $f(t_i)$ is an unknown smooth function, and ϵ_i's are i.i.d. random errors with mean 0, variance σ^2, which are independent of $(\boldsymbol{x}_i, t_i)$. PLMs are more flexible than standard linear models since they have both parametric and nonparametric components. They can be a suitable choice when one suspects that the response y linearly depends on x, but that it is nonlinearly related to t.

Surveys regarding the estimation and application of the model (7.1) can be found in the monograph of Hardle et al. (2000). Raheem et al. (2012) considered absolute penalty and shrinkage estimators in PLMs where the vector of coefficients $\boldsymbol{\beta}$ in the linear part can be partitioned as $(\boldsymbol{\beta}_1^\top, \boldsymbol{\beta}_2^\top)^\top$; $\boldsymbol{\beta}_1$ is the coefficient vector of the main effects, and $\boldsymbol{\beta}_2$ is the vector of the nuisance effects. For a more recent study about PLM, we refer to Roozbeh and Arashi (2016b). Since for estimation $\boldsymbol{\beta}$, we need to estimate the nonparametric component $f(.)$, we estimate it using the kernel smoothing method. Throughout, we do not further discuss the estimation of $f(.)$ in PLMs, since the main concern is the estimation of $\boldsymbol{\beta}$ in the ridge context.

To estimate $f(.)$, assume that $(y_i, \boldsymbol{x}_i, t_i)$, $i = 1, \dots, n$ satisfy the model (7.1). Since $\mathbb{E}(\varepsilon_i) = 0$, we have $f(t_i) = \mathbb{E}(y_i - \boldsymbol{x}_i^\top \boldsymbol{\beta})$ for $i = 1, \dots, n$. Hence, if we

Theory of Ridge Regression Estimation with Applications, First Edition.
A.K. Md. Ehsanes Saleh, Mohammad Arashi, and B.M. Golam Kibria.

know $\boldsymbol{\beta}$, a natural nonparametric estimator of $f(.)$ is given by

$$\hat{f}(t, \boldsymbol{\beta}) = \sum_{i=1}^{n} W_{ni}(t)(y_i - \boldsymbol{x}_i^\top \boldsymbol{\beta}), \tag{7.2}$$

where the positive weight function $W_{ni}(\cdot)$ satisfies the three regularity conditions given here:

(i) $\max_{1\le i\le n} \sum_{j=1}^{n} W_{ni}(t_j) = O(1)$,
(ii) $\max_{1\le i,j\le n} W_{ni}(t_j) = O\left(n^{-\frac{2}{3}}\right)$,
(iii) $\max_{1\le i\le n} \sum_{j=1}^{n} W_{ni}(t_j) I(|t_i - t_j| > c_n) = O(d_n)$, where $I(A)$ is the indicator function of the set A,

$$\lim_{n\to\infty} \sup nc_n^3 < \infty,$$
$$\lim_{n\to\infty} \sup nd_n^3 < \infty.$$

These assumptions guarantee the existence of $\hat{f}(t, \boldsymbol{\beta})$ at the optimal convergence rate $n^{-4/5}$, in PLMs with probability one. See Müller and Rönz (1999) for more details.

7.2 Partial Linear Model and Estimation

Consider the PLM defined by

$$y_j = \boldsymbol{x}_{j(1)}^\top \boldsymbol{\beta}_1 + \boldsymbol{x}_{j(2)}^\top \boldsymbol{\beta}_2 + g(l_j) + \epsilon_j, \qquad j = 1, \dots, n, \tag{7.3}$$

where y_j's are responses, $\boldsymbol{x}_{j(1)} = (x_{j1}, \dots, x_{jp_1})^\top$, $\boldsymbol{x}_{j(2)} = (x_{j(p_1+1)}, \dots, x_{jp})^\top$, are covariates of the unknown vectors, namely, $\boldsymbol{\beta}_1 = (\beta_1, \dots, \beta_{p_1})^\top$ and also $\boldsymbol{\beta}_2 = (\beta_{p_1+1}, \dots, \beta_p)^\top$, respectively, and $t_j \in [0, 1]$ are design points, $g(\cdot)$ is a real valued function defined on $[0, 1]$, and $\epsilon_1, \dots, \epsilon_n$ are independently and identically distributed (i.i.d.) errors with zero mean and constant variance, σ^2.

Our main objective is to estimate $(\boldsymbol{\beta}_1^\top, \boldsymbol{\beta}_2^\top)^\top$ when it is suspected that the p_2-dimensional sparsity condition $\boldsymbol{\beta}_2 = \mathbf{0}$ may hold. In this situation, we first estimate the parameter $\boldsymbol{\beta} = (\boldsymbol{\beta}_1^\top, \boldsymbol{\beta}_2^\top)^\top$ using's Speckman's (1988) approach of partial kernel smoothing method, which attains the usual parametric convergence rate $n^{-1/2}$ without undersmoothing the nonparametric component $g(\cdot)$.

Assume that $[\{(\boldsymbol{x}_{j(1)}^\top, \boldsymbol{x}_{j(2)}^\top), t_j, Y_j\} | j = 1, 2, \dots, n]$ satisfy the model (7.3). If $\boldsymbol{\beta} = (\boldsymbol{\beta}_1^\top, \boldsymbol{\beta}_2^\top)^\top$ is the true parameter, then $\mathbb{E}(\epsilon_j) = 0$. Also, for $j = 1, \dots, n$, we have

$$g(t_j) = \mathbb{E}(Y_j - \boldsymbol{x}_{j(1)}^\top \boldsymbol{\beta}_1 - \boldsymbol{x}_{j(2)}^\top \boldsymbol{\beta}_2).$$

Thus, a natural nonparametric estimator of $g(\cdot)$ given $\boldsymbol{\beta} = (\boldsymbol{\beta}_1^\top, \boldsymbol{\beta}_2^\top)^\top$ is

$$\tilde{g}(\boldsymbol{t}, \boldsymbol{\beta}_1, \boldsymbol{\beta}_2) = \sum_{j=1}^{n} W_{nj}(t_j)(Y_j - \boldsymbol{x}_{j(1)}^\top \boldsymbol{\beta}_1 - \boldsymbol{x}_{j(2)}^\top \boldsymbol{\beta}_2), \quad \boldsymbol{t} = (t_1, \dots, t_n) \tag{7.4}$$

with the probability weight functions $W_{nj}(\cdot)$ satisfies the three regularity conditions (i)–(iii).

Suppose that we partitioned the design matrix,

$$\hat{\boldsymbol{X}} = (\underset{n\times p_1}{\hat{\boldsymbol{X}}_1}, \quad \underset{n\times p_2}{\hat{\boldsymbol{X}}_2})^\top,$$

and others as

$$\hat{\boldsymbol{Y}} = (\hat{y}_1, \dots, \hat{y}_n)^\top, \quad \hat{\boldsymbol{X}} = (\hat{\boldsymbol{x}}_1, \dots, \hat{\boldsymbol{x}}_n)^\top,$$

where

$$\hat{y}_j = y_j - \sum_{i=1}^{n} W_{ni}(t_j) Y_i, \quad \hat{\boldsymbol{x}}_j = \boldsymbol{x}_j - \sum_{i=1}^{n} W_{ni}(t_j)\boldsymbol{x}_i, \quad j = 1, \dots, n$$

To estimate $\boldsymbol{\beta} = (\boldsymbol{\beta}_1^\top, \boldsymbol{\beta}_2^\top)^\top$, we minimize

$$\mathrm{SS}(\boldsymbol{\beta}) = \sum_{j=1}^{n} (y_j - \boldsymbol{x}_{j(1)}^\top \boldsymbol{\beta}_1 - \boldsymbol{x}_{j(2)}^\top \boldsymbol{\beta}_2 - \tilde{g}(t_j, \boldsymbol{\beta}_1, \boldsymbol{\beta}_2))^2 \tag{7.5}$$

with respect to $\boldsymbol{\beta} = (\boldsymbol{\beta}_1^\top, \boldsymbol{\beta}_2^\top)^\top$. This yields the estimator of $\boldsymbol{\beta}$ as

$$\tilde{\boldsymbol{\beta}}_n = (\hat{\boldsymbol{X}}^\top \hat{\boldsymbol{X}})^{-1} \hat{\boldsymbol{X}}^\top \hat{\boldsymbol{Y}} \tag{7.6}$$

$$\begin{pmatrix} \tilde{\boldsymbol{\beta}}_{1n} \\ \tilde{\boldsymbol{\beta}}_{2n} \end{pmatrix} = \begin{pmatrix} (\hat{\boldsymbol{X}}_1^\top \hat{\boldsymbol{M}}_2 \hat{\boldsymbol{X}}_1)^{-1} \hat{\boldsymbol{X}}_1^\top \hat{\boldsymbol{M}}_2 \hat{\boldsymbol{Y}} \\ (\hat{\boldsymbol{X}}_2^\top \hat{\boldsymbol{M}}_1 \hat{\boldsymbol{X}}_2)^{-1} \hat{\boldsymbol{X}}_2^\top \hat{\boldsymbol{M}}_1 \hat{\boldsymbol{Y}} \end{pmatrix}, \tag{7.7}$$

where

$$\hat{\boldsymbol{M}}_1 = \boldsymbol{I} - \hat{\boldsymbol{X}}_1^\top (\hat{\boldsymbol{X}}_1^\top \hat{\boldsymbol{X}}_1)^{-1} \hat{\boldsymbol{X}}_1^\top$$
$$\hat{\boldsymbol{M}}_2 = \boldsymbol{I} - \hat{\boldsymbol{X}}_2^\top (\hat{\boldsymbol{X}}_2^\top \hat{\boldsymbol{X}}_2)^{-1} \hat{\boldsymbol{X}}_2^\top,$$

respectively.

If $\boldsymbol{\beta} = \boldsymbol{0}$ is true, then the restricted estimator is

$$\hat{\boldsymbol{\beta}}_n = \begin{pmatrix} \hat{\boldsymbol{\beta}}_{1n} \\ \boldsymbol{0} \end{pmatrix}, \quad \text{where} \quad \hat{\boldsymbol{\beta}}_{1n} = (\hat{\boldsymbol{X}}_1^\top \hat{\boldsymbol{X}}_1)^{-1} \hat{\boldsymbol{X}}_1^\top \hat{\boldsymbol{Y}}. \tag{7.8}$$

Further, we assume that $(\boldsymbol{x}_j, t_j)$ are i.i.d. random variables and let

$$\mathbb{E}(x_{is} | t_i) = h_s(t_i), \qquad s = 1, \dots, p.$$

Then, the conditional variance of x_{is} given t_i is $\mathbb{E}(\boldsymbol{u}_j\boldsymbol{u}_j^\top) = \boldsymbol{C}$, where $\boldsymbol{u}_j = (u_{j1}, \dots, u_{jp})^\top$ with $u_{is} = x_{is} - h_s(t_i)$. Then, by law of large numbers (LLN),

$$\lim_n n^{-1} \sum_{k=1}^{n} u_{ik}u_{kj} = C_{ij} \qquad (i,j = 1, \dots, p),$$

and the matrix $\boldsymbol{C} = (C_{ij})$ is nonsingular with probability 1. Readers are referred to Shi and Lau (2000), Gao (1995, 1997), and Liang and Härdle (1999), among others. Moreover, for any permutation $(j_1, \dots, j_k)$ of $(1, \dots, n)$ as $n \to \infty$,

$$\left\| \max \sum_{i=1}^{n} W_{n_i}(t_j)u_i \right\| = O\left(n^{-\frac{1}{6}}\right). \tag{7.9}$$

Finally, we use the following assumption:

The functions $g(\cdot)$ and $h_s(\cdot)$ satisfy the Lipchitz condition of order 1 on [0, 1] for $s = 1, \dots, p$. Then, we have the following theorem.

Theorem 7.1 *Under the assumed regularity conditions as $n \to \infty$,*

$$\sqrt{n}(\tilde{\boldsymbol{\beta}}_n - \boldsymbol{\beta}) \xrightarrow{D} \mathcal{N}_p(\boldsymbol{0}, \sigma^2\boldsymbol{C}^{-1}).$$

As a result of Theorem 7.1, we have the following limiting cases:

(i) $\sqrt{n}(\tilde{\boldsymbol{\beta}}_1 - \boldsymbol{\beta}_1) \xrightarrow{D} \mathcal{N}_{p_1}(\boldsymbol{0}, \sigma^2\boldsymbol{C}_{11.2}^{-1})$

(ii) $\sqrt{n}(\tilde{\boldsymbol{\beta}}_2 - \boldsymbol{\beta}_2) \xrightarrow{D} \mathcal{N}_{p_2}(\boldsymbol{0}, \sigma^2\boldsymbol{C}_{22.1}^{-1})$

(iii) $\sqrt{n}(\tilde{\beta}_{jn} - \beta_j) \xrightarrow{D} \mathcal{N}(0, \sigma^2 C^{jj}), \quad j = 1, \dots, p$ where C^{jj} is the jth diagonal element of $\boldsymbol{C}^{-1}$.

7.3 Ridge Estimators of Regression Parameter

Let the PLM be given as the following form

$$\boldsymbol{Y} = \boldsymbol{X}\boldsymbol{\beta} + \boldsymbol{f}(t) + \boldsymbol{\epsilon}, \tag{7.10}$$

where

$$\begin{aligned}
\boldsymbol{Y} &= (y_1, \dots, y_n)^\top, \\
\boldsymbol{X} &= (\boldsymbol{x}_1, \dots, \boldsymbol{x}_n)^\top, \quad n \times p \text{ matrix}, \\
\boldsymbol{f}(t) &= (f(t_1), \dots, f(t_n))^\top, \\
\boldsymbol{\epsilon} &= (\epsilon_1, \dots, \epsilon_n)^\top.
\end{aligned}$$

In general, we assume that $\boldsymbol{\epsilon}$ is an n-vector of i.i.d. random variables with distribution $\mathcal{N}_n(\boldsymbol{0}, \sigma^2\boldsymbol{V})$, where $\boldsymbol{V}$ is a symmetric, positive definite known matrix and $\sigma^2 \in \mathbb{R}^+$ is an unknown parameter.

Using the model (7.20) and the error distribution of ϵ we obtain the maximum likelihood estimator (MLE) or generalized least squares estimator (GLSE) of $\boldsymbol{\beta}$ by minimizing

$$\min_{\boldsymbol{\beta}\in\mathbb{R}^p} (\tilde{\boldsymbol{Y}} - \tilde{\boldsymbol{X}}\boldsymbol{\beta})^\top \boldsymbol{V}^{-1}(\tilde{\boldsymbol{Y}} - \tilde{\boldsymbol{X}}\boldsymbol{\beta}) \tag{7.11}$$

to get

$$\tilde{\boldsymbol{\beta}}_n^{(G)} = \boldsymbol{C}^{-1}\tilde{\boldsymbol{X}}^\top \boldsymbol{V}^{-1}\tilde{\boldsymbol{Y}}, \quad \boldsymbol{C} = \tilde{\boldsymbol{X}}^\top \boldsymbol{V}^{-1}\tilde{\boldsymbol{X}}, \tag{7.12}$$

where $\tilde{\boldsymbol{Y}} = (\tilde{y}_1, \ldots, \tilde{y}_n)^\top, \tilde{\boldsymbol{X}} = (\tilde{\boldsymbol{x}}_1, \ldots, \tilde{\boldsymbol{x}}_n)^\top, \tilde{y}_i = y_i - \sum_{j=1}^n W_{nj}(t_i)y_j$, and $\tilde{\boldsymbol{x}}_i = \boldsymbol{x}_i - \sum_{j=1}^n W_{nj}(t_i)\boldsymbol{x}_j$, $i = 1, \ldots, n$.

Now, under a restricted modeling paradigm, suppose that $\boldsymbol{\beta}$ satisfies the following linear nonstochastic constraints

$$\boldsymbol{H}\boldsymbol{\beta} = \boldsymbol{h}, \tag{7.13}$$

where $\boldsymbol{H}$ is a $q \times p$ nonzero matrix with rank $q < p$ and $\boldsymbol{h}$ is a $q \times 1$ vector of prespecified values. We refer restricted partially linear model (RPLM) to (7.20). For the RPLM, one generally adopts the well-known GLSE given by

$$\hat{\boldsymbol{\beta}}_n^{G} = \tilde{\boldsymbol{\beta}}_n^{G} - \boldsymbol{C}^{-1}\boldsymbol{H}^\top(\boldsymbol{H}\boldsymbol{C}^{-1}\boldsymbol{H}^\top)^{-1}(\boldsymbol{H}\tilde{\boldsymbol{\beta}}_n^{G} - \boldsymbol{h}). \tag{7.14}$$

The GLSE is widely used as an unbiased estimator. We refer to Saleh (2006) and Saleh et al. (2014) for more details and applications of restricted parametric or nonparametric models.

Now, in the line of this book, if there exists multicollinearity in $\boldsymbol{C}$, the $\hat{\boldsymbol{\beta}}_n^{G}$ would be badly apart from the actual coefficient parameter in some directions of p-dimension space. Hence, instead of minimizing the GLSE objective function as in (7.21), following Roozbeh (2015), one may minimize the objective function where both $\boldsymbol{\beta}^\top\boldsymbol{\beta}$ and $\boldsymbol{H}\boldsymbol{\beta} - \boldsymbol{h}$ are penalty functions

$$\min_{\boldsymbol{\beta}\in\mathbb{R}^p} (\tilde{\boldsymbol{Y}} - \tilde{\boldsymbol{X}}\boldsymbol{\beta})^\top \boldsymbol{V}^{-1}(\tilde{\boldsymbol{Y}} - \tilde{\boldsymbol{X}}\boldsymbol{\beta}) + k\boldsymbol{\beta}^\top\boldsymbol{\beta} + \boldsymbol{s}(\boldsymbol{H}\boldsymbol{\beta} - \boldsymbol{h}), \tag{7.15}$$

where $\boldsymbol{s} = (s_1, \ldots, s_q)^\top$ is a vector of constants.

The resulting estimator is the restricted generalized ridge estimator (GRE), given by

$$\hat{\boldsymbol{\beta}}_n^{\mathrm{GRR(R)}}(k) = \hat{\boldsymbol{\beta}}_n^{\mathrm{GRR}}(k) - \boldsymbol{C}^{-1}(k)\boldsymbol{H}^\top(\boldsymbol{H}\boldsymbol{C}^{-1}(k)\boldsymbol{H}^\top)^{-1}(\boldsymbol{H}\hat{\boldsymbol{\beta}}_n^{\mathrm{GRR}}(k) - \boldsymbol{h}), \tag{7.16}$$

where

$$\begin{aligned} \hat{\boldsymbol{\beta}}_n^{\mathrm{GRR}}(k) &= \boldsymbol{R}(k)\tilde{\boldsymbol{\beta}}_n^{G}, \quad \boldsymbol{R}(k) = (\boldsymbol{I}_p + k\boldsymbol{C}^{-1})^{-1} \\ &= \boldsymbol{C}^{-1}(k)\tilde{\boldsymbol{X}}^\top \boldsymbol{V}^{-1}\tilde{\boldsymbol{Y}}, \quad \boldsymbol{C}(k) = \boldsymbol{C} + k\boldsymbol{I}_p \end{aligned} \tag{7.17}$$

is the GRE and $k \geq 0$ is the ridge parameter.

From Saleh (2006), the likelihood ratio criterion for testing the null hypothesis $\mathcal{H}_o : \boldsymbol{H\beta} = \boldsymbol{h}$, is given by

$$\mathcal{L}_n = \frac{(\boldsymbol{H}\tilde{\boldsymbol{\beta}}_n^{\rm G} - \boldsymbol{h})^\top (\boldsymbol{H}\boldsymbol{C}^{-1}\boldsymbol{H}^\top)^{-1}(\boldsymbol{H}\tilde{\boldsymbol{\beta}}_n^{\rm G} - \boldsymbol{h})}{qs^2}, \tag{7.18}$$

where $s^2 = \frac{1}{m}(\tilde{\boldsymbol{Y}} - \tilde{\boldsymbol{X}}\tilde{\boldsymbol{\beta}}_n^{\rm G})^\top \boldsymbol{V}^{-1}(\tilde{\boldsymbol{Y}} - \tilde{\boldsymbol{X}}\tilde{\boldsymbol{\beta}}_n^{\rm G})$, $m = n - p$, is an unbiased estimator of σ^2.

Note that $\mathcal{L}_n$ follows a noncentral F-distribution with $(q, n-p)$ DF and noncentrality parameter Δ^2 given by

$$\Delta^2 = \frac{(\boldsymbol{H\beta} - \boldsymbol{h})^\top (\boldsymbol{H}\boldsymbol{C}^{-1}\boldsymbol{H}^\top)^{-1}(\boldsymbol{H\beta} - \boldsymbol{h})}{\sigma^2}. \tag{7.19}$$

Following Saleh (2006), we define three sorts of estimators using the test statistic $\mathcal{L}_n$. First, we consider the preliminary test generalized ridge estimator (PTGRE) defined by

$$\hat{\boldsymbol{\beta}}_n^{\rm GRR(PT)}(k) = \hat{\boldsymbol{\beta}}_n^{\rm GRR(R)}(k) + [1 - I(\mathcal{L}_n \leq F_{q,n-p}(\alpha))](\hat{\boldsymbol{\beta}}_n^{\rm GRR}(k) - \hat{\boldsymbol{\beta}}_n^{\rm GRR(R)}(k)), \tag{7.20}$$

where $F_{q,n-p}(\alpha)$ is the upper α-level critical value for the test of $\mathcal{H}_o$ and $I(A)$ is the indicator function of the set A.

This estimator has been considered by Saleh and Kibria (1993). The PTGRE has the disadvantage that it depends on α, the level of significance, and also it yields the extreme results, namely, $\hat{\boldsymbol{\beta}}_n^{\rm GRR(R)}(k)$ and $\hat{\boldsymbol{\beta}}_n^{\rm GRR}(k)$ depending on the outcome of the test.

A nonconvex continuous version of the PTGRE is the Stein-type generalized ridge estimator (SGRE) defined by

$$\begin{aligned}\hat{\boldsymbol{\beta}}_n^{\rm GRR(S(G))}(k) &= \hat{\boldsymbol{\beta}}_n^{\rm GRR(R)}(k) + (1 - d\mathcal{L}_n^{-1})(\hat{\boldsymbol{\beta}}_n^{\rm GRR}(k) - \hat{\boldsymbol{\beta}}_n^{\rm GRR(R)}(k)),\\ d &= \frac{(q-2)(n-p)}{q(n-p+2)}, \quad q \geq 3.\end{aligned} \tag{7.21}$$

The Stein-type generalized ridge regression has the disadvantage that it has strange behavior for small values of $\mathcal{L}_n$. Also, the shrinkage factor $(1 - d\mathcal{L}_n^{-1})$ becomes negative for $\mathcal{L}_n < d$. Hence, we define the positive-rule Stein-type generalized ridge estimator (PRSGRE) given by

$$\hat{\boldsymbol{\beta}}_n^{\rm GRR(S+)}(k) = \hat{\boldsymbol{\beta}}_n^{\rm GRR(S)}(k) - (1 - d\mathcal{L}_n^{-1})I(\mathcal{L}_n \leq d)(\hat{\boldsymbol{\beta}}_n^{\rm GRR}(k) - \hat{\boldsymbol{\beta}}_n^{\rm GRR(R)}(k)). \tag{7.22}$$

Shrinkage estimators have been considered by Arashi and Tabatabaey (2009), Arashi et al. (2010, 2012), Arashi (2012), and extended to monotone functional estimation in multidimensional models just as the additive regression model, semi-parametric PLM, and generalized linear model by Zhang et al. (2008).

7.4 Biases and L$_2$ Risks of Shrinkage Estimators

In this section, we give exact expressions of the bias and L$_2$-risk functions for the estimators $\hat{\boldsymbol{\beta}}_n^{\text{GRR(S)}}(k)$ and $\hat{\boldsymbol{\beta}}_n^{\text{GRR(S+)}}(k)$. Since the properties of unrestricted, restricted, and preliminary test estimators have been widely investigated in the literature, we refer to Sarkar (1992) and Saleh and Kibria (1993) for the properties of unrestricted, restricted, and preliminary test generalized ridge regression estimator.

Theorem 7.2 *Biases of the Stein-type and positive-rule Stein-type of generalized ridge regression are, respectively, given by*

$$
\begin{aligned}
b(\hat{\boldsymbol{\beta}}_n^{\text{GRR(S)}}(k)) &= -qd\boldsymbol{R}(k)\boldsymbol{\delta}\mathbb{E}[\chi^{-2}_{q+2}(\Delta^*)] - k\boldsymbol{C}^{-1}(k)\boldsymbol{\beta} \\
b(\hat{\boldsymbol{\beta}}_n^{\text{GRR(S)}}(k)) &= \boldsymbol{R}(k)\boldsymbol{\delta}\left\{\frac{qd}{q+2}\mathbb{E}\left[F^{-1}_{q+2,n-p}(\Delta^*)I\left(F_{q+2,n-p}(\Delta^*) \le \frac{qd}{q+2}\right)\right]\right. \\
&\quad \left. - \frac{qd}{q+2}\mathbb{E}[F^{-1}_{q+2,n-p}(\Delta^*)] - G_{q+2,n-p}(x', \Delta^*)\right\} - k\boldsymbol{C}^{-1}(k)\boldsymbol{\beta},
\end{aligned}
\tag{7.23}
$$

where $\boldsymbol{\delta} = \boldsymbol{C}^{-1}\boldsymbol{H}^\top(\boldsymbol{H}\boldsymbol{C}^{-1}\boldsymbol{H}^\top)^{-1}(\boldsymbol{H}\boldsymbol{\beta} - \boldsymbol{h})$,

$$
G_{q+2i,m}(x', \Delta^2) = \sum_{r=0}^{\infty} \frac{e^{-\frac{1}{2}\Delta^2}\left(\frac{1}{2}\Delta^2\right)^r}{\Gamma(r+1)} I_{x'}\left[\frac{q+2i}{2} + r, \frac{m}{2}\right],
$$

$$
\mathbb{E}[\chi^{-2}_{q+s}(\Delta^2)]^n = \sum_{r=0}^{\infty} \frac{\exp\left\{-\frac{1}{2}\Delta^2\right\}\left(\frac{1}{2}\Delta^2\right)^r}{\Gamma(r+1)} \frac{\Gamma\left(\frac{q+s}{2} + r - n\right)}{2^n\Gamma\left(\frac{q+s}{2} + r\right)},
$$

$$
\begin{aligned}
\mathbb{E}\left[\begin{array}{c} F^{-j}_{q+s,m}(\Delta^*)I \\ \left(F_{q+s,m}(\Delta^*) < \dfrac{qd}{q+s}\right)\end{array}\right] &= \sum_{r=0}^{\infty} \frac{\exp\left\{-\frac{1}{2}\Delta^2\right\}\left(\frac{1}{2}\Delta^2\right)^r}{\Gamma(r+1)} \\
&\quad \times \left(\frac{q+s}{m}\right)^j \frac{B\left(\frac{q+s+2r-2j}{2} \frac{n-p+2j}{2}\right)}{B\left(\frac{q+s+2r}{2}, \frac{m}{2}\right)} \\
&\quad \times I_x\left(\frac{q+s+2r-2j}{2}, \frac{m+2j}{2}\right),
\end{aligned}
$$

and $x' = \frac{qF_\alpha}{m+qF_\alpha}$, $x = \frac{qd}{m+qd}$, *$B(a,b)$ is the beta function, and $I_{x'}$ is the Pearson regularized incomplete beta function* $I_{x'}(a,b) = \frac{B_{x'}(a,b)}{B(a,b)} = \frac{\int_0^{x'} t^{a-1}(1-t)^{b-1}}{\int_0^1 t^{a-1}(1-t)^{b-1}}$.

Theorem 7.3 *Risks of the Stein-type and its positive-rule Stein-type of generalized ridge regression are, respectively, given by*

$$\begin{aligned}
R(\hat{\boldsymbol{\beta}}_n^{\mathrm{GRR(S)}}(k)) &= \sigma^2 \operatorname{tr}(\boldsymbol{R}(k)\boldsymbol{C}^{-1}\boldsymbol{R}(k)) + k^2\boldsymbol{\beta}^\top\boldsymbol{C}^{-2}(k)\boldsymbol{\beta}\\
&\quad + 2qdk\boldsymbol{\delta}^\top\boldsymbol{R}(k)\boldsymbol{C}^{-1}(k)\boldsymbol{\beta}\mathbb{E}(\chi^{-2}_{q+2}(\Delta^2))\\
&\quad - dq\sigma^2 \operatorname{tr}(\boldsymbol{R}(k)\boldsymbol{C}^{-1}\boldsymbol{H}^\top(\boldsymbol{H}\boldsymbol{C}^{-1}\boldsymbol{H}^\top)^{-1}\boldsymbol{H}\boldsymbol{C}^{-1}\boldsymbol{R}(k))\\
&\quad \times \left\{(q-2)\mathbb{E}(\chi^{-4}_{q+2}(\Delta^2))\right.\\
&\quad + \left[1 - \frac{(q+2)\boldsymbol{\delta}^\top\boldsymbol{R}^2(k)\boldsymbol{\delta}}{2\sigma^2\Delta^2 \operatorname{tr}(\boldsymbol{R}_k\boldsymbol{C}^{-1}\boldsymbol{H}^\top(\boldsymbol{H}\boldsymbol{C}^{-1}\boldsymbol{H}^\top)^{-1}\boldsymbol{H}\boldsymbol{C}^{-1}\boldsymbol{R}(k))}\right]\\
&\quad \left.\times (2\Delta^2)\mathbb{E}(\chi^{-4}_{q+4}(\Delta^2))\right\},\\
R(\hat{\boldsymbol{\beta}}_n^{\mathrm{GRR(S+)}}(k)) &= R(\hat{\boldsymbol{\beta}}_n^{\mathrm{GRR(S)}}(k))\\
&\quad - \sigma^2 \operatorname{tr}(\boldsymbol{R}(k)\boldsymbol{C}^{-1}\boldsymbol{H}^\top(\boldsymbol{H}\boldsymbol{C}^{-1}\boldsymbol{H}^\top)^{-1}\boldsymbol{H}\boldsymbol{C}^{-1}\boldsymbol{R}(k))\\
&\quad \times \mathbb{E}\left[\left(1 - \frac{qd}{q+2}F^{-1}_{q+2,n-p}(\Delta^2)\right)^2 I\left(F_{q+2,n-p}(\Delta^2) \le \frac{qd}{q+2}\right)\right]\\
&\quad + \boldsymbol{\delta}^\top\boldsymbol{R}(k)^2\boldsymbol{\delta}\mathbb{E}\left[\left(1 - \frac{qd}{q+4}F^{-1}_{q+4,n-p}(\Delta^2)\right)^2 I\left(F_{q+4,n-p}(\Delta^*) \le \frac{qd}{q+4}\right)\right]\\
&\quad - 2\boldsymbol{\delta}^\top\boldsymbol{R}(k)^2\boldsymbol{\delta}\mathbb{E}\left[\left(\frac{qd}{q+2}F^{-1}_{q+2,n-p}(\Delta^2) - 1\right) I\left(F_{q+2,n-p}(\Delta^2) \le \frac{qd}{q+2}\right)\right]\\
&\quad - 2k\boldsymbol{\delta}^\top\boldsymbol{R}(k)\boldsymbol{C}^{-1}(k)\boldsymbol{\beta}\mathbb{E}\left[\left(\frac{qd}{q+2}F^{-1}_{q+2,n-p}(\Delta^2) - 1\right) I\left(F_{q+2,n-p}(\Delta^2) \le \frac{qd}{q+2}\right)\right].
\end{aligned}$$

For analytical comparison between the proposed estimators, we refer to Roozbeh (2015) and continue our study with numerical results.

7.5 Numerical Analysis

To examine the L_2-risk performance of the proposed estimators, we adopted the Monte Carlo simulation study of Roozbeh (2015) for illustration.

To achieve different degrees of collinearity, following Kibria (2003) the explanatory variables were generated using the given model for $n = 1000$:

$$x_{ij} = (1-\gamma^2)^{\frac{1}{2}} z_{ij} + \gamma z_{ip}, \qquad i = 1, \dots, n, \quad j = 1, \dots, p,$$

where z_{ij} are independent standard normal pseudorandom numbers, and γ is specified so that the correlation between any two explanatory variables is given by γ^2. These variables are then standardized so that $\boldsymbol{X}^\top\boldsymbol{X}$ and $\boldsymbol{X}^\top\boldsymbol{Y}$ are in correlation forms. Three different sets of correlation corresponding to $\gamma = 0.80, 0.90$, and 0.99 are considered. Then n observations for the dependent variable are determined by

$$y_i = \sum_{j=1}^{6} x_{ji}\beta_j + f(t_i) + \epsilon_i, \quad i = 1, \dots, n, \tag{7.24}$$

where $\boldsymbol{\beta} = (-1, -1, 2, 3, -5, 4)^\top$, and

$$f(t) = \frac{1}{5}[\phi(t; -7, 1.44) + \phi(t; -3.5, 1) + \phi(t; 0, 0.64) + \phi(t; 3.5, 0.36) + \phi(t; 7, 0.16)],$$

which is a mixture of normal densities for $t \in [-9, 9]$ and $\phi(x; \mu, \sigma^2)$ is the normal probability density function (p.d.f.) with mean μ and variance σ^2. The main reason for selecting such a structure for the nonlinear part is to check the efficiency of the nonparametric estimations for wavy function. This function is difficult to estimate and provides a good test case for the nonparametric regression method.

According to Roozbeh (2015), it is assumed $\boldsymbol{\epsilon} \sim \mathcal{N}_n(\mathbf{0}, \sigma^2 \boldsymbol{V})$, for which the elements of $\boldsymbol{V}$ are $v_{ij} = \left(\frac{1}{n}\right)^{|i-j|}$ and $\sigma^2 = 4.00$. The set of linear restrictions is assumed to have form $\boldsymbol{H}\boldsymbol{\beta} = \mathbf{0}$, where

$$\boldsymbol{H} = \begin{pmatrix} 1 & 5 & -3 & -1 & -3 & 0 \\ -2 & 0 & 0 & -2 & 0 & 1 \\ 2 & 2 & 1 & 3 & -1 & 3 \\ 4 & 2 & 2 & 2 & 0 & -1 \\ 5 & 3 & 4 & -5 & -3 & 0 \end{pmatrix}.$$

For the weight function $W_{ni}(t_j)$, we use

$$W_{ni}(t_j) = \frac{1}{nh_n} K\left(\frac{t_i - tj}{h_n}\right) = \frac{1}{nh_n} . \frac{1}{\sqrt{2\pi}} \exp\left\{-\frac{(t_i - t_j)^2}{2h_n^2}\right\}, \quad h_n = 0.01,$$

which is Priestley and Chao's weight with the Gaussian kernel. We also apply the cross-validation (CV) method to select the optimal bandwidth h_n, which minimizes the following CV function

$$\text{CV} = \frac{1}{n} \sum_{i=1}^{n} (\tilde{\boldsymbol{Y}}^{(-i)} - \tilde{\boldsymbol{X}}^{(-i)} \hat{\boldsymbol{\beta}}^{(-i)})^2,$$

where $\hat{\boldsymbol{\beta}}^{(-i)}$ obtains by replacing $\tilde{\boldsymbol{X}}$ and $\tilde{\boldsymbol{Y}}$ by

$$\tilde{\boldsymbol{X}}^{(-i)} = (\tilde{x}_{jk}^{(-i)}), \quad \text{where} \quad \tilde{x}_{sk}^{(-i)} = x_{sk} - \sum_{j \neq i}^{n} W_{nj}(t_i) x_{sj}$$

$$\boldsymbol{Y}^{(-i)} = (\tilde{y}_1^{(-i)}, \ldots, \tilde{y}_n^{(-i)}), \quad \text{where} \quad \tilde{y}_k^{(-i)} = y_k - \sum_{j \neq i}^{n} W_{nj}(t_i) y_j.$$

Here, $\boldsymbol{Y}^{(-i)}$ is the predicted value of $\boldsymbol{Y} = (y_1, \ldots, y_n)^\top$ at $\boldsymbol{x}_i = (x_{1i}, x_{2i}, \ldots, x_{pi})^\top$ with y_i and $\boldsymbol{x}_i$ left out of the estimation of the $\boldsymbol{\beta}$.

The ratio of largest eigenvalue to smallest eigenvalue of the design matrix in model (7.24) is approximately $\lambda_5/\lambda_1 = 672.11,\ 1549.80$, and $14\,495.28$ for $\gamma = 0.80,\ 0.90$, and 0.99, respectively.

In Tables 7.1–7.6, we computed the proposed estimators' L_2 risk,

$$\mathrm{MSE}(\hat{f}(\boldsymbol{t}), f(\boldsymbol{t})) = \frac{1}{n} \sum_{i=1}^{n} [\hat{f}(t_i) - f(t_i)]^2$$

for different values of k and γ.

We found the best values of k (k_{opt}) by plotting

$$\Delta = R(\hat{\boldsymbol{\beta}}_n^{\text{GRR(S) or GRR(S+)}}) - R(\hat{\boldsymbol{\beta}}_n^{\text{GRR(S) or GRR(S+)}})(k)$$

vs. k for each of the proposed estimators (see Figure 7.1).

Table 7.1 Evaluation of the Stein-type generalized RRE at different k values in model (7.24) with $\gamma = 0.80$.

Coefficients (k)	0	0.05	0.10	0.15	0.20	0.25	k_{opt} = 0.299	0.30
$\hat{\beta}_1$	−1.007	−1.009	−1.010	−1.012	−1.014	−1.015	−0.994	−1.017
$\hat{\beta}_2$	−0.900	−0.902	−0.903	−0.905	−0.906	−0.908	−0.995	−0.909
$\hat{\beta}_3$	1.986	1.986	1.986	1.986	1.987	1.987	2.004	1.987
$\hat{\beta}_4$	3.024	3.025	3.026	3.027	3.028	3.029	3.004	3.030
$\hat{\beta}_5$	−5.073	−5.075	−5.078	−5.081	−5.083	−5.085	−5.002	−5.087
$\hat{\beta}_6$	3.999	4.002	4.005	4.008	4.011	4.013	3.996	4.016
$\hat{R}(\hat{\boldsymbol{\beta}}_n^{\text{GRR(S)}}(k))$	0.274	0.274	0.273	0.273	0.273	0.273	0.273	0.273

Table 7.2 Evaluation of PRSGRE at different k values in model (7.24) with $\gamma = 0.80$.

Coefficients (k)	0	0.05	0.10	0.15	0.20	0.25	k_{opt} = 0.2635	0.30
$\hat{\beta}_1$	−1.004	−1.002	−1.000	−0.998	−0.996	−0.994	−0.993	−0.992
$\hat{\beta}_2$	−1.004	−1.002	−1.000	−0.999	−0.997	−0.995	−0.995	−0.994
$\hat{\beta}_3$	2.008	2.007	2.006	2.006	2.005	2.005	2.004	2.004
$\hat{\beta}_4$	3.012	3.010	3.009	3.007	3.006	3.004	3.004	3.003
$\hat{\beta}_5$	−5.020	−5.016	−5.013	−5.009	−5.006	−5.003	−5.0021	−4.999
$\hat{\beta}_6$	4.016	4.012	4.008	4.004	4.000	3.997	3.996	3.993
$\hat{R}(\hat{\boldsymbol{\beta}}_n^{\text{GRR(S+)}}(k))$	0.242	0.241	0.241	0.241	0.241	0.241	0.241	0.241
$\widehat{\mathrm{MSE}}(\hat{f}(\boldsymbol{t}), f(\boldsymbol{t}))$	0.027	0.027	0.027	0.028	0.028	0.028	0.028	0.028

Table 7.3 Evaluation of SGRE at different k values in model (7.24) with $\gamma = 0.90$.

Coefficients (k)	0	0.10	0.20	0.30	k_{opt} = 0.396	0.40	0.50	0.60
$\hat{\beta}_1$	−0.969	−0.968	−0.966	−0.965	−0.964	−0.964	−0.962	−0.961
$\hat{\beta}_2$	−0.968	−0.967	−0.966	−0.964	−0.963	−0.963	−0.962	−0.961
$\hat{\beta}_3$	2.025	2.024	2.024	2.0232	2.022	2.022	2.021	2.020
$\hat{\beta}_4$	3.009	3.008	3.006	3.005	3.003	3.003	3.002	3.001
$\hat{\beta}_5$	−5.061	−5.058	−5.055	−5.052 184	−5.049	−5.048	−5.045	−5.042
$\hat{\beta}_6$	3.992	3.989	3.985	3.982 498	3.979	3.979	3.975	3.972
$\hat{R}(\hat{\beta}_n^{\text{GRR(S)}}(k))$	0.540	0.537	0.535	0.534	0.534	0.534	0.534	0.536
$\widehat{\text{MSE}}(\hat{f}(\boldsymbol{t}),f(\boldsymbol{t}))$	0.090	0.090	0.089	0.089	0.089	0.089	0.088	0.088

Table 7.4 Evaluation of PRSGRE at different k values in model (7.24) with $\gamma = 0.90$.

Coefficients (k)	0	0.10	0.20	0.30	k_{opt} = 0.348	0.40	0.50	0.60
$\hat{\beta}_1$	−0.966	−0.963	−0.956	−0.950	−0.947	−0.943	−0.937	−0.931
$\hat{\beta}_2$	−0.968	−0.963	−0.957	−0.952	−0.949	−0.946	−0.941	−0.936
$\hat{\beta}_3$	2.025	2.023	2.021	2.018	2.017	2.016	2.014	2.012
$\hat{\beta}_4$	3.009	3.005	3.000	2.996	2.994	2.991	2.987	2.983
$\hat{\beta}_5$	−5.061	−5.050	−5.039	−5.027	−5.022	−5.016	−5.005	−4.994
$\hat{\beta}_6$	3.992	3.978	3.965	3.951	3.944	3.937	3.924	3.911
$\hat{R}(\hat{\beta}_n^{\text{GRR(SG+)}}(k))$	0.476	0.474	0.472	0.471 787	0.471	0.471	0.472	0.474
$\widehat{\text{MSE}}(\hat{f}(\boldsymbol{t}),f(\boldsymbol{t}))$	0.0903	0.0905	0.090	0.090	0.090	0.091	0.0912	0.091

Table 7.5 Evaluation of SGRE at different k values in model (7.24) with $\gamma = 0.99$.

Coefficients (k)	0	0.10	0.20	0.30	0.40	k_{opt} = 0.422	0.50	0.60
$\hat{\beta}_1$	−1.430	−1.366	−1.308	−1.257	−1.210	−1.200	−1.167	−1.127
$\hat{\beta}_2$	−1.267	−1.205	−1.150	−1.100	−1.056	−1.046	−1.015	−0.977
$\hat{\beta}_3$	1.487	1.500	1.510	1.517	1.522	1.523	1.525	1.527
$\hat{\beta}_4$	3.046	3.024	3.001	2.977	2.954	2.949	2.930	2.906
$\hat{\beta}_5$	−4.945	−4.876	−4.811	−4.747	−4.687	−4.674	−4.628	−4.572
$\hat{\beta}_6$	5.025	4.854	4.700	4.563	4.438	4.412	4.324	4.221
$\hat{R}(\hat{\beta}_n^{\text{GRR(S)}}(k))$	5.748	5.391	5.179	5.082	5.076	5.085	5.142	5.265
$\widehat{\text{MSE}}(\hat{f}(\boldsymbol{t}),f(\boldsymbol{t})])$	0.040	0.040	0.041	0.041	0.042	0.042	0.042	0.042

Table 7.6 Evaluation of PRSGRE at different k values in model (7.24) with $\gamma = 0.99$.

Coefficients (k)	0	0.10	0.20	0.30	k_{opt} = 0.365	0.40	0.50	0.60
$\hat{\beta}_1$	−1.430	−1.340	−1.260	−1.188	−1.144	−1.122	−1.062	−1.007
$\hat{\beta}_2$	−1.267	−1.179	−1.101	−1.030	−0.987	−0.965	−0.907	−0.854
$\hat{\beta}_3$	1.487	1.496	1.501	1.503	1.503	1.503	1.500	1.496
$\hat{\beta}_4$	3.046	3.007	2.967	2.927	2.901	2.887	2.847	2.808
$\hat{\beta}_5$	−4.945	−4.834	−4.728	−4.626	−4.562	−4.528	−4.433	−4.342
$\hat{\beta}_6$	5.025	4.786	4.572	4.379	4.264	4.205	4.046	3.901
$\hat{R}(\hat{\beta}_n^{GRR(S+)}(k))$	5.060	4.752	4.584	4.525	4.534	4.552	4.648	4.798
$\widehat{MSE}(\hat{f}(t), f(t))$	0.040	0.041	0.042	0.042	0.043	0.043	0.044	0.045

Figure 7.2 shows the fitted function by kernel smoothing after estimation of the linear part of the model using $\hat{\beta}_n^{GRR(S)}(k_{opt})$ and $\hat{\beta}_n^{GRR(S+)}(k_{opt})$, that are, $Y - X\hat{\beta}_n^{S(G) \text{ or } S(G)+}(k_{opt})$, respectively, for $\gamma = 0.80, 0.90$, and 0.99. The minimum of CV approximately obtained was $h_n = 0.82$ for the model (7.24) with $n = 1000$. The diagram of CV vs. h_n is also plotted in Figure 7.3.

From the simulation, we found that at most k for superiority of PRSGRE over SGLSE are 0.598, 0.792, and 0.844 for $\gamma = 0.80,\ 0.90$, and 0.99, respectively. It is realized that the ridge parameter k is an increasing function of γ for superiority of PRSGRE over SGRE in L_2-risk sense. These k values are 0.527, 0.697, and 0.730 for superiority of PRSGRE over PRSGLSE. Finally, the PRSGRE is better than SGRE for all values of k and γ in L_2-risk sense.

7.5.1 Example: Housing Prices Data

The housing prices data consist of 92 detached homes in the Ottawa area that were sold during 1987. The variables are defined as follows. The dependent variable is sale price (SP), the independent variables include lot size (lot area; LT), square footage of housing (SFH), average neighborhood income (ANI), distance to highway (DHW), presence of garage (GAR), and fireplace (FP). The full parametric model has the form

$$\begin{aligned}(\text{SP})_i = \beta_0 + \beta_1(\text{LT})_i + \beta_2(\text{SFH})_i + \beta_3(\text{FP})_i + \beta_4(\text{DHW})_i \\ + \beta_5(\text{GAR})_i + \beta_6(\text{ANI})_i + \epsilon_i. \qquad (7.25)\end{aligned}$$

Looking at the correlation matrix given in Table 7.7, there exists a potential multicollinearity between variables SFH & FP and DHW & ANI.

We find that the eigenvalues of the matrix $X^\top X$ are given by $\lambda_1 = 1.682\,93$, $\lambda_2 = 6.658\,370$, $\lambda_3 = 15.617\,10$, $\lambda_4 = 18.824\,93$, $\lambda_5 = 23.900\,21$,

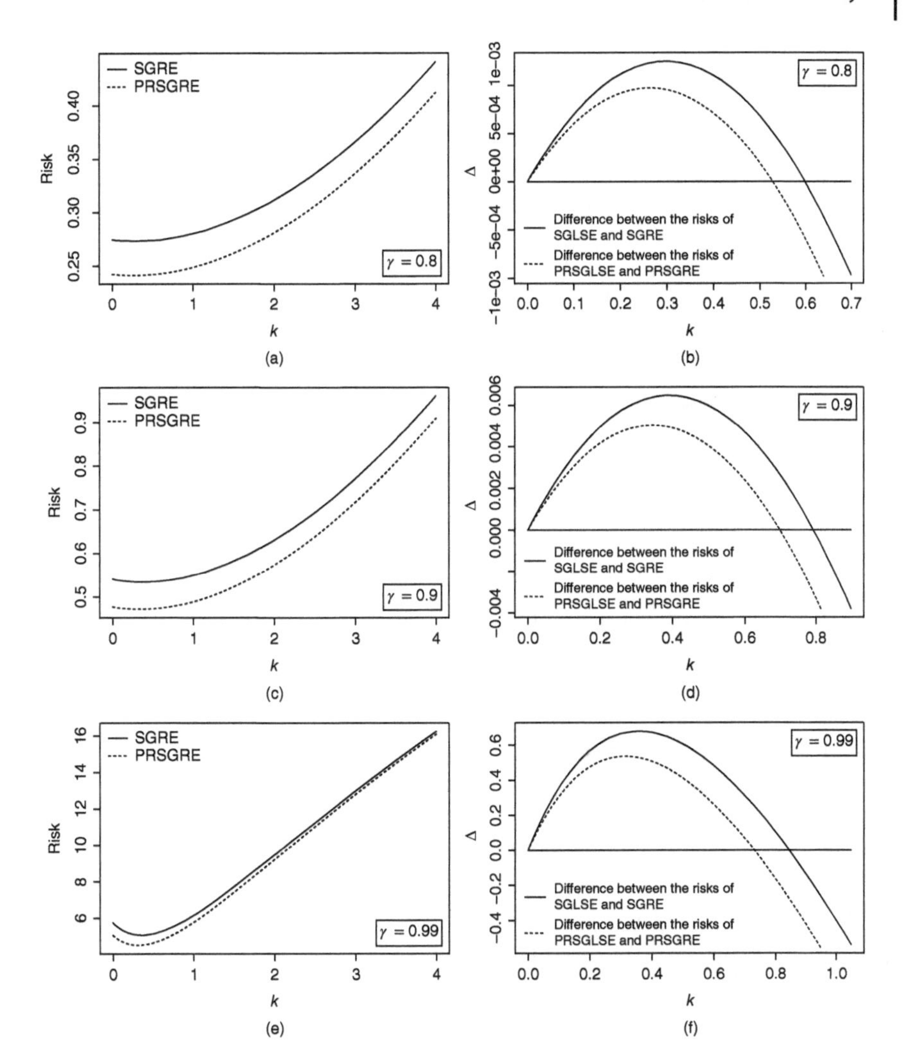

Figure 7.1 Plots of Δ and risks vs. k for different values of γ.

$\lambda_6 = 230.2869$, and $\lambda_7 = 238\ 556.3$. It is easy to see that the condition number is approximately equal to 141 750.325 00. So, $\boldsymbol{X}$ is morbidity badly.

An appropriate approach, as a remedy for multicollinearity, is to replace the pure parametric model with a PLM. Akdeniz and Tabakan (2009) proposed a PLM (here termed as semi-parametric model) for this data by taking the lot area as the nonparametric component.

To realize the type of relation from linearity/nonlinearity viewpoint between dependent variable (SP) and explanatory variables (except for the binary ones),

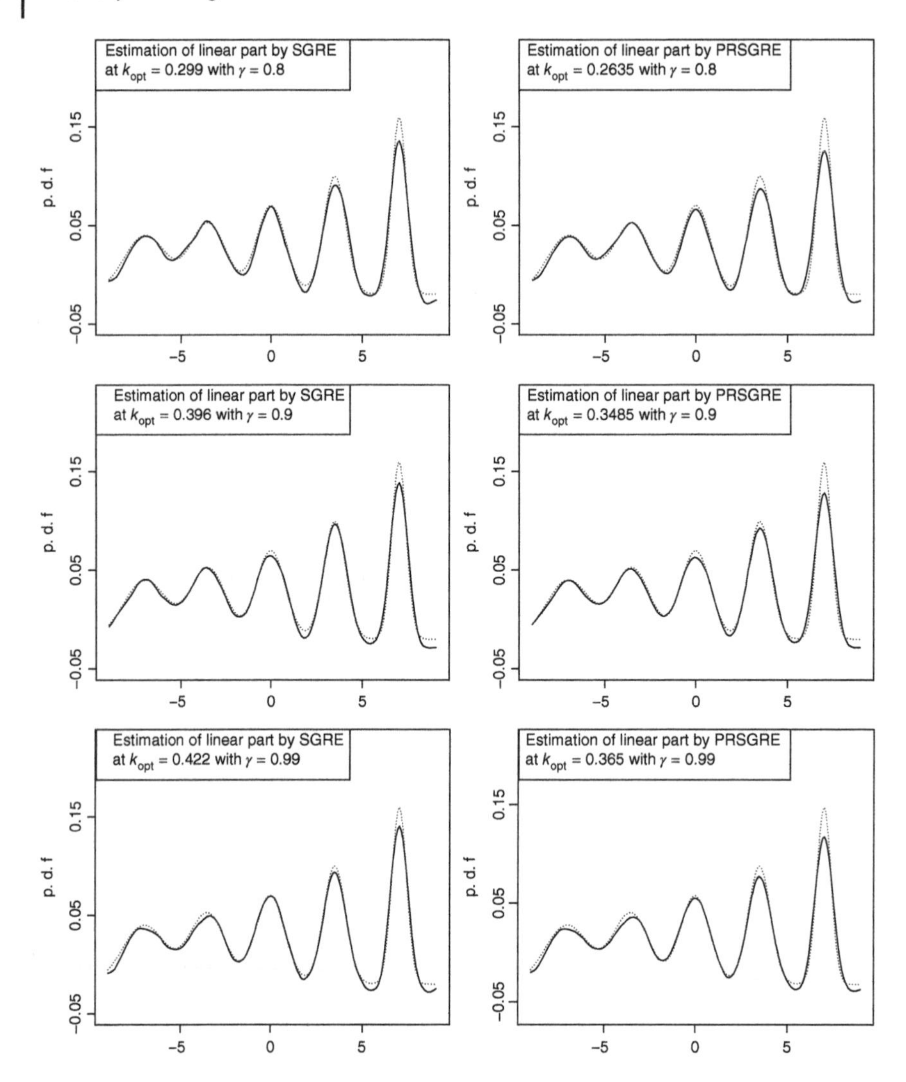

Figure 7.2 Estimation of the mixtures of normal p.d.fs by the kernel approach. Solid lines are the estimates and dotted lines are the true functions.

they are plotted in Figure 7.4. According to this figure, we consider the average neighborhood income (ANI) as a nonparametric part. So, the specification of the semi-parametric model is

$$\begin{aligned}(\mathrm{SP})_i = \boldsymbol{\beta}_0 + \boldsymbol{\beta}_1(\mathrm{LT})_i + \boldsymbol{\beta}_2(\mathrm{SFH})_i + \boldsymbol{\beta}_3(\mathrm{FP})_i + \boldsymbol{\beta}_4(\mathrm{DHW})_i \\ + \boldsymbol{\beta}_5(\mathrm{GAR})_i + f(\mathrm{ANI})_i + \epsilon_i.\end{aligned} \tag{7.26}$$

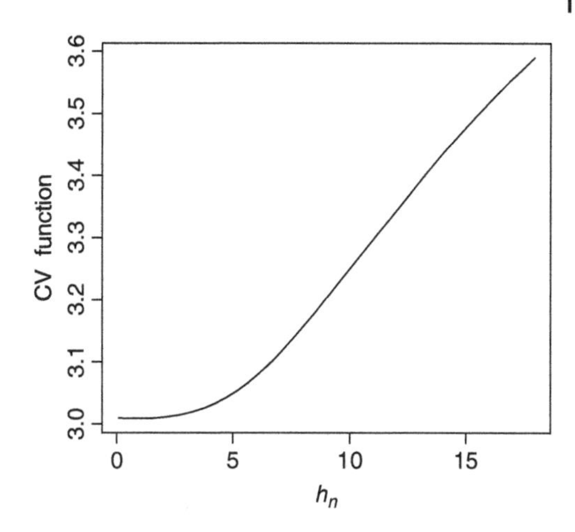

Figure 7.3 Plot of CV vs. h_n.

Table 7.7 Correlation matrix.

Variable	SP	LT	SFH	FP	DHW	GAR	ANI
SP	1.00	0.14	0.47	0.33	−0.10	0.29	0.34
LT	0.14	1.00	0.15	0.15	0.08	0.16	0.13
SFH	0.47	0.15	1.00	0.46	0.02	0.22	0.27
FP	0.33	0.15	0.46	1.00	0.10	0.14	0.38
DHW	−0.10	0.08	0.02	0.10	1.00	0.05	−0.10
GAR	0.29	0.16	0.22	0.14	0.05	1.00	0.02
ANI	0.34	0.13	0.27	0.38	−0.10	0.02	1.00

To compare the performance of the proposed restricted estimators, following Roozbeh (2015), we consider the parametric restriction $\boldsymbol{H\beta} = \boldsymbol{0}$, where

$$\boldsymbol{H} = \begin{pmatrix} -1 & 0 & -1 & -1 & 1 \\ 1 & 0 & -1 & 2 & 0 \\ 0 & -1 & 0 & -2 & 8 \end{pmatrix}.$$

The test statistic for testing $\mathcal{H}_o : \boldsymbol{H\beta} = \boldsymbol{0}$, given our observations, is

$$\chi^2_{\text{rank}(\boldsymbol{H})} \simeq (\boldsymbol{H}\hat{\boldsymbol{\beta}}_n^{\text{G}} - \boldsymbol{h})^\top (\boldsymbol{H}\hat{\boldsymbol{\Sigma}}_{\hat{\beta}}\boldsymbol{H}^\top)^{-1} (\boldsymbol{H}\hat{\boldsymbol{\beta}}_n^{\text{G}} - \boldsymbol{h}) = 0.4781,$$

where $\hat{\boldsymbol{\Sigma}}_{\hat{\beta}} = \hat{\sigma}^2(\tilde{\boldsymbol{X}}^\top \tilde{\boldsymbol{X}})^{-1}$. Thus, we conclude that the null-hypothesis $\mathcal{H}_o$ is accepted.

Table 7.8 summarizes the results. The "parametric estimates" refer to a model in which ANI enters. In the "semiparametric estimates," we have used the kernel regression procedure with optimal bandwidth $h_n = 3.38$ for estimating

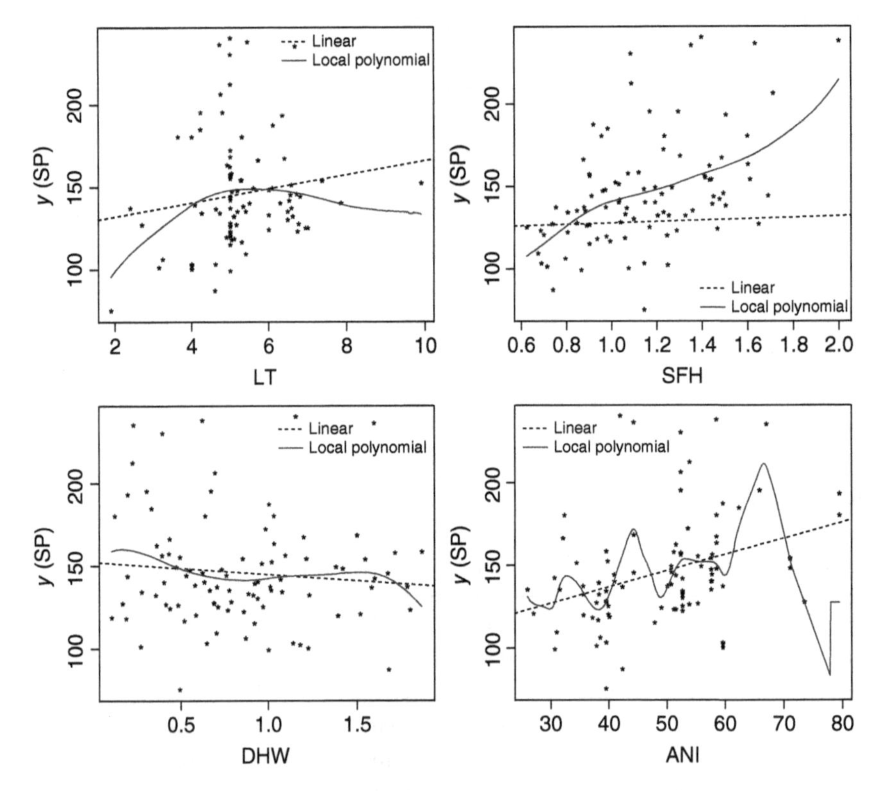

Figure 7.4 Plots of individual explanatory variables vs. dependent variable, linear fit (dash line), and local polynomial fit (solid line).

Table 7.8 Fitting of parametric and semi-parametric models to housing prices data.

	Parametric estimates		**Semiparametric estimates**	
Variable	**Coef.**	**s.e.**	**Coef.**	**s.e.**
Intercept	62.67	20.69	—	—
LT	0.81	2.65	9.40	2.25
SFH	39.27	11.99	73.45	11.15
FP	6.05	7.68	5.33	8.14
DHW	−8.21	6.77	−2.22	7.28
GAR	14.47	6.38	12.98	7.20
ANI	0.56	0.28	—	—
s^2	798.0625		149.73	
RSS	33 860.06		88 236.41	
R^2	0.3329		0.86	

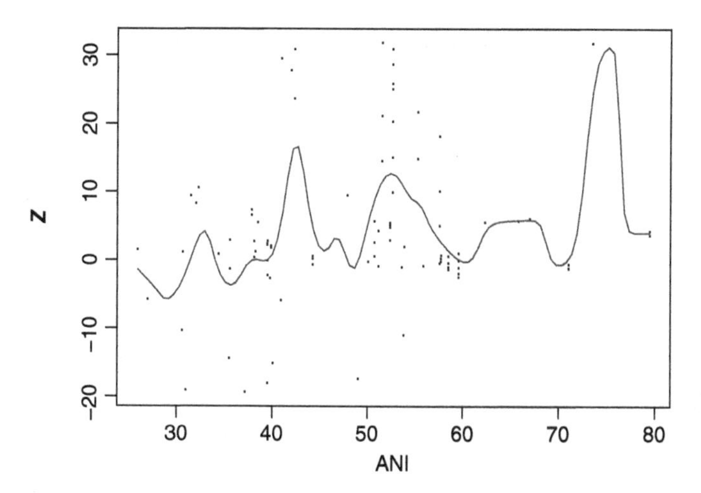

Figure 7.5 Estimation of nonlinear effect of ANI on dependent variable by kernel fit.

f(ANI). For estimating the nonparametric effect, first we estimated the parametric effects and then applied the kernel approach to fit $\boldsymbol{Z}_i = \text{SP}_i - \boldsymbol{x}_i\tilde{\hat{\boldsymbol{\beta}}}_n^{\text{G}}$ on ANI$_i$ for $i = 1, \dots, n$, where $\boldsymbol{x}_i = (\text{LT}_i, \text{SFH}_i, \text{FP}_i, \text{DHW}_i, \text{GAR}_i)$ (Figure 7.5).

The ratio of largest to smallest eigenvalues for the new design matrix in model (7.25) is approximately $\lambda_5/\lambda_1 = 427.9926$, and so there exists a potential multicollinearity between the columns of the design matrix. Now, in order to overcome the multicollinearity for better performance of the estimators, we used the proposed estimators for model (7.25).

The SGRRE and PRSGRE for different values of the ridge parameter are given in Tables 7.9 and 7.10, respectively. As it can be seen, $\hat{\boldsymbol{\beta}}_n^{\text{GRR(S+)}}(k_{\text{opt}})$ is the best estimator for the linear part of the semi-parametric regression model in the sense of risk.

Table 7.9 Evaluation of SGRRE at different k values for housing prices data.

Variable (k)	0.0	0.10	0.20	k_{opt} = 0.216	0.30	0.40	0.50	0.60
Intercept	—	—	—	—	—	—	—	—
LT	9.043	9.139	9.235	9.331	9.427	9.250	9.524	9.620
SFH	84.002	84.106	84.195	84.270	84.334	84.208	84.385	84.425
FP	0.460	0.036	0.376	0.777	1.168	0.441	1.549	1.919
DHW	−4.086	−4.382	−4.673	−4.960	−5.241	−4.719	−5.519	−5.792
GAR	4.924	4.734	4.548	4.367	4.190	4.519	4.017	3.848
ANI	—	—	—	—	—	—	—	—
$\hat{R}(\hat{\boldsymbol{\beta}}_n^{\text{GRR(S)}}(k))$	257.89	254.09	252.62	252.59	253.35	256.12	260.82	267.31
$\widehat{\text{MSE}}(\hat{f}(.), f(.))$	2241.81	2241.08	2240.50	2240.42	2240.12	2240.03	2240.02	2240.18

Table 7.10 Evaluation of PRSGRRE at different k values for housing prices data.

Variable (k)	0.0	0.10	k_{opt} = 0.195	0.20	0.30	0.40	0.50	0.60
Intercept	—	—	—	—	—	—	—	—
LT	9.226	9.355	9.476	9.481	9.604	9.725	9.843	9.958
SFH	78.614	77.705	76.860	76.820	75.959	75.120	74.303	73.506
FP	2.953	3.229	3.483	3.495	3.751	3.998	4.237	4.466
DHW	−3.136	−3.033	−2.937	−2.933	−2.835	−2.739	−2.645	−2.554
GAR	9.042	9.112	9.176	9.179	9.242	9.303	9.362	9.417
ANI	—	—	—	—	—	—	—	—
$\hat{R}(\hat{\beta}_n^{GRR(S+)}(k);\beta)$	234.02	230.68	229.67	230.82	232.68	234.02	239.12	246.00
$\widehat{MSE}(\hat{f}(.),f(.))$	2225.762	2233.685	2240.480	2240.838	2248.805	2256.785	2264.771	2272.758

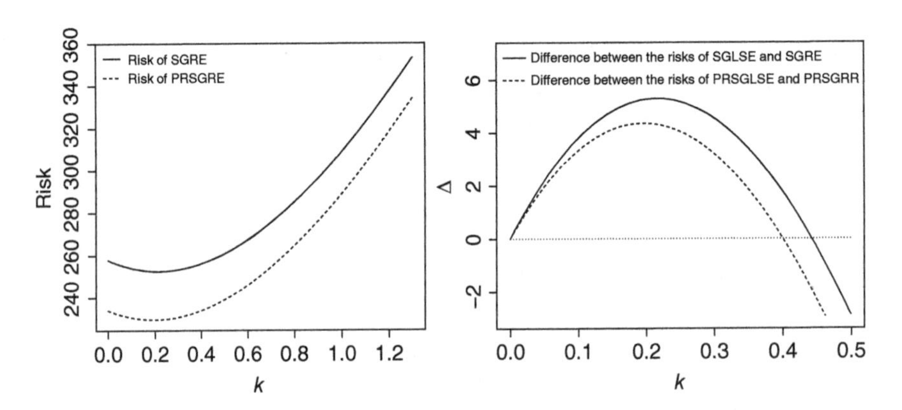

Figure 7.6 The diagrams of Δ and risk vs. k for housing prices data.

In Figure 7.6, the Δ and $R(.)$ of SGRE (solid line) and PRSGDRE (dash line) vs. ridge parameter k are plotted. Finally, we estimated the nonparametric effect (f(ANI)) after estimation of the linear part by $\hat{\beta}_n^{\text{GRR(S) or GRR(S+)}}(k_{opt})$ in Figure 7.7, i.e. we used kernel fit to regress $Z^{\text{GRR(S) or GRR(S+)}} = SP - X\hat{\beta}_n^{\text{GRR(S) or GRR(S+)}}(k_{opt})$ on ANI.

7.6 High-Dimensional PLM

In this section, we develop the theory proposed in the previous section for the high-dimensional case in which $p > n$. For our purpose, we adopt the structure

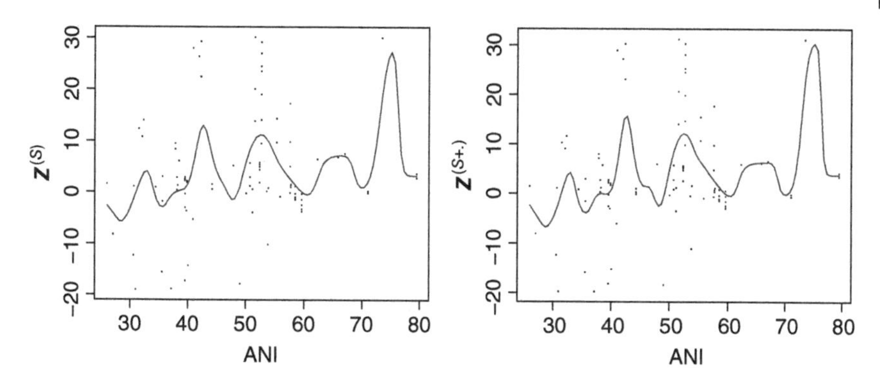

Figure 7.7 Estimation of *f*(ANI) by kernel regression after removing the linear part by the proposed estimators in housing prices data.

of Arashi and Roozbeh (2016), i.e. we partition the regression parameter $\boldsymbol{\beta}$ as $\boldsymbol{\beta} = (\boldsymbol{\beta}_1^\top, \boldsymbol{\beta}_2^\top)^\top$, where the sub-vector $\boldsymbol{\beta}_i$ has dimension p_i, $i = 1, 2$ and $p_1 + p_2 = p$ with $p_1 < n$. Hence, the model (7.20) has the form

$$\boldsymbol{Y} = \boldsymbol{X}_1\boldsymbol{\beta}_1 + \boldsymbol{X}_2\boldsymbol{\beta}_2 + \boldsymbol{f}(t) + \boldsymbol{\epsilon}, \tag{7.27}$$

where $\boldsymbol{X}$ is partitioned according to $(\boldsymbol{X}_1, \boldsymbol{X}_2)$ in such a way that $\boldsymbol{X}_i$ is a $n \times p_i$ sub-matrix, $i = 1, 2$.

In this section, we consider the estimation of $\boldsymbol{\beta}_1$ under the sparsity assumption $\boldsymbol{\beta}_2 = \boldsymbol{0}$. Hence, we minimize the objective function

$$\min_{\boldsymbol{\beta} \in \mathbb{R}^p} (\tilde{\boldsymbol{Y}} - \tilde{\boldsymbol{X}}\boldsymbol{\beta})^\top \boldsymbol{V}^{-1}(\tilde{\boldsymbol{Y}} - \tilde{\boldsymbol{X}}\boldsymbol{\beta}) + k_n\boldsymbol{\beta}^\top\boldsymbol{\beta} + \boldsymbol{r}\boldsymbol{\beta}_2, \tag{7.28}$$

where k_n is a function of sample size n and $\boldsymbol{r} = (r_1, \dots, r_{p_2})^\top$ is a vector of constants.

The resulting estimator of $\boldsymbol{\beta}_1$ is the generalized ridge regression estimator (RRE), given by

$$\begin{aligned}\hat{\boldsymbol{\beta}}_{n1}^{\mathrm{GRR}}(\omega_n, k_n) &= (\boldsymbol{X}_1^\top \boldsymbol{V}^{-1}\boldsymbol{X}_1 + k_n\boldsymbol{I}_{p_1})^{-1}\boldsymbol{X}_1^\top\boldsymbol{V}^{-1}\boldsymbol{Y}\\ &= (\boldsymbol{I}_{p_1} + k_n(\boldsymbol{X}_1^\top\boldsymbol{V}^{-1}\boldsymbol{X}_1)^{-1})^{-1}\hat{\boldsymbol{\beta}}_{n1}^{\mathrm{G}}(\omega_n)\\ &= \boldsymbol{T}_1(\omega_n, k_n)\hat{\boldsymbol{\beta}}_{n1}^{\mathrm{G}}(\omega_n),\end{aligned} \tag{7.29}$$

where $\boldsymbol{T}_1(\omega_n, k_n) = (\boldsymbol{I}_{p_1} + k_n(\boldsymbol{X}_1^\top\boldsymbol{V}^{-1}\boldsymbol{X}_1)^{-1})^{-1}$ and $\hat{\boldsymbol{\beta}}_{n1}^{\mathrm{GRR}}(\omega_n)$ is the generalized restricted estimator of $\boldsymbol{\beta}_1$ given by

$$\hat{\boldsymbol{\beta}}_{n1}^{\mathrm{GRR}}(\omega_n) = (\tilde{\boldsymbol{X}}_1^\top\boldsymbol{V}^{-1}\tilde{\boldsymbol{X}}_1)^{-1}\tilde{\boldsymbol{X}}_1^\top\boldsymbol{V}^{-1}\tilde{\boldsymbol{Y}}. \tag{7.30}$$

Here, ω_n is the bandwidth parameter as a function of sample size n, where we estimate $\boldsymbol{f}(t)$ by $\hat{\boldsymbol{f}}(t) = \boldsymbol{k}(t)\boldsymbol{Y}$, $\boldsymbol{k}(t) = (K_{\omega_n}(t, t_1), \dots, K_{\omega_n}(t, t_n))$ with $K_{\omega_n}(\cdot)$ is a kernel function of order m with bandwidth parameter ω_n (see Arashi and Roozbeh 2016).

Note that the generalized (unrestricted) ridge estimators of $\boldsymbol{\beta}_1$ and $\boldsymbol{\beta}_2$, respectively, have the forms

$$\hat{\boldsymbol{\beta}}_{n1}^{\text{GRR}}(\omega_n, k_n) = \boldsymbol{R}_1(\omega_n, k_n)\hat{\boldsymbol{\beta}}_{n1}^{\text{GRR}}(\omega_n),$$
$$\hat{\boldsymbol{\beta}}_{n2}^{\text{GRR}}(\omega_n, k_n) = \boldsymbol{R}_2(\omega_n, k_n)\hat{\boldsymbol{\beta}}_{n2}^{\text{GRR}}(\omega_n), \tag{7.31}$$

where

$$\boldsymbol{R}_1(\omega_n, k_n) = (\boldsymbol{I}_{p_1} + k_n(\boldsymbol{X}_1^\top \boldsymbol{\Sigma}_2^{-1}(\omega_n, k_n)\boldsymbol{X}_1)^{-1})^{-1}$$
$$\boldsymbol{R}_2(\omega_n, k_n) = (\boldsymbol{I}_{p_2} + k_n(\boldsymbol{X}_2^\top \boldsymbol{\Sigma}_1^{-1}(\omega_n, k_n)\boldsymbol{X}_2)^{-1})^{-1}$$
$$\boldsymbol{\Sigma}_i^{-1}(\omega_n, k_n) = \boldsymbol{V}^{-1} - \boldsymbol{V}^{-1}\boldsymbol{X}_i(\boldsymbol{X}_i^\top \boldsymbol{V}^{-1}\boldsymbol{X}_i + k_n\boldsymbol{I}_{p_i})^{-1}\boldsymbol{X}_i^\top \boldsymbol{V}^{-1}, \quad i = 1, 2.$$

In order to derive the forms of shrinkage estimators, we need to find the test statistic for testing the null-hypothesis $\mathcal{H}_o : \boldsymbol{\beta}_2 = \boldsymbol{0}$. Thus, we make use of the following statistic

$$\mathcal{L}_n^* = \frac{\hat{\boldsymbol{\beta}}_{n2}^{\text{GRR}}(\omega_n, k_n)^\top \boldsymbol{C}_2(\omega_n)\hat{\boldsymbol{\beta}}_{n2}^{\text{GRR}}(\omega_n, k_n)}{(n - p_1)s_*^2}, \tag{7.32}$$

where $\boldsymbol{C}_2(\omega_n) = \tilde{\boldsymbol{X}}_2^\top[\boldsymbol{V}^{-1} - \boldsymbol{V}^{-1}\tilde{\boldsymbol{X}}_1(\tilde{\boldsymbol{X}}_1^\top \boldsymbol{V}^{-1}\tilde{\boldsymbol{X}}_1)^{-1}\tilde{\boldsymbol{X}}_1^\top \boldsymbol{V}^{-1}]\tilde{\boldsymbol{X}}_2$ and

$$s_*^2 = \frac{1}{n - p_1}(\tilde{\boldsymbol{Y}} - \tilde{\boldsymbol{X}}_1\hat{\boldsymbol{\beta}}_1^{\text{GRR}}(\omega_n, k_n))^\top \boldsymbol{V}^{-1}(\tilde{\boldsymbol{Y}} - \tilde{\boldsymbol{X}}_1\hat{\boldsymbol{\beta}}_1^{\text{GRR}}(\omega_n, k_n)). \tag{7.33}$$

To find the asymptotic distribution of $\mathcal{L}_n^*$, we need to make some regularity conditions. Hence, we suppose the following assumptions hold

(A1) $\max_{1\le i\le n} \boldsymbol{x}_i^\top(\boldsymbol{X}^\top \boldsymbol{V}^{-1}\boldsymbol{X} + k_n\boldsymbol{I}_p)^{-1}\boldsymbol{x}_i = o(n)$, where $\boldsymbol{x}_i^\top$ is the ith row of $\boldsymbol{X}$.
(A2) $\frac{k_n}{n} \to k_o$ as $n \to \infty$,
(A3) Let

$$\begin{aligned}
\boldsymbol{A}_n &= \boldsymbol{X}^\top \boldsymbol{V}^{-1}\boldsymbol{X} + k_n\boldsymbol{I}_p \\
&= \begin{pmatrix} \boldsymbol{X}_1^\top \\ \boldsymbol{X}_2^\top \end{pmatrix} \boldsymbol{V}^{-1}(\boldsymbol{X}_1\ \boldsymbol{X}_2) + k_n\boldsymbol{I}_p \\
&= \begin{pmatrix} \boldsymbol{X}_1^\top \boldsymbol{V}^{-1}\boldsymbol{X}_1 & \boldsymbol{X}_1^\top \boldsymbol{V}^{-1}\boldsymbol{X}_2 \\ \boldsymbol{X}_2^\top \boldsymbol{V}^{-1}\boldsymbol{X}_1 & \boldsymbol{X}_2^\top \boldsymbol{V}^{-1}\boldsymbol{X}_2 \end{pmatrix} + k_n \begin{pmatrix} \boldsymbol{I}_{p_1} & \boldsymbol{0}_{p_1\times p_2} \\ \boldsymbol{0}_{p_2\times p_1} & \boldsymbol{I}_{p_2} \end{pmatrix} \\
&= \begin{pmatrix} \boldsymbol{A}_{n11} & \boldsymbol{A}_{n12} \\ \boldsymbol{A}_{n21} & \boldsymbol{A}_{n22} \end{pmatrix}.
\end{aligned}$$

Then, there exists a positive definite matrix $\boldsymbol{A}$ such that

$$\frac{1}{n}\boldsymbol{A}_n \to \boldsymbol{A} = \begin{pmatrix} \boldsymbol{A}_{11} & \boldsymbol{A}_{12} \\ \boldsymbol{A}_{21} & \boldsymbol{A}_{22} \end{pmatrix}, \quad \text{as} \quad n \to \infty.$$

(A4) $\boldsymbol{X}_2^\top \boldsymbol{V}^{-1}\boldsymbol{X}_2 = o(\sqrt{n})$.
(A5) $\boldsymbol{x}_i^\top \boldsymbol{x}_j = o(\sqrt{n})$, $i, j = 1, \dots, n$.

Note that by (A2), (A4), and (A5), one can directly conclude that as $n \to \infty$

$$\boldsymbol{\Sigma}_2^{-1}(\omega_n, k_n) \to \boldsymbol{V}^{-1}. \tag{7.34}$$

The test statistic $\mathcal{L}_n^*$ diverges as $n \to \infty$, under any fixed alternatives $A_\xi : \boldsymbol{\beta}_2 = \boldsymbol{\xi}$. To overcome this difficulty, in sequel, we consider the local alternatives

$$\mathcal{K}_{(n)} : \boldsymbol{\beta}_2 = \boldsymbol{\beta}_{2(n)} = n^{-\frac{1}{2}}\boldsymbol{\xi},$$

where $\boldsymbol{\xi} = (\xi_1, \dots, \xi_{p_2})^\top \in \mathbb{R}^{p_2}$ is a fixed vector.

Then, we have the following result about the asymptotic distribution of $\mathcal{L}_n^*$. The proof is left as an exercise.

Theorem 7.4 *Under the regularity conditions (A1)–(A3) and local alternatives $\{\mathcal{K}_{(n)}\}$, $\pounds_n$ is asymptotically distributed according to a noncentral chi-square distribution with p_2 degrees of freedom and noncentrality parameter $\frac{1}{2}\Delta^*$, where*

$$\Delta^* = \frac{1}{\sigma^2}\boldsymbol{\xi}'\boldsymbol{A}_{22.1}\boldsymbol{\xi}, \quad \boldsymbol{A}_{22.1} = \boldsymbol{A}_{22} - \boldsymbol{A}_{21}\boldsymbol{A}_{11}^{-1}\boldsymbol{A}_{12}.$$

Then, in a similar manner to that in Section 5.2, the PTGRE, SGRE, and PRSGRE of $\boldsymbol{\beta}_1$ are defined respectively as

$$\begin{aligned}
\hat{\boldsymbol{\beta}}_{n1}^{\mathrm{GRR(PT)}}(\omega_n, k_n) &= \hat{\boldsymbol{\beta}}_{n1}^{\mathrm{GRR(R)}}(\omega_n, k_n) + [1 - I(\mathcal{L}_n^* \le \chi_n^2 p_2(\alpha))] \\
&\quad \times (\hat{\boldsymbol{\beta}}_{n1}^{\mathrm{GRR}}(\omega_n, k_n) - \hat{\boldsymbol{\beta}}_{n1}^{\mathrm{GRR(R)}}(\omega_n, k_n)) \\
&= \boldsymbol{T}_1(\omega_n, k_n)\hat{\boldsymbol{\beta}}_{n1}^{\mathrm{GRR(R)}}(\omega_n) + [1 - I(\mathcal{L}_n^* \le \chi_{p_2}^2(\alpha))] \\
&\quad \times (\boldsymbol{R}_1(\omega_n, k_n)\hat{\boldsymbol{\beta}}_{n1}^{\mathrm{GRR}} - \boldsymbol{T}_1(\omega_n, k_n)\hat{\boldsymbol{\beta}}_{n1}^{\mathrm{GRR(R)}}(\omega_n))
\end{aligned} \tag{7.35}$$

$$\begin{aligned}
\hat{\boldsymbol{\beta}}_{n1}^{\mathrm{GRR(S)}}(\omega_n, k_n) &= \hat{\boldsymbol{\beta}}_{n1}^{\mathrm{G}}(\omega_n, k_n) + (1 - d\mathcal{L}_n^{*\,-1}) \\
&\quad \times (\hat{\boldsymbol{\beta}}_{n1}^{\mathrm{GRR}}(\omega_n, k_n) - \hat{\boldsymbol{\beta}}_{n1}^{\mathrm{GRR(R)}}(\omega_n, k_n)), \quad d = p_2 - 2 > 0 \\
&= \boldsymbol{T}_1(\omega_n, k_n)\hat{\boldsymbol{\beta}}_{n1}^{\mathrm{GRR(R)}}(\omega_n) + (1 - d\mathcal{L}_n^{*\,-1}) \\
&\quad \times (\boldsymbol{R}_1(\omega_n, k_n)\hat{\boldsymbol{\beta}}_{n1}^{\mathrm{GRR}} - \boldsymbol{T}_1(\omega_n, k_n)\hat{\boldsymbol{\beta}}_{n1}^{\mathrm{GRR(R)}}(\omega_n))
\end{aligned} \tag{7.36}$$

$$\begin{aligned}
\hat{\boldsymbol{\beta}}_{n1}^{\mathrm{GRR(S+)}}(\omega_n, k_n) &= \hat{\boldsymbol{\beta}}_{n1}^{\mathrm{GRR(S)}}(\omega_n, k_n) - (1 - d\mathcal{L}_n^{*\,-1})I(\mathcal{L}_n^* \le d) \\
&\quad \times (\hat{\boldsymbol{\beta}}_{n1}^{\mathrm{GRR}}(\omega_n, k_n) - \hat{\boldsymbol{\beta}}_{n1}^{\mathrm{GRR(R)}}(\omega_n, k_n)).
\end{aligned}$$

Under the foregoing regularity conditions (A1)–(A5) and local alternatives $\{\mathcal{K}_{(n)}\}$, one can derive asymptotic properties of the proposed estimators, which are left as exercise. To end this section, we present in the next section a real data example to illustrate the usefulness of the suggested estimators for high-dimensional data.

7.6.1 Example: Riboflavin Data

Here, we consider the data set about riboflavin (vitamin B_2) production in *Bacillus subtilis,* which can be found in the R package "hdi". There is a single real-valued response variable which is the logarithm of the riboflavin production rate. Furthermore, there are $p = 4088$ explanatory variables measuring the logarithm of the expression level of 4088 genes. There is one rather homogeneous data set from $n = 71$ samples that were hybridized repeatedly during a fed batch fermentation process where different engineered strains and strains grown under different fermentation conditions were analyzed. Based on 100-fold cross-validation, the LASSO shrinks 4047 parameters to zero and remains $p_1 = 41$ significant explanatory variables.

To detect the nonparametric part of the model, for $i = 1, \dots, 41$, we calculate

$$s_i^2 = \frac{1}{n - p_1 - 1}(\boldsymbol{Y} - \boldsymbol{X}_1[, -i]\hat{\boldsymbol{\beta}}_{n1}^{\mathrm{GRR}}(\omega_n))^{\top}(\boldsymbol{Y} - \boldsymbol{X}_1[, -i]\hat{\boldsymbol{\beta}}_{n1}^{\mathrm{GRR(R)}}(\omega_n)),$$

where $\boldsymbol{X}_1[, -i]$ is obtained by deleting the ith column of matrix $\boldsymbol{X}_1$. Among all 41 remaining genes, "*DNAJ_at*" had minimum s_*^2 value. We also use the added-variable plots to identify the parametric and nonparametric components of the model. Added-variable plots enable us to visually assess the effect of each predictor, having adjusted for the effects of the other predictors. By looking at the added-variable plot (Figure 7.8), we consider "*DNAJ_at*" as a nonparametric part. As it can be seen from this figure, the slope of the least

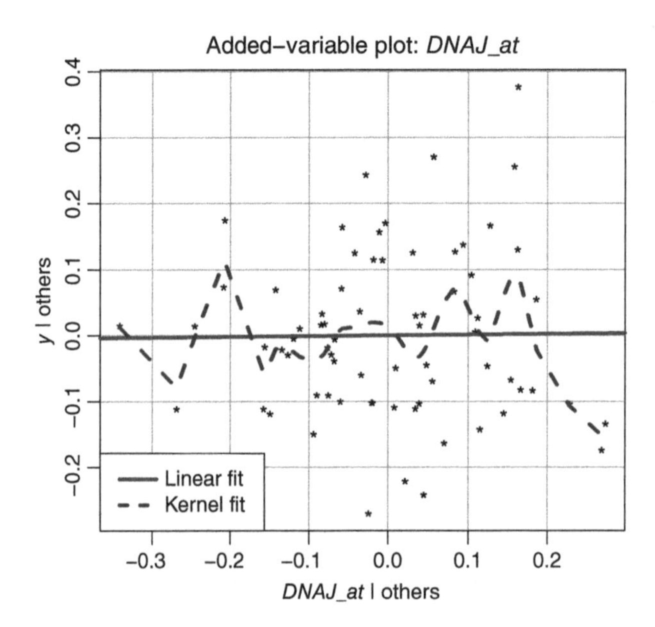

Figure 7.8 Added-variable plot of explanatory variables *DNAJ_at* vs. dependent variable, linear fit (solid line) and kernel fit (dashed line).

Table 7.11 Evaluation of proposed estimators for real data set.

Method	GRE	GRRE	SGRRE	PRSGRRE
RSS	16.1187	1.9231	1.3109	1.1509
R^2	0.7282	0.9676	0.9779	0.9806

squares line is equal to zero, approximately, and so, the specification of the sparse semi-parametric regression model is

$$Y = X_1\beta_1 + X_2\beta_2 + f(t) + \epsilon, \quad t = DNAJ\_at, \tag{7.37}$$

where $p_1 = 40$ and $p_2 = 4047$.

For estimating the nonparametric part of the model, $f(t)$, we use

$$W_{\omega_n}(t_j) = \frac{1}{n\omega_n} K\left(\frac{t - tj}{\omega_n}\right) = \frac{1}{n\omega_n} \cdot \frac{1}{\sqrt{2\pi}} \exp\left\{-\frac{(t_i - t_j)^2}{2\omega_n^2}\right\},$$

which is Priestley and Chao's weight with the Gaussian kernel.

We also apply the CV method to select the optimal bandwidth ω_n and k_n, which minimizes the following CV function

$$\mathrm{CV}(\omega_n, k_n) = \frac{1}{n}\sum_{i=1}^{n} (\tilde{Y}^{(-i)} - \tilde{X}^{(-i)}\hat{\beta}^{(-i)}(\omega_n, k_n))^2,$$

where $\hat{\beta}^{(-i)}$ is obtained by replacing $\tilde{X}$ and $\tilde{Y}$ with $\tilde{X}^{(-i)} = (\tilde{x}_{jk}^{(-i)})$, $1 \le k \le n$, $1 \le j \le p$, $\tilde{Y}^{(-i)} = (\tilde{y}_1^{(-i)}, \ldots, \tilde{y}_n^{(-i)})$, $\tilde{x}_{sk}^{(-i)} = x_{sk} - \sum_{j\neq i}^{n} W_{nj}(t_i)x_{sj}$, $\tilde{y}_k^{(-i)} = y_k - \sum_{j\neq i}^{n} W_{nj}(t_i)y_j$.

Table 7.11 shows a summary of the results. In this table, the RSS and R^2, respectively, are the residual sum of squares and coefficient of determination of the model, i.e. $\mathrm{RSS} = \sum_{i=1}^{n} (y_i - \hat{y}_i)^2$, $\hat{y}_i = \boldsymbol{x}_i\hat{\boldsymbol{e}} + \hat{f}(t_i)$, and $R^2 = 1 - \mathrm{RSS}/S_{yy}$. For estimation of the nonparametric effect, at first we estimated the parametric effects by one of the proposed methods and then the local polynomial approach was applied to fit $y_i - \boldsymbol{x}_i^\top\hat{\boldsymbol{\beta}}$ on t_i, $i = 1, \ldots, n$.

In this example, as can be seen from Figure 7.8, the nonlinear relation between dependent variable and *DNAJ_at* can be detected, and so the pure parametric model does not fit the data; the semi-parametric regression model fits more significantly. Further, from Table 7.11, it can be deduced that PRSGRE is quite efficient, in the sense that it has significant value of goodness of fit.

7.7 Summary and Concluding Remarks

In this chapter, apart from the preliminary test and penalty estimators, we mainly concentrated on the Stein-type shrinkage ridge estimators, for estimating the regression parameters of the partially linear regression models. We developed the problem of ridge estimation for the low/high-dimensional case. We analyzed the performance of the estimators extensively in Monte Carlo simulations, where the nonparametric component of the model was estimated by the kernel method. The housing prices data used the low-dimensional part and the riboflavin data used for the high-dimensional part for practical illustrations, in which the cross-validation function was applied to obtain the optimal bandwidth parameters.

Problems

7.1 Prove that the test statistic for testing $\mathcal{H}_o : \boldsymbol{\beta}_2 = \mathbf{0}$ is

$$\mathcal{L}_n = \frac{1}{\sigma^2} \tilde{\boldsymbol{\beta}}_{2n}^\top (\boldsymbol{X}_2^\top \hat{\boldsymbol{M}}_1 \boldsymbol{X}_2) \tilde{\boldsymbol{\beta}}_{2n}$$

and c_α is the α-level critical value. What would be the distribution of $\mathcal{L}_n$ under both null and alternative hypotheses?

7.2 Prove that the biases of the Stein-type and positive-rule Stein-type of generalized RRE are, respectively, given by

$$b(\hat{\boldsymbol{\beta}}_n^{\text{GRR(S)}}(k)) = -qd\boldsymbol{R}(k)\boldsymbol{\delta}\mathbb{E}[\chi_{q+2}^{-2}(\Delta^*)] - kC^{-1}(k)\boldsymbol{\beta}$$

$$\begin{aligned} b(\hat{\boldsymbol{\beta}}_n^{\text{GRR(S+)}}(k)) = \boldsymbol{R}(k)\boldsymbol{\delta} & \\ \times \Bigg\{ \frac{qd}{q+2} \mathbb{E}\left[F_{q+2,n-p}^{-1}(\Delta^*) I\left(F_{q+2,n-p}(\Delta^*) \leq \frac{qd}{q+2} \right) \right] & \\ - \frac{qd}{q+2} \mathbb{E}[F_{q+2,n-p}^{-1}(\Delta^*)] - G_{q+2,n-p}(x', \Delta^*) \Bigg\} & \\ - kC^{-1}(k)\boldsymbol{\beta}, & \end{aligned}$$

where $\boldsymbol{\delta}$ and $G_{q+2i,m}(x', \Delta^2)$ are defined in Theorem 7.2.

7.3 Show that the risks of the Stein-type and positive-rule Stein-type of generalized RRE are, respectively, given by

$$\begin{aligned} R(\hat{\boldsymbol{\beta}}_n^{\text{GRR(S)}}(k)) = & \ \sigma^2 \operatorname{tr}(\boldsymbol{R}(k)\boldsymbol{C}^{-1}\boldsymbol{R}(k)) + k^2 \boldsymbol{\beta}^\top \boldsymbol{C}^{-2}(k)\boldsymbol{\beta} \\ & + 2qdk\boldsymbol{\delta}^\top \boldsymbol{R}(k)\boldsymbol{C}^{-1}(k)\boldsymbol{\beta}\mathbb{E}(\chi_{q+2}^{-2}(\Delta^2)) \\ & - dq\sigma^2 \operatorname{tr}(\boldsymbol{R}(k)\boldsymbol{C}^{-1}\boldsymbol{H}^\top(\boldsymbol{H}\boldsymbol{C}^{-1}\boldsymbol{H}^\top)^{-1}\boldsymbol{H}\boldsymbol{C}^{-1}\boldsymbol{R}(k)) \\ & \times \Bigg\{ (q-2)\mathbb{E}(\chi_{q+2}^{-4}(\Delta^2)) \\ & + \left[1 - \frac{(q+2)\boldsymbol{\delta}^\top \boldsymbol{R}^2(k)\boldsymbol{\delta}}{2\sigma^2\Delta^2 \operatorname{tr}(\boldsymbol{R}_k \boldsymbol{C}^{-1}\boldsymbol{H}^\top(\boldsymbol{H}\boldsymbol{C}^{-1}\boldsymbol{H}^\top)^{-1}\boldsymbol{H}\boldsymbol{C}^{-1}\boldsymbol{R}(k))} \right] \\ & \times (2\Delta^2)\mathbb{E}(\chi_{q+4}^{-4}(\Delta^2)) \Bigg\}, \end{aligned}$$

$$
\begin{aligned}
R(\hat{\boldsymbol{\beta}}_n^{\text{GRR(S+)}}(k)) &= R(\hat{\boldsymbol{\beta}}_n^{\text{GRR(S)}}(k)) \\
&\quad - \sigma^2 \operatorname{tr}(\boldsymbol{R}(k)\boldsymbol{C}^{-1}\boldsymbol{H}^\top(\boldsymbol{H}\boldsymbol{C}^{-1}\boldsymbol{H}^\top)^{-1}\boldsymbol{H}\boldsymbol{C}^{-1}\boldsymbol{R}(k)) \\
&\quad \times \mathbb{E}\left[\left(1 - \frac{qd}{q+2} F_{q+2,n-p}^{-1}(\Delta^2)\right)^2 I\left(F_{q+2,n-p}(\Delta^2) \leq \frac{qd}{q+2}\right)\right] \\
&\quad + \boldsymbol{\delta}^\top \boldsymbol{R}(k)^2 \boldsymbol{\delta} \mathbb{E}\left[\left(1 - \frac{qd}{q+4} F_{q+4,n-p}^{-1}(\Delta^2)\right)^2 I\left(F_{q+4,n-p}(\Delta^*) \leq \frac{qd}{q+4}\right)\right] \\
&\quad - 2\boldsymbol{\delta}^\top \boldsymbol{R}(k)^2 \boldsymbol{\delta} \mathbb{E}\left[\left(\frac{qd}{q+2} F_{q+2,n-p}^{-1}(\Delta^2) - 1\right) I\left(F_{q+2,n-p}(\Delta^2) \leq \frac{qd}{q+2}\right)\right] \\
&\quad - 2k\boldsymbol{\delta}^\top \boldsymbol{R}(k)\boldsymbol{C}^{-1}(k)\boldsymbol{\beta} \mathbb{E}\left[\left(\frac{qd}{q+2} F_{q+2,n-p}^{-1}(\Delta^2) - 1\right) I\left(F_{q+2,n-p}(\Delta^2) \leq \frac{qd}{q+2}\right)\right].
\end{aligned}
$$

7.4 Prove that the test statistic for testing $\mathcal{H}_o : \boldsymbol{\beta}_2 = \boldsymbol{0}$ is

$$
\mathcal{L}_n^* = \frac{\hat{\boldsymbol{\beta}}_{n2}^{\text{GRR}}(\omega_n, k_n)^\top \boldsymbol{C}_2(\omega_n) \hat{\boldsymbol{\beta}}_{n2}^{\text{GRR}}(\omega_n, k_n)}{(n - p_1)s_*^2}
$$

where $\boldsymbol{C}_2(\omega_n) = \tilde{\boldsymbol{X}}_2^\top[\boldsymbol{V}^{-1} - \boldsymbol{V}^{-1}\tilde{\boldsymbol{X}}_1(\tilde{\boldsymbol{X}}_1^\top \boldsymbol{V}^{-1}\tilde{\boldsymbol{X}}_1)^{-1}\tilde{\boldsymbol{X}}_1^\top \boldsymbol{V}^{-1}]\tilde{\boldsymbol{X}}_2$ and

$$
s_*^2 = \frac{1}{n - p_1}(\tilde{\boldsymbol{Y}} - \tilde{\boldsymbol{X}}_1 \hat{\boldsymbol{\beta}}_{G1}(\omega_n, k_n))^\top \boldsymbol{V}^{-1}(\tilde{\boldsymbol{Y}} - \tilde{\boldsymbol{X}}_1 \hat{\boldsymbol{\beta}}_{G1}(\omega_n, k_n)).
$$

7.5 Prove Theorem 7.2.

7.6 Consider a real data set, where the design matrix elements are moderate to highly correlated, then find the efficiency of the estimators using unweighted risk functions. Find parallel formulas for the efficiency expressions and compare the results with that of the efficiency using weighted risk function. Are the two results consistent?

8

Logistic Regression Model

In this chapter, we discuss the shrinkage and penalty estimators related to logistic regression models. Logistic regression is often used when the response variable encodes a binary outcome. Logistic regression is a widely used tool for the statistical analysis of observed proportions or rates. Methods for fitting logistic multiple regression models have been available through standard statistical packages for the past several years.

8.1 Introduction

Logistic regression is a popular method to model binary data in biostatistics and health sciences. It has extensive applications in many different spheres of life. A classic area of application of logistic regression is biomedical studies. There are other areas like the prediction of loan returning behavior of bank clients or whether a company would or would not reach a bankruptcy situation in the future (Beaver 1966; Martin 1977; Tam and Kiang 1992). Another popular example is to investigate the occurrence of a disease as related to different characteristics of the patients, such as age, sex, food habit, daily exercise, and others. Fitting a model with appropriate significant predictors is a challenging job. When selecting the variables for a linear model, we generally look at individual p-values. This procedure can be deceptive. If the variables are highly correlated, the p-values may also be high, motivating the researcher to mistakenly deduce that those variables are not important predictors. On the other hand, for the same reason, some irrelevant variables that are not associated with the response variable may be included in the model. Unstable parameter estimates occur when the number of covariates is relatively large or when the covariates are highly correlated. This chapter deals with the estimation of the parameters for the logistic regression model when the covariates are highly correlated. To describe the problem, let $Y_i \in \{0, 1\}$ denote the dichotomous dependent variable and let $\boldsymbol{x}_i = (1, x_{1i}, x_{2i}, \dots, x_{pi})^\top$ be a $(p+1)$-dimensional

Theory of Ridge Regression Estimation with Applications, First Edition.
A.K. Md. Ehsanes Saleh, Mohammad Arashi, and B.M. Golam Kibria.

vector of explanatory variables for the ith observation. The conditional probability of $Y_i = 1$ given $\boldsymbol{x}_i$ is given by

$$P(Y_i = 1|\boldsymbol{x}_i) = \pi(\boldsymbol{x}_i) = (1 + \exp\{-\beta_0 + \boldsymbol{x}_i^\top \boldsymbol{\beta}\})^{-1}, \tag{8.1}$$

where $\boldsymbol{\beta} = (\beta_0, \beta_1, \dots, \beta_p)^\top$ is the $(p+1)$-vector regression parameter of interest. The logit transformation in terms of $\pi(\boldsymbol{x}_i)$ is given by

$$\ln\left(\frac{\pi(\boldsymbol{x}_i)}{1 - \pi(\boldsymbol{x}_i)}\right) = \beta_0 + \boldsymbol{x}_i^\top \boldsymbol{\beta} = (\beta_0 + \beta_1 x_{1i} + \cdots + \beta_p x_{pi}). \tag{8.2}$$

Our primary objective is to estimate $\boldsymbol{\beta}$ when it is suspected that it belongs to the linear subspace defined by $\boldsymbol{H}\boldsymbol{\beta} = \boldsymbol{h}$, where $\boldsymbol{H}$ is a $q \times p$ matrix of known real values with rank q and $\boldsymbol{h}$ is a q-vector of known real values. The most common method of estimation of $\boldsymbol{\beta}$ is the maximum likelihood (ML) method. That is to say, we have to minimize the log-likelihood function

$$\sum_{i=1}^{n} y_i \log \pi(x_i) + (1 - y_i) \log(1 - \pi(x_i))$$

with respect to β_j's ($j = 1, \dots, p$). The ML equation is given by

$$\boldsymbol{X}^\top(\boldsymbol{Y} - \pi(\boldsymbol{x})) = \boldsymbol{0}, \tag{8.3}$$

where $\boldsymbol{X}$ is the $n \times (p+1)$ matrix of covariates, $\boldsymbol{Y} = (Y_1, \dots, Y_n)^\top$ is an n-vector of binary variables and $\pi(\boldsymbol{x}) = (\pi(x_1), \dots, \pi(x_n))^\top$. Since (8.1) is nonlinear in $\boldsymbol{\beta}$, one must use the iterative method to solve it. The Hosmer and Lemeshow (1989) iterative method finds $(t+1)$th iteration based on tth iterative results as

$$\boldsymbol{\beta}^{(t+1)} = \boldsymbol{\beta}^{(t)} + \{\boldsymbol{X}^\top \operatorname{Diag}(\pi(\boldsymbol{x}_i)^{(t)}(1 - \pi(\boldsymbol{x}_i)^{(t)}))\boldsymbol{X}\}^{-1}\boldsymbol{X}^\top(\boldsymbol{Y} - \boldsymbol{m}^{(t)}), \tag{8.4}$$

where $\boldsymbol{m}^{(t)} = (\pi(x_1)^{(t)}, \dots, \pi(x_p)^{(t)})^\top$.

The solution of (8.1) is obtained using the following iterative weighted least-squares algorithm given by

$$\hat{\boldsymbol{\beta}}_n^{\mathrm{ML}} = (\boldsymbol{X}^\top \hat{\boldsymbol{W}}_n \boldsymbol{X})^{-1}(\boldsymbol{X}^\top \hat{\boldsymbol{W}}_n \boldsymbol{Z}), \tag{8.5}$$

where $\hat{\boldsymbol{W}}_n = \operatorname{Diag}(\hat{\pi}(x_1)(1 - \hat{\pi}(x_1)), \dots, \hat{\pi}(x_n)(1 - \hat{\pi}(x_n)))$, $\boldsymbol{Z} = (Z_1, \dots, Z_n)^\top$ and

$$Z_i = \ln\left(\frac{\hat{\pi}(x_i)}{1 - \hat{\pi}(x_i)}\right) + \frac{Y_i - \hat{\pi}(x_i)}{\hat{\pi}(x_i)(1 - \hat{\pi}(x_i))}. \tag{8.6}$$

The estimated information matrix may be written as

$$\boldsymbol{I}(\hat{\boldsymbol{\beta}}_n^{\mathrm{ML}}) = \boldsymbol{X}^\top \boldsymbol{W} \boldsymbol{X} = \operatorname{Cov}(\hat{\boldsymbol{\beta}}_n^{\mathrm{ML}}).$$

Hence, one can show that

$$\sqrt{n}(\hat{\boldsymbol{\beta}}_n^{\mathrm{ML}} - \boldsymbol{\beta}) \sim \mathcal{N}_{p+1}(\boldsymbol{0}, \boldsymbol{I}(\hat{\boldsymbol{\beta}}_n^{\mathrm{ML}})^{-1}). \tag{8.7}$$

Now, consider the partition $\boldsymbol{\beta} = (\boldsymbol{\beta}_1^\top, \boldsymbol{\beta}_2^\top)^\top$ and $\boldsymbol{X} = (\boldsymbol{X}_1, \boldsymbol{X}_2)$. Then, we may write the estimators of $(\boldsymbol{\beta}_1^\top, \boldsymbol{\beta}_2^\top)^\top$ as

$$\begin{pmatrix} \hat{\boldsymbol{\beta}}_{1n}^{\mathrm{ML}} \\ \hat{\boldsymbol{\beta}}_{2n}^{\mathrm{ML}} \end{pmatrix} = \begin{pmatrix} \boldsymbol{X}_1^\top \hat{\boldsymbol{W}} \boldsymbol{X}_1 & \boldsymbol{X}_1^\top \hat{\boldsymbol{W}} \boldsymbol{X}_2 \\ \boldsymbol{X}_2^\top \hat{\boldsymbol{W}} \boldsymbol{X}_1 & \boldsymbol{X}_2^\top \hat{\boldsymbol{W}} \boldsymbol{X}_2 \end{pmatrix}^{-1} \begin{pmatrix} \boldsymbol{X}_1^\top \hat{\boldsymbol{W}} \boldsymbol{Z}_{(1)} \\ \boldsymbol{X}_1^\top \hat{\boldsymbol{W}} \boldsymbol{Z}_{(2)} \end{pmatrix}, \tag{8.8}$$

where

$$\begin{aligned} \boldsymbol{Z}_{(1)} &= (Z_1, \ldots, Z_{(p_1)})^\top \\ \boldsymbol{Z}_{(2)} &= (Z_{p_1+1}, \ldots, Z_p)^\top, \end{aligned}$$

respectively.

If we suspect *sparsity* in $\boldsymbol{\beta}$, i.e. $\boldsymbol{\beta}_2 = \boldsymbol{0}$, then the restricted maximum likelihood estimator (RMLE) of $\boldsymbol{\beta}_R = \begin{pmatrix} \boldsymbol{\beta}_1 \\ \boldsymbol{0} \end{pmatrix}$ is given by $\begin{pmatrix} \hat{\boldsymbol{\beta}}_{1n}^{\mathrm{ML}} \\ \boldsymbol{0} \end{pmatrix}$ where

$$\hat{\boldsymbol{\beta}}_{1n}^{\mathrm{ML}(R)} = (\boldsymbol{X}_1^\top \hat{\boldsymbol{W}}_n \boldsymbol{X}_1)^{-1} \boldsymbol{X}_1^\top \hat{\boldsymbol{W}}_n \boldsymbol{Z}_{(1)}. \tag{8.9}$$

8.1.1 Penalty Estimators

We consider two basic penalty estimators, namely, (i) ridge regression estimators (RREs) and the (ii) least absolute shrinkage and selection operator (LASSO) estimator of $\boldsymbol{\beta}$.

As for the RRE, we use $\min_{\boldsymbol{\beta} \in \mathbb{R}^p} \{l(\boldsymbol{\beta}) + k\boldsymbol{\beta}^\top \boldsymbol{\beta}\}$, yielding a normal equation as

$$\sum_{i=1}^{s} y_i x_{i\alpha} - \sum_{i=1}^{s} n_i x_{i\alpha} \hat{\pi}(x_i) + k\beta_\alpha = 0, \quad \alpha = 0, 1, \ldots, p. \tag{8.10}$$

In matrix form, we may write (8.10) as

$$(\boldsymbol{X}^\top \hat{\boldsymbol{W}} \boldsymbol{X} + k\boldsymbol{I}_p)\boldsymbol{\beta} = \boldsymbol{X}^\top \hat{\boldsymbol{W}} \boldsymbol{Z} \tag{8.11}$$

so that the RRE of $\boldsymbol{\beta}$ is given by

$$\hat{\boldsymbol{\beta}}_n^{\mathrm{RRML}}(k) = (\boldsymbol{X}^\top \hat{\boldsymbol{W}} \boldsymbol{X} + k\boldsymbol{I}_p)^{-1} \boldsymbol{X}^\top \hat{\boldsymbol{W}} \boldsymbol{Z}. \tag{8.12}$$

To obtain the RRE) of $\boldsymbol{\beta}$, we consider the asymptotic marginal distribution of $(\hat{\boldsymbol{\beta}}_{1n}^{\mathrm{ML}^\top}, \hat{\boldsymbol{\beta}}_{2n}^{\mathrm{ML}^\top})^\top$ given by

$$\sqrt{n}(\hat{\beta}_{in}^{\mathrm{ML}} - \beta_i) \sim \mathcal{N}_{p_i}(\boldsymbol{0}, \sigma^2 \boldsymbol{C}_{ii\cdot j}^{-1}), \quad i = 1, 2; j = 1 \text{ if } i = 1; \; j = 2 \text{ if } i = 2.$$

Then, the RRE of $\boldsymbol{\beta}$ may be defined as

$$\begin{pmatrix} \hat{\boldsymbol{\beta}}_{1n}^{\mathrm{ML}} \\ (\boldsymbol{I}_{p_2} + k\boldsymbol{C}_{22\cdot1}^{-1})\hat{\boldsymbol{\beta}}_{2n}^{\mathrm{ML}}, \end{pmatrix}$$

where

$$C_{11\cdot 2} = X_1^\top \hat{W} X_1 - (X_1^\top \hat{W} X_2)(X_2^\top \hat{W} X_2)^{-1}(X_2^\top \hat{W} X_1)$$
$$C_{22\cdot 1} = X_2^\top \hat{W} X_2 - (X_2^\top \hat{W} X_1)(X_1^\top \hat{W} X_1)^{-1}(X_1^\top \hat{W} X_2),$$

respectively.

To obtain the RRE of $\boldsymbol{\beta}$, we consider the asymptotic marginal distribution of $(\hat{\boldsymbol{\beta}}_{1n}^\top, \hat{\boldsymbol{\beta}}_{2n}^\top)^\top$ given by

$$\sqrt{n}(\hat{\beta}_{in} - \beta_i) \sim \mathcal{N}_{p_i}(\mathbf{0}, \sigma^2 C_{ii\cdot j}^{-1}), \quad i = 1, 2,\ j = 2 \text{ if } i = 1;\ j = 1 \text{ if } i = 2.$$

Then, the RRE of $\boldsymbol{\beta}$ may be defined as

$$\begin{pmatrix} \hat{\boldsymbol{\beta}}_{1n} \\ (\boldsymbol{I}_{p_2} + k C_{22\cdot 1}^{-1})^{-1} \tilde{\boldsymbol{\beta}}_{2n}, \end{pmatrix}$$

where

$$C_{11\cdot 2} = X_1^\top \hat{W} X_1 - (X_1^\top \hat{W} X_2)(X_2^\top \hat{W} X_2)^{-1}(X_2^\top \hat{W} X_1)$$
$$C_{22\cdot 1} = X_2^\top \hat{W} X_2 - (X_2^\top \hat{W} X_1)(X_1^\top \hat{W} X_1)^{-1}(X_1^\top \hat{W} X_2),$$

respectively, with

$$C = \begin{pmatrix} C_{11} & C_{12} \\ C_{21} & C_{22} \end{pmatrix} \tag{8.13}$$

$$C_{11} = X_1^\top \hat{W} X_1, \quad C_{22} = X_2^\top \hat{W} X_2, \quad C_{12} = X_1^\top \hat{W} X_2, \quad C_{21} = X_2^\top \hat{W} X_1$$

and

$$C^{-1} = \begin{pmatrix} C_{11\cdot 2}^{-1} & -C_{11\cdot 2}^{-1} C_{12} C_{22}^{-1} \\ -C_{22}^{-1} C_{21} C_{11\cdot 2}^{-1} & C_{22\cdot 1}^{-1} \end{pmatrix}. \tag{8.14}$$

Following and Donoho and Johnstone (1994), the modified least absolute shrinkage and selection operator (MLASSO) estimator may be defined by

$$\hat{\beta}_n^{\text{MLASSO}}(\lambda) = (\text{sgn}(\tilde{\beta}_{jn})(|\tilde{\beta}_{jn}| - \lambda\sqrt{C^{jj}})^+ | j = 0, 1, \ldots, p)^\top, \tag{8.15}$$

where C^{jj} is the jth diagonal element of C^{-1}. This estimator forces some coefficient exactly equal to zero.

8.1.2 Shrinkage Estimators

Since we suspect that the sparsity condition $\boldsymbol{\beta}_2 = \mathbf{0}$ may hold, we use the Wald statistic, $\mathcal{L}_n$, for testing the hypothesis $\mathcal{H}_o : \boldsymbol{\beta}_2 = \mathbf{0}$, where

$$\mathcal{L}_n = \hat{\boldsymbol{\beta}}_{2n}^{\text{ML}^\top} C_{22} \hat{\boldsymbol{\beta}}_{2n}^{\text{ML}} \xrightarrow{\mathcal{D}} \chi_{p_2}^2. \tag{8.16}$$

For large samples and under $\boldsymbol{\beta}_2 = \mathbf{0}$, $\mathcal{L}_n$ has a chi-square distribution with p_2 degree of freedom (DF). Let $\chi_{p_2}^2(\alpha) = c_\alpha$, (say) be the α-level critical value from the null distribution of $\mathcal{L}_n$. Then, the preliminary test estimator (PTE) of $\boldsymbol{\beta}$ may be written using the marginal distribution of $(\hat{\boldsymbol{\beta}}_{1n}^{\text{ML(PT)}^\top}, \hat{\boldsymbol{\beta}}_{2n}^{\text{ML(PT)}^\top})^\top$ as

$$\hat{\boldsymbol{\beta}}_n^{\text{ML(PT)}} = \begin{pmatrix} \hat{\boldsymbol{\beta}}_{1n}^{\text{ML}} \\ \hat{\boldsymbol{\beta}}_{2n}^{\text{ML}} I(\mathcal{L}_n > c_\alpha) \end{pmatrix}, \tag{8.17}$$

where $I(A)$ is the indicator function of the set A.

We notice that PTE is a discrete function and loses some optimality properties. Thus, we consider the continuous version of the PTE which mimics the Stein-type estimator (SE) given by

$$\hat{\boldsymbol{\beta}}_n^{\text{ML(S)}} = \begin{pmatrix} \hat{\boldsymbol{\beta}}_{1n}^{\text{ML}} \\ \hat{\boldsymbol{\beta}}_{2n}^{\text{ML}} (1 - (p_2 - 2)\mathcal{L}_n^{-1})^+ \end{pmatrix}. \tag{8.18}$$

Since the SE may yield estimators with incorrect signs, we consider the positive-rule Stein-type estimator (PRSE) of $\boldsymbol{\beta} = (\boldsymbol{\beta}_1^\top, \boldsymbol{\beta}_2^\top)^\top$ given by

$$\hat{\boldsymbol{\beta}}_n^{\text{ML(S+)}} = \begin{pmatrix} \hat{\boldsymbol{\beta}}_{1n}^{\text{ML}} \\ \hat{\boldsymbol{\beta}}_{2n}^{\text{ML}} (1 - (p_2 - 2)\mathcal{L}_n^{-1})^+ \end{pmatrix}. \tag{8.19}$$

8.1.3 Results on MLASSO

In this section, we present the results on the MLASSO estimators defined by (8.15) repeated here.

$$\hat{\boldsymbol{\beta}}_n^{\text{MLASSO}}(\lambda) = (\text{sgn}(\tilde{\beta}_{jn})(|\tilde{\beta}_{jn}| - \lambda\sqrt{C^{jj}})^+ | j = 0, \ldots, p)^\top.$$

By now we know that the asymptotic marginal distribution of

$$\sqrt{n}(\hat{\beta}_{jn}^{\text{ML}} - \beta_j) \xrightarrow{\mathcal{D}} \mathcal{N}(0, C^{jj}), \quad j = 0, 1, \ldots, p. \tag{8.20}$$

Consider the family of diagonal linear projections,

$$\boldsymbol{T}_{\text{DP}}(\boldsymbol{\beta}, k) = (k_1 \hat{\beta}_{1n}^{\text{MLASSO}}, \ldots, k_r \hat{\beta}_{rn}^{\text{MLASSO}})^\top, \quad k_j \in (0, 1). \tag{8.21}$$

This estimator *kills* or *keeps* a parameter β_j, i.e. it does *subset selection*. Under the rule, we incur a risk C^{jj} if we use $\tilde{\beta}_{jn}$, and β_j^2 if we estimate 0 instead. Hence, the ideal choice is $I(|\beta_j| > \sqrt{C^{jj}})$, i.e. predictors whose true coefficients exceed the noise level of $\sqrt{C^{jj}}$. This yields the ideal L_2 risk

$$R(\boldsymbol{T}_{\text{Dp}}) = \sum_{j=1}^{p} \min\{\beta_j^2, C^{jj}\}. \tag{8.22}$$

Next, if p_1 coefficients exceed the noise level and p_2 coefficients are 0's, then the L_2 risk is given by

$$(\boldsymbol{C}_{11\cdot2}^{-1} + \boldsymbol{\delta}_2^\top \boldsymbol{\delta}_2), \quad \boldsymbol{\delta}_2 = (\boldsymbol{\delta}_{21}, \ldots, \boldsymbol{\delta}_{2p_2})^\top. \tag{8.23}$$

The configuration described yields the lower bound of L_2 risk of $\hat{\boldsymbol{\beta}}_n^{\text{MLASSO}}(\lambda)$ as

$$R(\hat{\boldsymbol{\beta}}_n^{\text{MLASSO}}(\lambda):\boldsymbol{I}_p) \geq \text{tr}(\boldsymbol{C}_{11\cdot2}^{-1}) + \boldsymbol{\delta}_2^\top\boldsymbol{\delta}_2. \tag{8.24}$$

The weighted asymptotic distributional risk (ADR) is then given by

$$R(\hat{\boldsymbol{\beta}}_n^{\text{MLASSO}}(\lambda):\boldsymbol{C}_{11\cdot2}, \boldsymbol{C}_{22\cdot1}) = p_1 + \Delta^2.$$

We shall use this lower bound to compare the MLASSO with other estimators.

The asymptotic distributional bias (ADB) of an estimator $\hat{\boldsymbol{\beta}}^*$ is defined as

$$\text{ADB}(\hat{\boldsymbol{\beta}}^*) = \lim_{n\to\infty} \mathbb{E}[\sqrt{n}(\hat{\boldsymbol{\beta}}^* - \boldsymbol{\beta})]. \tag{8.25}$$

8.1.4 Results on PTE and Stein-Type Estimators

In this section, we present the ADB and asymptotic distributional L_2-risk (ADR) of the shrinkage estimators.

Theorem 8.1 *Under the assumed regularity conditions and a sequence of local alternatives $\{\mathcal{K}_{(n)}\}$ defined by $\mathcal{K}_{(n)}:\boldsymbol{\beta}_{(n)} = n^{-1/2}(\boldsymbol{\delta}_1^\top, \boldsymbol{\delta}_2^\top)^\top$, the following holds:*

$$\text{ADB}(\hat{\boldsymbol{\beta}}_n^{\text{ML}}) = \begin{pmatrix} \mathbf{0} \\ \mathbf{0} \end{pmatrix}$$

$$\text{ADB}(\hat{\boldsymbol{\beta}}_n^{\text{ML}(R)}) = \begin{pmatrix} \mathbf{0} \\ -\boldsymbol{\delta}_2 \end{pmatrix}$$

$$\text{ADB}(\hat{\boldsymbol{\beta}}_n^{\text{ML(PT)}}(\alpha)) = \begin{pmatrix} \mathbf{0} \\ -\boldsymbol{\delta}_2 H_{p_2+2}(c_\alpha; \Delta^2) \end{pmatrix}$$

$$\text{ADB}(\hat{\boldsymbol{\beta}}_n^{\text{ML(S)}}) = \begin{pmatrix} \mathbf{0} \\ -(p_2-2)\boldsymbol{\delta}_2\mathbb{E}[\chi_{p_2+2}^{-2}(\Delta^2)] \end{pmatrix}$$

$$\text{ADB}(\hat{\boldsymbol{\beta}}_n^{\text{ML(S+)}}) = \begin{pmatrix} \mathbf{0} \\ -(p_2-2)[\boldsymbol{\delta}_2\mathbb{E}[\chi_{p_2+2}^{-2}(\Delta^2)] + (p_2-2)\boldsymbol{\delta}_2\mathbb{E}[\chi_{p_2+2}^{-2}(\Delta^2)]] \end{pmatrix}$$

$$\text{ADB}(\hat{\boldsymbol{\beta}}_n^{\text{MLASSO}}) = (\sqrt{C^{jj}}[(2\Phi(\Delta_j - 1) - \Delta_j H_3(k^2; \Delta_j^2)], j = 1, \ldots, p.$$

$$\text{ADB}(\hat{\boldsymbol{\beta}}_n^{\text{RRML}}(k)) = \begin{pmatrix} \mathbf{0} \\ -k(\boldsymbol{C}_{22\cdot1} + k\boldsymbol{I}_{p_2})^{-1}\boldsymbol{\delta}_2 \end{pmatrix},$$

where H-function is the cumulative distribution function (c.d.f.) of the noncentral χ^2 distribution with noncentrality parameter $\Delta^2 = \boldsymbol{\delta}_2^\top \boldsymbol{C}_{22\cdot1}\boldsymbol{\delta}_2$ and ν DF

Now, respectively, the expressions of L_2-risk expressions are given by

$$\text{ADR}(\hat{\boldsymbol{\beta}}_n^{\text{ML}}) = \begin{pmatrix} \text{tr}(\boldsymbol{C}_{11\cdot2}^{-1}) \\ \text{tr}(\boldsymbol{C}_{22\cdot1}^{-1}) \end{pmatrix}. \tag{8.26}$$

That means the risk function for the maximum likelihood estimator is

$$\text{ADR}(\hat{\boldsymbol{\beta}}_n^{\text{ML}}) = \text{tr}(\boldsymbol{C}_{11\cdot 2}^{-1}) + \text{tr}(\boldsymbol{C}_{22\cdot 1}^{-1}) = \text{tr}(\boldsymbol{C}^{-1}) \tag{8.27}$$

and the asymptotic weighted L_2-risk will be

$$\text{ADR}(\hat{\boldsymbol{\beta}}_n^{\text{ML}} : \boldsymbol{C}_{11\cdot 2}, \boldsymbol{C}_{22\cdot 1}) = \sigma^2(p_1 + p_2). \tag{8.28}$$

The risk function of restricted estimator is

$$\text{ADR}(\hat{\boldsymbol{\beta}}_n^{\text{ML}(R)}) = \text{tr}(\boldsymbol{C}_{11}^{-1}) + \boldsymbol{\delta}_2^\top \boldsymbol{\delta}_2 \tag{8.29}$$

and the weighted L_2-risk will be

$$\text{ADR}(\hat{\boldsymbol{\beta}}_n^{\text{ML}(R)} : \boldsymbol{C}_{11\cdot 2}, \boldsymbol{C}_{22\cdot 1}) = [\text{tr}(\boldsymbol{C}_{11}^{-1}\boldsymbol{C}_{11\cdot 2}) + \Delta^2]. \tag{8.30}$$

Then the risk function for $\hat{\boldsymbol{\beta}}_n^{\text{ML(PT)}}$ is

$$\begin{aligned}\text{ADR}(\hat{\boldsymbol{\beta}}_n^{\text{ML(PT)}}(\alpha)) = \{&\text{tr}(\boldsymbol{C}_{11\cdot 2}^{-1}) + \text{tr}(\boldsymbol{C}_{22\cdot 1}^{-1})[1 - H_{p_2+2}(c_\alpha : \Delta^2)] \\ &+ \boldsymbol{\delta}_2^\top \boldsymbol{\delta}_2 \{2H_{p_2+2}(c_\alpha : \Delta^2) - H_{p_2+4}(c_\alpha : \Delta^2)\}\end{aligned} \tag{8.31}$$

and the weighted L_2-risk will be

$$\begin{aligned}\text{ADR}(\hat{\boldsymbol{\beta}}_n^{\text{ML(PT)}}(\alpha); \boldsymbol{C}_{11\cdot 2}, \boldsymbol{C}_{22\cdot 1}) = \{&p_1 + p_2[1 - H_{p_2+2}(c_\alpha : \Delta^2)] \\ &+ \Delta^2\{2H_{p_2+2}(c_\alpha : \Delta^2) - H_{p_2+4}(c_\alpha : \Delta^2)\}\}.\end{aligned} \tag{8.32}$$

The risk function for $\hat{\boldsymbol{\beta}}_n^{\text{ML(S)}}$ is

$$\begin{aligned}\text{ADR}(\hat{\boldsymbol{\beta}}_n^{\text{ML(S)}} : \boldsymbol{I}_{p_1}) = \sigma^2\{&\text{tr}(\boldsymbol{C}_{11\cdot 2}^{-1}) + \text{tr}(\boldsymbol{C}_{22\cdot 1}^{-1}) \\ &- (p_2 - 2)\,\text{tr}(\boldsymbol{C}_{22\cdot 1}^{-1}) \\ &\times (2\mathbb{E}[\chi_{p_2+2}^{-2}(\Delta^2)] - (p_2 - 2)\mathbb{E}[\chi_{p_2+2}^{-4}(\Delta^2)]) \\ &+ (p_2^2 - 4)\boldsymbol{\delta}_2^\top \boldsymbol{\delta}_2 \mathbb{E}[\chi_{p_2+4}^{-4}(\Delta^2)]\}.\end{aligned} \tag{8.33}$$

The weight L_2-risk function is

$$R(\hat{\boldsymbol{\beta}}_n^{\text{ML(S)}}; \boldsymbol{C}_{11\cdot 2}, \boldsymbol{C}_{22\cdot 1}) = \sigma^2\{p_1 + p_2 - (p_2 - 2)^2\mathbb{E}[\chi_{p_2}^{-2}(\Delta^2)]\}. \tag{8.34}$$

The risk function for $\hat{\boldsymbol{\beta}}_n^{\text{ML(S+)}}$ is

$$\begin{aligned}R(\hat{\boldsymbol{\beta}}_n^{\text{ML(S+)}}) = R(\hat{\boldsymbol{\beta}}_n^{\text{ML(S)}}) &- (p_2 - 2)\,\text{tr}(\boldsymbol{C}_{22\cdot 1}^{-1}) \\ &\times \mathbb{E}[(1 - (p_2 - 2)^2 \chi_{p_2+2}^{-2}(\Delta^2))^2 I(\chi_{p_2+2}^2(\Delta^2) \le p_2 - 2)] \\ &+ \boldsymbol{\delta}_2^\top \boldsymbol{\delta}_2 \{2\mathbb{E}[(1 - (p_2 - 2)\chi_{p_2+2}^{-2}(\Delta^2)) I(\chi_{p_2+2}^2(\Delta^2) < p_2 - 2)] \\ &- \mathbb{E}[(1 - (p_2 - 2)\chi_{p_2+4}^{-2}(\Delta^2))^2 I(\chi_{p_2+4}^2(\Delta^2) \le p_2 - 2)]\}.\end{aligned}$$

The weighted L_2-risk function is

$$\begin{aligned}
R(\hat{\boldsymbol{\beta}}_n^{\text{ML(S+)}} : \boldsymbol{C}_{11\cdot 2}, \boldsymbol{C}_{22\cdot 1}) = {} & \sigma^2\{p_1 + p_2 - (p_2-2)^2\mathbb{E}[\chi_{p_2}^{-2}(\Delta^2)]\} \\
& - (p_2-2)p_2 \\
& \times \mathbb{E}[(1-(p_2-2)^2\chi_{p_2+2}^{-2}(\Delta^2))^2 I(\chi_{p_2+2}^2(\Delta^2) \le p_2-2)] \\
& + \Delta^2\{2\mathbb{E}[(1-(p_2-2)\chi_{p_2+2}^{-2}(\Delta^2))I(\chi_{p_2+2}^2(\Delta^2) \le p_2-2)] \\
& - \mathbb{E}[(1-(p_2-2)\chi_{p_2+4}^{-2}(\Delta^2))^2 I(\chi_{p_2+4}^2(\Delta^2) \le p_2-2)]\}.
\end{aligned}$$

8.1.5 Results on Penalty Estimators

First, we consider the RRE as defined before.

$$\hat{\boldsymbol{\beta}}_n^{\text{RRML}}(k) = \begin{pmatrix} \hat{\boldsymbol{\beta}}_{1n}^{\text{ML}} \\ (\boldsymbol{I}_{p_2} + k\boldsymbol{C}_{22\cdot 1}^{-1})^{-1}\hat{\boldsymbol{\beta}}_{2n}^{\text{ML}} \end{pmatrix}. \tag{8.35}$$

The ADR for RRE is

$$\text{ADR}(\hat{\boldsymbol{\beta}}_n^{\text{RRML}}(k)) = \text{tr}(\boldsymbol{\Sigma}) + \frac{1}{(1+k)^2}[\text{tr}(\boldsymbol{C}_{22\cdot 1}^{-1}) + k^2\boldsymbol{\delta}_2^{\top}(\boldsymbol{C}_{22\cdot 1} + k\boldsymbol{I}_{p_2})^{-2}\boldsymbol{\delta}_2], \tag{8.36}$$

where $\boldsymbol{\Sigma} = (\boldsymbol{C}_{22\cdot 1} + k\boldsymbol{I}_{p_2})^{-1}\boldsymbol{C}_{22\cdot 1}(\boldsymbol{C}_{22\cdot 1} + k\boldsymbol{I}_{p_2})^{-1}$.

The weighted L_2-risk function is

$$\begin{aligned}
\text{ADR}(\hat{\boldsymbol{\beta}}_n^{\text{RRML}}(k) : \boldsymbol{C}_{11\cdot 2}, \boldsymbol{C}_{22\cdot 1}) &= \left\{p_1 + \frac{1}{(1+k)^2}[p_2 + k^2\Delta^2]\right\} \\
&= \left\{p_1 + \frac{p_2\Delta^2}{(p_2+\Delta^2)}\right\} \quad \text{if} \quad k_{\text{opt}} = p_2\Delta^{-2}.
\end{aligned} \tag{8.37}$$

$$\text{ADR}(\hat{\boldsymbol{\beta}}_n^{\text{MLASSO}}(\lambda)) = \text{tr}(\boldsymbol{C}_{11\cdot 2}^{-1}) + \boldsymbol{\delta}_2^{\top}\boldsymbol{\delta}_2. \tag{8.38}$$

The weighted L_2-risk function is

$$\text{ADR}(\hat{\boldsymbol{\beta}}_n^{\text{MLASSO}} : \boldsymbol{C}_{11\cdot 2}, \boldsymbol{C}_{22\cdot 1}) = p_1 + \Delta^2. \tag{8.39}$$

8.2 Asymptotic Distributional L_2 Risk Efficiency Expressions of the Estimators

In this section, we study the comparative properties of the estimators relative to the MLEs. We consider the weighted L_2 risk to compare the performance of the estimators.

The asymptotic distributional L_2-risk efficiency (ADRE) expressions are given by

$$\text{ADRE}(\hat{\boldsymbol{\beta}}_n^{\text{ML}(R)} : \hat{\boldsymbol{\beta}}_n^{\text{ML}}) = \left(1 + \frac{p_2}{p_1}\right)\left(1 + \frac{\Delta^2}{p_1}\right)^{-1}$$

$$\text{ADRE}(\hat{\boldsymbol{\beta}}_n^{\text{MLASSO}} : \hat{\boldsymbol{\beta}}_n^{\text{ML}}) = \left(1 + \frac{p_2}{p_1}\right)\left(1 + \frac{\Delta^2}{p_1}\right)^{-1}$$

$$\text{ADRE}(\hat{\boldsymbol{\beta}}_n^{\text{ML(PT)}} : \hat{\boldsymbol{\beta}}_n^{\text{ML}}) = \left(1 + \frac{p_2}{p_1}\right)\left\{1 + \frac{p_2}{p_1}(1 - H_{p_2+2}(c_\alpha; \Delta^2)) + \frac{\Delta^2}{p_1}(2H_{p_2+2}(c_\alpha : \Delta^2) - H_{p_2+4}(c_\alpha : \Delta^2))\right\}^{-1}$$

$$\text{ADRE}(\hat{\boldsymbol{\beta}}^{\text{ML(S)}} : \hat{\boldsymbol{\beta}}_n^{\text{ML}}) = \left(1 + \frac{p_2}{p_1}\right)\left(1 + \frac{p_2}{p_1} - \frac{1}{p_1}(p_2 - 2)^2 \mathbb{E}[\chi_{p_2}^{-2}(\Delta^2)]\right)^{-1}$$

$$\text{ADRE}(\hat{\boldsymbol{\beta}}_n^{\text{ML(S+)}} : \hat{\boldsymbol{\beta}}_n^{\text{ML}}) = \left(1 + \frac{p_2}{p_1}\right)\left\{1 + \frac{p_2}{p_1} - \frac{1}{p_1}(p_2 - 2)\mathbb{E}[\chi_{p_2}^{-2}(\Delta^2)] - \frac{p_2}{p_1}\mathbb{E}[(1 - (p_2 - 2\chi_{p_2+2}^{-2}(\Delta^2))^2 I(\chi_{p_2+2}^2(\Delta^2)) < p_2 - 2)] - R^*\right\}^{-1}$$

$$\text{ADRE}(\hat{\boldsymbol{\beta}}^{\text{RR}}(k_{\text{opt}}) : \hat{\boldsymbol{\beta}}_n^{\text{ML}}) = \left(1 + \frac{p_2}{p_1}\right)\left(\frac{p_2\Delta^2}{p_1(p_2 + \Delta^2)}\right)^{-1}.$$

8.2.1 MLASSO vs. MLE

The ADR is

$$\text{ADR}(\hat{\boldsymbol{\beta}}_n^{\text{MLASSO}} : \hat{\boldsymbol{\beta}}_n^{\text{ML}}) = \frac{p_1 + p_2}{p_1 + \Delta^2}. \tag{8.40}$$

This expression is a decreasing function of $\Delta^2 = 0$ and is given by

$$\text{ADR}(\hat{\boldsymbol{\beta}}_n^{\text{MLASSO}} : \hat{\boldsymbol{\beta}}_n^{\text{ML}}) = 1 + \frac{p_2}{p_1}.$$

Table 8.1 is given for $p_1 = 2, 3, 5, 7, p = 10, 20, 30, 40, 60$, and 100.

It is observed from Table 8.1 that, for fixed p, when p_1 increases the ADR decreases; and, for fixed p_1, when p increases the ADR also increases. That means p_1 has a negative effect, while p has a positive effect on the ADR.

Table 8.1 Relative efficiency table for different values of p and p_1.

p_1	$p = 10$	$p = 20$	$p = 30$	$p = 40$	$p = 60$	$p = 100$
2.00	5.00	10.00	15.00	20.00	30.00	50.00
3.00	3.33	6.67	10.00	13.33	20.00	33.33
5.00	2.00	4.00	6.00	8.00	12.00	20.00
7.00	1.43	2.86	4.29	5.71	8.57	14.29

8.2.2 MLASSO vs. RMLE

Next, we consider the comparison of MLASSO vs. RMLE. Here, the ADR risk difference is given by

$$\begin{aligned}\text{ADR}(\hat{\boldsymbol{\beta}}_n^{\text{ML}(R)}) - \text{ADR}(\hat{\boldsymbol{\beta}}_n^{\text{MLASSO}}(\lambda)) &= [\text{tr}(\boldsymbol{C}_{11}^{-1}\boldsymbol{C}_{11\cdot2}) + \Delta^2] - \{p_1 + \Delta^2\} \\ &= \{\text{tr}(\boldsymbol{C}_{11}^{-1}\boldsymbol{C}_{11\cdot2}) - p_1\}. \end{aligned} \tag{8.41}$$

The L_2-risk difference is nonnegative whenever

$$\text{tr}(\boldsymbol{C}_{11}^{-1}\boldsymbol{C}_{11\cdot2}) \geq p_1. \tag{8.42}$$

Thus, MLE outperforms LASSO when $\text{tr}(\boldsymbol{C}_{11}^{-1}\boldsymbol{C}_{11\cdot2}) \geq p_1$. Otherwise, LASSO will outperform when $\text{tr}(\boldsymbol{C}_{11}^{-1}\boldsymbol{C}_{11\cdot2}) < p_1$. Hence, neither LASSO nor RMLE outperforms the others uniformly.

The $\text{ADR}(\hat{\boldsymbol{\beta}}_n^{\text{ML}(R)} : \hat{\boldsymbol{\beta}}_n^{\text{ML}})$ equals

$$\frac{(p_1 + p_2)}{\text{tr}(\boldsymbol{C}_{11}^{-1}\boldsymbol{C}_{11\cdot2}) + \Delta^2} \tag{8.43}$$

has its maximum at $\Delta^2 = 0$ given

$$\frac{p_1 + p_2}{\text{tr}(\boldsymbol{C}_{11}^{-1}\boldsymbol{C}_{11\cdot2})}. \tag{8.44}$$

Data was generated from multivariate normal distribution with correlation among X's of 0.6, 0.8, and 0.9. Some tabular values for $p_1 = 2, 3, 5, 7$ and $p = 10$ are given in Table 8.2.

From Table 8.2, it appears that the efficiency of the RMLE decreases as the correlation among regressors increases. Also, for given correlations, the efficiency increases and p_1 increases while p_2 decreases.

8.2.3 Comparison of MLASSO vs. PTE

In this case, the ADR difference is given by

$$\begin{aligned}&\text{ADR}(\hat{\boldsymbol{\beta}}_n^{\text{ML(PT)}}(\alpha)) - \text{ADR}(\hat{\boldsymbol{\beta}}_n^{\text{MLASSO}}(\lambda)) \\ &\quad = \sigma^2\{p_1 + p_2[1 - H_{p_2+2}(c_\alpha : \Delta^2)] + \Delta^2\{2H_{p_2+2}(c_\alpha : \Delta^2) - H_{p_2+4}(c_\alpha : \Delta^2)\}\} \\ &\qquad -\sigma^2\{p_1 + \Delta^2\}.\end{aligned}$$

Table 8.2 Relative efficiency table for different values of p_1.

p	p_1	RMLE ($\rho = 0.6$)	RMLE ($\rho = 0.7$)	RMLE ($\rho = 0.8$)
10	2	4.14	3.92	3.86
10	3	4.73	4.47	4.41
10	5	5.91	5.59	5.52
10	7	7.10	6.71	6.62

The given expression is nonnegative whenever

$$0 \le \Delta^2 \le \frac{p_2[1 - H_{p_2+2}(c_\alpha : \Delta^2)]}{1 - \{2H_{p_2+2}(c_\alpha : \Delta^2) - H_{p_2+4}(c_\alpha : \Delta^2)\}}. \tag{8.45}$$

Hence, the MLASSO estimator outperforms PTE whenever (8.45) holds; and for

$$\Delta^2 \ge \frac{p_2[1 - H_{p_2+2}(c_\alpha : \Delta^2)]}{1 - \{2H_{p_2+2}(c_\alpha : \Delta^2) - H_{p_2+4}(c_\alpha : \Delta^2)\}}, \tag{8.46}$$

the PTE outperforms the MLASSO estimator.

8.2.4 PT and MLE

The relative efficiency (REff) is

$$\text{ADR}(\hat{\boldsymbol{\beta}}_n^{\text{ML(PT)}} : \hat{\boldsymbol{\beta}}_n^{\text{ML}}) = \frac{\sigma^2(p_1 + p_2)}{R(\hat{\boldsymbol{\beta}}_n^{\text{ML(PT)}}(\alpha))} \tag{8.47}$$

The maximum ADR is achieved at $\Delta^2 = 0$.

$$\text{ADR}(\hat{\boldsymbol{\beta}}_n^{\text{ML(PT)}} : \hat{\boldsymbol{\beta}}_n^{\text{ML}}) = \left[1 + \frac{p_1 - p_2}{p_1 + p_2} H_{p_2+2}(c_\alpha : 0)\right]^{-1}. \tag{8.48}$$

Some tabular values of ADR are given for $p_1 = 2, 3, 5$, and 7 and $p = 10, 20, 30, 40, \textit{and}\ 60$ in Table 8.3. From Table 8.3, it appears that for fixed p_1, as the value

Table 8.3 Relative efficiency for $\alpha = 0.05$.

		p					
α	p_1	10	20	30	40	60	100
0.05	2	2.43	4.53	6.30	7.79	10.13	13.18
	3	1.65	3.15	4.48	5.66	7.65	10.52
	5	1.00	1.95	2.84	3.66	5.13	7.49
	7	0.72	1.41	2.08	2.70	3.86	5.82
0.10	2	2.35	4.08	5.33	6.24	7.46	8.72
	3	1.62	2.95	4.00	4.84	6.07	7.52
	5	1.00	1.90	2.67	3.34	4.42	5.89
	7	0.72	1.40	2.01	2.55	3.47	4.84
0.20	2.00	2.18	3.34	3.98	4.35	4.76	5.08
	3	1.57	2.59	3.24	3.68	4.21	4.69
	5	1.00	1.79	2.37	2.80	3.41	4.06
	7	0.73	1.36	1.86	2.27	2.86	3.58
0.50	2.00	1.70	2.01	2.10	2.13	2.15	2.15
	3	1.39	1.79	1.94	2.01	2.07	2.10
	5	1.00	1.47	1.69	1.80	1.92	2.01
	7	0.76	1.24	1.49	1.63	1.79	1.92

of p increases, the REff also increases. For a given p, the REff decreases as p_1 increases and p_2 decreases.

8.2.5 Comparison of MLASSO vs. SE

In this case, the L_2-risk difference of the SE and MLASSO is given by

$$\begin{aligned}
&\text{ADR}(\hat{\boldsymbol{\beta}}_n^{\text{ML(S)}}) - \text{ADR}(\hat{\boldsymbol{\beta}}_n^{\text{MLASSO}}(\lambda_p)) \\
&\quad = \sigma^2\{p_1 + p_2 - (p_2-2)^2\mathbb{E}[\chi_{p_2}^{-2}(\Delta^2)] - (p_1 + \Delta^2)\}.
\end{aligned} \tag{8.49}$$

The expression (8.49) is nonnegative whenever

$$0 \le \Delta^2 \le (p_2-2)^2\mathbb{E}[\chi_{p_2+2}^{-2}(\Delta^2)]. \tag{8.50}$$

This implies that MLASSO SE whenever (8.53) is satisfied. While for

$$\Delta^2 > (p_2-2)^2\mathbb{E}[\chi_{p_2+2}^{-2}(\Delta^2)], \tag{8.51}$$

we find that the SE outperforms MLASSO. Hence neither MLASSO nor SE outperforms the other uniformly in Δ^2.

8.2.6 Comparison of MLASSO vs. PRSE

In this case, the L_2-risk difference of PRSE and MLASSO is given by

$$\begin{aligned}
&\text{ADR}(\hat{\boldsymbol{\beta}}_n^{\text{ML(S+)}}) - \text{ADR}(\hat{\boldsymbol{\beta}}_n^{\text{MLASSO}}(\lambda_p)) \\
&\quad = \{p_1 + p_2 - (p_2-2)^2\mathbb{E}[\chi_{p_2}^{-2}(\Delta^2)]\} \\
&\qquad - (p_2-2)p_2\mathbb{E}[(1-(p_2-2)^2\chi_{p_2+2}^{-2}(\Delta^2))^2 I(\chi_{p_2+2}^{2}(\Delta^2) < p_2-2)] \\
&\qquad - \Delta^2\{2\mathbb{E}[(1-(p_2-2)\chi_{p_2+2}^{-2}(\Delta^2))I(\chi_{p_2+2}^{2}(\Delta^2) < p_2-2)] \\
&\qquad - \mathbb{E}[(1-(p_2-2)\chi_{p_2+4}^{-2}(\Delta^2))^2 I(\chi_{p_2+4}^{2}(\Delta^2) < p_2-2)]\} \\
&\qquad - \{p_1 + \Delta^2\} \\
&\quad = \{p_2 - (p_2-2)^2\mathbb{E}[\chi_{p_2}^{-2}(\Delta^2)]\} \\
&\qquad - (p_2-2)p_2\mathbb{E}[(1-(p_2-2)^2\chi_{p_2+2}^{-2}(\Delta^2))^2 I(\chi_{p_2+2}^{2}(\Delta^2) < p_2-2)] \\
&\qquad - \Delta^2\{2\mathbb{E}[(1-(p_2-2)\chi_{p_2+2}^{-2}(\Delta^2))I(\chi_{p_2+2}^{2}(\Delta^2) < p_2-2)] \\
&\qquad - \mathbb{E}[(1-(p_2-2)\chi_{p_2+4}^{-2}(\Delta^2))^2 I(\chi_{p_2+4}^{2}(\Delta^2) < p_2-2)]\} - \sigma^2\Delta^2.
\end{aligned} \tag{8.52}$$

The expression (8.52) is nonnegative whenever

$$0 \leq \Delta^2 \leq \frac{f_1(\Delta^2)}{f_2(\Delta^2)}, \tag{8.53}$$

where

$$\begin{aligned} f_1(\Delta^2) &= \{p_2 - (p_2-2)^2\mathbb{E}[\chi_{p_2}^{-2}(\Delta^2)]\} \\ &\quad - (p_2-2)p_2\mathbb{E}[(1-(p_2-2)^2\chi_{p_2+2}^{-2}(\Delta^2))^2 I(\chi_{p_2+2}^2(\Delta^2) < p_2-2)] \end{aligned} \tag{8.54}$$

$$\begin{aligned} f_2(\Delta^2) &= 1 - \{2\mathbb{E}[(1-(p_2-2)\chi_{p_2+2}^{-2}(\Delta^2))I(\chi_{p_2+2}^2(\Delta^2) < p_2-2)] \\ &\quad - \mathbb{E}[(1-(p_2-2)\chi_{p_2+4}^{-2}(\Delta^2))^2 I(\chi_{p_2+4}^2(\Delta^2) < p_2-2)]\}. \end{aligned} \tag{8.55}$$

This implies that MLASSO outperforms PRSE whenever (8.53) is satisfied. While for

$$\Delta^2 > \frac{f_1(\Delta^2)}{f_2(\Delta^2)}, \tag{8.56}$$

the PRSE outperforms MLASSO.

8.2.7 RRE vs. MLE

The asymptotic distributional REff of the RRE compared to MLE is given by

$$\begin{aligned} \text{ADR}(\hat{\boldsymbol{\beta}}_n^{\text{RRML}}(k):\hat{\boldsymbol{\beta}}_n^{\text{ML}}) &= \frac{p_1+p_2}{p_1 + \frac{p_2+k^2\Delta^2}{(1+k)^2}} \\ &= \frac{p_1+p_2}{p_1+\frac{p_2\Delta^2}{p_2+\Delta^2}}, \quad k = p_2/\Delta^2. \end{aligned} \tag{8.57}$$

For $\Delta^2 = 0$, the given RMLE becomes $1 + \frac{p_2}{p_1} > 1$; for $\Delta^2 = \infty$, it becomes p_2. Thus, for any values of Δ^2, $\text{ADR}(\hat{\boldsymbol{\beta}}_n^{\text{RRML}}(k):\hat{\boldsymbol{\beta}}_n^{\text{ML}}) > 1$, and the RRE uniformly dominates MLE.

8.2.7.1 RRE vs. RMLE

Next, we consider comparison of the RRE vs. RMLE. Here, the ADR difference is given by

$$\begin{aligned} &\text{ADR}(\hat{\boldsymbol{\beta}}_n^{\text{ML(R)}}) - \text{ADR}(\hat{\boldsymbol{\beta}}_n^{\text{RRML}}(k)) \\ &\quad = \sigma^2\left\{\text{tr}(\boldsymbol{C}_{11}^{-1}\boldsymbol{C}_{11\cdot2}) + \Delta^2 - p_1 - \frac{p_2+k^2\Delta^2}{(1+k)^2}\right\}. \end{aligned} \tag{8.58}$$

Now, if we consider optimum $k = p_2\Delta^{-2}$, then the risk difference becomes

$$\begin{aligned}
&\mathrm{ADR}(\hat{\boldsymbol{\beta}}_n^{\mathrm{ML(R)}}) - \mathrm{ADR}(\hat{\boldsymbol{\beta}}_n^{\mathrm{RRML}}(k)) \\
&\quad = \sigma^2 \left\{ \mathrm{tr}(\boldsymbol{C}_{11}^{-1}\boldsymbol{C}_{11\cdot 2}) + \Delta^2 - p_1 - \frac{p_2\Delta^2}{(p_2 + \Delta^2)} \right\} \\
&\quad = \sigma^2 \left\{ \frac{\Delta^4}{(p_2 + \Delta^2)} - (p_1 - \mathrm{tr}(\boldsymbol{C}_{11}^{-1}\boldsymbol{C}_{11\cdot 2})) \right\} \\
&\quad = \sigma^2 \left\{ \frac{\Delta^4}{(p_2 + \Delta^2)} - \mathrm{tr}(\boldsymbol{M}_0) \right\},
\end{aligned} \tag{8.59}$$

where

$$\begin{aligned}
\boldsymbol{M}_0 &= (\boldsymbol{C}_{11}^{-1}\boldsymbol{C}_{11} - \boldsymbol{C}_{11}^{-1}\boldsymbol{C}_{11\cdot 2}) = \boldsymbol{C}_{11}^{-1}[\boldsymbol{C}_{11} - \boldsymbol{C}_{11\cdot 2}] \\
&= \boldsymbol{C}_{11}^{-1}[\boldsymbol{C}_{11} - \boldsymbol{C}_{11} + \boldsymbol{C}_{12}\boldsymbol{C}_{22}^{-1}\boldsymbol{C}_{21}] = \boldsymbol{C}_{11}^{-1}\boldsymbol{C}_{12}\boldsymbol{C}_{22}^{-1}\boldsymbol{C}_{21}.
\end{aligned} \tag{8.60}$$

Solving the equation

$$\frac{\Delta^4}{(p_2 + \Delta^2)} = \mathrm{tr}(\boldsymbol{M}_0). \tag{8.61}$$

Then using $ax^2 + bx + c, x = \frac{-b \pm \sqrt{b^2 - 4ac}}{2a}$, we obtain

$$\begin{aligned}
\Delta_0^2 &= \frac{\mathrm{tr}(\boldsymbol{M}_0) \pm \sqrt{\mathrm{tr}\ (\boldsymbol{M}_0)^2 + 4p_2\ \mathrm{tr}(\boldsymbol{M}_0)}}{2} \\
&= \frac{\mathrm{tr}(\boldsymbol{M}_0) + \sqrt{\mathrm{tr}\ (\boldsymbol{M}_0)^2 + 4p_2\ \mathrm{tr}(\boldsymbol{M}_0)}}{2} \\
&= \frac{1}{2} \left\{ \mathrm{tr}(\boldsymbol{M}_0) + \mathrm{tr}(\boldsymbol{M}_0)\sqrt{1 + \frac{4p_2}{\mathrm{tr}(\boldsymbol{M}_0)}} \right\} \\
&= \frac{1}{2}\ \mathrm{tr}(\boldsymbol{M}_0) \left\{ 1 + \sqrt{1 + \frac{4p_2}{\mathrm{tr}(\boldsymbol{M}_0)}} \right\}.
\end{aligned} \tag{8.62}$$

Now, if $0 \le \Delta^2 \le \Delta_0^2$, then RMLE outperforms RRE; and if $\Delta^2 \in (\Delta_0^2, \infty)$, RRE outperforms RMLE.

The asymptotic distributional REff equals

$$\mathrm{ADR}(\hat{\boldsymbol{\beta}}_n^{\mathrm{RRML}}(k) : \hat{\boldsymbol{\beta}}_n^{\mathrm{ML}}) = \frac{\mathrm{tr}(\boldsymbol{C}_{11}^{-1}\boldsymbol{C}_{11\cdot 2}) + \Delta^2}{p_1 + \frac{p_2 + k^2\Delta^2}{(1+k)^2}} \tag{8.63}$$

and has its maximum at $\Delta^2 = 0$ given by

$$\frac{\mathrm{tr}(\boldsymbol{C}_{11}^{-1}\boldsymbol{C}_{11\cdot 2})}{p_1}. \tag{8.64}$$

8.2.8 Comparison of RRE vs. PTE

In this case, the ADR difference is given by

$$\begin{aligned}
&\mathrm{ADR}(\hat{\boldsymbol{\beta}}_n^{\mathrm{ML(PT)}}(\alpha)) - \mathrm{ADR}(\hat{\boldsymbol{\beta}}_n^{\mathrm{ML}}(k)) \\
&= \sigma^2\{p_2[1 - H_{p_2+2}(c_\alpha : \Delta^2)] + \Delta^2\{2H_{p_2+2}(c_\alpha : \Delta^2) - H_{p_2+4}(c_\alpha : \Delta^2)\} \\
&\quad - \left.\frac{p_2\Delta^2}{(p_2 + \Delta^2)}\right\}.
\end{aligned} \tag{8.65}$$

Note that the risk of $\hat{\boldsymbol{\beta}}_n^{\mathrm{ML(PT)}}(\alpha)$ is an increasing function of Δ^2 crossing the p_2-line to a maximum and then dropping monotonically toward the p_2-line as $\Delta^2 \to \infty$. The value of the risk is $p_2(1 - H_{p_2+2}(\chi^2_{p_2}(\alpha); 0))(< p_2)$ at $\Delta^2 = 0$. On the other hand, $\frac{p_2\Delta^2}{p_2+\Delta^2}$ is an increasing function of Δ^2 below the p_2-line with a minimum value 0 at $\Delta^2 = 0$; and as $\Delta^2 \to \infty$, $\frac{p_2\Delta^2}{p_2+\Delta^2} \to p_2$. Hence, the risk difference in Eq. (8.65) is nonnegative for $\Delta^2 \in \mathbb{R}^+$. Thus, the RRE uniformly performs better than PTE. The given expression is nonnegative whenever

$$0 \le \Delta^2 \le \frac{p_2[1 - H_{p_2+2}(c_\alpha : \Delta^2)] + p_1 H_{p_2+2}(c_\alpha : \Delta^2) - \frac{p_2}{(1+k)^2}}{\frac{k^2}{(1+k)^2} - \{2H_{p_2+2}(c_\alpha : \Delta^2) - H_{p_2+4}(c_\alpha : \Delta^2)\}}, \tag{8.66}$$

Hence, RRE outperforms PTE whenever (8.66) holds; and for

$$\Delta^2 > \frac{p_2[1 - H_{p_2+2}(c_\alpha : \Delta^2)] + p_1 H_{p_2+2}(c_\alpha : \Delta^2) - \frac{p_2}{(1+k)^2}}{\frac{k^2}{(1+k)^2} - \{2H_{p_2+2}(c_\alpha : \Delta^2) - H_{p_2+4}(c_\alpha : \Delta^2)\}}, \tag{8.67}$$

the PTE outperforms RRE.

8.2.9 Comparison of RRE vs. SE

In this case, the L_2-risk difference of SE and RRE is given by

$$\begin{aligned}
&\mathrm{ADR}(\hat{\boldsymbol{\beta}}_n^{\mathrm{ML(S)}}) - \mathrm{ADR}(\hat{\boldsymbol{\beta}}_n^{\mathrm{RRML}}(k)) \\
&= \sigma^2\left\{p_1 + p_2 - (p_2 - 2)^2\mathbb{E}[\chi^{-2}_{p_2}(\Delta^2)] - \left(p_1 + \frac{1}{(1+k)^2}[p_2 + k^2\Delta^2]\right)\right\} \\
&= \{p_2 - (p_2 - 2)^2\mathbb{E}[\chi^{-2}_{p_2}(\Delta^2)]\} - \left\{\frac{p_2\Delta^2}{(p_2 + \Delta^2)}\right\}.
\end{aligned} \tag{8.68}$$

Note that the first function is increasing in Δ^2 with a value 2 at $\Delta^2 = 0$; and as $\Delta^2 \to \infty$, it tends to p_2. The second function is also increasing in Δ^2 with a value 0 at $\Delta^2 = 0$, and approaches the value p_2 as $\Delta^2 \to \infty$. Hence, the risk difference is nonnegative for all $\Delta^2 \in \mathbb{R}^+$. Consequently, RRE outperforms SE uniformly.

8.2.10 Comparison of RRE vs. PRSE

In this case, the L_2-risk difference of RRE and PRSE is given by

$$\begin{aligned}
&\mathrm{ADR}(\hat{\boldsymbol{\beta}}_n^{\mathrm{ML(S)}}) - \mathrm{ADR}(\hat{\boldsymbol{\beta}}_n^{\mathrm{RRML}}(k)) \\
&= \{p_1 + p_2 - (p_2-2)^2\mathbb{E}[\chi_{p_2}^{-2}(\Delta^2)]\} \\
&\quad - (p_2-2)p_2\mathbb{E}[(1-(p_2-2)^2\chi_{p_2+2}^{-2}(\Delta^2))^2 I(\chi_{p_2+2}^{2}(\Delta^2) < p_2-2)] \\
&\quad + \Delta^2\{2\mathbb{E}[(1-(p_2-2)\chi_{p_2+2}^{-2}(\Delta^2))I(\chi_{p_2+2}^{2}(\Delta^2) < p_2-2)] \\
&\quad - \mathbb{E}[(1-(p_2-2)\chi_{p_2+4}^{-2}(\Delta^2))^2 I(\chi_{p_2+4}^{2}(\Delta^2) < p_2-2)]\} \\
&\quad - \left\{p_1 + \frac{1}{(1+k)^2}[p_2 + k^2\Delta^2]\right\} \\
&= \{p_2 - (p_2-2)^2\mathbb{E}[\chi_{p_2}^{-2}(\Delta^2)]\} \\
&\quad - (p_2-2)p_2\mathbb{E}[(1-(p_2-2)^2\chi_{p_2+2}^{-2}(\Delta^2))^2 I(\chi_{p_2+2}^{2}(\Delta^2) < p_2-2)] \\
&\quad + \Delta^2\{2\mathbb{E}[(1-(p_2-2)\chi_{p_2+2}^{-2}(\Delta^2))I(\chi_{p_2+2}^{2}(\Delta^2) < p_2-2)] \\
&\quad - \mathbb{E}[(1-(p_2-2)\chi_{p_2+4}^{-2}(\Delta^2))^2 I(\chi_{p_2+4}^{2}(\Delta^2) < p_2-2)]\} \\
&\quad - \frac{p_2\Delta^2}{p_2+\Delta^2}.
\end{aligned} \tag{8.69}$$

Consider the $\mathrm{R}(\hat{\boldsymbol{\beta}}_n^{\mathrm{ML(S+)}})$. It is a monotonically increasing function of Δ^2. At the point $\Delta^2 = 0$, its value is

$$\sigma^2(p_1+2) - \sigma^2 p_2\mathbb{E}[(1-(p_2-2)\chi_{p_2+2}^{-2}(0))^2 I(\chi_{p_2+2}^{-2}(0) < p_2-2)] \geq 0;$$

and as $\Delta^2 \to \infty$, it tends to $\sigma^2(p_1+p_2)$. For $\mathrm{ADR}(\hat{\boldsymbol{\beta}}_n^{\mathrm{RRML}}(k))$, at $\Delta^2 = 0$, the value is $\sigma^2 p_1$; and as $\Delta^2 \to \infty$, it tends to $\sigma^2(p_1+p_2)$. Hence, the L_2-risk difference in (8.69) is nonnegative and RRE uniformly outperforms PRSE.

Note that the risk difference of $\hat{\boldsymbol{\beta}}_n^{\mathrm{ML(S+)}}$ and $\hat{\boldsymbol{\beta}}_n^{\mathrm{RRML}}(k_{\mathrm{opt}})$ at $\Delta^2 = 0$ is

$$\begin{aligned}
&(p_1+2) - p_2\mathbb{E}[(1-(p_2-2)\chi_{p_2+2}^{-2}(0))^2 I(\chi_{p_2+2}^{-2}(0) < p_2-2)] - \sigma^2 p_1 \\
&= (2 - p_2\mathbb{E}[(1-(p_2-2)\chi_{p_2+2}^{-2}(0))^2 I(\chi_{p_2+2}^{-2}(0) < p_2-2)]) \geq 0.
\end{aligned} \tag{8.70}$$

Because the expected value in Eq. (8.32) is a decreasing function of DF, and

$$2 > p_2\mathbb{E}[(1-(p_2-2)\chi_{p_2+2}^{-2}(0))^2 I(\chi_{p_2+2}^{-2}(0) < p_2-2)].$$

8.2.11 PTE vs. SE and PRSE

Following the previous sections, we can make a comparison of PTE with SEs and PRSEs and find the value of Δ^2, for which one estimator dominates others.

8.2.12 Numerical Comparison Among the Estimators

Since the weighted asymptotic risk of the restricted estimator depends on the characteristics of the $\boldsymbol{X}$ matrix, we have generated 1000 samples from a multivariate normal distribution with fixed correlation between p explanatory variables as 0.6, 0.8, and 0.9. We compared relative efficiencies and presented them in Tables 8.4–8.6.

We also note that the RRE depends on the unknown parameter k. Following Kibria (2003, 2012) and Kibria and Banik (2016), we consider several k values and compute the efficiency tables. However, rather than showing the exhaustive work involved using different k values, we consider the optimal value of k, which is described here.

It can be shown that $\text{ADR}(\hat{\boldsymbol{\beta}}_n^{\text{ML(R)}}; \boldsymbol{C}_{11\cdot 2}, \boldsymbol{C}_{22\cdot 1}) \leq \text{ADR}(\hat{\boldsymbol{\beta}}_n^{\text{ML}}; \boldsymbol{C}_{11\cdot 2}, \boldsymbol{C}_{22\cdot 1})$, whenever, $k \in (0, k_{\text{opt}}]$ with $k_{\text{opt}} = \frac{p_2}{\boldsymbol{\beta}_2^\top \boldsymbol{\beta}_2} = p_2 \Delta^{-2}$. Thus, $\text{ADR}(\hat{\boldsymbol{\beta}}_n^{\text{RRML}}(k))$ in Eq. (8.37) at $k = k_{\text{opt}}$ becomes,

$$\text{ADR}(\hat{\boldsymbol{\beta}}_n; \boldsymbol{C}_{11\cdot 2}, \boldsymbol{C}_{22\cdot 1}) = \left\{ p_1 + \frac{p_2 \Delta^2}{(p_2 + \Delta^2)} \right\}.$$

For some selected values of p_1, p, and n, some selected tabular values of efficiency of SE vs. MLE are given in Figures 8.1–8.3 and Tables 8.4–8.6 for $p_1 = 3$, $p_2 = 5$, and $p_5 = 7$.

If we review Tables 8.4–8.6, we observe that the restricted estimator outperformed all the estimators for $\Delta = 0$ or near it. For any value of Δ^2, LASSO, RRE, PTE, SE, and PRE dominate MLE. However, RMLE dominates RRE when $\Delta^2 \in (0, 5.91)$ for any value of p_1 or p_1. Also, PRSE uniformly dominates SE for any values of Δ^2. The RRE dominates the rest of the estimators for $\Delta^2 > \Delta_0^2$, where Δ_0^2 depends on the values of n, p_1, and p.

8.3 Summary and Concluding Remarks

This chapter considers MLE, RMLE, PTE, SE, PRSE, RRE, and, finally, MLASSO for estimating the parameters for the logistic regression model. The performances of the estimators are compared on the basis of quadratic risk functions under both null and alternative hypotheses, which specify certain restrictions on the regression parameters. Under the restriction $\mathcal{H}_o$, the RMLE performs the best compared to other estimators; however, it performs the worst when RMLE moves away from its origin. RRE uniformly dominates all estimators except the RMLE. PRSE uniformly dominates PTE, SE, and MLE.

Table 8.4 Relative efficiency for $n = 20, p = 10$, and $p_1 = 3$.

		RMLE					PTE				
		ρ					α				
Δ^2	MLE	0.6	0.8	0.9	MLASSO	RRE	0.15	0.2	0.25	SE	PRSE
0.00	1.00	6.78	6.37	6.23	3.33	3.33	2.47	2.27	2.11	1.56	1.74
0.10	1.00	6.35	5.99	5.86	3.23	3.23	2.39	2.21	2.05	1.55	1.72
0.20	1.00	5.97	5.65	5.54	3.12	3.13	2.32	2.15	2.00	1.54	1.71
0.30	1.00	5.63	5.35	5.25	3.03	3.04	2.26	2.09	1.95	1.53	1.70
0.50	1.00	5.06	4.83	4.75	2.86	2.88	2.14	1.98	1.86	1.51	1.67
0.70	1.00	4.60	4.41	4.34	2.70	2.75	2.03	1.89	1.78	1.49	1.64
0.90	1.00	4.21	4.05	3.99	2.56	2.63	1.94	1.81	1.71	1.48	1.62
1.00	1.00	4.04	3.89	3.84	2.50	2.58	1.89	1.77	1.67	1.47	1.61
1.50	1.00	3.36	3.26	3.22	2.22	2.36	1.71	1.61	1.53	1.44	1.56
1.80	1.00	3.05	2.97	2.94	2.08	2.26	1.62	1.53	1.46	1.42	1.53
2.00	1.00	2.88	2.80	2.77	2.00	2.20	1.56	1.49	1.42	1.41	1.51
2.86	1.00	2.31	2.26	2.24	1.71	1.99	1.38	1.33	1.29	1.37	1.44
3.00	1.00	2.23	2.19	2.17	1.67	1.96	1.36	1.31	1.27	1.36	1.43
3.88	1.00	1.87	1.84	1.82	1.45	1.82	1.24	1.20	1.18	1.32	1.38
4.90	1.00	1.57	1.55	1.54	1.27	1.70	1.14	1.12	1.10	1.29	1.33
5.00	1.00	1.54	1.52	1.51	1.25	1.69	1.13	1.11	1.10	1.29	1.33
5.91	1.00	1.35	1.34	1.33	1.12	1.61	1.07	1.06	1.05	1.26	1.29
9.93	1.00	0.88	0.87	0.87	0.77	1.41	0.96	0.97	0.98	1.19	1.20
10.00	1.00	0.87	0.86	0.86	0.77	1.40	0.96	0.97	0.98	1.19	1.20
15.00	1.00	0.61	0.60	0.60	0.56	1.29	0.96	0.97	0.98	1.14	1.14
19.94	1.00	0.47	0.46	0.46	0.44	1.22	0.98	0.99	0.99	1.11	1.11
20.00	1.00	0.47	0.46	0.46	0.43	1.22	0.98	0.99	0.99	1.11	1.11
25.00	1.00	0.38	0.38	0.38	0.36	1.18	0.99	1.00	1.00	1.09	1.09
30.00	1.00	0.32	0.32	0.32	0.30	1.15	1.00	1.00	1.00	1.08	1.08
35.00	1.00	0.27	0.27	0.27	0.26	1.13	1.00	1.00	1.00	1.07	1.07
39.95	1.00	0.24	0.24	0.24	0.23	1.12	1.00	1.00	1.00	1.06	1.06
40.00	1.00	0.24	0.24	0.24	0.23	1.12	1.00	1.00	1.00	1.06	1.06
50.00	1.00	0.19	0.19	0.19	0.19	1.09	1.00	1.00	1.00	1.05	1.05
59.96	1.00	0.16	0.16	0.16	0.16	1.08	1.00	1.00	1.00	1.04	1.04
60.00	1.00	0.16	0.16	0.16	0.16	1.08	1.00	1.00	1.00	1.04	1.04
100.00	1.00	0.10	0.10	0.10	0.10	1.05	1.00	1.00	1.00	1.02	1.02

Table 8.5 Relative efficiency for $n = 20, p = 10$, and $p_1 = 5$.

		RMLE					PTE				
		ρ					α				
Δ^2	MLE	0.6	0.8	0.9	MLASSO	RRE	0.15	0.2	0.25	SE	PRSE
0.00	1.00	8.17	7.52	7.28	2.00	2.00	1.74	1.67	1.60	1.22	1.32
0.10	1.00	7.55	6.99	6.79	1.96	1.96	1.70	1.63	1.57	1.22	1.31
0.20	1.00	7.02	6.53	6.35	1.92	1.93	1.67	1.60	1.54	1.21	1.30
0.30	1.00	6.56	6.13	5.98	1.89	1.89	1.64	1.57	1.51	1.21	1.30
0.50	1.00	5.80	5.46	5.34	1.82	1.83	1.58	1.52	1.46	1.20	1.29
0.70	1.00	5.20	4.93	4.82	1.75	1.78	1.52	1.47	1.42	1.20	1.27
0.90	1.00	4.71	4.48	4.40	1.69	1.74	1.47	1.42	1.38	1.19	1.26
1.00	1.00	4.50	4.29	4.21	1.67	1.71	1.45	1.40	1.36	1.19	1.26
1.50	1.00	3.67	3.53	3.48	1.54	1.63	1.35	1.31	1.28	1.17	1.23
1.80	1.00	3.31	3.19	3.15	1.47	1.58	1.30	1.27	1.24	1.17	1.22
2.00	1.00	3.10	3.00	2.96	1.43	1.56	1.27	1.24	1.21	1.16	1.21
2.86	1.00	2.45	2.39	2.36	1.27	1.47	1.17	1.15	1.13	1.14	1.18
3.00	1.00	2.37	2.31	2.29	1.25	1.45	1.15	1.13	1.12	1.14	1.18
3.88	1.00	1.96	1.92	1.90	1.13	1.39	1.08	1.07	1.06	1.13	1.16
4.90	1.00	1.63	1.61	1.59	1.01	1.34	1.02	1.02	1.02	1.11	1.13
5.00	1.00	1.61	1.58	1.57	1.00	1.33	1.02	1.02	1.02	1.11	1.13
5.91	1.00	1.40	1.38	1.37	0.92	1.30	0.99	0.99	0.99	1.10	1.12
9.93	1.00	0.90	0.89	0.88	0.67	1.20	0.94	0.95	0.96	1.07	1.08
10.00	1.00	0.89	0.88	0.88	0.67	1.20	0.94	0.95	0.96	1.07	1.07
15.00	1.00	0.62	0.61	0.61	0.50	1.14	0.96	0.97	0.98	1.05	1.05
19.94	1.00	0.47	0.47	0.47	0.40	1.11	0.99	0.99	0.99	1.04	1.04
20.00	1.00	0.47	0.47	0.47	0.40	1.11	0.99	0.99	0.99	1.04	1.04
25.00	1.00	0.38	0.38	0.38	0.33	1.09	1.00	1.00	1.00	1.03	1.03
30.00	1.00	0.32	0.32	0.32	0.29	1.08	1.00	1.00	1.00	1.03	1.03
35.00	1.00	0.28	0.28	0.27	0.25	1.07	1.00	1.00	1.00	1.02	1.02
39.95	1.00	0.24	0.24	0.24	0.22	1.06	1.00	1.00	1.00	1.02	1.02
40.00	1.00	0.24	0.24	0.24	0.22	1.06	1.00	1.00	1.00	1.02	1.02
50.00	1.00	0.20	0.19	0.19	0.18	1.05	1.00	1.00	1.00	1.02	1.02
59.96	1.00	0.16	0.16	0.16	0.15	1.04	1.00	1.00	1.00	1.01	1.01
60.00	1.00	0.16	0.16	0.16	0.15	1.04	1.00	1.00	1.00	1.01	1.01
100.00	1.00	0.10	0.10	0.10	0.10	1.02	1.00	1.00	1.00	1.01	1.01

Table 8.6 Relative efficiency for $n = 20, p = 10$, and $p_1 = 7$.

		RMLE					PTE				
		ρ					α				
Δ^2	MLE	0.6	0.8	0.9	MLASSO	RRE	0.15	0.2	0.25	SE	PRSE
0.00	1.00	6.78	6.37	6.23	3.33	3.33	2.47	2.27	2.11	1.56	1.74
0.10	1.00	6.35	5.99	5.86	3.23	3.23	2.39	2.21	2.05	1.55	1.72
0.20	1.00	5.97	5.65	5.54	3.12	3.13	2.32	2.15	2.00	1.54	1.71
0.30	1.00	5.63	5.35	5.25	3.03	3.04	2.26	2.09	1.95	1.53	1.70
0.50	1.00	5.06	4.83	4.75	2.86	2.88	2.14	1.98	1.86	1.51	1.67
0.70	1.00	4.60	4.41	4.34	2.70	2.75	2.03	1.89	1.78	1.49	1.64
0.90	1.00	4.21	4.05	3.99	2.56	2.63	1.94	1.81	1.71	1.48	1.62
1.00	1.00	4.04	3.89	3.84	2.50	2.58	1.89	1.77	1.67	1.47	1.61
1.50	1.00	3.36	3.26	3.22	2.22	2.36	1.71	1.61	1.53	1.44	1.56
1.80	1.00	3.05	2.97	2.94	2.08	2.26	1.62	1.53	1.46	1.42	1.53
2.00	1.00	2.88	2.80	2.77	2.00	2.20	1.56	1.49	1.42	1.41	1.51
2.86	1.00	2.31	2.26	2.24	1.71	1.99	1.38	1.33	1.29	1.37	1.44
3.00	1.00	2.23	2.19	2.17	1.67	1.96	1.36	1.31	1.27	1.36	1.43
3.88	1.00	1.87	1.84	1.82	1.45	1.82	1.24	1.20	1.18	1.32	1.38
4.90	1.00	1.57	1.55	1.54	1.27	1.70	1.14	1.12	1.10	1.29	1.33
5.00	1.00	1.54	1.52	1.51	1.25	1.69	1.13	1.11	1.10	1.29	1.33
5.91	1.00	1.35	1.34	1.33	1.12	1.61	1.07	1.06	1.05	1.26	1.29
9.93	1.00	0.88	0.87	0.87	0.77	1.41	0.96	0.97	0.98	1.19	1.20
10.00	1.00	0.87	0.86	0.86	0.77	1.40	0.96	0.97	0.98	1.19	1.20
15.00	1.00	0.61	0.60	0.60	0.56	1.29	0.96	0.97	0.98	1.14	1.14
19.94	1.00	0.47	0.46	0.46	0.44	1.22	0.98	0.99	0.99	1.11	1.11
20.00	1.00	0.47	0.46	0.46	0.43	1.22	0.98	0.99	0.99	1.11	1.11
25.00	1.00	0.38	0.38	0.38	0.36	1.18	0.99	1.00	1.00	1.09	1.09
30.00	1.00	0.32	0.32	0.32	0.30	1.15	1.00	1.00	1.00	1.08	1.08
35.00	1.00	0.27	0.27	0.27	0.26	1.13	1.00	1.00	1.00	1.07	1.07
39.95	1.00	0.24	0.24	0.24	0.23	1.12	1.00	1.00	1.00	1.06	1.06
40.00	1.00	0.24	0.24	0.24	0.23	1.12	1.00	1.00	1.00	1.06	1.06
50.00	1.00	0.19	0.19	0.19	0.19	1.09	1.00	1.00	1.00	1.05	1.05
59.96	1.00	0.16	0.16	0.16	0.16	1.08	1.00	1.00	1.00	1.04	1.04
60.00	1.00	0.16	0.16	0.16	0.16	1.08	1.00	1.00	1.00	1.04	1.04
100.00	1.00	0.10	0.10	0.10	0.10	1.05	1.00	1.00	1.00	1.02	1.02

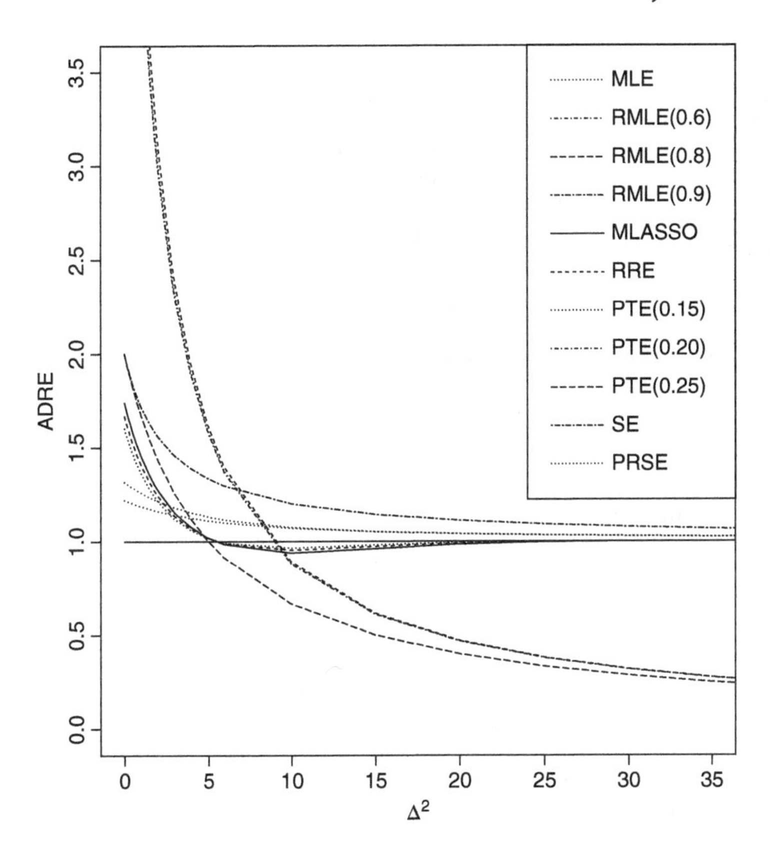

Figure 8.1 Relative efficiency for $p = 10$ and $p_1 = 3$.

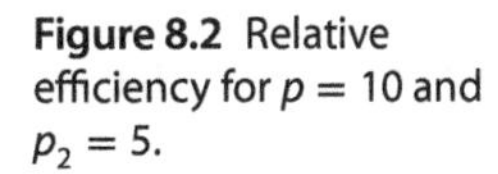

Figure 8.2 Relative efficiency for $p = 10$ and $p_2 = 5$.

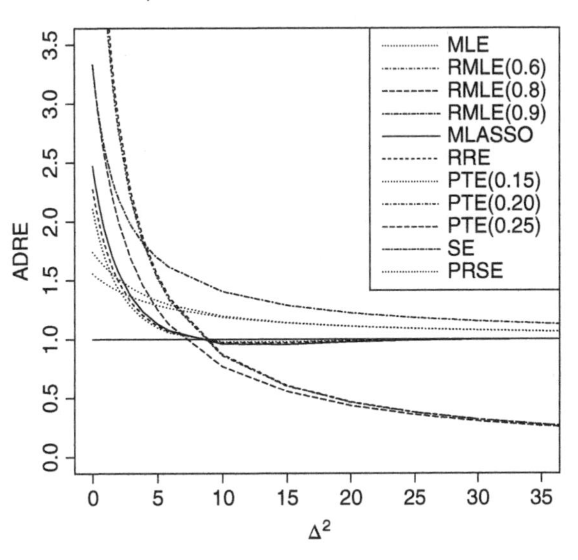

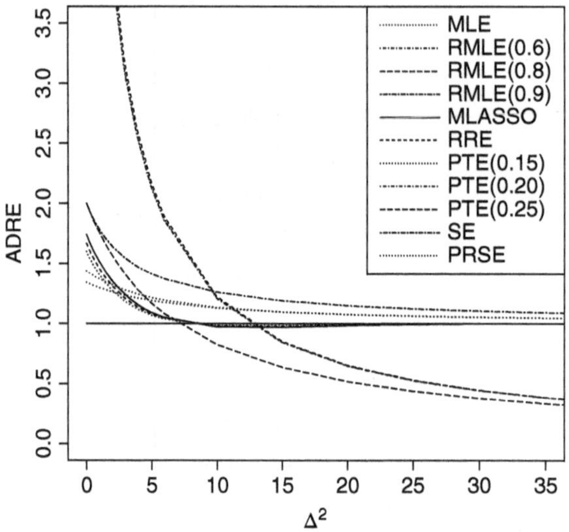

Figure 8.3 Relative efficiency for $p = 10$ and $p_2 = 7$.

Problems

8.1 Under usual notations show that

$$\sqrt{n}(\hat{\boldsymbol{\beta}}_n^{\text{ML}} - \boldsymbol{\beta}) \sim \mathcal{N}_p(\mathbf{0}, I(\hat{\boldsymbol{\beta}}_n^{\text{ML}})^{-1}).$$

8.2 Show that for testing the hypothesis $\mathcal{H}_o : \boldsymbol{\beta}_2 = \mathbf{0}$ vs. $\mathcal{H}_A : \boldsymbol{\beta}_2 \neq \mathbf{0}$, the test statistic is

$$\mathcal{L}_n = \hat{\boldsymbol{\beta}}_{2n}^{\text{ML}^\top} C_{22} \hat{\boldsymbol{\beta}}_{2n}^{\text{ML}}$$

and for a large sample set and under the null hypothesis, $\mathcal{L}_n$ has a chi-square distribution with p_2 DF

8.3 Prove Theorem 8.1.

8.4 Show that the risk function for RRE is

$$\text{ADR}(\hat{\boldsymbol{\beta}}_n^{\text{RRML}}) = \left\{ \text{tr}(\boldsymbol{C}_{11\cdot2}^{-1}) + \frac{1}{(1+k)^2}[\text{tr}(\boldsymbol{C}_{22\cdot1}^{-1}) + k^2 \boldsymbol{\delta}_2^\top \boldsymbol{\delta}_2] \right\}.$$

8.5 Show that the risk function for $\hat{\boldsymbol{\beta}}_n^{\text{ML(PT)}}$ is

$$\begin{aligned}\text{ADR}(\hat{\boldsymbol{\beta}}_n^{\text{ML(PT)}}) = \{&\text{tr}(\boldsymbol{C}_{11\cdot2}^{-1}) + \text{tr}(\boldsymbol{C}_{22\cdot1}^{-1})[1 - H_{p_2+2}(c_\alpha : \Delta^2)] \\ &+ \text{tr}(\boldsymbol{C}_{11}^{-1}) H_{p_2+2}(c_\alpha : \Delta^2)\} \\ &+ \boldsymbol{\delta}_2^\top \boldsymbol{\delta}_2 \{2H_{p_2+2}(c_\alpha : \Delta^2) - H_{p_2+4}(c_\alpha : \Delta^2)\}\end{aligned}$$

and that the weighted L_2-risk will be

$$\begin{aligned}\mathrm{ADR}(\hat{\boldsymbol{\beta}}_n^{\mathrm{ML(PT)}}; \boldsymbol{C}_{11\cdot 2}, \boldsymbol{C}_{22\cdot 1}) = \{ & p_1 + p_2[1 - H_{p_2+2}(c_\alpha : \Delta^2)] \\ & + p_1 H_{p_2+2}(c_\alpha : \Delta^2) \\ & + \Delta^2\{2H_{p_2+2}(c_\alpha : \Delta^2) - H_{p_2+4}(c_\alpha : \Delta^2)\}\}.\end{aligned}$$

8.6 Verify that the risk function for $\hat{\boldsymbol{\beta}}_n^{\mathrm{ML(S)}}$ is

$$\begin{aligned}\mathrm{ADR}\,\hat{\boldsymbol{\beta}}_n^{\mathrm{ML(S)}}) = \{ & \mathrm{tr}(\boldsymbol{C}_{11\cdot 2}^{-1}) + \mathrm{tr}(\boldsymbol{C}_{22\cdot 1}^{-1}) - (p_2 - 2)\,\mathrm{tr}(\boldsymbol{C}_{22\cdot 1}^{-1}) \\ & \times (2\mathbb{E}[\chi_{p_2+2}^{-2}(\Delta^2)] - (p_2 - 2)\mathbb{E}[\chi_{p_2+2}^{-4}(\Delta^2)]) \\ & + (p_2^2 - 4)\boldsymbol{\delta}_2^\top \boldsymbol{\delta}_2 \mathbb{E}[\chi_{p_2+4}^{-4}(\Delta^2)]\}.\end{aligned}$$

8.7 Show that the risk function for $\hat{\boldsymbol{\beta}}_n^{\mathrm{ML(S+)}}$ is

$$\begin{aligned}\mathrm{ADR}(\hat{\boldsymbol{\beta}}_n^{\mathrm{ML(S+)}}) = \mathrm{ADR}(\hat{\boldsymbol{\beta}}_n^{\mathrm{ML(S)}}) & - (p_2 - 2)\,\mathrm{tr}(\boldsymbol{C}_{22\cdot 1}^{-1}) \\ & \times \mathbb{E}[(1 - (p_2 - 2)^2 \chi_{p_2+2}^{-2}(\Delta^2))^2 I(\chi_{p_2+2}^2(\Delta^2) \le p_2 - 2)] \\ & + \boldsymbol{\delta}_2^\top \boldsymbol{\delta}_2 \{2\mathbb{E}[(1 - (p_2 - 2)\chi_{p_2+2}^{-2}(\Delta^2))I(\chi_{p_2+2}^2(\Delta^2) < p_2 - 2)] \\ & - \mathbb{E}[(1 - (p_2 - 2)\chi_{p_2+4}^{-2}(\Delta^2))^2 I(\chi_{p_2+4}^2(\Delta^2) \le p_2 - 2)]\}.\end{aligned}$$

8.8 Compare the risk function of PTE and SE and find the value of Δ^2 for which PTE dominates SE or vice versa.

8.9 Compare the risk function of PTE and PRSE and find the value of Δ^2 for which PTE dominates PRSE or vice versa.

8.10 Show that MLASSO out performs PTE when

$$\Delta^2 \ge \frac{p_2[1 - H_{p_2+2}(c_\alpha : \Delta^2)] + p_1 H_{p_2+2}(c_\alpha : \Delta^2)}{1 - \{2H_{p_2+2}(c_\alpha : \Delta^2) - H_{p_2+4}(c_\alpha : \Delta^2)\}}.$$

8.11 Consider a real data set, where the design matrix elements are moderate to highly correlated, then find the efficiency of the estimators using unweighted risk functions. Find parallel formulas for the efficiency expressions and compare the results with that of the efficiency using weighted risk function. Are the two results consistent?

9

Regression Models with Autoregressive Errors

9.1 Introduction

One of the important assumptions of the linear model is that observed responses of the model are independent. However, in reality, significant serial correlation might occur when data are collected sequentially in time. Autocorrelation, also known as serial correlation, occurs when successive items in a series are correlated so that their covariance is not zero and they are not independent. The main objective of this chapter is to develop some penalty and improved estimators, namely, ridge regression estimator (RRE) and the least absolute shrinkage and selection operator (LASSO) and the Stein-type estimators for the linear regression model with AR(1) errors when some of the coefficients are not statistically significant.

To describe the problem of autocorrelation, we consider the following regression model,

$$y_i = \boldsymbol{x}_i^\top \beta + \xi_i, \quad i = 1, 2, \ldots, n, \tag{9.1}$$

where y_i's are n responses, $\boldsymbol{x}_i = (x_{i1}, \ldots, x_{ip})^\top$ is a $p \times 1$ known vector of regressors, $\boldsymbol{\beta} = (\beta_1, \ldots, \beta_p)^\top$ is an unknown $p \times 1$ vector of unknown regression parameters, and ξ_i is an $n \times 1$ disturbance vector.

Under the assumption of independence and $\boldsymbol{X}$ is of full rank, the errors are not autocorrelated and $\boldsymbol{X}$ and ξ are independently distributed. In that case, the least squares estimator (LSE) of $\boldsymbol{\beta}$ is obtained as

$$\tilde{\boldsymbol{\beta}}_n = (\boldsymbol{X}^\top \boldsymbol{X})^{-1} \boldsymbol{X}^\top \boldsymbol{Y},$$

with the covariance matrix of $\tilde{\boldsymbol{\beta}}_n$ as

$$\text{Cov}(\tilde{\boldsymbol{\beta}}_n) = \sigma^2 (\boldsymbol{X}^\top \boldsymbol{X})^{-1}.$$

In a real-life situation, the necessary assumptions for the LSE may not be met. That means, both regressors and responses may be correlated instead of independent. In that case, the LSE does not possess the optimum property.

Theory of Ridge Regression Estimation with Applications, First Edition.
A.K. Md. Ehsanes Saleh, Mohammad Arashi, and B.M. Golam Kibria.

Now we will assume that the errors of the model (9.1) has an AR error term of order 1. That is

$$\xi_i = \phi\xi_{i-1} + u_i, \tag{9.2}$$

where ϕ is the autoregressive parameter and $\{u_i\}$ are independently and identically distributed white noise random variables with zero mean and σ_u^2 variance. The variance covariance matrix of $\boldsymbol{Y}$ is

$$\text{Var}(\boldsymbol{Y}) = \text{Var}(\boldsymbol{\xi}) = \sigma^2\boldsymbol{V}(\phi),$$

where

$$\boldsymbol{V}(\phi) = \frac{1}{1-\phi^2}\begin{pmatrix} 1 & \phi & \phi^2 & \dots & \phi^{n-1} \\ \phi & 1 & \phi & \dots & \phi^{n-2} \\ \phi^2 & \phi & 1 & \dots & \\ & & & & \\ \phi^{n-1} & \phi^{n-2} & \phi & \dots & 1 \end{pmatrix}. \tag{9.3}$$

This $\boldsymbol{V}(Y)$ matrix is function of AR parameters ϕ and σ^2, which need to be estimated from data.
Since the covariance matrix of ξ is nonspherical, the LSE of $\boldsymbol{\beta}$ will be inefficient compared to generalized least squares estimator (GLSE). Thus we need to estimate the parameters using the GLSE. The GLSE of $\boldsymbol{\beta}$ is

$$\tilde{\boldsymbol{\beta}}_n^{\text{G}} = \left(\boldsymbol{X}^\top\boldsymbol{V}(\phi)^{-1}\boldsymbol{X}\right)^{-1}\boldsymbol{X}^\top\boldsymbol{V}(\phi)^{-1}\boldsymbol{y}, \tag{9.4}$$

with covariance matrix

$$\text{Cov}(\tilde{\boldsymbol{\beta}}_n^{\text{G}}) = \sigma^2\left(\boldsymbol{X}^\top\boldsymbol{V}(\phi)^{-1}\boldsymbol{X}\right)^{-1},$$

where

$$\boldsymbol{V}(\phi)^{-1} = (1-\phi^2)\begin{pmatrix} 1 & -\phi & 0 & 0 & \dots 0 \\ -\phi & 1+\phi^2 & -\phi & 0 & \dots 0 \\ 0 & -\phi & 1+\phi^2 & -\phi & \dots 0 \\ & & & & \\ 0 & 0 & \dots 0 & -\phi & 1 \end{pmatrix} \tag{9.5}$$

and ϕ is a parameter and can be estimated using the least squares principle as

$$\hat{\phi} = \frac{\sum_{i=2}^{n}\hat{\xi}_i\hat{\xi}_{i-1}}{\sum_{i=2}^{n}\hat{\xi}_i^2}.$$

Here, $\hat{\xi}_i$ is the ith residual and $\hat{\xi}_i = y_i - \boldsymbol{x}_i^\top\tilde{\boldsymbol{\beta}}$, where

$$\tilde{\boldsymbol{\beta}} = \left(\boldsymbol{X}^\top\boldsymbol{X}\right)^{-1}\boldsymbol{X}^\top\boldsymbol{y}$$

and

$$\widehat{\sigma}^2 = \frac{1}{n-p-1}(y - X\tilde{\beta})^\top V^{-1}(\phi)\left(y - X\widehat{\beta}\right).$$

Now, we are interested in the estimation of subvector β_1 when one suspects from previous studies/experiences that β_2 is also equal to zero. As such, consider the partition of $\beta = (\beta_1^\top, \beta_2^\top)^\top$, where β_1 and β_2 have dimensions p_1 and p_2, respectively, and $p_1 + p_2 = p$ and $X = (X_1, X_2)$. Then, we may write the estimators of $(\beta_1^\top, \beta_2^\top)^\top$ as

$$\begin{pmatrix} \tilde{\beta}_{1n}^{\mathrm{G}} \\ \tilde{\beta}_{2n}^{\mathrm{G}} \end{pmatrix} = \begin{pmatrix} X_1^\top V(\phi)X_1 & X_1^\top V(\phi)X_2 \\ X_2^\top V(\phi)X_1 & X_2^\top V(\phi)X_2 \end{pmatrix}^{-1} \begin{pmatrix} X_1^\top V(\phi) \\ X_2^\top V(\phi)y \end{pmatrix}. \tag{9.6}$$

If we suspect sparsity in β, i.e. $\beta_2 = \mathbf{0}$, then the estimator of $\beta_{\mathrm{R}} = (\beta_1^\top, \mathbf{0}^\top)^\top$ is given by

$$\widehat{\beta}_n^{\mathrm{G}} = \begin{pmatrix} \widehat{\beta}_1^{\mathrm{G}} \\ \mathbf{0} \end{pmatrix}$$

where

$$\widehat{\beta}_{1n}^{\mathrm{G}} = (X_1^\top V(\widehat{\phi})X_1)^{-1}(X_1^\top V(\widehat{\phi})y). \tag{9.7}$$

9.1.1 Penalty Estimators

We consider two basic penalty estimators, namely, (i) RREs and the (ii) LASSO estimator of β.

As for the RREs, we $\underset{\beta \in \mathbb{R}^p}{\text{minimize}}\{l(\beta) + k\beta^\top\beta\}$, yielding a normal equation as

$$\left(X^\top V(\phi)X + kI_p\right)\beta = X^\top V(\phi)y \tag{9.8}$$

so that the RRE of β is obtained as

$$\begin{aligned} \widehat{\beta}_n^{\mathrm{GRR}}(k) &= \left(X^\top V(\phi)X + kI_p\right)^{-1} X^\top V(\phi)y \\ &= \left(I_p + k(X^\top V(\phi)X)^{-1}\right)^{-1}\left(X^\top V(\phi)X\right)^{-1} X^\top V(\phi)y \\ &= \left(I_p + k(X^\top V(\phi)X)^{-1}\right)^{-1}\widehat{\beta} \end{aligned} \tag{9.9}$$

with the covariance matrix of $\widehat{\beta}_n^{\mathrm{GRR}}(k)$

$$\begin{aligned} \mathrm{Cov}\left(\widehat{\beta}_n^{\mathrm{GRR}}(k)\right) &= \sigma^2\left(X^\top V(\phi)X + kI_p\right)^{-1}\left(X^\top V(\phi)X\right) \\ &\quad \times\left(X^\top V(\phi)X + kI_p\right)^{-1}. \end{aligned} \tag{9.10}$$

Here, both parameters k and ϕ are unknown and need to be estimated from data. RRE for the autoregressive model has been considered by several researchers such as Ismail and Suvarna (2016). To obtain the RRE of $(\boldsymbol{\beta}_1^\top, \boldsymbol{\beta}_2^\top)^\top$, we consider the asymptotic marginal distribution of

$$\begin{aligned}
\tilde{\boldsymbol{\beta}}_{1n}^{\mathrm{G}} &\sim \mathcal{N}_{p_1}(\beta_1, \sigma^2 \boldsymbol{B}_{11\cdot 2}^{-1}(\phi)) \\
\tilde{\boldsymbol{\beta}}_{2n}^{\mathrm{G}} &\sim \mathcal{N}_{p_2}(\beta_2, \sigma^2 \boldsymbol{B}_{22\cdot 1}^{-1}(\phi)),
\end{aligned} \tag{9.11}$$

where

$$\begin{aligned}
\boldsymbol{B}_{11\cdot 2}(\phi) &= \boldsymbol{X}_1^\top \boldsymbol{V}(\phi)\boldsymbol{X}_1 - (\boldsymbol{X}_1^\top \boldsymbol{V}(\phi)\boldsymbol{X}_2)(\boldsymbol{X}_2^\top \boldsymbol{V}(\phi)\boldsymbol{X}_2)^{-1}(\boldsymbol{X}_2^\top \boldsymbol{V}(\phi)\boldsymbol{X}_1) \\
\boldsymbol{B}_{22\cdot 1}(\phi) &= \boldsymbol{X}_2^\top \boldsymbol{V}(\phi)\boldsymbol{X}_2 - (\boldsymbol{X}_2^\top \boldsymbol{V}(\phi)\boldsymbol{X}_1)(\boldsymbol{X}_1^\top \boldsymbol{V}(\phi)\boldsymbol{X}_1)^{-1}(\boldsymbol{X}_1^\top \boldsymbol{V}(\phi)\boldsymbol{X}_2),
\end{aligned}$$

respectively, with

$$\boldsymbol{B}(\phi) = \begin{pmatrix} \boldsymbol{B}_{11}(\phi) & \boldsymbol{B}_{12}(\phi) \\ \boldsymbol{B}_{21}(\phi) & \boldsymbol{B}_{22}(\phi) \end{pmatrix}. \tag{9.12}$$

Also,

$$\begin{aligned}
\boldsymbol{B}_{11}(\phi) &= \boldsymbol{X}_1^\top \boldsymbol{V}(\phi)\boldsymbol{X}_1 \\
\boldsymbol{B}_{22}(\phi) &= \boldsymbol{X}_2^\top \boldsymbol{V}(\phi)\boldsymbol{X}_2 \\
\boldsymbol{B}_{12}(\phi) &= \boldsymbol{X}_1^\top \boldsymbol{V}(\phi)\boldsymbol{X}_2 \\
\boldsymbol{B}_{21}(\phi) &= \boldsymbol{X}_2^\top \boldsymbol{V}(\phi)\boldsymbol{X}_1
\end{aligned}$$

and

$$\boldsymbol{B}^{-1}(\phi) = \begin{pmatrix} \boldsymbol{B}_{11\cdot 2}^{-1}(\phi) & -\boldsymbol{B}_{11\cdot 2}^{-1}(\phi)\boldsymbol{B}_{12}(\phi)\boldsymbol{B}_{22}^{-1}(\phi) \\ -\boldsymbol{B}_{22}^{-1}(\phi)\boldsymbol{B}_{21}(\phi)\boldsymbol{B}_{11\cdot 2}^{-1}(\phi) & \boldsymbol{B}_{22\cdot 1}^{-1}(\phi) \end{pmatrix}. \tag{9.13}$$

From (9.9), we define the RREs of $(\boldsymbol{\beta}_1^\top, \boldsymbol{\beta}_2^\top)^\top$ as

$$\begin{aligned}
\hat{\boldsymbol{\beta}}_{1n}^{\mathrm{GRR}}(k) &= \tilde{\boldsymbol{\beta}}_{1n}^{\mathrm{G}} \\
\hat{\boldsymbol{\beta}}_{2n}^{\mathrm{GRR}}(k) &= \left(\boldsymbol{I}_{p_2} + k\boldsymbol{B}_{22\cdot 1}(\phi)\right)^{-1} \tilde{\boldsymbol{\beta}}_{2n}^{\mathrm{G}}.
\end{aligned} \tag{9.14}$$

9.1.2 Shrinkage Estimators

9.1.2.1 Preliminary Test Estimator

Since, we suspect that the sparsity condition, $\boldsymbol{\beta}_2 = \boldsymbol{0}$, may hold, we use the Wald test statistic, $\mathcal{L}_n$, for testing the hypothesis $\mathcal{H}_o : \boldsymbol{\beta}_2 = \boldsymbol{0}$, where

$$\mathcal{L}_n = \frac{n\tilde{\boldsymbol{\beta}}_{2n}^{\mathrm{G}\top} \boldsymbol{B}_{22\cdot 1}(\phi)\tilde{\boldsymbol{\beta}}_{2n}^{\mathrm{G}}}{\sigma^2} \sim \chi_{p_2}^2. \tag{9.15}$$

For large samples and under $\boldsymbol{\beta}_2 = \boldsymbol{0}$, $\mathcal{L}_n$ has a chi-square distribution with p_2 degrees of freedom (D.F.) Let $\chi_{p_2}^2(\alpha) = c_\alpha$, (say) be the α-level critical value from the null distribution of $\mathcal{L}_n$. Then, the preliminary test estimator (PTE) of $\boldsymbol{\beta}$ may be written using the marginal distribution of $\left(\hat{\boldsymbol{\beta}}_{1n}^{\mathrm{G(PT)}}, \hat{\boldsymbol{\beta}}_{2n}^{\mathrm{G(PT)}}\right)^\top$ as

$$\hat{\boldsymbol{\beta}}_n^{\mathrm{G(PT)}} = \begin{pmatrix} \tilde{\boldsymbol{\beta}}_{1n}^{\mathrm{G}} \\ \tilde{\boldsymbol{\beta}}_{2n}^{\mathrm{G}} I(\mathcal{L}_n > c_\alpha) \end{pmatrix}, \tag{9.16}$$

where $I(A)$ is the indicator function of the set A.

9.1.2.2 Stein-Type and Positive-Rule Stein-Type Estimators

We notice that PTE is a discrete function and loses some optimality properties. Thus, we consider the continuous version of the PTE which mimics the Stein-type estimators given by

$$\hat{\boldsymbol{\beta}}_n^{\mathrm{G(S)}} = \begin{pmatrix} \tilde{\boldsymbol{\beta}}_{1n}^{\mathrm{G}} \\ \tilde{\boldsymbol{\beta}}_{2n}^{\mathrm{G}} (1 - (p_2 - 2)(\mathcal{L}_n^{-1})) \end{pmatrix}. \tag{9.17}$$

Since the SE may yield estimators with wrong signs, we consider the positive-rule Stein-type estimator (PRSE) of $\beta = (\beta_1, \beta_2)^\top$ given by

$$\hat{\boldsymbol{\beta}}_n^{\mathrm{G(S+)}} = \begin{pmatrix} \tilde{\boldsymbol{\beta}}_{1n}^{\mathrm{G}} \\ \tilde{\boldsymbol{\beta}}_{2n}^{\mathrm{G}} (1 - (p_2 - 2)\mathcal{L}_n^{-1})^+ \end{pmatrix}. \tag{9.18}$$

9.1.3 Results on Penalty Estimators

A modified least absolute shrinkage and selection operator (MLASSO) estimator may be defined by the vector

$$\hat{\boldsymbol{\beta}}_n^{\mathrm{MLASSO}}(\lambda) = \left(\mathrm{sgn}(\tilde{\boldsymbol{\beta}}_{jn}^{\mathrm{G}})(|\tilde{\boldsymbol{\beta}}_{jn}^{\mathrm{G}}| - \lambda\sqrt{\boldsymbol{B}^{jj}(\phi)})|j = 0, 1, \ldots, p \right)^\top, \tag{9.19}$$

where $\boldsymbol{B}^{jj}(\phi)$ is the jth diagonal element of $\boldsymbol{B}^{-1}(\phi)$. This estimator puts some coefficient exactly equal to zero.

Now, we know that the asymptotic marginal distribution of

$$\sqrt{n}(\tilde{\boldsymbol{\beta}}_{jn}^{\mathrm{G}} - \beta_j) \sim \mathcal{N}(\mathbf{0}, \sigma^2 \boldsymbol{B}^{jj}(\phi)), \quad j = 1, \ldots, p. \tag{9.20}$$

We consider the estimation of $\boldsymbol{\beta}$ by any estimator $\hat{\boldsymbol{\beta}}_n^*$ under the L_2-risk function

$$\mathrm{ADR}(\hat{\boldsymbol{\beta}}_n^*) = \mathbb{E}\left[\|\hat{\boldsymbol{\beta}}_n^* - \boldsymbol{\beta}\|_2^2 \right]. \tag{9.21}$$

Consider the family of diagonal linear projections

$$\boldsymbol{T}_{\mathrm{DP}}(\beta, k) = \left(k_1 \hat{\boldsymbol{\beta}}_{1n}^{\mathrm{MLASSO}}, \ldots, k_r \hat{\boldsymbol{\beta}}_{rn}^{\mathrm{MLASSO}} \right)^\top, \quad k_j \in (0, 1). \tag{9.22}$$

Following Chapter 8, we obtain the ideal L_2 risk

$$\mathrm{ADR}(\boldsymbol{T}_{\mathrm{DP}}) = \sum_{j=1}^{p} \min(\beta_j^2, B^{jj}(\phi)). \tag{9.23}$$

Now, if p_1 coefficients exceed the noise level and p_2 coefficients are 0's, then the L_2 risk is given by

$$\mathrm{R}(\boldsymbol{T}_{\mathrm{DP}}) = \sigma^2 \operatorname{tr}(\boldsymbol{B}_{11\cdot 2}^{-1}(\phi)) + \boldsymbol{\beta}_2^\top \boldsymbol{\beta}_2, \quad \boldsymbol{\beta}_2 = (\beta_{21}, \dots, \beta_{2p_2})^\top. \tag{9.24}$$

Consequently, the weighted L_2-risk lower bound is given by $\hat{\boldsymbol{\beta}}_n^{\mathrm{MLASSO}}(\lambda)$ as

$$\mathrm{R}(\hat{\boldsymbol{\beta}}_n^{\mathrm{MLASSO}}(\lambda) : \boldsymbol{B}_{11\cdot 2}^{-1}(\phi), \boldsymbol{B}_{22\cdot 1}^{-1}(\phi)) = \sigma^2(p_1 + \Delta^2), \quad \Delta^2 = \frac{\boldsymbol{\beta}_2^\top \boldsymbol{B}_{22\cdot 1}(\phi)\boldsymbol{\beta}_2}{\sigma^2}. \tag{9.25}$$

We shall use this lower bound to compare MLASSO with other estimators. For details on LASSO estimators, see Chapter 5.

9.1.4 Results on PTE and Stein-Type Estimators

In this section, we present the asymptotic distributional bias (ADB) and asymptotic distributional L_2-risk (ADR) of the shrinkage estimators.

Theorem 9.1 *Under the assumed regularity conditions and a sequence of local alternatives $\{\mathcal{K}_{(n)}\}$ defined by $\mathcal{K}_{(n)} : \boldsymbol{\beta}_2(n) = n^{-1/2}\boldsymbol{\delta}_2$, $\boldsymbol{\delta}_2 \neq 0$, the following holds:*

$$\mathrm{ADB}(\tilde{\boldsymbol{\beta}}_n^{\mathrm{G}}) = \begin{pmatrix} \mathrm{ADB}(\tilde{\boldsymbol{\beta}}_{1n}^{\mathrm{G}}) \\ \mathrm{ADB}(\tilde{\boldsymbol{\beta}}_{1n}^{\mathrm{G}}) \end{pmatrix} = \begin{pmatrix} \mathbf{0} \\ \mathbf{0} \end{pmatrix} \tag{9.26}$$

$$\mathrm{ADB}(\hat{\boldsymbol{\beta}}_n^{\mathrm{G}}) = \begin{pmatrix} \mathrm{ADB}(\tilde{\boldsymbol{\beta}}_{1n}^{\mathrm{G}}) \\ \mathrm{ADB}(\hat{\boldsymbol{\beta}}_{1n}^{\mathrm{G}}) \end{pmatrix} = \begin{pmatrix} \mathbf{0} \\ -\boldsymbol{\beta}_2 \end{pmatrix} \tag{9.27}$$

$$\mathrm{ADB}(\hat{\boldsymbol{\beta}}_n^{\mathrm{G(PT)}}) = \begin{pmatrix} \mathrm{ADB}(\hat{\boldsymbol{\beta}}_{1n}^{\mathrm{G(PT)}}) \\ \mathrm{ADB}(\hat{\boldsymbol{\beta}}_{1n}^{\mathrm{G(PT)}}) \end{pmatrix} = \begin{pmatrix} \mathbf{0} \\ -\boldsymbol{\beta}_2 H_{p_2+2}(c_\alpha; \Delta^2) \end{pmatrix}, \tag{9.28}$$

where $\Delta^2 = \boldsymbol{\delta}_2^\top \boldsymbol{B}_{22\cdot 1}^{-1}\boldsymbol{\delta}_2/\sigma^2$ and H-function are the cumulative distribution function (c.d.f.) of the chi-square distribution.

$$\mathrm{ADB}(\hat{\boldsymbol{\beta}}_n^{\mathrm{G(S)}}) = \begin{pmatrix} \mathrm{ADB}(\hat{\boldsymbol{\beta}}_{1n}^{\mathrm{G(S)}}) \\ \mathrm{ADB}(\hat{\boldsymbol{\beta}}_{1n}^{\mathrm{G(S)}}) \end{pmatrix} = \begin{pmatrix} 0 \\ -(p_2 - 2)\boldsymbol{\beta}_2 \mathbb{E}[\chi_{p_2+2}^{-2}(\Delta^2)] \end{pmatrix} \tag{9.29}$$

$$\text{ADB}(\hat{\boldsymbol{\beta}}_n^{\text{G(S)+}}) = \begin{pmatrix} \text{ADB}(\hat{\boldsymbol{\beta}}_{1n}^{\text{G(S+)}}) \\ \text{ADB}(\hat{\boldsymbol{\beta}}_{2n}^{\text{G(S+)}}) \end{pmatrix} = \begin{pmatrix} \mathbf{0} \\ \text{ADB}(\hat{\boldsymbol{\beta}}_{2n}^{\text{G(S)}}) \\ -(p_2-2)\boldsymbol{\beta}_2 H_{p_2+2}(c_\alpha;\Delta^2) \end{pmatrix} \tag{9.30}$$

Theorem 9.2 *Under the assumed regularity conditions and a sequence of local alternatives $\{\mathcal{K}_{(n)}\}$ defined by $\mathcal{K}_{(n)} : \boldsymbol{\beta}_{2(n)} = n^{-1/2}\boldsymbol{\delta}_2$, $\boldsymbol{\delta}_2 \neq \mathbf{0}$, the expressions of asymptotic distributional L_2-risk expressions are, respectively, given by*

$$\text{ADR}(\tilde{\boldsymbol{\beta}}_n^{\text{G}}) = \begin{pmatrix} \text{ADR}(\tilde{\boldsymbol{\beta}}_{1n}^{\text{G}}) \\ \text{ADR}(\tilde{\boldsymbol{\beta}}_{2n}^{\text{G}}) \end{pmatrix} = \begin{pmatrix} \sigma^2 \operatorname{tr}(\boldsymbol{B}_{11\cdot2}^{-1}(\phi)) \\ \sigma^2 \operatorname{tr}(\boldsymbol{B}_{22\cdot1}^{-1}(\phi)) \end{pmatrix}. \tag{9.31}$$

That means the ADR function for GLSE is

$$\begin{aligned}\text{ADR}(\tilde{\boldsymbol{\beta}}_n^{\text{G}}) &= \sigma^2\{\operatorname{tr}(\boldsymbol{B}_{11\cdot2}^{-1}(\phi)) + \operatorname{tr}(\boldsymbol{B}_{22\cdot1}^{-1}(\phi))\} \\ &= \sigma^2 \operatorname{tr}(\boldsymbol{B}^{-1}(\phi)),\end{aligned} \tag{9.32}$$

and the weighted L_2-risk will be

$$\text{ADR}(\tilde{\boldsymbol{\beta}}_n^{\text{G}} : \boldsymbol{B}_{11\cdot2}(\phi), \boldsymbol{B}_{22\cdot1}(\phi)) = \sigma^2(p_1+p_2). \tag{9.33}$$

The risk function of the restricted GLSE estimator, $\hat{\boldsymbol{\beta}}_n^{\text{G}}$, is

$$\text{ADR}(\hat{\boldsymbol{\beta}}_n^{\text{G(R)}}) = \begin{pmatrix} \text{ADR}(\hat{\boldsymbol{\beta}}_{1n}^{\text{G(R)}}) \\ \text{ADR}(\hat{\boldsymbol{\beta}}_{2n}^{\text{G(R)}}) \end{pmatrix} = \begin{pmatrix} \sigma^2 \operatorname{tr}(\boldsymbol{B}_{11}^{-1}(\phi)) \\ \boldsymbol{\beta}_2^\top \boldsymbol{\beta}_2 \end{pmatrix}. \tag{9.34}$$

That means, the risk function for restricted generalized least squares estimator (RGLSE) is

$$\text{ADR}(\hat{\boldsymbol{\beta}}_n^{\text{G(R)}}) = \sigma^2 \operatorname{tr}(\boldsymbol{B}_{11}^{-1}(\phi)) + \boldsymbol{\beta}_2^\top \boldsymbol{\beta}_2, \tag{9.35}$$

and the weighted L_2-risk will be

$$\text{ADR}(\hat{\boldsymbol{\beta}}_n^{\text{G(R)}} : \boldsymbol{B}_{11\cdot2}(\phi), \boldsymbol{B}_{22\cdot1}(\phi)) = \sigma^2\{\operatorname{tr}(\boldsymbol{B}_{11}^{-1}(\phi)\boldsymbol{B}_{11\cdot2}(\phi)) + \Delta^2\}. \tag{9.36}$$

The risk of PTGLSE is

$$\text{ADR}(\hat{\boldsymbol{\beta}}_n^{\text{G(PT)}}) = \begin{pmatrix} \text{ADR}(\hat{\boldsymbol{\beta}}_{1n}^{\text{G(PT)}}) \\ \text{ADR}(\hat{\boldsymbol{\beta}}_{2n}^{\text{G(PT)}}) \end{pmatrix}, \tag{9.37}$$

where

$$\text{ADR}(\hat{\boldsymbol{\beta}}_{1n}^{\text{G(PT)}}) = \sigma^2 \operatorname{tr}(\boldsymbol{B}_{11\cdot2}^{-1}(\phi)) \tag{9.38}$$

and

$$\begin{aligned}\text{ADR}(\hat{\boldsymbol{\beta}}_{2n}^{\text{G(PT)}}) &= \sigma^2\{\operatorname{tr}(\boldsymbol{B}_{22\cdot1}^{-1}(\phi))[1 - H_{p_2+2}(c_\alpha : \Delta^2)]\} \\ &\quad + \boldsymbol{\delta}_2^\top \boldsymbol{\delta}_2\{2H_{p_2+2}(c_\alpha : \Delta^2) - H_{p_2+4}(c_\alpha : \Delta^2)\}.\end{aligned} \tag{9.39}$$

Then the final risk function for $\widehat{\boldsymbol{\beta}}_n^{\mathrm{G(PT)}}$ *is*

$$\mathrm{ADR}(\widehat{\boldsymbol{\beta}}_n^{\mathrm{G(PT)}}) = \sigma^2\{\mathrm{tr}(\boldsymbol{B}_{11\cdot 2}^{-1}(\phi)) + \mathrm{tr}(\boldsymbol{B}_{22\cdot 1}^{-1}(\phi))[1 - H_{p_2+2}(c_\alpha : \Delta^2)] + \boldsymbol{\delta}_2^\top\boldsymbol{\delta}_2\{2H_{p_2+2}(c_\alpha : \Delta^2) - H_{p_2+4}(c_\alpha : \Delta^2)\}\}, \tag{9.40}$$

and the weighted L_2*-risk will be*

$$\begin{aligned}&\mathrm{ADR}(\widehat{\boldsymbol{\beta}}_n^{\mathrm{G(PT)}}; \boldsymbol{B}_{11\cdot 2}(\phi), \boldsymbol{B}_{22\cdot 1}(\phi))\\ &= \sigma^2\{p_1 + p_2[1 - H_{p_2+2}(c_\alpha : \Delta^2)] + \Delta^2\{2H_{p_2+2}(c_\alpha : \Delta^2) - H_{p_2+4}(c_\alpha : \Delta^2)\}\}.\end{aligned} \tag{9.41}$$

The risk of Stein-type GLSE,

$$\mathrm{ADR}(\widehat{\boldsymbol{\beta}}_n^{\mathrm{G(S)}}) = \begin{pmatrix}\mathrm{ADR}(\widehat{\boldsymbol{\beta}}_{1n}^{\mathrm{G(S)}})\\ \mathrm{ADR}(\widehat{\boldsymbol{\beta}}_{2n}^{\mathrm{G(S)}})\end{pmatrix}, \tag{9.42}$$

where

$$\mathrm{ADR}(\widehat{\boldsymbol{\beta}}_{1n}^{\mathrm{G(S)}}) = \sigma^2\ \mathrm{tr}(\boldsymbol{B}_{11\cdot 2}^{-1}(\phi)) \tag{9.43}$$

and

$$\begin{aligned}\mathrm{ADR}(\widehat{\boldsymbol{\beta}}_{2n}^{\mathrm{G(S)}}) &= \sigma^2\{\mathrm{tr}(\boldsymbol{B}_{22\cdot 1}^{-1}(\phi)) - (p_2 - 2)\ \mathrm{tr}(\boldsymbol{B}_{22\cdot 1}^{-1}(\phi))\\ &\quad\times(2\mathbb{E}[\chi_{p_2+2}^{-2}(\Delta^2)] - (p_2 - 2)\mathbb{E}[\chi_{p_2+2}^{-4}(\Delta^2)])\}\\ &\quad+ (p_2^2 - 4)\boldsymbol{\delta}_2^\top\boldsymbol{\delta}_2\mathbb{E}[\chi_{p_2+4}^{-4}(\Delta^2)].\end{aligned} \tag{9.44}$$

Then, the final risk function for $\widehat{\boldsymbol{\beta}}_n^{\mathrm{G(S)}}$ *is*

$$\begin{aligned}\mathrm{ADR}(\widehat{\boldsymbol{\beta}}_n^{\mathrm{G(S)}}) &= \sigma^2\{(\mathrm{tr}(\boldsymbol{B}_{11\cdot 2}^{-1}(\phi)) + \mathrm{tr}(\boldsymbol{B}_{22\cdot 1}^{-1}(\phi))) - (p_2 - 2)\ \mathrm{tr}(\boldsymbol{B}_{22\cdot 1}^{-1})\\ &\quad\times(2\mathbb{E}[\chi_{p_2+2}^{-2}(\Delta^2)] - (p_2 - 2)\mathbb{E}[\chi_{p_2+2}^{-4}(\Delta^2)])\}\\ &\quad+ (p_2^2 - 4)\boldsymbol{\delta}_2^\top\boldsymbol{\delta}_2\mathbb{E}[\chi_{p_2+4}^{-4}(\Delta^2)].\end{aligned} \tag{9.45}$$

The weighted L_2*-risk is*

$$\mathrm{ADR}(\widehat{\boldsymbol{\beta}}_n^{\mathrm{G(S)}}; \boldsymbol{B}_{11\cdot 2}(\phi), \boldsymbol{B}_{22\cdot 1}(\phi)) = \sigma^2\{p_1 + p_2 - (p_2 - 2)^2\mathbb{E}[\chi_{p_2}^{-2}(\Delta^2)]\}. \tag{9.46}$$

Finally, the risk function of the positive-rule Stein-type GLSE ($\widehat{\boldsymbol{\beta}}_n^{\mathrm{G(S+)}}$) *is obtained as follows:*

$$\begin{aligned}\mathrm{ADR}(\widehat{\boldsymbol{\beta}}_{1n}^{\mathrm{G(S+)}}) &= \mathrm{ADR}(\widehat{\boldsymbol{\beta}}_n^{\mathrm{G(S)}}) - \sigma^2(p_2 - 2)\ \mathrm{tr}(\boldsymbol{B}_{22\cdot 1}^{-1}(\phi))\\ &\quad\times\mathbb{E}[(1 - (p_2 - 2)^2\chi_{p_2+2}^{-2}(\Delta^2))^2 I(\chi_{p_2+2}^2(\Delta^2) < p_2 - 2)]\\ &\quad+ \boldsymbol{\delta}_2^\top\boldsymbol{\delta}_2\{2\mathbb{E}[(1 - (p_2 - 2)\chi_{p_2+2}^{-2}(\Delta^2))I(\chi_{p_2+2}^2(\Delta^2) < p_2 - 2)]\\ &\quad- \mathbb{E}[(1 - (p_2 - 2)\chi_{p_2+4}^{-2}(\Delta^2))^2 I(\chi_{p_2+4}^2(\Delta^2) < p_2 - 2)]\}.\end{aligned} \tag{9.47}$$

The weighted L_2-risk function is

$$\begin{aligned}\mathrm{ADR}(\widehat{\boldsymbol{\beta}}_n^{\mathrm{G(S+)}} : \boldsymbol{B}_{11\cdot2}(\phi), \boldsymbol{B}_{22\cdot1}(\phi)) = {}& \sigma^2\{p_1 + p_2 - (p_2-2)^2\mathbb{E}[\chi_{p_2}^{-2}(\Delta^2)]\}\\ &- \sigma^2 p_2\mathbb{E}[(1-(p_2-2)^2\chi_{p_2+2}^{-2}(\Delta^2))^2\\ &\quad I(\chi_{p_2+2}^2(\Delta^2) \le p_2-2)]\\ &+ \Delta^2\{2\mathbb{E}[(1-(p_2-2)\chi_{p_2+2}^{-2}(\Delta^2))\\ &\quad I(\chi_{p_2+2}^2(\Delta^2) \le p_2-2)\\ &- \mathbb{E}[(1-(p_2-2)\chi_{p_2+4}^{-2}(\Delta^2))^2\\ &\quad I(\chi_{p_2+4}^2(\Delta^2) \le p_2-2)].\end{aligned} \tag{9.48}$$

9.1.5 Results on Penalty Estimators

First, we consider the simplest version of the RRE given by

$$\widehat{\boldsymbol{\beta}}_n^{\mathrm{GRR}}(k) = \begin{pmatrix} \tilde{\boldsymbol{\beta}}_{1n}^{\mathrm{G}} \\ \frac{1}{1+k}\tilde{\boldsymbol{\beta}}_{2n}^{\mathrm{G}} \end{pmatrix}. \tag{9.49}$$

The risk function for RRE is

$$\mathrm{ADR}(\widehat{\boldsymbol{\beta}}_n(k)) = \left\{\sigma^2\ \mathrm{tr}(\mathbf{B}_{11\cdot2}^{-1}(\phi)) + \frac{1}{(1+k)^2}[\sigma^2\ \mathrm{tr}(\mathbf{B}_{22\cdot1}^{-1}(\phi)) + k^2\boldsymbol{\delta}_2^{\top}\boldsymbol{\delta}_2]\right\}. \tag{9.50}$$

The weighted L_2-risk function is

$$\mathrm{ADR}(\widehat{\boldsymbol{\beta}}_n(k) : \boldsymbol{B}_{11\cdot2}(\phi), \boldsymbol{B}_{22\cdot1}(\phi)) = \sigma^2\left\{p_1 + \frac{1}{(1+k)^2}[p_2 + k^2\Delta^2]\right\}. \tag{9.51}$$

The optimum value of k is obtained as $k_{\mathrm{opt}} = p_2\Delta^{-2}$. So that

$$\mathrm{ADR}(\widehat{\boldsymbol{\beta}}_n(k) : \boldsymbol{B}_{11\cdot2}(\phi), \boldsymbol{B}_{22\cdot1}(\phi)) = \sigma^2\left\{p_1 + \frac{p_2\Delta^2}{p_2+\Delta^2}\right\}. \tag{9.52}$$

The risk function of the MLASSO estimator is

$$\mathrm{ADR}(\widehat{\boldsymbol{\beta}}_n^{\mathrm{MLASSO}}(\lambda)) = \sigma^2\ \mathrm{tr}(\boldsymbol{B}_{11\cdot2}^{-1}(\phi)) + \boldsymbol{\delta}_2^{\top}\boldsymbol{\delta}_2. \tag{9.53}$$

The weighted L_2-risk function is

$$\mathrm{ADR}(\widehat{\boldsymbol{\beta}}_n^{\mathrm{MLASSO}} : \boldsymbol{B}_{11\cdot2}(\phi), \boldsymbol{B}_{22\cdot1}(\phi)) = \sigma^2\{p_1 + \Delta^2\}. \tag{9.54}$$

9.2 Asymptotic Distributional L_2-risk Efficiency Comparison

In this section, we study the comparative properties of the estimators relative to the GLSEs. We consider the weighted L_2-risk to compare the performance of the estimators. However, it is noted that all of the weighted L_2-risk functions in this chapter except for the restricted LSEs are identical to those of the L_2 risk functions in Chapter 8. Thus, we should skip the finite sample comparison among the estimators except for the RGLSE.

9.2.1 Comparison of GLSE with RGLSE

In this case, the asymptotic distributional relative weighted L_2-risk efficiency of RGLSE vs. GLSE is given by

$$\begin{aligned}\text{ADRE}(\hat{\boldsymbol{\beta}}_n^{\text{G}} : \tilde{\boldsymbol{\beta}}_n^{\text{G}}) &= \frac{p_1 + p_2}{\text{tr}(\boldsymbol{B}_{11}^{-1}(\phi)\boldsymbol{B}_{11\cdot2}(\phi)) + \Delta^2} \\ &= \left(1 + \frac{p_2}{p_1}\right)\left(1 - \frac{\text{tr}(\boldsymbol{N}_0(\phi))}{p_1} + \frac{\Delta^2}{p_1}\right)^{-1}, \end{aligned} \tag{9.55}$$

where

$$\boldsymbol{N}_0(\phi) = \boldsymbol{B}_{11}^{-1}(\phi)\boldsymbol{B}_{12}(\phi)\boldsymbol{B}_{22}^{-1}(\phi)\boldsymbol{B}_{21}(\phi).$$

The $\text{ADRE}(\hat{\boldsymbol{\beta}}_n^{\text{G}} : \tilde{\boldsymbol{\beta}}_n^{\text{G}})$ is a decreasing function of Δ^2. At $\Delta^2 = 0$, its value is

$$\left(1 + \frac{p_2}{p_1}\right)\left(1 - \frac{\text{tr}(\boldsymbol{N}_0(\phi))}{p_1}\right)^{-1}$$

and as $\Delta^2 \to \infty$, its value is 0. It crosses the 1-line at $\Delta^2 = p_2 + \boldsymbol{N}_0(\phi)$. So,

$$0 \leq \text{ADRE}(\hat{\boldsymbol{\beta}}_n^{\text{G}} : \tilde{\boldsymbol{\beta}}_n^{\text{G}}) \leq \left(1 + \frac{p_2}{p_1}\right)\left(1 - \frac{\text{tr}(\boldsymbol{N}_0(\phi))}{p_1}\right)^{-1}.$$

In order to compute $\text{tr}(\boldsymbol{N}_0)$, we need to find $\boldsymbol{B}_{11}(\phi)$, $\boldsymbol{B}_{22}(\phi)$, and $\boldsymbol{B}_{12}(\phi)$. These are obtained by generating explanatory variables by the following equation based on McDonald and Galarneau (1975),

$$x_{ij} = \sqrt{1 - \rho^2} z_{ij} + \rho z_{ip}, \quad i = 1, 2, \ldots, n; j = 1, \ldots, p. \tag{9.56}$$

where z_{ij} are independent $\mathcal{N}(0, 1)$ pseudorandom numbers and ρ^2 is the correlation between any two explanatory variables. In this study, we take $\rho^2 = 0.1, 0.2, 0.8$, and 0.9 which shows the variables are lightly collinear and severely collinear. In our case, we chose $n = 100$ and various (p_1, p_2). The resulting output is then used to compute $\text{tr}(\boldsymbol{N}_0(\phi))$.

9.2.2 Comparison of GLSE with PTE

Here, the relative weighted L_2-risk efficiency expression for PTE vs. LSE is given by

$$\text{ADRE}(\hat{\boldsymbol{\beta}}_n^{\text{G(PT)}} : \tilde{\boldsymbol{\beta}}_n^{\text{G}}) = \frac{p_1 + p_2}{g(\Delta^2, \alpha)}, \tag{9.57}$$

where

$$g(\Delta^2, \alpha) = p_1 + p_2(1 - H_{p_2+2}(c_\alpha; \Delta^2)) + \Delta^2[2H_{p_2+2}(c_\alpha; \Delta^2) - H_{p_2+4}(c_\alpha; \Delta^2)]. \tag{9.58}$$

Then, the PTE outperforms the LSE for

$$0 \le \Delta^2 \le \frac{p_2 H_{p_2+2}(c_\alpha; \Delta^2)}{2H_{p_2+2}(c_\alpha; \Delta^2) - H_{p_2+4}(c_\alpha; \Delta^2)} = \Delta^2_{\text{G(PT)}}. \tag{9.59}$$

Otherwise, LSE outperforms the PTE in the interval $(\Delta^2_{\text{G(PT)}}, \infty)$. We may mention that $\text{ADRE}(\hat{\boldsymbol{\beta}}_n^{\text{G(PT)}} : \tilde{\boldsymbol{\beta}}_n^{\text{G}})$ is a decreasing function of Δ^2 with a maximum at the point $\Delta^2 = 0$, then decreases crossing the 1-line to a minimum at the point $\Delta^2 = \Delta^2_{\text{G(PT)}}(\min)$ with a value $M_{\text{G(PT)}}(\alpha)$, and then increases toward 1-line.

The $\text{ADRE}(\hat{\boldsymbol{\beta}}_n^{\text{G(PT)}}; \tilde{\boldsymbol{\beta}}_n^{\text{G}})$ belongs to the interval

$$M_{\text{G(PT)}}(\alpha) \le \text{ADRE}(\hat{\boldsymbol{\beta}}_n^{\text{G(PT)}}; \tilde{\boldsymbol{\beta}}_n^{\text{G}}) \le \left(1 + \frac{p_2}{p_1}\right)\left(1 + \frac{p_2}{p_1}[1 - H_{p_2+2}(c_\alpha; 0)]\right)^{-1},$$

where $M_{\text{G(PT)}}(\alpha)$ depends on the size α and given by

$$\begin{aligned} M_{\text{G(PT)}}(\alpha) = \left(1 + \frac{p_2}{p_1}\right)\Bigg\{ & 1 + \frac{p_2}{p_1}[1 - H_{p_2+2}(c_\alpha; \Delta^2_{\text{G(PT)}}(\min))] \\ & + \frac{\Delta^2_{\text{G(PT)}}(\min)}{p_1} \\ & \times [2H_{p_2+2}(c_\alpha; \Delta^2_{\text{G(PT)}}(\min)) - H_{p_2+4}(c_\alpha; \Delta^2_{\text{G(PT)}}(\min))]\Bigg\}^{-1}. \end{aligned}$$

The quantity $\Delta^2_{\text{G(PT)}}(\min)$ is the value Δ^2 at which the relative weighted L_2-risk efficiency value is minimum.

9.2.3 Comparison of LSE with SE and PRSE

Since SE and PRSE need $p_2 \ge 3$ to express their weighted L_2-risk expressions, we assume always $p_2 \ge 3$. We have

$$\text{ADRE}(\hat{\boldsymbol{\beta}}_n^{\text{G(S)}}; \tilde{\boldsymbol{\beta}}_n^{\text{G}}) = \left(1 + \frac{p_2}{p_1}\right)\left(1 + \frac{p_2}{p_1} - \frac{(p_2 - 2)^2}{p_1}\mathbb{E}[\chi^{-2}_{p_2}(\Delta^2)]\right)^{-1}. \tag{9.60}$$

It is a decreasing function of Δ^2. At $\Delta^2 = 0$, its value is $\left(1+\frac{p_2}{p_1}\right)\left(1+\frac{2}{p_1}\right)^{-1}$ and when $\Delta^2 \to \infty$, its value goes to 1. Hence, for $\Delta^2 \in \mathbb{R}^+$,

$$1 \le \left(1+\frac{p_2}{p_1}\right)\left(1+\frac{p_2}{p_1}-\frac{(p_2-2)^2}{p_1}\mathbb{E}[\chi_{p_2}^{-2}(\Delta^2)]\right)^{-1} \le \left(1+\frac{p_2}{p_1}\right)\left(1+\frac{2}{p_1}\right)^{-1}.$$

Also,

$$\begin{aligned}
\text{ADRE}(\hat{\boldsymbol{\beta}}_n^{\text{G(S+)}};\tilde{\boldsymbol{\beta}}_n^{\text{G}}) = \left(1+\frac{p_2}{p_1}\right)\Bigg\{ & 1+\frac{p_2}{p_1}-\frac{(p_2-2)^2}{p_1}\mathbb{E}[\chi_{p_2}^{-2}(\Delta^2)] \\
& -\frac{p_2}{p_1}\mathbb{E}[(1-(p_2-2)\chi_{p_2+2}^{-2}(\Delta^2))^2 I(\chi_{p_2+2}^2(\Delta^2) < (p_2-2))] \\
& +\frac{\Delta^2}{p_1}(2\mathbb{E}[(1-(p_2-2)\chi_{p_2+2}^{-2}(\Delta^2)) I(\chi_{p_2+2}^2(\Delta^2) < (p_2-2))] \\
& -\mathbb{E}[(1-(p_2-2)\chi_{p_2+4}^{-2}(\Delta^2))^2 I(\chi_{p_2+4}^2(\Delta^2) < (p_2-2))])\Bigg\}^{-1}.
\end{aligned} \tag{9.61}$$

So that,

$$\text{ADRE}(\hat{\boldsymbol{\beta}}_n^{\text{G(S+)}};\tilde{\boldsymbol{\beta}}_n^{\text{G}}) \ge \text{ADRE}(\hat{\boldsymbol{\beta}}_n^{\text{G(S)}};\tilde{\boldsymbol{\beta}}_n^{\text{G}}) \ge 1 \quad \forall \Delta^2 \in \mathbb{R}^+.$$

9.2.4 Comparison of LSE and RLSE with RRE

First, we consider weighted L_2-risk difference of GLSE and RRE given by

$$\begin{aligned}
\sigma^2(p_1+p_2) - \sigma^2 p_1 - \sigma^2\frac{p_2\Delta^2}{p_2+\Delta^2} &= \sigma^2 p_2\left(1-\frac{\Delta^2}{p_2+\Delta^2}\right) \\
&= \frac{\sigma^2 p_2^2}{p_2+\Delta^2}, \quad \forall\, \Delta^2 \in \mathbb{R}^+.
\end{aligned} \tag{9.62}$$

Hence, RRE outperforms the GLSE uniformly. Similarly, for the RGLSE and RRE, the weighted L_2-risk difference is given by

$$\begin{aligned}
\sigma^2(\text{tr}(\boldsymbol{C}_{11}^{-1}\boldsymbol{C}_{11\cdot2}) + \Delta^2) - \left(\sigma^2 p_1 + \frac{\sigma^2 p_2\Delta^2}{p_2+\Delta^2}\right) & \\
= \sigma^2\left\{[\text{tr}\boldsymbol{C}_{11}^{-1}\boldsymbol{C}_{11\cdot2} - p_1] + \frac{\Delta^4}{p_2+\Delta^2}\right\} & \\
= \sigma^2\left(\frac{\Delta^4}{p_2+\Delta^2} - \text{tr}(\boldsymbol{N}_0(\phi))\right). &
\end{aligned} \tag{9.63}$$

If $\Delta^2 = 0$, then (9.63) is negative. Hence, RGLSE outperforms RRE at this point. Solving the equation

$$\frac{\Delta^4}{p_2+\Delta^2} = \text{tr}(\boldsymbol{N}_0(\phi)) \tag{9.64}$$

for Δ^2, we get

$$\Delta_0^2 = \frac{1}{2}\,\mathrm{tr}(\boldsymbol{N}_0(\phi))\left\{1 + \sqrt{1 + \frac{4p_2}{\mathrm{tr}(\boldsymbol{N}_0(\phi))}}\right\}. \tag{9.65}$$

If $0 \le \Delta^2 \le \Delta_0^2$, RGLSE outperforms better than the RRE, and if $\Delta^2 \in (\Delta_0^2, \infty)$, RRE performs better than RGLSE; Thus, RGLSE nor RRE outperforms the other uniformly.

In addition, the relative weighted L_2-risk efficiency of RRE vs. GLSE equals

$$\mathrm{ADRE}(\hat{\boldsymbol{\beta}}_n^{\mathrm{GRR}}(k_{\mathrm{opt}}) : \tilde{\boldsymbol{\beta}}_n) = \frac{p_1 + p_2}{p_1 + \frac{p_2\Delta^2}{p_2+\Delta^2}} = \left(1 + \frac{p_2}{p_1}\right)\left(1 + \frac{p_2\Delta^2}{p_1(p_2 + \Delta^2)}\right)^{-1}, \tag{9.66}$$

which is a decreasing function of Δ^2 with maximum $\left(1 + \frac{p_2}{p_1}\right)$ at $\Delta^2 = 0$ and minimum 1 as $\Delta^2 \to \infty$. So,

$$1 \le \left(1 + \frac{p_2}{p_1}\right)\left(1 + \frac{p_2}{p_1\left(1 + \frac{p_2}{\Delta^2}\right)}\right)^{-1} \le 1 + \frac{p_2}{p_1}; \quad \forall \Delta^2 \in \mathbb{R}^+.$$

9.2.5 Comparison of RRE with PTE, SE and PRSE

9.2.5.1 Comparison Between $\hat{\boldsymbol{\beta}}_n^{\mathrm{GRR}}(k_{\mathrm{opt}})$ and $\hat{\boldsymbol{\beta}}_n^{\mathrm{G(PT)}}$

Here, the weighted L_2-risk difference of $\hat{\boldsymbol{\beta}}_n^{\mathrm{G(PT)}}$ and $\hat{\boldsymbol{\beta}}_n^{\mathrm{GRR}}(k_{\mathrm{opt}})$ is given by

$$\begin{aligned}
&\mathrm{ADR}(\hat{\boldsymbol{\beta}}_n^{\mathrm{G(PT)}}(\alpha) : \boldsymbol{B}_{11\cdot 2}(\phi), \boldsymbol{B}_{22\cdot 1}(\phi)) - \mathrm{ADR}(\hat{\boldsymbol{\beta}}_n^{\mathrm{GRR}}(k_{\mathrm{opt}}) : \boldsymbol{B}_{11\cdot 2}(\phi), \boldsymbol{B}_{22\cdot 1}(\phi)) \\
&= \sigma^2[p_2(1 - H_{p_2+2}(c_\alpha; \Delta^2)) + \Delta^2\{2H_{p_2+2}(c_\alpha; \Delta^2) - H_{p_2+4}(c_\alpha; \Delta^2)\}] \\
&\quad - \frac{\sigma^2 p_2 \Delta^2}{p_2 + \Delta^2}
\end{aligned} \tag{9.67}$$

Note that the risk of $\hat{\boldsymbol{\beta}}_{2n}^{\mathrm{G(PT)}}$ is an increasing function of Δ^2 crossing the p_2-line to a maximum and then drops monotonically toward the p_2-line as $\Delta^2 \to \infty$. The value of the risk is $p_2(1 - H_{p_2+2}(\chi^2_{p_2}(\alpha); 0))(< p_2)$ at $\Delta^2 = 0$. On the other hand, $\frac{p_2\Delta^2}{p_2+\Delta^2}$ is an increasing function of Δ^2 below the p_2-line with a minimum value 0 at $\Delta^2 = 0$ and as $\Delta^2 \to \infty$, $\frac{p_2\Delta^2}{p_2+\Delta^2} \to p_2$. Hence, the risk difference in Eq. (9.67) is nonnegative for $\Delta^2 \in \mathbb{R}^+$. Thus, the RRE uniformly performs better than PTE.

9.2.5.2 Comparison Between $\hat{\beta}_n^{\text{GRR}}(k_{\text{opt}})$ and $\hat{\beta}_n^{\text{G(S)}}$

The weighted L_2-risk difference of $\hat{\boldsymbol{\beta}}_n^{\text{G(S)}}$ and $\hat{\boldsymbol{\beta}}_n^{\text{GRR}}(k_{\text{opt}})$ is given by

$$\begin{aligned}
&\text{ADR}(\hat{\boldsymbol{\beta}}_n^{\text{G(S)}} : \boldsymbol{B}_{11\cdot 2}(\phi), \boldsymbol{B}_{22\cdot 1}(\phi)) - \text{ADR}(\hat{\boldsymbol{\beta}}_n^{\text{GRR}}(k_{\text{opt}}) : \boldsymbol{B}_{11\cdot 2}(\phi), \boldsymbol{B}_{22\cdot 1}(\phi)) \\
&= \sigma^2(p_1 + p_2 - (p_2 - 2)^2\mathbb{E}[\chi_{p_2}^{-2}(\Delta^2)]) - \sigma^2\left(p_1 + \frac{p_2\Delta^2}{p_2 + \Delta^2}\right) \\
&= \sigma^2\left[p_2 - (p_2 - 2)^2\mathbb{E}[\chi_{p_2}^{-2}(\Delta^2)] - \frac{p_2\Delta^2}{p_2 + \Delta^2}\right].
\end{aligned} \tag{9.68}$$

Note that the first function is increasing in Δ^2 with a value 2 at $\Delta^2 = 0$; and as $\Delta^2 \to \infty$, it tends to p_2. The second function is also increasing in Δ^2 with a value 0 at $\Delta^2 = 0$ and approaches the value p_2 as $\Delta^2 \to \infty$. Hence, the risk difference is nonnegative for all $\Delta^2 \in \mathbb{R}^+$. Consequently, RRE outperforms SE uniformly.

9.2.5.3 Comparison of $\hat{\beta}_n^{\text{GRR}}(k_{\text{opt}})$ with $\hat{\beta}_n^{\text{G(S+)}}$

The L_2 risk of $\hat{\boldsymbol{\beta}}_n^{\text{G(S+)}}$ is

$$\text{ADR}(\hat{\boldsymbol{\beta}}_n^{\text{G(S+)}} : \boldsymbol{B}_{11\cdot 2}(\phi), \boldsymbol{B}_{22\cdot 1}(\phi)) = \text{ADR}(\hat{\boldsymbol{\beta}}_n^{\text{G(S)}} : \boldsymbol{B}_{11\cdot 2}(\phi), \boldsymbol{B}_{22\cdot 1}(\phi)) - \text{R}^*, \tag{9.69}$$

where

$$\begin{aligned}
\text{R}^* = {} & \sigma^2 p_2 \mathbb{E}[(1 - (p_2 - 2)\chi_{p_2+2}^{-2}(\Delta^2))^2 I(\chi_{p_2+2}^{-2}(\Delta^2) < p_2 - 2)] \\
& - \sigma^2\Delta^2\{2\mathbb{E}[(1 - (p_2 - 2)\chi_{p_2+2}^{-2}(\Delta^2))I(\chi_{p_2+2}^{-2}(\Delta^2) < p_2 - 2)] \\
& - \mathbb{E}[(1 - (p_2 - 2)\chi_{p_2+4}^{-2}(\Delta^2))^2 I(\chi_{p_2+4}^{-2}(\Delta^2) < p_2 - 2)]\};
\end{aligned} \tag{9.70}$$

and $\text{ADR}(\hat{\boldsymbol{\beta}}_n^{\text{G(S)}} : \boldsymbol{B}_{11\cdot 2}(\phi), \boldsymbol{B}_{22\cdot 1}(\phi))$ is

$$\text{ADR}(\hat{\boldsymbol{\beta}}_n^{\text{G(S)}} : \boldsymbol{B}_{11\cdot 2}(\phi), \boldsymbol{B}_{22\cdot 1}(\phi)) = \sigma^2(p_1 + p_2 - (p_2 - 2)^2\mathbb{E}[\chi_{p_2}^{-2}(\Delta^2)])$$

The weighted L_2-risk difference of PRSE and RRE is given by

$$\begin{aligned}
\text{ADR}(\hat{\boldsymbol{\beta}}_n^{\text{G(S+)}} &: \boldsymbol{B}_{11\cdot 2}(\phi), \boldsymbol{B}_{22\cdot 1}(\phi)) - \text{ADR}(\hat{\boldsymbol{\beta}}_n^{\text{GRR}}(k_{\text{opt}}) : \boldsymbol{B}_{11\cdot 2}(\phi), \boldsymbol{B}_{22\cdot 1}(\phi)) \\
&= [\text{ADR}(\hat{\boldsymbol{\beta}}_n^{\text{G(S)}} : \boldsymbol{B}_{11\cdot 2}(\phi), \boldsymbol{B}_{22\cdot 1}(\phi)) - \text{R}^*] \\
&\quad - \text{ADR}(\hat{\boldsymbol{\beta}}_n^{\text{GRR}}(k_{\text{opt}}) : \boldsymbol{B}_{11\cdot 2}(\phi), \boldsymbol{B}_{22\cdot 1}(\phi)) \geq 0,
\end{aligned} \tag{9.71}$$

where

$$\text{ADR}(\hat{\boldsymbol{\beta}}_n^{\text{GRR}}(k_{\text{opt}}) : \boldsymbol{B}_{11\cdot 2}(\phi), \boldsymbol{B}_{22\cdot 1}(\phi)) = \sigma^2\left(p_1 + \frac{p_2\Delta^2}{p_2 + \Delta^2}\right)$$

Consider the ADR($\hat{\boldsymbol{\beta}}_n^{\text{G(S+)}}$). It is a monotonically increasing function of Δ^2. At $\Delta^2 = 0$, its value is $\sigma^2(p_1 + 2) - \sigma^2 p_2 \mathbb{E}[(1 - (p_2 - 2)\chi^{-2}_{p_2+2}(0))^2 I(\chi^{-2}_{p_2+2}(0) < p_2 - 2)] \geq 0$; and as $\Delta^2 \to \infty$, it tends to $\sigma^2(p_1 + p_2)$. For ADR($\hat{\boldsymbol{\beta}}_n^{\text{GRR}}(k_{\text{opt}}) : \boldsymbol{B}_{11\cdot2}(\phi), \boldsymbol{B}_{22\cdot1}(\phi)$), at $\Delta^2 = 0$, the value is $\sigma^2 p_1$; and as $\Delta^2 \to \infty$, it tends to $\sigma^2(p_1 + p_2)$. Hence, the L_2-risk difference in (9.71) is nonnegative and RRE uniformly outperforms PRSE.

Note that the risk difference of $\hat{\boldsymbol{\beta}}_n^{\text{G(S+)}}$ and $\hat{\boldsymbol{\beta}}_n^{\text{GRR}}(k_{\text{opt}})$ at $\Delta^2 = 0$ is

$$
\begin{aligned}
&\sigma^2(p_1 + 2) - \sigma^2 p_2 \mathbb{E}[(1 - (p_2 - 2)\chi^{-2}_{p_2+2}(0))^2 I(\chi^{-2}_{p_2+2}(0) < p_2 - 2)] - \sigma^2 p_1 \\
&= \sigma^2(2 - p_2 \mathbb{E}[(1 - (p_2 - 2)\chi^{-2}_{p_2+2}(0))^2 I(\chi^{-2}_{p_2+2}(0) < p_2 - 2)]) \geq 0 \qquad (9.72)
\end{aligned}
$$

because the expected value in Eq. (9.72) is a decreasing function of D.F., and

$$
2 > p_2 \mathbb{E}[(1 - (p_2 - 2)\chi^{-2}_{p_2+2}(0))^2 I(\chi^{-2}_{p_2+2}(0) < p_2 - 2)].
$$

9.2.6 Comparison of MLASSO with GLSE and RGLSE

First, note that if p_1 coefficients $|\beta_j| > \sigma\sqrt{C^{jj}}$ and p_2 coefficients are zero in a sparse solution, the lower bound of the weighted L_2 risk is given by $\sigma^2(p_1 + \Delta^2)$. Thereby, we compare all estimators relative to this quantity. Hence, the weighted L_2-risk difference between LSE and MLASSO is given by

$$
\sigma^2(p_1 + p_2) - \sigma^2(p_1 + \Delta^2 - \text{tr}(\boldsymbol{N}_0(\phi))) = \sigma^2[(p_2 + \text{tr}(\boldsymbol{N}_0(\phi))) - \Delta^2]. \qquad (9.73)
$$

Hence, if $\Delta^2 \in (0, p_2 + \text{tr}(\boldsymbol{M}_0))$, the MLASSO performs better than the LSE, while if $\Delta^2 \in (p_2 + \text{tr}(\boldsymbol{N}_0(\phi)), \infty)$ the LSE performs better than the MLASSO. Consequently, neither LSE nor the MLASSO performs better than the other uniformly.

Next we compare the RGLSE and MLASSO. In this case, the weighted L_2-risk difference is given by

$$
\sigma^2(p_1 + \Delta^2 - \text{tr}(\boldsymbol{N}_0(\phi))) - \sigma^2(p_1 + \Delta^2) = -\sigma^2(\text{tr}(\boldsymbol{N}_0(\phi))) < 0. \qquad (9.74)
$$

Hence, the RGLSE uniformly performs better than the MLASSO. If $\text{tr}(\boldsymbol{N}_0(\phi)) = 0$, MLASSO and RGLSE are L_2-risk equivalent. If the GLSEs are independent, then $\text{tr}(\boldsymbol{N}_0(\phi)) = 0$. Hence, MLASSO satisfies the oracle properties.

9.2.7 Comparison of MLASSO with PTE, SE, and PRSE

We first consider the PTE vs. MLASSO. In this case, the weighted L_2-risk difference is given by

$$\begin{aligned}
&\text{ADR}(\hat{\boldsymbol{\beta}}_n^{\text{G(PT)}}; \boldsymbol{B}_{11\cdot 2}(\phi), \boldsymbol{B}_{22\cdot 1}(\phi)) - \text{ADR}(\hat{\boldsymbol{\beta}}_n^{\text{MLASSO}}(k_{\text{opt}}); \boldsymbol{B}_{11\cdot 2}(\phi), \boldsymbol{B}_{22\cdot 1}(\phi)) \\
&\quad = \sigma^2[p_2(1 - H_{p_2+2}(c_\alpha; \Delta^2)) - \Delta^2\{1 - 2H_{p_2+2}(c_\alpha; \Delta^2) + H_{p_2+4}(c_\alpha; \Delta^2)\}] \\
&\quad \geq \sigma^2 p_2(1 - H_{p_2+2}(c_\alpha; 0)) \geq 0, \quad \text{if } \Delta^2 = 0 \qquad (9.75)
\end{aligned}$$

Hence, the MLASSO outperforms the PTE when $\Delta^2 = 0$. But, when $\Delta^2 \neq 0$, then the MLASSO outperforms the PTE for

$$0 \leq \Delta^2 \leq \frac{p_2[1 - H_{p_2+2}(c_\alpha; \Delta^2)]}{1 - 2H_{p_2+2}(c_\alpha; \Delta^2) + H_{p_2+4}(c_\alpha; \Delta^2)}. \qquad (9.76)$$

Otherwise, PTE outperforms the MLASSO. Hence, neither PTE nor MLASSO outperforms the other uniformly.

Next, we consider SE and PRSE vs. the MLASSO. In these two cases, we have weighted L_2-risk differences given by

$$\begin{aligned}
&\text{ADR}(\hat{\boldsymbol{\beta}}_n^{\text{G(S)}}; \boldsymbol{B}_{11\cdot 2}(\phi), \boldsymbol{B}_{22\cdot 1}(\phi)) - \text{ADR}(\hat{\boldsymbol{\beta}}_n^{\text{MLASSO}}(k_{\text{opt}}); \boldsymbol{B}_{11\cdot 2}(\phi), \boldsymbol{B}_{22\cdot 1}(\phi)) \\
&\quad = \sigma^2[p_1 + p_2 - (p_2 - 2)^2 \mathbb{E}[\chi_{p_2+2}^{-2}(\Delta^2)] - (p_1 + \Delta^2)] \\
&\quad = \sigma^2[p_2 - (p_2 - 2)^2 \mathbb{E}[\chi_{p_2+2}^{-2}(\Delta^2)] - \Delta^2] \qquad (9.77)
\end{aligned}$$

and from (9.69),

$$\begin{aligned}
&\text{ADR}(\hat{\boldsymbol{\beta}}_n^{\text{G(S+)}}; \boldsymbol{B}_{11\cdot 2}(\phi), \boldsymbol{B}_{22\cdot 1}(\phi)) - \text{ADR}(\hat{\boldsymbol{\beta}}_n^{\text{MLASSO}}(\lambda); \boldsymbol{B}_{11\cdot 2}(\phi), \boldsymbol{B}_{22\cdot 1}(\phi)) \\
&\quad = \text{ADR}(\hat{\boldsymbol{\beta}}_n^{\text{G(S)}}; \boldsymbol{C}_{11\cdot 2}, \boldsymbol{C}_{22\cdot 1}) - \text{ADR}(\hat{\boldsymbol{\beta}}_n^{\text{MLASSO}}(\lambda); \boldsymbol{C}_{11\cdot 2}, \boldsymbol{C}_{22\cdot 1}) - \text{R}^*, \qquad (9.78)
\end{aligned}$$

where R^* is given by (9.70). Hence, the MLASSO outperforms the SE as well as the PRSE in the interval

$$0 \leq \Delta^2 \leq p_2 - (p_2 - 2)^2 \mathbb{E}[\chi_{p_2}^{-2}(\Delta^2)]. \qquad (9.79)$$

Thus, neither the SE nor the PRSE outperforms the MLASSO, uniformly.

9.2.8 Comparison of MLASSO with RRE

Here, the weighted L_2-risk difference is given by

$$\begin{aligned}
&\text{ADR}(\hat{\boldsymbol{\beta}}_n^{\text{MLASSO}}(\lambda); \boldsymbol{B}_{11\cdot 2}(\phi), \boldsymbol{B}_{22\cdot 1}(\phi)) - \text{ADR}(\hat{\boldsymbol{\beta}}_n^{\text{GRR}}(k_{\text{opt}}); \boldsymbol{B}_{11\cdot 2}(\phi), \boldsymbol{B}_{22\cdot 1}(\phi)) \\
&= \sigma^2 \left[(p_1 + \Delta^2) - \left(p_1 + \frac{p_2\Delta^2}{p_2 + \Delta^2}\right)\right] = \frac{\sigma^2\Delta^2}{p_2 + \Delta^2} \geq 0. \qquad (9.80)
\end{aligned}$$

Hence, the RRE outperforms the MLASSO uniformly.

9.3 Example: Sea Level Rise at Key West, Florida

In this section, we want to illustrate the methodology of this chapter by using a real data application. In this regard, we want to see whether there is any relationship between sea level rise at Key West, Florida with the following regressors: time (year); atmospheric carbon dioxide concentration (CO_2); ocean heat content (OHC); global mean temperature (Temp); RF; PC; sunspots (SP); Pacific decadal oscillation (PDO); Southern Oscillation Index (SOI), which measures the strength of the Southern Oscillation.

The sources of data are: (i) National Oceanographic and Atmospheric Administration (NOAA), (ii) Australian Government Bureau of Meteorology, and (iii) NASA Goddard Institute for Space Studies (NASA GISS). Since there are a lot of missing values and some variables are measured at a later time, we consider data between 1959 and 2016 so that we can consider nine (9) independent variables (regressors). All variables are standardized so that all regression coefficients are comparable.

We consider the following linear regression model:

$$\begin{aligned}\text{SL} = {} & \beta_1 * \text{CO}_2 + \beta_2 * \text{OHC} + \beta_3 * \text{Year} + \beta_4 \\ & * \text{Temp} + \beta_5 * \text{RF} + \beta_6 * \text{PC} + \beta_7 \\ & * \text{SP} + \beta_8 * \text{PDO} + \beta_9 * \text{SOI}.\end{aligned}$$

We first apply the LSE and fit the following linear regression model:

$$\begin{aligned}\text{SL} = {} & 1.9035 * \text{CO}_2 - 0.1869 * \text{OHC} - 0.7909 * \text{Year} - 0.0488 \\ & * \text{Temp} - 0.2710 * \text{RF} + 0.2158 * \text{PC} - 0.0277 \\ & * \text{SP} - 0.0084 * \text{PDO} - 0.0995 * \text{SOI}\end{aligned}$$

with MSE $= 0.2198$, $\text{R}^2 = 0.815$, and $F = 23.49$ with D.F. $= (9, 48)$ and P-value $= 1.042 \times 10^{-14}$. This indicates that overall the regression model is significant.

The correlation matrix among the variables is given in Table 9.1. If we review Table 9.1, we can see that there are moderate to strong relationships among some of the regressors.

9.3.1 Estimation of the Model Parameters

9.3.1.1 Testing for Multicollinearity

We can see from Table 9.2 that among the nine regressors, six have variance inflation factor (VIF) greater than 10. So there is moderate to strong multicollinearity existing in the data.

Condition index: $\kappa = 1366.27220$. Since the condition number exceeds 1000, we may conclude that at least one of the regressors is responsible for the multicollinearity problem in the data.

Thus, from the correlation matrix, VIF, and the condition number, we conclude that these data suffer from the problem of multicollinearity.

Table 9.1 Correlation coefficients among the variables.

	SL	CO_2	OHC	Year	Temp	RF	PC	SP	PDO	SOI
SL	1.000	0.894	0.846	0.885	0.828	−0.020	−0.007	−0.142	−0.045	−0.109
CO_2	0.894	1.000	0.952	0.993	0.920	0.010	0.008	−0.182	−0.083	−0.044
OHC	0.846	0.952	1.000	0.920	0.912	−0.035	−0.051	−0.195	−0.120	0.020
Year	0.885	0.993	0.920	1.000	0.898	0.008	0.012	−0.174	−0.061	−0.067
Temp	0.828	0.920	0.912	0.898	1.000	0.080	0.063	−0.081	−0.011	−0.158
RF	−0.020	0.010	−0.035	0.008	0.080	1.000	0.967	0.119	0.045	−0.241
PC	−0.007	0.008	−0.051	0.012	0.063	0.967	1.000	0.121	0.025	−0.243
SP	−0.142	−0.182	−0.195	−0.174	−0.081	0.119	0.121	1.000	0.002	−0.043
PDO	−0.045	−0.083	−0.120	−0.061	−0.011	0.045	0.025	0.002	1.000	−0.587
SOI	−0.109	−0.044	0.020	−0.067	−0.158	−0.241	−0.243	−0.043	−0.587	1.000

Table 9.2 VIF values related to sea level rise at Key West, Florida data set.

Variables	VIF
CO_2	239.285
OHC	21.678
Year	139.294
Temp	9.866
RF	16.371
PC	16.493
SP	1.135
PDO	1.624
SOI	1.877

9.3.1.2 Testing for Autoregressive Process

Using the following R command, we found that the residual of the model follow AR(1) process. The estimated AR(1) coefficient, $\hat{\phi} = -0.334$.

```
> phi=Arima(Res, order=c(1,1,0))$coe
> phi
       ar1
-0.3343068
```

Since the data follow AR(1) process and the regressors are correlated, this data will be the most appropriate to analyze for this chapter.

9.3.1.3 Estimation of Ridge Parameter k

Following Kibria (2003) and Kibria and Banik (2016), we may estimate the ridge coefficient k. However, since different methods produce different values of k, we consider the

```
> rstats2(ridge1.lm)$PRESS
 K=0.001   K=0.01   K=0.05    K=0.1    K=0.5    K=0.9      K=1
15.40242 15.26157 14.80704 14.59746 14.82246 15.82000 16.12388
```

We use $k = 0.10$ because this value gives a smaller predicted residual error sum of squares (PRESS).

Based on the full fitted model, we consider the following hypotheses: $p_1 = 2$ and $p_2 = 7$. We test the following hypothesis,

$$\boldsymbol{\beta}_2 = \mathbf{0},$$

where

$$\boldsymbol{\beta}_2 = (\beta_2, \beta_4, \beta_5, \beta_6, \beta_7, \beta_8, \beta_9)^{\top}$$

The fitted reduced linear regression model is

$$\text{SL} = 1.056 * \text{CO}_2 + 10.166 * \text{Year}.$$

Now, using $k = 0.10$, and $\phi = -0.334$, the estimated values of the regression coefficient are provided in Table 9.3.

If we review Table 9.3, we can see that LASSO kept variable CO_2 and kicked out the rest of the regressors. The sign of the temperature has been changed from negative to positive. This is true because as the temperature go up, the sea level should go up too. Table 9.4 also indicates that the LSE gave the wrong sign for the temperature variable and only CO_2 is marginally significant at 5% significance level.

Table 9.3 Estimation of parameter using different methods ($p_1 = 2, p_2 = 7, \hat{\phi} = -0.34$, and $k = 0.10$).

	GLSE	RGLSEE	RRE	MLASSO	PTGLSE	SGLSE	PRSLSE
CO_2	2.533	2.533	2.533	2.533	2.533	2.533	2.533
OHC	−1.173	−1.173	−1.173	0.000	−1.173	−1.173	−1.173
Year	−0.313	−0.313	−0.298	0.000	−0.313	−0.269	−0.269
Temp	0.208	0.208	0.198	0.000	0.208	0.178	0.178
RF	0.046	0.046	0.044	0.000	0.046	0.039	0.039
PC	0.021	0.021	0.020	0.000	0.021	0.018	0.018
SP	−0.321	−0.321	−0.306	0.000	−0.321	−0.276	−0.276
PDO	−0.178	−0.178	−0.169	0.000	−0.178	−0.152	−0.152
SOI	−0.105	−0.105	−0.100	0.000	−0.105	−0.090	−0.090

Table 9.4 Estimation of parameter using LSE ($p_1 = 2, p_2 = 7, \hat{\phi} = -0.34$, and $k = 0.10$).

	LSE	Standard error	*t*-Value	Pr(> |*t*|)
Intercept	5.379×10^{-15}	0.061	0.000	1.0000
CO_2	1.903	0.960	1.982	0.0532
OHC	−0.186	−0.289	−0.647	0.521
Year	−0.790	−0.732	−1.079	0.285
Temp	−0.048	0.195	−0.250	0.8033
RF	−0.271	0.251	−1.079	0.286
PC	0.215	0.252	0.856	0.396
SP	−0.027	0.061	−0.418	0.677
PDO	−0.008	0.079	−0.106	0.916
SOI	−0.085	0.085	−1.170	0.247

Now, we consider the relative efficiency (REff) criterion to compare the performance of the estimators.

9.3.2 Relative Efficiency

9.3.2.1 Relative Efficiency (REff)

The REff of $\hat{\boldsymbol{\beta}}^*$ compared to GLSE is defined as

$$\text{REff}(\hat{\boldsymbol{\beta}}^*) = \frac{\text{R}(\tilde{\boldsymbol{\beta}}_n^{\text{G}})}{\text{R}(\hat{\boldsymbol{\beta}}^*)}, \tag{9.81}$$

where $\hat{\boldsymbol{\beta}}^*$ could be any of the proposed estimators. If we review the risk functions (except GLSE) under Theorem 9.1, all the terms contain $\boldsymbol{\delta}^\top\boldsymbol{\delta}$. To write these risk functions in terms of Δ, we adopt the following procedure.

We obtain from Anderson (1984, Theorem A.2.4, p. 590) that

$$\gamma_p \le \frac{\boldsymbol{\delta}^\top\boldsymbol{\delta}/\sigma^2}{\boldsymbol{\delta}^\top\boldsymbol{B}_{22\cdot1}^{-1}\boldsymbol{\delta}/\sigma^2} \le \gamma_1. \tag{9.82}$$

Since, $\Delta^2 = \boldsymbol{\delta}^\top\boldsymbol{B}_{22\cdot1}^{-1}\boldsymbol{\delta}/\sigma^2$, the above equation can be written as

$$\sigma^2\Delta\gamma_p \le \boldsymbol{\delta}^\top\boldsymbol{\delta} \le \sigma^2\Delta^2\gamma_1, \tag{9.83}$$

where γ_1 and γ_p are, respectively, the largest and the smallest characteristic roots of the matrix, $\boldsymbol{B}_{22\cdot1}$. Using this result, the risk functions can be written as follows:

The risk function for GLSE is

$$\text{R}(\tilde{\boldsymbol{\beta}}_n^{\text{G}}) = \sigma^2\{\text{tr}(\boldsymbol{B}_{11\cdot2}^{-1})(\phi) + \text{tr}(\boldsymbol{B}_{22\cdot1}^{-1}(\phi))\} = \sigma^2\ \text{tr}(\boldsymbol{B}^{-1}(\phi)). \tag{9.84}$$

The risk function for RGLSE is

$$R(\widehat{\boldsymbol{\beta}}_n^{\mathrm{G}}) = \sigma^2 \operatorname{tr}(\boldsymbol{B}_{11}^{-1}(\phi)) + \sigma^2\Delta^2\gamma_1. \tag{9.85}$$

The risk of PTGLSE is

$$\begin{aligned} R(\widehat{\boldsymbol{\beta}}_n^{\mathrm{G(PT)}}) &= \sigma^2\{\operatorname{tr}(\boldsymbol{B}_{11\cdot2}^{-1}) + \operatorname{tr}(\boldsymbol{B}_{22\cdot1}^{-1}(\phi))[1 - H_{p_2+2}(c_\alpha : \Delta^2)] \\ &\quad + \operatorname{tr}(\boldsymbol{B}_{11}^{-1}(\phi))H_{p_2+2}(c_\alpha : \Delta^2)\} \\ &\quad + \sigma^2\Delta^2\gamma_1\{2H_{p_2+2}(c_\alpha : \Delta^2) - H_{p_2+4}(c_\alpha : \Delta^2)\}. \end{aligned} \tag{9.86}$$

The risk function for SGLSE is

$$\begin{aligned} R(\widehat{\boldsymbol{\beta}}_n^{\mathrm{G(S)}}) &= \sigma^2\{(\operatorname{tr}(\boldsymbol{B}_{11\cdot2}^{-1}(\phi)) + \operatorname{tr}(\boldsymbol{B}_{22\cdot1}^{-1}(\phi))) - (p_2 - 2)\operatorname{tr}(\boldsymbol{B}_{22\cdot1}^{-1}) \\ &\quad \times (2\mathbb{E}[\chi_{p_2+2}^{-2}(\Delta^2)] - (p_2 - 2)\mathbb{E}[\chi_{p_2+2}^{-4}(\Delta^2)])\} \\ &\quad + (p_2^2 - 4)\sigma^2\Delta^2\gamma_1\mathbb{E}[\chi_{p_2+4}^{-4}(\Delta^2)]. \end{aligned} \tag{9.87}$$

The risk function of PRSGLSE is

$$\begin{aligned} R(\widehat{\boldsymbol{\beta}}_{1n}^{\mathrm{G(S)}}) &= R(\widehat{\boldsymbol{\beta}}_n^{\mathrm{G(S)}}) - \sigma^2(p_2 - 2)\operatorname{tr}(\boldsymbol{B}_{22\cdot1}^{-1}(\phi)) \\ &\quad \times \mathbb{E}[(1 - (p_2 - 2)^2\chi_{p_2+2}^{-2}(\Delta^2))^2 I(\chi_{p_2+2}^{2}(\Delta^2) < p_2 - 2)] \\ &\quad + \sigma^2\Delta^2\gamma_1\{2\mathbb{E}[(1 - (p_2 - 2)\chi_{p_2+2}^{-2}(\Delta^2))I(\chi_{p_2+2}^{2}(\Delta^2) < p_2 - 2)] \\ &\quad - \mathbb{E}[(1 - (p_2 - 2)\chi_{p_2+4}^{-2}(\Delta^2))^2 I(\chi_{p_2+4}^{2}(\Delta^2) < p_2 - 2)]\}. \end{aligned} \tag{9.88}$$

The risk function for generalized RRE is

$$R(\widehat{\boldsymbol{\beta}}_n^{\mathrm{GRR}}(k)) = \sigma^2\left\{\operatorname{tr}(B_{11\cdot2}^{-1}(\phi)) + \frac{1}{(1+k)^2}[\operatorname{tr}(B_{22\cdot1}^{-1}(\phi)) + k^2\sigma^2\Delta\gamma_1]\right\}. \tag{9.89}$$

The risk function of MLASSO is

$$R(\widehat{\boldsymbol{\beta}}_n^{\mathrm{MLASSO}}(\lambda)) = \sigma^2 \operatorname{tr}(B_{11\cdot2}^{-1}(\phi)) + \sigma^2\Delta\gamma_1. \tag{9.90}$$

Using the risk functions in (9.84)–(9.90), the REff of the estimators for different values of Δ^2 are given in Table 9.5.

The relative efficiencies of the estimators for different values of Δ^2 are given in Figure 9.1 and Table 9.6.

If we review Tables 9.5 and 9.6, we can see that the performance of the estimators except the GLSE depend on the values of Δ^2, p_1, p_2, k, and ϕ. We immediately see that the restricted estimator outperforms all the estimators when the restriction is at 0. However, as Δ^2 goes away from the null hypothesis, the REff of RGLSE goes down and performs the worst when Δ^2 is large. Both PRSGLSE and SGLSE uniformly dominate the RRE and MLASSO estimator. We also observe that the GLSE uniformly dominates the PTGLSE for $\Delta^2 \in (0, 40)$. The REff of

Table 9.5 Relative efficiency of the proposed estimators ($p_1 = 2, p_2 = 7, \widehat{\phi} = -0.34$, and $k = 0.10$).

Δ^2	GLSE	RGLSE	RRE	MLASSO	PTGLSE	SGLSE	PRSGLSE
0.000	1.000	2.605	1.022	1.144	0.815	1.099	1.177
0.100	1.000	2.512	1.022	1.125	0.806	1.087	1.164
0.500	1.000	2.198	1.022	1.058	0.776	1.045	1.115
1.000	1.000	1.901	1.021	0.984	0.743	1.004	1.066
2.000	1.000	1.497	1.020	0.863	0.690	0.947	0.995
5.000	1.000	0.914	1.016	0.631	0.612	0.871	0.893
10.000	1.000	0.554	1.010	0.436	0.626	0.849	0.855
15.000	1.000	0.398	1.004	0.333	0.726	0.857	0.858
20.000	1.000	0.310	0.998	0.269	0.841	0.870	0.870
25.000	1.000	0.254	0.993	0.226	0.926	0.882	0.882
30.000	1.000	0.215	0.987	0.195	0.971	0.893	0.893
40.000	1.000	0.165	0.975	0.153	0.997	0.911	0.911
50.000	1.000	0.134	0.964	0.125	1.000	0.923	0.923
60.000	1.000	0.112	0.954	0.106	1.000	0.933	0.933
100.000	1.000	0.069	0.913	0.066	1.000	0.999	0.999

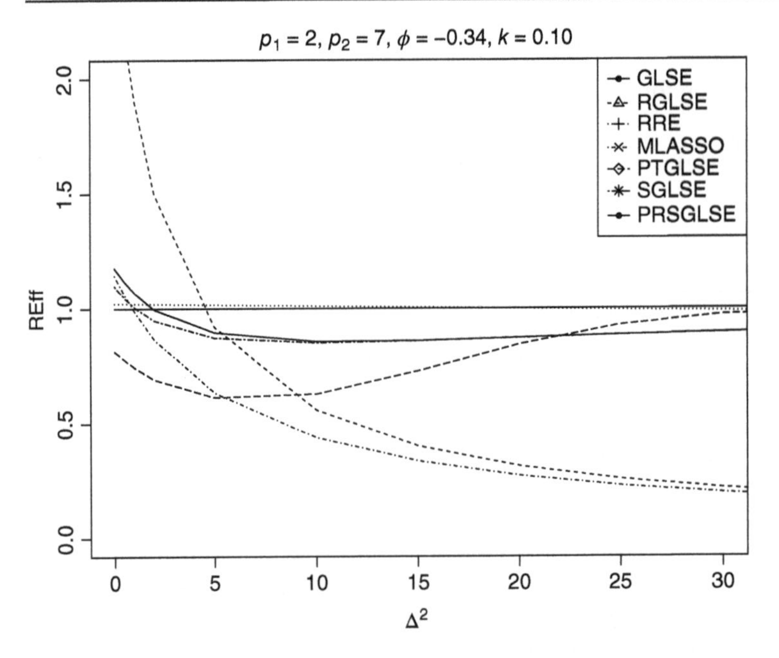

Figure 9.1 Relative efficiency of the estimators for $p_1 = 2, p_2 = 7, \phi = -0.34$, and $k = 0.10$.

Table 9.6 The relative efficiency of the proposed estimators ($p_1 = 6, p_2 = 3, \hat{\phi} = -0.34$, and $k = 0.10$).

Δ^2	GLSE	RGLSE	RRE	MLASSO	PTGLSE	SGLSE	PRSGLSE
0.000	1.000	1.117	1.004	1.023	0.579	1.008	1.011
0.100	1.000	1.062	1.004	0.977	0.566	0.993	0.998
0.500	1.000	0.889	1.002	0.828	0.523	0.944	0.956
1.000	1.000	0.738	1.000	0.696	0.482	0.902	0.918
2.000	1.000	0.551	0.996	0.527	0.429	0.854	0.871
5.000	1.000	0.313	0.985	0.305	0.386	0.825	0.834
10.000	1.000	0.182	0.967	0.179	0.477	0.856	0.858
15.000	1.000	0.128	0.950	0.127	0.664	0.887	0.888
20.000	1.000	0.099	0.933	0.098	0.843	0.909	0.909
25.000	1.000	0.081	0.917	0.080	0.945	0.924	0.924
30.000	1.000	0.068	0.901	0.068	0.984	0.935	0.935
40.000	1.000	0.052	0.871	0.052	0.999	0.949	0.949
50.000	1.000	0.042	0.843	0.042	1.000	0.959	0.959
60.000	1.000	0.035	0.817	0.035	1.000	0.965	0.965
100.000	1.000	0.021	0.727	0.021	1.000	0.996	0.996

PTGLSE increases as Δ^2 increases. These conclusions looks unusual, probably due to autocorrelated data. However, when ϕ is large, we have the usual conclusions (see Tables 9.7 and 9.8).

9.3.2.2 Effect of Autocorrelation Coefficient ϕ

To see the effect of the autocorrelation coefficient ϕ on the performance of the proposed shrinkage, LASSO and RREs, we evaluated the REff for various values of $\phi = 0.35, 0.55$, and 0.75 and presented them, respectively, in Figures 9.2–9.4 and Tables 9.6–9.9. If we review these figures and tables, under $\Delta^2 = 0$, we observe that the RGLSE performed the best followed by PRSGLSE, LASSO, SGLSE, and RRE; and PTGLSE (with $\alpha = 0.05$) performed the worst. The performance of the RGLSE becomes worse when Δ^2 increases and becomes inefficient for large Δ^2.

To see the opposite effect of the autocorrelation coefficient on the proposed estimators, we evaluated the REff of the estimators for $\phi = 0.85$ and $\phi = -0.85$ and presented them in Figures 9.5 and 9.6 and Tables 9.10 and 9.11. If we review these two figures and tables, we can see that the proposed estimators perform better for positive value of ϕ than for the negative value of ϕ.

Table 9.7 The relative efficiency of the proposed estimators ($p_1 = 2, p_2 = 7, \phi = 0.35$, and $k = 0.10$).

Δ^2	GLSE	RGLSE	RRE	MLASSO	PTGLSE	SGLSE	PRSGLSE
0.000	1.000	4.334	1.056	1.279	0.989	1.185	1.355
0.100	1.000	3.871	1.056	1.236	0.963	1.158	1.321
0.500	1.000	2.711	1.054	1.087	0.874	1.070	1.208
1.000	1.000	1.973	1.051	0.945	0.787	0.990	1.104
2.000	1.000	1.277	1.046	0.750	0.666	0.887	0.968
5.000	1.000	0.621	1.031	0.462	0.510	0.766	0.800
10.000	1.000	0.334	1.006	0.282	0.494	0.736	0.744
15.000	1.000	0.229	0.983	0.203	0.596	0.749	0.751
20.000	1.000	0.174	0.961	0.159	0.743	0.769	0.770
25.000	1.000	0.140	0.940	0.130	0.871	0.789	0.789
30.000	1.000	0.117	0.919	0.110	0.947	0.807	0.807
40.000	1.000	0.089	0.881	0.085	0.994	0.836	0.836
50.000	1.000	0.071	0.846	0.069	1.000	0.858	0.858
60.000	1.000	0.060	0.814	0.058	1.000	0.875	0.875
100.000	1.000	0.036	0.706	0.035	1.000	0.995	0.995

Table 9.8 The relative efficiency of the proposed estimators ($p_1 = 2, p_2 = 7, \phi = 0.55$, and $k = 0.10$).

Δ^2	GLSE	RGLSE	RRE	MLASSO	PTGLSE	SGLSE	PRSGLSE
0.000	1.000	5.512	1.076	1.410	1.107	1.262	1.537
0.100	1.000	4.829	1.076	1.361	1.076	1.234	1.492
0.500	1.000	3.228	1.074	1.194	0.970	1.141	1.348
1.000	1.000	2.282	1.071	1.035	0.868	1.055	1.218
2.000	1.000	1.439	1.066	0.818	0.728	0.945	1.052
5.000	1.000	0.683	1.052	0.502	0.548	0.814	0.853
10.000	1.000	0.364	1.028	0.305	0.522	0.776	0.785
15.000	1.000	0.248	1.005	0.219	0.619	0.784	0.786
20.000	1.000	0.188	0.984	0.171	0.760	0.801	0.802
25.000	1.000	0.152	0.963	0.140	0.881	0.818	0.818
30.000	1.000	0.127	0.943	0.119	0.951	0.834	0.834
40.000	1.000	0.096	0.906	0.091	0.995	0.859	0.859
50.000	1.000	0.077	0.871	0.074	1.000	0.878	0.878
60.000	1.000	0.064	0.840	0.062	1.000	0.893	0.893

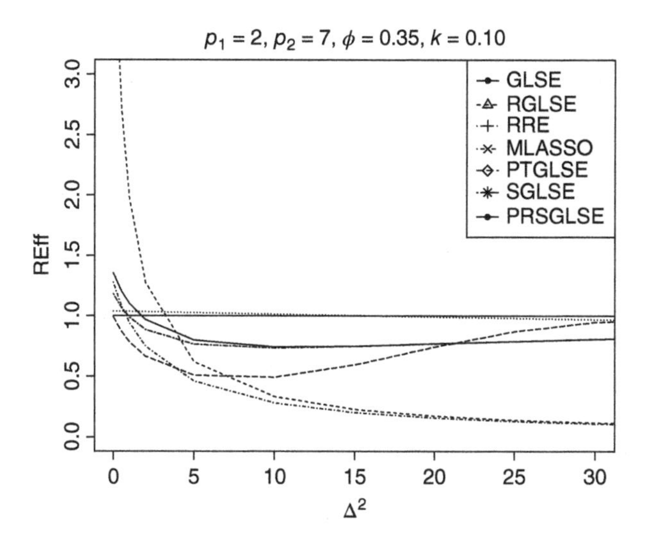

Figure 9.2 Relative efficiency of the estimators for $p_1 = 2, p_2 = 7, \phi = 0.35$, and $k = 0.10$.

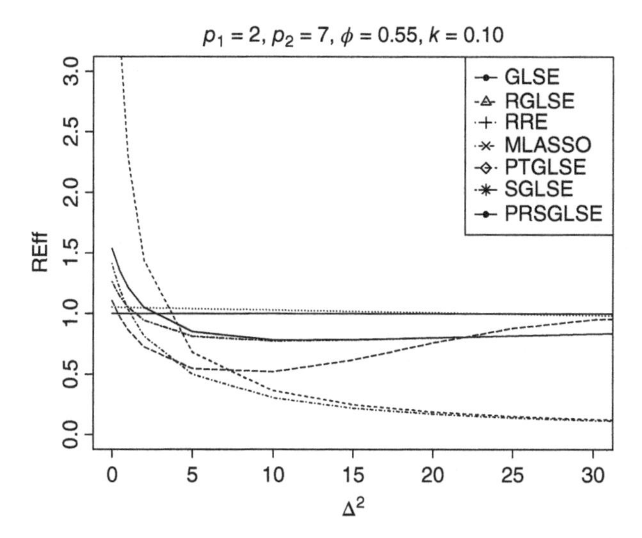

Figure 9.3 Relative efficiency of the estimators for $p_1 = 2, p_2 = 7, \phi = 0.55$, and $k = 0.10$.

9.4 Summary and Concluding Remarks

In this chapter, we considered the multiple linear regression model when the regressors are not independent and errors follow an AR(1) process. We proposed some shrinkage, namely, restricted estimator, PTE, Stein-type estimator,

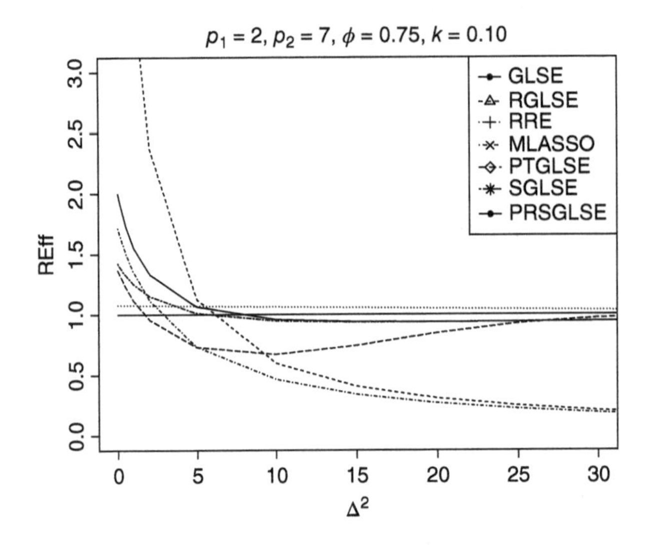

Figure 9.4 Relative efficiency of the estimators for $p_1 = 2, p_2 = 7, \phi = 0.75$, and $k = 0.10$.

Table 9.9 The relative efficiency of the proposed estimators ($p_1 = 2, p_2 = 7, \phi = 0.75$, and $k = 0.10$).

Δ^2	GLSE	RGLSE	RRE	MLASSO	PTGLSE	SGLSE	PRSGLSE
0.000	1.000	8.994	1.113	1.712	1.366	1.422	1.996
0.100	1.000	7.887	1.112	1.667	1.335	1.401	1.933
0.500	1.000	5.287	1.111	1.510	1.225	1.325	1.731
1.000	1.000	3.744	1.110	1.351	1.115	1.252	1.552
2.000	1.000	2.364	1.106	1.116	0.956	1.150	1.330
5.000	1.000	1.123	1.097	0.733	0.731	1.010	1.064
10.000	1.000	0.599	1.081	0.467	0.671	0.948	0.958
15.000	1.000	0.408	1.066	0.342	0.743	0.936	0.938
20.000	1.000	0.310	1.051	0.270	0.846	0.937	0.937
25.000	1.000	0.249	1.036	0.223	0.927	0.940	0.940
30.000	1.000	0.209	1.022	0.190	0.971	0.945	0.945
40.000	1.000	0.158	0.995	0.147	0.997	0.952	0.952
50.000	1.000	0.126	0.970	0.119	1.000	0.959	0.959
60.000	1.000	0.106	0.945	0.101	1.000	0.964	0.964
100.000	1.000	0.064	0.859	0.062	1.000	1.025	1.025

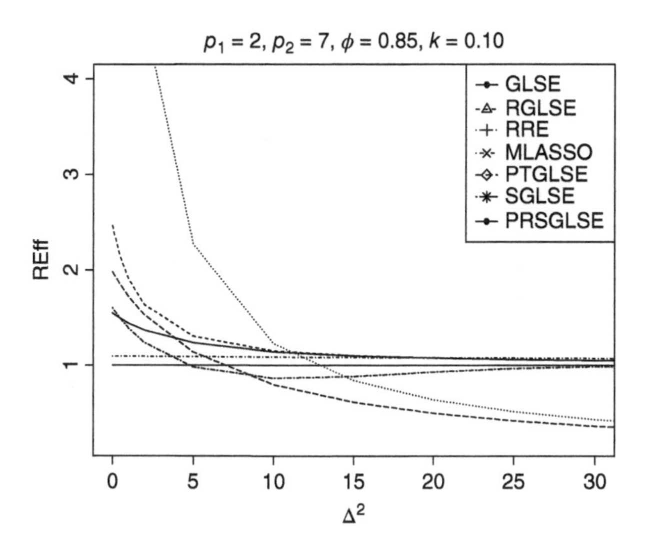

Figure 9.5 Relative efficiency of the estimators for $p_1 = 2, p_2 = 7, \phi = 0.85$, and $k = 0.10$.

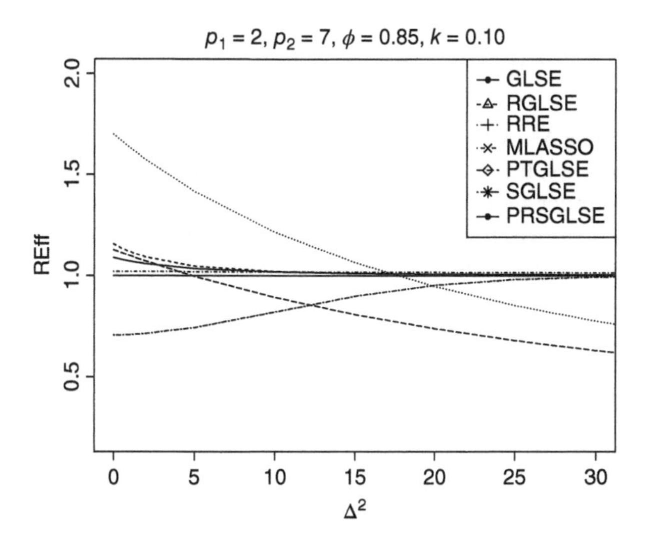

Figure 9.6 Relative efficiency of the estimators for $p_1 = 2, p_2 = 7, \phi = -0.85$, and $k = 0.10$.

PRSE as well as penalty estimators, namely, LASSO and RRE for estimating the regression parameters. We obtained the asymptotic distributional L_2 risk of the estimators and compared them in the sense of smaller risk and noncentrality parameter Δ^2. We found that the performance of the estimators depend on the value of the autocorrelation coefficient, ϕ, number of regressors,

Table 9.10 The relative efficiency of the proposed estimators ($p_1 = 2, p_2 = 7, \phi = 0.85$, and $k = 0.10$).

Δ^2	GLSE	RGLSE	RRE	MLASSO	PTGLSE	SGLSE	PRSGLSE
0.000	1.000	14.861	1.094	1.978	1.602	1.546	2.459
0.100	1.000	13.895	1.094	1.960	1.585	1.538	2.389
0.500	1.000	11.028	1.094	1.891	1.518	1.508	2.161
1.000	1.000	8.767	1.093	1.811	1.443	1.475	1.957
2.000	1.000	6.218	1.093	1.669	1.317	1.419	1.699
5.000	1.000	3.321	1.092	1.352	1.084	1.305	1.376
10.000	1.000	1.869	1.089	1.027	0.950	1.205	1.215
15.000	1.000	1.301	1.087	0.828	0.940	1.153	1.154
20.000	1.000	0.997	1.085	0.694	0.962	1.121	1.122
25.000	1.000	0.809	1.082	0.597	0.982	1.100	1.100
30.000	1.000	0.680	1.080	0.524	0.993	1.085	1.085
40.000	1.000	0.516	1.076	0.421	0.999	1.066	1.066
50.000	1.000	0.416	1.071	0.352	1.000	1.054	1.054
60.000	1.000	0.348	1.067	0.302	1.000	1.045	1.045
100.000	1.000	0.211	1.049	0.193	1.000	1.044	1.044

and noncentrality parameter Δ^2. A real-life data was analyzed to illustrate the performance of the estimators. It is shown that the ridge estimator dominates the rest of the estimators under a correctly specified model. However, it shows poor performance when Δ^2 moves from the null hypothesis. We also observed that the proposed estimators perform better for a positive value of ϕ than for a negative value of ϕ.

Problems

9.1 Under usual notations, show that

$$\tilde{\boldsymbol{\beta}}_n^{\mathrm{G}} = (\boldsymbol{X}^\top \boldsymbol{V}(\phi)^{-1}\boldsymbol{X})^{-1}\boldsymbol{X}^\top \boldsymbol{V}(\phi)^{-1}\boldsymbol{y}.$$

9.2 Show that for testing the hypothesis $\mathcal{H}_o : \boldsymbol{\beta}_2 = \boldsymbol{0}$ vs. $\mathcal{H}_A : \boldsymbol{\beta}_2 \neq \boldsymbol{0}$, the test statistic is

$$\mathcal{L}_n = \frac{n\tilde{\boldsymbol{\beta}}_{2n}^{\mathrm{G}\top}\boldsymbol{B}_{22\cdot1}\tilde{\boldsymbol{\beta}}_{2n}^{\mathrm{G}}}{\sigma^2}$$

and for large sample and under the null hypothesis, $\mathcal{L}_n$ has chi-square distribution with p_2 D.F.

Table 9.11 The relative efficiency of the proposed estimators ($p_1 = 2, p_2 = 7, \phi = -0.85$, and $k = 0.10$).

Δ^2	GLSE	RGLSE	RRE	MLASSO	PTGLSE	SGLSE	PRSGLSE
0.000	1.000	1.699	1.020	1.127	0.705	1.088	1.157
0.100	1.000	1.695	1.020	1.126	0.706	1.087	1.153
0.500	1.000	1.679	1.020	1.118	0.708	1.082	1.139
1.000	1.000	1.658	1.020	1.109	0.712	1.077	1.124
2.000	1.000	1.619	1.020	1.091	0.721	1.068	1.100
5.000	1.000	1.511	1.019	1.041	0.760	1.049	1.060
10.000	1.000	1.361	1.019	0.968	0.842	1.033	1.035
15.000	1.000	1.237	1.018	0.903	0.915	1.025	1.025
20.000	1.000	1.134	1.017	0.847	0.962	1.020	1.020
25.000	1.000	1.047	1.017	0.798	0.985	1.016	1.016
30.000	1.000	0.973	1.016	0.754	0.995	1.014	1.014
40.000	1.000	0.851	1.015	0.679	1.000	1.011	1.011
50.000	1.000	0.757	1.014	0.617	1.000	1.009	1.009
60.000	1.000	0.681	1.013	0.566	1.000	1.007	1.007
100.000	1.000	0.487	1.008	0.425	1.000	1.009	1.009

9.3 Prove Theorem 9.1.

9.4 Show that the risk function for RRE is

$$R(\hat{\boldsymbol{\beta}}_n^{\text{GRR}}) = \sigma^2 \left\{ \text{tr}(\boldsymbol{B}_{11\cdot 2}^{-1}(\phi)) + \frac{1}{(1+k)^2}[\text{tr}(\boldsymbol{B}_{22\cdot 1}^{-1}(\phi)) + k^2 \boldsymbol{\delta}_2^\top \boldsymbol{\delta}_2] \right\}.$$

9.5 Show that the risk function for $\hat{\boldsymbol{\beta}}_n^{\text{G(PT)}}(\alpha)$ is

$$\begin{aligned} R(\hat{\boldsymbol{\beta}}_n^{\text{G(PT)}}(\alpha)) = \sigma^2 \{ & \text{tr}(\boldsymbol{B}_{11\cdot 2}^{-1}) + \text{tr}(\boldsymbol{B}_{22\cdot 1}^{-1}(\phi))[1 - H_{p_2+2}(c_\alpha : \Delta^2)] \\ & + \text{tr}(\boldsymbol{B}_{11}^{-1}(\phi)) H_{p_2+2}(c_\alpha : \Delta^2) \} \\ & + \boldsymbol{\delta}_2^\top \boldsymbol{\delta}_2 \{ 2H_{p_2+2}(c_\alpha : \Delta^2) - H_{p_2+4}(c_\alpha : \Delta^2) \}. \end{aligned}$$

9.6 Verify that the risk function for $\hat{\boldsymbol{\beta}}_n^{\text{G(S)}}$ is

$$\begin{aligned} R(\hat{\boldsymbol{\beta}}_n^{\text{G(S)}}) = \sigma^2 \{ & (\text{tr}(\boldsymbol{B}_{11\cdot 2}^{-1}(\phi)) + \text{tr}(\boldsymbol{B}_{22\cdot 1}^{-1}(\phi))) - (p_2 - 2)\,\text{tr}(\boldsymbol{B}_{22\cdot 1}^{-1}) \\ & \times (2\mathbb{E}[\chi_{p_2+2}^{-2}(\Delta^2)] - (p_2 - 2)\mathbb{E}[\chi_{p_2+2}^{-4}(\Delta^2)]) \} \\ & + (p_2^2 - 4)\boldsymbol{\delta}_2^\top \boldsymbol{\delta}_2 \mathbb{E}[\chi_{p_2+4}^{-4}(\Delta^2)]. \end{aligned}$$

9.7 Show that the risk function for $\hat{\boldsymbol{\beta}}_n^{\mathrm{G(S+)}}$ is

$$\begin{aligned}
\mathrm{R}(\hat{\boldsymbol{\beta}}_{1n}^{\mathrm{G(S+)}}) = \mathrm{R}(\hat{\boldsymbol{\beta}}_n^{\mathrm{G(S)}})\\
&- \sigma^2(p_2-2)tr[\boldsymbol{B}_{22\cdot1}^{-1}(\phi)]\\
&\times \mathbb{E}[(1-(p_2-2)^2\chi_{p_2+2}^{-2}(\Delta^2))^2 I(\chi_{p_2+2}^2(\Delta^2) < p_2-2)]\\
&+ \boldsymbol{\delta}_2^{\top}\boldsymbol{\delta}_2\{2\mathbb{E}[(1-(p_2-2)\chi_{p_2+2}^{-2}(\Delta^2))I(\chi_{p_2+2}^2(\Delta^2) < p_2-2)]\\
&- \mathbb{E}[(1-(p_2-2)\chi_{p_2+4}^{-2}(\Delta^2))^2 I(\chi_{p_2+4}^2(\Delta^2) < p_2-2)]\}.
\end{aligned}$$

9.8 Compare the risk function of PTGLSE and SGLSE and find the value of Δ^2 for which PTGLSE dominates SGLSE or vice versa.

9.9 Compare the risk function of PTGLSE and PRSGLSE and find the value of Δ^2 for which PTGLSE dominates PRSGLSE or vice versa.

9.10 Show that MLASSO outperforms PTGLSE when

$$\Delta^2 \geq \frac{p_2[1-H_{p_2+2}(c_\alpha : \Delta^2)] + p_1 H_{p_2+2}(c_\alpha : \Delta^2)}{1-\{2H_{p_2+2}(c_\alpha : \Delta^2) - H_{p_2+4}(c_\alpha : \Delta^2)\}}. \tag{9.91}$$

9.11 Consider a real data set where the design matrix elements are moderate to highly correlated, and then find the efficiency of the estimators using unweighted risk functions. Find parallel formulas for the efficiency expressions and compare the results with that of the efficiency using weighted risk function. Are the two results consistent?

10

Rank-Based Shrinkage Estimation

This chapter introduces the R-estimates and provides a comparative study of ridge regression estimator (RRE), least absolute shrinkage and selection operator (LASSO), preliminary test estimator (PTE) and the Stein-type estimator based on the theory of rank-based statistics and the nonorthogonality design matrix of a given linear model.

10.1 Introduction

It is well known that the usual rank estimators (REs) are robust in the linear regression models, asymptotically unbiased with minimum variance. But, the data analyst may point out some deficiency with the R-estimators when one considers the "prediction accuracy" and "interpretation." To overcome these concerns, we propose the rank-based least absolute shrinkage and selection operator (RLASSO) estimator. It defines a continuous shrinking operation that can produce coefficients that are exactly "zero" and competitive with the rank-based "subset selection" and RRE, retaining the good properties of both the R-estimators. RLASSO simultaneously estimates and selects the coefficients of a given linear regression model.

However, there are rank-based PTEs and Stein-type estimators (see Saleh 2006; Jureckova and Sen 1996, and Puri and Sen 1986). These R-estimators provide estimators which shrink toward the target value and do not select coefficients for appropriate prediction and interpretations. Hoerl and Kennard (1970) introduced ridge regression based on the Tikhonov (1963) regularization, and Tibshirani (1996) introduced the LASSO estimators in a parametric formulation. The methodology is minimization of least squares objective function subject to L_2- and L_1-penalty restrictions. However, the L_2 penalty does not produce a sparse solution, but the L_1-penalty does.

This chapter points to the useful aspects of RLASSO and the rank-based ridge regression R-estimators as well as the limitations. Conclusions are based on

Theory of Ridge Regression Estimation with Applications, First Edition.
A.K. Md. Ehsanes Saleh, Mohammad Arashi, and B.M. Golam Kibria.

asymptotic L_2-risk lower bound of RLASSO with the actual asymptotic L_2 risk of other R-estimators.

10.2 Linear Model and Rank Estimation

Consider the multiple linear model,

$$Y = \theta \mathbf{1}_n + X\boldsymbol{\beta} + \epsilon = \theta \mathbf{1}_n + X_1\boldsymbol{\beta}_1 + X_2\boldsymbol{\beta}_2 + \epsilon, \tag{10.1}$$

where $Y = (Y_1, Y_2, \ldots, Y_n)^\top$, $X = (X_1, X_2)$ is the $n \times p$ matrix of real numbers, θ is the intercept parameter, and $\boldsymbol{\beta} = (\boldsymbol{\beta}_1, \boldsymbol{\beta}_2)^\top$ is the p-vector of regression parameters. We assume that:

(1) Errors $\epsilon = (\epsilon_1, \ldots, \epsilon_n)^\top$ are independently and identically distributed (i.i.d.) random variables with (unknown) cumulative distributional function (c.d.f.) F having absolutely continuous probability density function (p.d.f.) f with finite and nonzero Fisher information

$$0 < I(f) = \int_{-\infty}^{+\infty} \left[-\frac{f'(x)}{f(x)}\right]^2 f(x)\mathrm{d}x < \infty. \tag{10.2}$$

(2) For the definition of the linear rank statistics, we consider the score generating function $\varphi : (0,1) \to \mathbb{R}$ which is assumed to be nonconstant, non-decreasing, and square integrable on $(0,1)$ so that

$$A_\varphi^2 = \int_0^1 \varphi^2(u)\mathrm{d}u - \left(\int_0^1 \varphi(u)\mathrm{d}u\right)^2. \tag{10.3}$$

The scores are defined in either of the following ways:

$$a_n(i) = \mathbb{E}\varphi(U_{i:n}) \quad \text{or} \quad a_n(i) = \varphi\left(\frac{i}{n+1}\right), \qquad i = 1, \ldots, n$$

for $n \geq 1$, where $U_{1:n} \leq \cdots \leq U_{n:n}$ are order statistics from a sample of size n from $\mathcal{U}(0,1)$.

(3) Let

$$C_n = \frac{1}{n}\sum_{i=1}^{n}(\boldsymbol{x}_i - \overline{\boldsymbol{x}}_n)(\boldsymbol{x}_i - \overline{\boldsymbol{x}}_n)^\top, \tag{10.4}$$

where $\boldsymbol{x}_i$ is the ith row of X and $\overline{\boldsymbol{x}} = \frac{1}{n}\sum_{i=1}^{n} x_i$, $n > p$. We assume that

(a) $\lim_{n\to\infty} C_n = C$

(b) $\lim_{n\to\infty} \max_{1\leq i\leq n} (\boldsymbol{x}_i - \overline{\boldsymbol{x}}_n)^\top C_n^{-1}(\boldsymbol{x}_i - \overline{\boldsymbol{x}}_n) = 0.$

For the R-estimation of the parameter $\boldsymbol{\beta}$, define for $\boldsymbol{b} \in \mathbb{R}^p$ the rank of $y_i - \boldsymbol{x}_i^\top \boldsymbol{b}$ among $y_1 - \boldsymbol{x}_1^\top \boldsymbol{b}, \ \dots, \ y_n - \boldsymbol{x}_n^\top \boldsymbol{b}$ by $R_{ni}(\boldsymbol{b})$. Then for each $n \geq 1$, consider the set of scores $a_n(1) \leq \dots \leq a_n(n)$ and define the vector of linear rank statistics

$$\boldsymbol{L}_n(\boldsymbol{b}) = (L_{n1}(\boldsymbol{b}), \dots, L_{nn}(\boldsymbol{b}))^\top = \frac{1}{\sqrt{n}} \sum_{i=1}^{n} (\boldsymbol{x}_i - \overline{\boldsymbol{x}}_n) a_n(R_{ni}(\boldsymbol{b})). \tag{10.5}$$

Since $R_{ni}(\boldsymbol{b})$ are translation invariant, there is no need of adjustment for the intercept parameter θ.

If we set $\|\boldsymbol{a}\| = \sum_{i=1}^{p} |a_j|$ for $\boldsymbol{a} = (a_1, \dots, a_n)^\top$, then the unrestricted RE is defined by any central point of the set

$$\tilde{\boldsymbol{\beta}}_n^{\mathrm{R}} = \{\boldsymbol{b} \in \mathbb{R}^p \, : \, \|\boldsymbol{L}_n(\boldsymbol{b})\| = \min\}. \tag{10.6}$$

Let the RE be as $\tilde{\boldsymbol{\beta}}_n^{\mathrm{R}}$. Then, using the uniform asymptotic linearity of Jureckova (1971),

$$\lim_{n\to\infty} \mathbb{P}\left(\sup_{\|\boldsymbol{\omega}\|<k} \left\| \boldsymbol{L}_n\left(\boldsymbol{\beta} + \frac{\boldsymbol{\omega}}{\sqrt{n}}\right) - \boldsymbol{L}_n(\boldsymbol{\beta}) + \gamma \boldsymbol{C}\boldsymbol{\omega} \right\| > \epsilon \right) = 0 \tag{10.7}$$

for any $k > 0$ and $\epsilon > 0$. Then, it is well-known that

$$\sqrt{n}(\tilde{\boldsymbol{\beta}}_n^{\mathrm{R}} - \boldsymbol{\beta}) \xrightarrow{\mathcal{D}} \mathcal{N}_p(\boldsymbol{0}, \eta^2 \boldsymbol{C}^{-1}), \tag{10.8}$$

where

$$\eta^2 = \frac{A_\varphi^2}{\gamma^2}, \qquad \gamma = \int_0^1 \varphi(u) \left\{ -\frac{f'(F^{-1}(u))}{f(F^{-1}(u))} \right\} \mathrm{d}u \tag{10.9}$$

and

$$\lim_{n\to\infty} \frac{1}{n} \boldsymbol{C}_n = \boldsymbol{C} = \begin{pmatrix} \boldsymbol{C}_{11} & \boldsymbol{C}_{12} \\ \boldsymbol{C}_{21} & \boldsymbol{C}_{22} \end{pmatrix}.$$

Similarly, when $\boldsymbol{\beta}_2 = \boldsymbol{0}$, the model reduces to

$$\underbrace{\boldsymbol{Y}}_{n\times 1} = \underbrace{\boldsymbol{X}_1}_{n\times p_1} \underbrace{\boldsymbol{\beta}_1}_{p_1\times 1} + \underbrace{\boldsymbol{\epsilon}}_{n\times 1}, \quad \boldsymbol{X}_1 = (\boldsymbol{x}_1^{(1)}, \cdots, \boldsymbol{x}_n^{(1)}); \ \boldsymbol{x}_j \in \mathbb{R}^p. \tag{10.10}$$

Accordingly, for the R-estimation of $\boldsymbol{\beta}_1$, in this case, we define rank of $y_i - \boldsymbol{x}_i^{(1)\top} \boldsymbol{\beta}_1$ among

$$y_1 - \boldsymbol{x}_1^{(1)\top} \boldsymbol{b}_1, \dots, y_n - \boldsymbol{x}_n^{(1)\top} \boldsymbol{b}_1$$

as $R_{n_i}(\boldsymbol{b}_1)$. Then, for each $n \geq 1$, consider the set of scores $a_n(1) \leq \ldots \leq a_n(n)$ and define the vector of linear rank statistics

$$L_n(\boldsymbol{b}_1) = (L_{n_1}(\boldsymbol{b}_1), \ldots, L_{n_{p_1}}(\boldsymbol{b}_1)), \tag{10.11}$$

where

$$L_{n_j}(\boldsymbol{b}_1) = \frac{1}{\sqrt{n}} \sum_{i=1}^{n} (\boldsymbol{x}_i^{(1)} - \overline{\boldsymbol{x}}_n^{(1)}) a_n(R_{n_i}(\boldsymbol{b}_1)). \tag{10.12}$$

Then, define the restricted RE of $\boldsymbol{\beta}_1$ as

$$\begin{aligned}\hat{\boldsymbol{\beta}}_{1n}^{\mathrm{R(R)}} &= \{\boldsymbol{b}_1 \in \mathbb{R}^{p_1} : \|L_n(\boldsymbol{b}_1)\| = \min\} \\ &= (\hat{\beta}_{1n}, \ldots, \hat{\beta}_{p_1 n})^\top,\end{aligned} \tag{10.13}$$

which satisfy the following equality

$$\lim_{n\to\infty} \mathbb{P}\left(\sup_{\|\boldsymbol{w}_1^{(1)} < k\|} \left\| L_n\left(\boldsymbol{\beta}_1 + \frac{\boldsymbol{w}_1^{(1)}}{\sqrt{n}}\right) - L_n(\boldsymbol{\beta}_1) + \gamma \boldsymbol{C}_{11} \boldsymbol{w}_1^{(1)} \right\| > \epsilon \right) = 0, \tag{10.14}$$

where $\boldsymbol{w}_1^{(1)} = (w_1, \ldots, w_{p_1})^\top$, for any $k > 0$ and $\epsilon > 0$. Then, it follows that

$$\sqrt{n}(\hat{\boldsymbol{\beta}}_1^{\mathrm{R(R)}} - \boldsymbol{\beta}_1) \xrightarrow{\mathcal{D}} \mathcal{N}_p(\boldsymbol{0}, \eta^2 \boldsymbol{C}_{11}^{-1}). \tag{10.15}$$

We are basically interested in the R-estimation of $\boldsymbol{\beta}$ when it is suspected the sparsity condition $\boldsymbol{\beta}_2 = \boldsymbol{0}$ may hold. Under the given setup, the RE is written as

$$\tilde{\boldsymbol{\beta}}_n = \begin{pmatrix} \tilde{\boldsymbol{\beta}}_{1n}^{\mathrm{R}} \\ \tilde{\boldsymbol{\beta}}_{2n}^{\mathrm{R}}, \end{pmatrix} \tag{10.16}$$

where $\tilde{\boldsymbol{\beta}}_{1n}^{\mathrm{R}}$ and $\tilde{\boldsymbol{\beta}}_{2n}^{\mathrm{R}}$ are p_1 and p_2 vectors, respectively; and if $\boldsymbol{\beta}_{\mathrm{R}} = (\boldsymbol{\beta}_1^\top, \boldsymbol{0}^\top)^\top$ is satisfied, then the restricted RE of $\boldsymbol{\beta}_{\mathrm{R}}$ is

$$\hat{\boldsymbol{\beta}}_n^{\mathrm{R(R)}} = \begin{pmatrix} \hat{\boldsymbol{\beta}}_{1n}^{\mathrm{R(R)}}, \\ \boldsymbol{0} \end{pmatrix}, \tag{10.17}$$

where $\hat{\boldsymbol{\beta}}_{1n}^{\mathrm{R(R)}}$ is defined by (10.13).

In this section, we are interested in the study of some shrinkage estimators stemming from $\tilde{\boldsymbol{\beta}}_n^{\mathrm{R}}$ and $\hat{\boldsymbol{\beta}}_n^{\mathrm{R(R)}}$. In order to look at the performance characteristics of several R-estimators, we use the two components asymptotic weighted L_2-risk function

$$\mathrm{ADR}(\hat{\boldsymbol{\beta}}_n^* : \boldsymbol{W}_1, \boldsymbol{W}_2) = \lim_{n\to\infty} n\mathbb{E}\left[\left\|\hat{\boldsymbol{\beta}}_{1n}^* - \boldsymbol{\beta}_1\right\|_{\boldsymbol{W}_1}^2\right] + \lim_{n\to\infty} \mathbb{E}\left[\left\|\hat{\boldsymbol{\beta}}_{2n}^* - \boldsymbol{\beta}_2\right\|_{\boldsymbol{W}_2}^2\right] \tag{10.18}$$

where $\hat{\boldsymbol{\beta}}_n^*$ is any estimator of the form $(\hat{\boldsymbol{\beta}}_{1n}^{*\top}, \hat{\boldsymbol{\beta}}_{2n}^{*\top})^\top$ of $\boldsymbol{\beta} = (\boldsymbol{\beta}_1^\top, \boldsymbol{\beta}_2^\top)^\top$.

Note that $\sqrt{n}(\tilde{\boldsymbol{\beta}}_n^{\rm R} - \boldsymbol{\beta}) \sim \mathcal{N}_p(\mathbf{0}, \eta^2 \boldsymbol{C}^{-1})$ so that the marginal distribution of $\tilde{\boldsymbol{\beta}}_{1n}^{\rm R}$ and $\tilde{\boldsymbol{\beta}}_{2n}^{\rm R}$ are given by

$$\begin{aligned}&\sqrt{n}(\tilde{\boldsymbol{\beta}}_{1n}^{\rm R} - \boldsymbol{\beta}_1) \sim \mathcal{N}_{p_1}(\mathbf{0}, \eta^2 \boldsymbol{C}_{11\cdot 2}^{-1})\\&\sqrt{n}(\tilde{\boldsymbol{\beta}}_{2n}^{\rm R} - \boldsymbol{\beta}_2) \sim \mathcal{N}_{p_2}(\mathbf{0}, \eta^2 \boldsymbol{C}_{22\cdot 1}^{-1}).\end{aligned} \tag{10.19}$$

Hence, asymptotic distributional weighted $\rm L_2$ risk of $\sqrt{n}(\tilde{\boldsymbol{\beta}}_n^{\rm R} - \boldsymbol{\beta})$ is given by

$$\text{ADR}(\tilde{\boldsymbol{\beta}}_n^{\rm R}, \boldsymbol{C}_{11\cdot 2}, \boldsymbol{C}_{22\cdot 1}) = \eta^2(p_1 + p_2). \tag{10.20}$$

For the test of sparsity $\boldsymbol{\beta}_2 = \mathbf{0}$, we define the aligned rank statistic

$$\mathcal{L}_n = A_n^{-2}\{\hat{L}_{n(2)}^{\top} \boldsymbol{C}_{n22.1} \hat{L}_{n(2)}\}, \quad \hat{L}_{n(2)} = L_{n(2)}(\hat{\boldsymbol{\beta}}_{1n}, \mathbf{0}) \tag{10.21}$$

and $L_n(b_1, b_2) = (L_{n(2)}^{\top}(b_1, b_2), L_{n(2)}^{\top}(b_1, b_2))^{\top}$. Further, it is shown (see, Puri and Sen 1986) that

$$\mathcal{L}_n = \frac{1}{\eta^2} \tilde{\boldsymbol{\beta}}_{2n}^{\rm R^{\top}} \boldsymbol{C}_{22\cdot 1} \tilde{\boldsymbol{\beta}}_{2n}^{\rm R} + o_p(1). \tag{10.22}$$

It is easy to see that under $\mathcal{H}_0 : \boldsymbol{\beta}_2 = \mathbf{0}$, $\mathcal{L}_n$ has asymptotically the chi-square distribution with p_2 degrees of freedom (DF) and under the sequence of local alternatives $\{\mathcal{K}_{n(2)}\}$ defined by

$$\mathcal{K}_{n(2)} : \boldsymbol{\beta}_{2(n)} = n^{-\frac{1}{2}}(\boldsymbol{\delta}_1^{\top}, \boldsymbol{\delta}_2^{\top})^{\top}, \quad \boldsymbol{\delta}_1 = (\delta_1, \ldots, \delta_{p_1})^{\top}; \boldsymbol{\delta}_2 = (\delta_{p_1+1}, \ldots, \delta_p)^{\top}. \tag{10.23}$$

For a suitable estimator $\hat{\boldsymbol{\beta}}_n^*$ of $\boldsymbol{\beta}$, we denote by

$$G^*(\boldsymbol{x}) = \lim_{n\to\infty} P\left\{ n^{\frac{1}{2}}(\boldsymbol{\beta}_n^* - \boldsymbol{\beta}) \le \boldsymbol{x} \middle| \mathcal{K}_{n(2)} \right\}, \tag{10.24}$$

where we assume that G^* is non-degenerate.

One may find the asymptotic distributional weighted $\rm L_2$ risks are

$$\text{ADR}(\hat{\boldsymbol{\beta}}_{1n}^{\rm R(R)}; \boldsymbol{C}_{11\cdot 2}, \boldsymbol{C}_{22\cdot 1}) = \eta^2(p_1 - \text{tr}(\boldsymbol{M}_0) + \Delta^2), \tag{10.25}$$

where

$$\boldsymbol{M}_0 = \boldsymbol{C}_{11}^{-1}\boldsymbol{C}_{12}\boldsymbol{C}_{22}^{-1}\boldsymbol{C}_{21} \quad \text{and} \quad \Delta^2 = \frac{1}{\eta^2}\boldsymbol{\delta}_2^{\top}\boldsymbol{C}_{22\cdot 1}\boldsymbol{\delta}_2. \tag{10.26}$$

Our focus in this chapter is the comparative study of performance properties of these rank-based penalty R-estimators and PTEs and Stein-type R-estimators. We refer to Saleh (2006) for the comparative study of the PTEs and Stein-type estimators. We extend the study to include penalty estimators which have not been done yet.

Now, we recall several results that form the asymptotic distribution of

$$\sqrt{n}(\tilde{\boldsymbol{\beta}}^{\rm R} - \boldsymbol{\beta}) \sim \mathcal{N}_p(\mathbf{0}, \eta^2 \boldsymbol{C}^{-1}).$$

For the RE given by $(\tilde{\boldsymbol{\beta}}_{1n}^{\mathrm{R}^\top}, \tilde{\boldsymbol{\beta}}_{2n}^{\mathrm{R}^\top})^\top$, we see under regularity condition and Eq. (10.22),

$$\begin{pmatrix} \sqrt{n}(\tilde{\boldsymbol{\beta}}_{1n}^{\mathrm{R}} - \boldsymbol{\beta}) \\ \sqrt{n}\left(\tilde{\boldsymbol{\beta}}_{2n}^{\mathrm{R}} - n^{-\frac{1}{2}}\boldsymbol{\delta}_2\right) \end{pmatrix} = \boldsymbol{C}^{-1}\gamma^{-1}n^{-\frac{1}{2}}L_n(\boldsymbol{\beta}) + o_p(1), \tag{10.27}$$

where $n^{-\frac{1}{2}}\gamma^{-1}L_n(\boldsymbol{\beta}) \xrightarrow{\mathcal{D}} \mathcal{N}_p(0, \eta^2\boldsymbol{C})$.

As a consequence, the asymptotic marginal distribution of $\tilde{\boldsymbol{\beta}}_{jn}^{\mathrm{R}}$ $(j = 1, \dots, p)$ under local alternatives

$$\mathcal{K}_{(n)} : \boldsymbol{\beta}_{jn} = n^{-\frac{1}{2}}\delta_j, \quad j = 1, \dots, p$$

is given by

$$Z_j = \frac{\sqrt{n}\tilde{\boldsymbol{\beta}}_{jn}}{\eta\sqrt{\boldsymbol{C}_n^{jj}}} \xrightarrow{\mathcal{D}} \mathcal{N}(\Delta_j, 1), \qquad \Delta_j = \frac{\delta_j}{\eta\sqrt{\boldsymbol{C}^{jj}}}, \tag{10.28}$$

where $\boldsymbol{C}_n^{jj}$ is jth diagonal of $\left(\frac{1}{n}\boldsymbol{C}_n\right)^{-1}$.

10.2.1 Penalty R-Estimators

In this section, we consider three basic penalty estimators, namely:

(i) The hard threshold estimator (HTE) (Donoho and Johnstone 1994),
(ii) The LASSO by Tibshirani (1996),
(iii) The RRE by Hoerl and Kennard (1970).

Motivated by the idea that only few regression coefficients contribute signal, we consider threshold rules that retain only observed data that exceed a multiple of the noise level. Accordingly, we consider the "subset selection" rule given by Donoho and Johnstone (1994) known as the "hard threshold" rule as given by

$$\begin{aligned} \hat{\boldsymbol{\beta}}_n^{\mathrm{HT}}(\kappa) &= (\sqrt{n}\tilde{\beta}_{jn}^{\mathrm{R}} I(|\sqrt{n}\tilde{\beta}_{jn}^{\mathrm{R}}| > \kappa\eta\sqrt{C_n^{jj}}) | j = 1, \dots, p)^\top, \\ &\stackrel{\mathcal{D}}{=} (\eta\sqrt{C_n^{jj}} Z_j I(|Z_j| > \kappa) | j = 1, \dots, p)^\top \quad \text{as} \quad n \to \infty, \end{aligned} \tag{10.29}$$

where $\tilde{\beta}_{jn}^{\mathrm{R}}$ is the jth element of $\tilde{\boldsymbol{\beta}}_n^{\mathrm{R}}$ and $I(A)$ is an indicator function of the set A. The quantity κ is called the threshold parameter. The components of $\hat{\boldsymbol{\beta}}_n^{\mathrm{HT}}(\kappa)$ are kept as $\tilde{\beta}_{jn}$ if they are significant and zero, otherwise. It is apparent that each component of $\hat{\boldsymbol{\beta}}_n^{\mathrm{HT}}(\kappa)$ is a PTE of the predictor concerned. The components of $\hat{\boldsymbol{\beta}}_n^{\mathrm{HT}}(\kappa)$ are PTEs and discrete variables and lose some optimality properties.

Hence, one may define a continuous version of (10.29) based on marginal distribution of $\sqrt{n}\tilde{\beta}_{jn}^{\mathrm{R}}$ $(j = 1, \ldots, p)$.

In accordance with the principle of the PTE approach (see Saleh 2006), we define the Stein-type estimator as the continuous version of PTE based on the marginal distribution of

$$\tilde{\beta}_{jn}^{\mathrm{R}} \sim \mathcal{N}(\beta, \sigma^2 C^{jj}) \quad j = 1, \ldots, p$$

given by

$$\begin{aligned}\hat{\boldsymbol{\beta}}_n^{\mathrm{S}}(\kappa) &= \left(\sqrt{n}\tilde{\beta}_{jn}^{\mathrm{R}} - \kappa\eta\sqrt{C_n^{jj}}\frac{\sqrt{n}\tilde{\beta}_{jn}^{\mathrm{R}}}{\sqrt{n}|\tilde{\beta}_{jn}^{\mathrm{R}}|} \middle| j = 1, \ldots, p \right)^{\top} \\ &\stackrel{\mathcal{D}}{=} (\eta\sqrt{C^{jj}}\mathrm{sgn}(Z_j)(|Z_j| - \kappa)|j = 1, \ldots, p)^{\top} \quad \text{as} \quad n \to \infty \\ &= (\hat{\beta}_{1n}^{\mathrm{S}}(\kappa), \ldots, \hat{\beta}_{pn}^{\mathrm{S}}(\kappa))^{\top}.\end{aligned}$$

See Saleh (2006, p. 83) for more details.

Another continuous version proposed by Tibshirani (1996) and Donoho and Johnstone (1994) is called the LASSO. In order to develop LASSO for our case, we propose the following modified R LASSO (MRL) given by

$$\hat{\boldsymbol{\beta}}_n^{\mathrm{MRL}}(\lambda) = (\hat{\beta}_{1n}^{\mathrm{MRL}}(\lambda), \ldots, \hat{\beta}_{pn}^{\mathrm{MRL}}(\lambda))^{\top}, \tag{10.30}$$

where for $j = 1, \ldots, p$,

$$\begin{aligned}\hat{\beta}_{jn}^{\mathrm{MRL}}(\lambda) &= \mathrm{sgn}(\tilde{\beta}_{jn}^{\mathrm{R}})(|\tilde{\beta}_{jn}^{\mathrm{R}}| - \lambda\eta\sqrt{\boldsymbol{C}_n^{jj}})^{+} \\ &\stackrel{\mathcal{D}}{=} \eta\sqrt{\boldsymbol{C}^{jj}}\mathrm{sgn}(Z_j)(|Z_j| - \lambda)^{+} \quad \text{as} \quad n \to \infty.\end{aligned} \tag{10.31}$$

The estimator $\hat{\boldsymbol{\beta}}_n^{\mathrm{MRL}}(\lambda)$ defines a continuous shrinkage operation that produces a sparse solution which may be derived as follows:

One may show that $\hat{\boldsymbol{\beta}}_n^{\mathrm{MRL}}(\lambda)$ is the solution of the equation

$$-\tilde{\boldsymbol{\beta}}_m^{\mathrm{R}} + \hat{\boldsymbol{\beta}}_n^{\mathrm{MRL}}(\lambda) + \lambda\eta\mathrm{Diag}\left((\boldsymbol{C}^{11})^{\frac{1}{2}}, \ldots, (\boldsymbol{C}^{pp})^{\frac{1}{2}}\right)\,\mathrm{sgn}(\hat{\boldsymbol{\beta}}_n^{\mathrm{MRL}}(\lambda)) = \boldsymbol{0}. \tag{10.32}$$

Thus, the jth component of (10.32) equals

$$-\tilde{\beta}_{jn}^{\mathrm{R}} + \hat{\beta}_{jn}^{\mathrm{MRL}}(\lambda) + \lambda\eta\boldsymbol{C}^{jj}\ \mathrm{sgn}(\hat{\boldsymbol{\beta}}_{jn}^{\mathrm{MRL}}(\lambda)) = \boldsymbol{0}, \quad j = 1, \ldots, p. \tag{10.33}$$

Now, consider three cases, $\mathrm{sgn}(\hat{\boldsymbol{\beta}}_{jn}^{\mathrm{MRL}}(\lambda)) = \pm 1$ and 0. We can show that the asymptotic distributional expressions for $\hat{\boldsymbol{\beta}}_{jn}^{\mathrm{MRL}}(\lambda)$ is given by

$$\eta\sqrt{\boldsymbol{C}^{ii}}\mathrm{sgn}(Z_j)(|Z_j| - \kappa)^{+}. \tag{10.34}$$

Finally, we consider the R ridge regression estimator (RRRE) of $(\boldsymbol{\beta}_1^\top, \boldsymbol{\beta}_2^\top)$. They are obtained using marginal distributions of $\sqrt{n}\tilde{\beta}_{jn}^{\mathrm{R}} \sim \mathcal{N}(\delta_j, \sigma^2 C^{jj})$, $j = 1, \ldots, p$, as

$$\tilde{\boldsymbol{\beta}}_n^{\mathrm{RRR}}(k) = \begin{pmatrix} \tilde{\boldsymbol{\beta}}_{1n}^{\mathrm{R}} \\ \frac{1}{1+k}\tilde{\boldsymbol{\beta}}_{2n}^{\mathrm{R}} \end{pmatrix}, \tag{10.35}$$

to accommodate sparsity condition; see Tibshirani (1996) on the summary of properties discussed earlier.

Our problem is to compare the performance characteristics of these penalty estimators with that of the Stein-type and preliminary test R-estimators (PTREs) with respect to asymptotic distributional mean squared error criterion. We present the PTREs and Stein-type R-estimators in the next section.

10.2.2 PTREs and Stein-type R-Estimators

For the model (10.1), if we suspect sparsity condition, i.e. $\boldsymbol{\beta}_2 = \mathbf{0}$, then the restricted R-estimator of $(\boldsymbol{\beta}_1^\top, \boldsymbol{\beta}_2^\top)^\top$ is $\hat{\boldsymbol{\beta}}_n^{\mathrm{R(R)}} = (\tilde{\boldsymbol{\beta}}_{1n}^{\mathrm{R}^\top}, \mathbf{0}^\top)^\top$. For the test of the null-hypothesis $\mathcal{H}_o : \boldsymbol{\beta}_2 = \mathbf{0}$ vs. $\mathcal{H}_A : \boldsymbol{\beta}_2 \neq \mathbf{0}$, the rank statistic for the test of $\mathcal{H}_o$ is given by

$$\mathcal{L}_n = A_n^{-2}\{\hat{L}_{n(2)}^\top \boldsymbol{C}_{n22.1}\hat{L}_{n(2)}\}, \quad \hat{L}_{n(2)} = L_{n(2)}(\tilde{\boldsymbol{\beta}}_{1n}^{\mathrm{R}}, \mathbf{0}), \tag{10.36}$$

where $L_n(b_1, b_2) = (L_{n(2)}^\top(b_1, b_2), L_{n(2)}^\top(b_1, b_2))^\top$, $\boldsymbol{C}_{n22\cdot1} = \boldsymbol{C}_{n22} - \boldsymbol{C}_{n21}\boldsymbol{C}_{n11}^{-1}\boldsymbol{C}_{n12}$, and

$$A_n^2 = \frac{1}{n-1}\sum_{i=1}^{n}(a_n(i) - \overline{a}_n)^2, \quad \overline{a} = \frac{1}{n}\sum_{i=1}^{n} a_n(i).$$

It is well known that under the model (10.1) and assumptions (10.2)–(10.4) as $n \to \infty$, $\mathcal{L}_n$ follows the chi-squared distribution with p_2 DF Then, we define the PTE of $(\boldsymbol{\beta}_1^\top, \boldsymbol{\beta}_2^\top)^\top$ as

$$\begin{pmatrix} \hat{\boldsymbol{\beta}}_{1n}^{\mathrm{R(PT)}}(\alpha) \\ \hat{\boldsymbol{\beta}}_{2n}^{\mathrm{R(PT)}} \end{pmatrix} = \begin{pmatrix} \tilde{\boldsymbol{\beta}}_{1n}^{\mathrm{R}} \\ \tilde{\boldsymbol{\beta}}_{2n}^{\mathrm{R}} - \tilde{\boldsymbol{\beta}}_{2n}^{\mathrm{R}} I(\mathcal{L}_n < \chi_{p_2}^2(\alpha)) \end{pmatrix}. \tag{10.37}$$

where $I(A)$ is the indicator function of the set A.

Similarly, we define the Stein-type R-estimator as

$$\begin{pmatrix} \hat{\boldsymbol{\beta}}_{1n}^{\mathrm{R(S)}} \\ \hat{\boldsymbol{\beta}}_{2n}^{\mathrm{R(S)}} \end{pmatrix} = \begin{pmatrix} \tilde{\boldsymbol{\beta}}_{1n}^{\mathrm{R}} \\ \tilde{\boldsymbol{\beta}}_{2n}^{\mathrm{R}}(1 - (p_2 - 2)\mathcal{L}_n^{-1}) \end{pmatrix}. \tag{10.38}$$

Finally, the positive rule Stein-type estimator is given by

$$\begin{pmatrix} \hat{\boldsymbol{\beta}}_{1n}^{\mathrm{R(S+)}} \\ \hat{\boldsymbol{\beta}}_{2n}^{\mathrm{R(S+)}} \end{pmatrix} = \begin{pmatrix} \tilde{\boldsymbol{\beta}}_{1n}^{\mathrm{R}} \\ \tilde{\boldsymbol{\beta}}_{2n}^{\mathrm{R}}(1 - (p_2 - 2)\mathcal{L}_n^{-1})I(\mathcal{L}_n > p_2 - 2) \end{pmatrix}. \tag{10.39}$$

10.3 Asymptotic Distributional Bias and L_2 Risk of the R-Estimators

First, we consider the asymptotic distribution bias (ADB) and ADR of the penalty estimators.

10.3.1 Hard Threshold Estimators (Subset Selection)

It is easy to see that

$$\begin{aligned}\text{ADB}(\hat{\boldsymbol{\beta}}_n^{\text{HT}}(\kappa)) &= \lim_{n\to\infty} \mathbb{E}\left[\hat{\boldsymbol{\beta}}_n^{\text{HT}}(\kappa) - \eta^{-\frac{1}{2}}\boldsymbol{\delta}\right] \\ &= (-\delta_j H_3(\kappa^2, \Delta_j^2)|j = 1, \dots, p)^\top, \end{aligned} \tag{10.40}$$

where $H_\nu(\kappa^2, \Delta_j^2)$ is the c.d.f. of a noncentral chi-square distribution with ν DF and noncentrality parameter, $\frac{1}{2}\Delta_j^2$.

The ADR of $\hat{\boldsymbol{\beta}}_n^{\text{HT}}(\kappa)$ is given by

$$\begin{aligned}\text{ADR}(\hat{\boldsymbol{\beta}}_n^{\text{HT}}(\kappa)) &= \sum_{j=1}^{p} \lim_{n\to\infty} \mathbb{E}[\sqrt{n}(\tilde{\beta}_{jn} - \beta_j) - \sqrt{n}\tilde{\beta}_{jn} I(|\sqrt{n}\tilde{\beta}_{jn}| < \kappa\eta\sqrt{C^{jj}})]^2 \\ &= \eta^2 \sum_{j=1}^{p} C^{jj}\{(1 - H_3(\kappa^2; \Delta_j^2)) + \Delta_j^2(2H_3(\kappa^2; \Delta_j^2) - H_5(\kappa^2; \Delta_j^2))\}.\end{aligned} \tag{10.41}$$

Since

$$[\sqrt{n}\tilde{\beta}_{jn} I(\sqrt{n}|\tilde{\beta}_{jn}| > \kappa\eta\sqrt{C^{jj}}) - \beta_j]^2 \le \sqrt{n}\left(\tilde{\beta}_{jn}^2 - n^{-\frac{1}{2}\delta_j}\right)^2 + \delta_j^2 \tag{10.42}$$

for $j = 1, \dots, p$. Hence,

$$\text{ADR}(\hat{\boldsymbol{\beta}}_n^{\text{HT}}(\kappa)) \le \eta^2 \operatorname{tr}(\boldsymbol{C}^{-1}) + \boldsymbol{\delta}^\top\boldsymbol{\delta}, \quad \forall \boldsymbol{\delta} \in \mathbb{R}^p. \tag{10.43}$$

Note that (10.43) is free of the threshold parameter, κ.

Thus, we have Lemma 10.1, which gives the asymptotic distributional upper bound (ADUB) of $\text{ADR}(\hat{\boldsymbol{\beta}}_n^{\text{HT}}(\kappa))$ as

$$\begin{aligned}&\text{ADR}(\hat{\boldsymbol{\beta}}_n^{\text{HT}}(\kappa); \boldsymbol{C}_{11\cdot 2}, \boldsymbol{C}_{22\cdot 1}) \\ &\le \begin{cases} (i)\eta^2(1+\kappa^2)(p_1+p_2), \kappa > 1 \\ (ii)\eta^2(p_1+p_2) + \boldsymbol{\delta}_1^\top \boldsymbol{C}_{11\cdot 2}\boldsymbol{\delta}_1 + \boldsymbol{\delta}_2^\top \boldsymbol{C}_{22\cdot 1}\boldsymbol{\delta}_2 \quad \boldsymbol{\delta}_1 \in \mathbb{R}^{p_1}, \ \boldsymbol{\delta}_2 \in \mathbb{R}^{p_2} \\ (iii)\eta^2 \rho_{\text{HT}}(\kappa, 0)(p_1+p_2) + 1.2(\boldsymbol{\delta}_1^\top \boldsymbol{C}_{11\cdot 2}\boldsymbol{\delta}_1 + \boldsymbol{\delta}_2^\top \boldsymbol{C}_{22\cdot 1}\boldsymbol{\delta}_2), \quad 0 < \boldsymbol{\delta} < \kappa \mathbf{1}_p^\top. \end{cases}\end{aligned} \tag{10.44}$$

Lemma 10.1 *Under the assumed regularity conditions, the ADUB of* L_2 *risk is given by*

$$\mathrm{ADR}(\hat{\boldsymbol{\beta}}_n^{\mathrm{HT}}(\kappa)) \leq \begin{cases} (\mathrm{i})\eta^2(1+\kappa^2)\ \mathrm{tr}\ \boldsymbol{C}^{-1}, \kappa > 1 \\ (\mathrm{ii})\eta^2\ \mathrm{tr}(\boldsymbol{C}^{-1}) + \boldsymbol{\delta}^\top\boldsymbol{\delta} \quad \boldsymbol{\delta} \in \mathbb{R}^p \\ (\mathrm{iii})\eta^2\rho_{\mathrm{HT}}(\kappa, 0)\ \mathrm{tr}(\boldsymbol{C}^{-1}) + 1.2\boldsymbol{\delta}^\top\boldsymbol{\delta}, \quad 0 < \boldsymbol{\delta} < \kappa\mathbf{1}_p^\top, \end{cases} \tag{10.45}$$

where

$$\rho_{\mathrm{HT}}(\kappa, 0) = 2[1 - \Phi(\kappa) + \lambda\phi(\lambda)]. \tag{10.46}$$

If we have a sparse solution with p_1 coefficients $|\delta_j| > \kappa\eta\sqrt{C^{jj}}$ $(j = 1, \ldots, p)$ and p_2 zero coefficients, then

$$\hat{\boldsymbol{\beta}}_n^{\mathrm{HT}}(\kappa) = \begin{pmatrix} \tilde{\boldsymbol{\beta}}_n^{\mathrm{R}} \\ \mathbf{0} \end{pmatrix}. \tag{10.47}$$

Then, the ADUB of the weighted L_2-risk is given by

$$\mathrm{ADR}(\hat{\boldsymbol{\beta}}_n^{\mathrm{HT}}(\kappa); \boldsymbol{C}_{11\cdot 2}, \boldsymbol{C}_{22\cdot 1}) \leq \eta^2(p_1 + \Delta^2), \quad \delta^2 = \frac{1}{\eta^2}\boldsymbol{\delta}_2^\top\boldsymbol{C}_{22\cdot 1}\boldsymbol{\delta}_2 \tag{10.48}$$

is independent of the threshold parameter, κ.

10.3.2 Rank-based LASSO

The ADB and ADR of $\hat{\boldsymbol{\beta}}_n^{\mathrm{MRL}}(\lambda)$ are given by

$$\begin{aligned} &\mathrm{ADB}(\hat{\boldsymbol{\beta}}_n^{\mathrm{MRL}}(\lambda)) \\ &\quad = \lim_{n\to\infty} \mathbb{E}\left[\sqrt{n}\left(\hat{\boldsymbol{\beta}}_n^{\mathrm{MRL}}(\lambda) - n^{-\frac{1}{2}}\boldsymbol{\delta}\right)\right] \\ &\quad = (\eta\sqrt{C^{jj}}[(2\Phi(\Delta_j) - 1)(1 - H_3(\lambda^2; \Delta_j^2)) + \Delta_j H_3(\lambda^2; \Delta_j^2)]|j = 1, \ldots, p)^\top \end{aligned} \tag{10.49}$$

and

$$\mathrm{ADR}(\hat{\boldsymbol{\beta}}_n^{\mathrm{MRL}}(\kappa)) = \eta^2 \sum_{j=1}^{p} C^{jj}\rho_{\mathrm{ST}}(\lambda, \Delta_j), \tag{10.50}$$

where

$$\begin{aligned} \rho_{\mathrm{ST}}(\lambda, \Delta_j^2) = {} & 1 + \lambda^2 + (\Delta_j^2 - \lambda^2 - 1)\{\Phi(\lambda - \Delta_j) - \Phi(-\lambda - \Delta_j)\} \\ & -\{(\lambda - \Delta_j)\phi(\lambda + \Delta_j)\phi(\lambda - \Delta_j)\} \end{aligned} \tag{10.51}$$

Hence, the following Lemma 10.2 gives the ADUB of L_2-risk of $\hat{\boldsymbol{\beta}}_n^{\mathrm{MRL}}(\lambda)$.

Lemma 10.2 *Under the assumed regularity conditions, the ADUB of the* L_2 *risk is given by*

$$\text{ADR}(\hat{\boldsymbol{\beta}}_n^{\text{MRL}}(\lambda)) \leq \begin{cases} \eta^2(1+\lambda^2)\,\text{tr}(\boldsymbol{C}^{-1}), & \lambda > 1 \\ \eta^2\,\text{tr}(\boldsymbol{C}^{-1}) + \boldsymbol{\delta}^\top\boldsymbol{\delta}, & \boldsymbol{\delta} \in \mathbb{R}^p \\ \eta^2\rho_{\text{ST}}(\lambda,0)\,\text{tr}(\boldsymbol{C}^{-1}) + 1.2\boldsymbol{\delta}^\top\boldsymbol{\delta}, & 0 < \boldsymbol{\delta} < \lambda\boldsymbol{1}_p^\top, \end{cases} \tag{10.52}$$

where $\rho_{\text{ST}}(\lambda,0) = (1+\lambda^2)(2\Phi(\lambda)-1) - 2\lambda\phi(\lambda)$.

As for the sparse solution, the weighted L_2-risk upper bound are given by

$$\text{ADR}(\hat{\boldsymbol{\beta}}_n^{\text{MRL}}(\lambda); \boldsymbol{C}_{11\cdot2}; \boldsymbol{C}_{22\cdot1}) \leq \eta^2(p_1 + \Delta^2) \tag{10.53}$$

independent of λ.

Next, we consider "asymptotic oracle for orthogonal linear projection" (AOOLP) in the following section.

10.3.3 Multivariate Normal Decision Theory and Oracles for Diagonal Linear Projection

Consider the following problem in multivariate normal decision theory. We are given the least-squares estimator (LSE) of $\boldsymbol{\beta}$, namely, $\tilde{\boldsymbol{\beta}}_n = (\tilde{\beta}_{1n}, \ldots, \tilde{\beta}_{pn})^\top$ according to

$$\sqrt{n}\tilde{\beta}_{jn} \stackrel{\mathcal{D}}{=} \delta_j + \eta\sqrt{C^{jj}}Z_j, \quad Z_j \sim \mathcal{N}(0,1), \tag{10.54}$$

where $\eta^2 C^{jj}$ is the marginal variance of $\tilde{\beta}_{jn}$, $j = 1, \ldots, p$, and noise level and $\{\beta_j\}_{j=1}^p$ are the object of interest. We measure the quality of the estimator based on L_2-loss and define the risk as

$$R(\hat{\boldsymbol{\beta}}_n^*) = \mathbb{E}[n\|\boldsymbol{\beta}_n^* - \boldsymbol{\beta}\|]^2. \tag{10.55}$$

If there is a sparse solution, then use the (10.18) formulation. We consider a family of diagonal linear projections,

$$\boldsymbol{T}_{\text{DP}}(\hat{\boldsymbol{\beta}}_n^{\text{MRL}}, \boldsymbol{\delta}) = (\lambda_1\hat{\boldsymbol{\beta}}_{1n}^{\text{MRL}}, \ldots, \lambda_2\hat{\boldsymbol{\beta}}_{pn}^{\text{MRL}})^\top, \quad \delta_j \in \{0,1\}. \tag{10.56}$$

Such estimators "keep" or "kill" the coordinate. The ideal diagonal coefficients, λ_j are in this case $I(|\delta_j| > \eta\sqrt{C^{jj}})$. These coefficients estimate those β_j's which are larger than the noise level $\eta\sqrt{C^{jj}}$, yielding the asymptotic lower bound on the risk as

$$R(\boldsymbol{T}_{\text{DP}}) = \sum_{j=1}^{p} \min(\delta_j^2, \eta^2 C^{jj}). \tag{10.57}$$

As a special case of (10.57), we obtain

$$R(\boldsymbol{T}_{\text{DP}}) = \begin{cases} \eta^2 \operatorname{tr}(\boldsymbol{C}^{-1}) & \text{if all } |\sqrt{n}\beta_j| \geq \eta\sqrt{C^{jj}}, j = 1, \ldots, p \\ \boldsymbol{\delta}^{\top}\boldsymbol{\delta} & \text{if all } |\sqrt{n}\beta_j| < \eta\sqrt{C^{jj}}, j = 1, \ldots, p. \end{cases} \tag{10.58}$$

In general, the risk $R(\boldsymbol{T}_{\text{DP}})$ cannot be attained for all $\boldsymbol{\beta}$ by any estimator, linear or nonlinear. However, for the sparse case, if p_1 is the number of nonzero coefficients, $|\sqrt{n}\beta_j| > \eta\sqrt{C^{jj}}$; $(j = 1, \ldots, p_1)$ and p_2 is the number of zero coefficients, then (10.58) reduces to the lower bound given by

$$R(\boldsymbol{T}_{\text{DP}}) = \eta^2 \operatorname{tr}(\boldsymbol{C}_{11\cdot 2}^{-1}) + \boldsymbol{\delta}_2^{\top}\boldsymbol{\delta}_2. \tag{10.59}$$

Consequently, the weighted L_2-risk lower bound is given by

$$R(\boldsymbol{T}_{\text{DP}}; \boldsymbol{C}_{11\cdot 2}, \boldsymbol{C}_{22\cdot 1}) = \eta^2(p_1 + \Delta^2), \quad \Delta^2 = \frac{1}{\eta^2}\boldsymbol{\delta}^2\boldsymbol{C}_{22\cdot 1}\boldsymbol{\delta}_2. \tag{10.60}$$

10.4 Comparison of Estimators

In this section, we compare various estimators with respect to the unrestricted R-estimator (URE), in terms of relative weighted L_2-risk efficiency (RWRE).

10.4.1 Comparison of RE with Restricted RE

In this case, the RWRE of restricted R-estimator versus RE is given by

$$\begin{aligned} \text{RWRE}(\hat{\boldsymbol{\beta}}_n^{\text{R}} : \tilde{\boldsymbol{\beta}}_n^{\text{R}}) &= \frac{p_1 + p_2}{\operatorname{tr}(\boldsymbol{C}_{11}^{-1}\boldsymbol{C}_{11\cdot 2}) + \Delta^2} \\ &= \left(1 + \frac{p_2}{p_1}\right)\left(1 - \frac{\operatorname{tr}(\boldsymbol{M}_0)}{p_1} + \frac{\Delta^2}{p_1}\right)^{-1}; \end{aligned} \tag{10.61}$$

where $\boldsymbol{M}_0 = \boldsymbol{C}_{11}^{-1}\boldsymbol{C}_{12}\boldsymbol{C}_{22}^{-1}\boldsymbol{C}_{21}$. The $\text{RWRE}(\hat{\boldsymbol{\beta}}_n^{\text{R}} : \tilde{\boldsymbol{\beta}}_n^{\text{R}})$ is a decreasing function of Δ^2. So,

$$0 \leq \text{RWRE}(\hat{\boldsymbol{\beta}}_n^{\text{R}} : \tilde{\boldsymbol{\beta}}_n^{\text{R}}) \leq \left(1 + \frac{p_2}{p_1}\right)\left(1 - \frac{\operatorname{tr} \boldsymbol{M}_0}{p_1}\right)^{-1}.$$

In order to compute $\operatorname{tr}(\boldsymbol{M}_0)$, we need to find $\boldsymbol{C}_{11}$, $\boldsymbol{C}_{22}$, and $\boldsymbol{C}_{12}$. These are obtained by generating explanatory variables using the following equation following McDonald and Galarneau (1975),

$$x_{ij} = \sqrt{1 - \rho^2} z_{ij} + \rho z_{ip}, \quad i = 1, \ldots, n; j = 1, \ldots, p, \tag{10.62}$$

where z_{ij} are independent $\mathcal{N}(0, 1)$ pseudorandom numbers and ρ^2 is the correlation between any two explanatory variables. In this study, we take $\rho^2 = 0.1, 0.2, 0.8$, and 0.9, which shows the variables are lightly collinear and severely collinear. In our case, we chose $n = 100$ and various (p_1, p_2). The resulting output is then used to compute $\operatorname{tr}(\boldsymbol{M}_0)$.

10.4.2 Comparison of RE with PTRE

Here, the RWRE expression for PTRE vs. RE is given by

$$\mathrm{RWRE}(\hat{\boldsymbol{\beta}}_n^{\mathrm{R(PT)}}(\alpha) : \tilde{\boldsymbol{\beta}}_n^{\mathrm{R}}) = \frac{p_1 + p_2}{g(\Delta^2, \alpha)}, \tag{10.63}$$

where

$$g(\Delta^2, \alpha) = p_1 + p_2(1 - H_{p_2+2}(c_\alpha; \Delta^2)) + \Delta^2[2H_{p_2+2}(c_\alpha; \Delta^2) - H_{p_2+4}(c_\alpha; \Delta^2)].$$

Then, the PTRE outperforms the RE for

$$0 \leq \Delta^2 \leq \frac{p_2 H_{p_2+2}(c_\alpha; \Delta^2)}{2H_{p_2+2}(c_\alpha; \Delta^2) - H_{p_2+4}(c_\alpha; \Delta^2)} = \Delta^2_{\mathrm{R(PT)}}. \tag{10.64}$$

Otherwise, RE outperforms the PTRE in the interval $(\Delta^2_{\mathrm{R(PT)}}, \infty)$. We may mention that $\mathrm{RWRE}(\hat{\boldsymbol{\beta}}_n^{\mathrm{R(PT)}}(\alpha) : \tilde{\boldsymbol{\beta}}_n^{\mathrm{R}})$ is a decreasing function of Δ^2 with a maximum at $\Delta^2 = 0$, then it decreases crossing the 1-line to a minimum at $\Delta^2 = \Delta^2_{\mathrm{R(PT)}}(\min)$ with a value $M_{\mathrm{R(PT)}}(\alpha)$, and then increases toward 1-line.

The RWRE expression for PTRE vs. RE belongs to the interval

$$M_{\mathrm{R(PT)}}(\alpha) \leq \mathrm{RWRE}(\hat{\boldsymbol{\beta}}_n^{\mathrm{R(PT)}}; \tilde{\boldsymbol{\beta}}_n^{\mathrm{R}}) \leq \left(1 + \frac{p_2}{p_1}\right)\left(1 + \frac{p_2}{p_1}[1 - H_{p_2+2}(c_\alpha; 0)]\right)^{-1},$$

where $M_{\mathrm{R(PT)}}(\alpha)$ depends on the size α and given by

$$\begin{aligned} M_{\mathrm{R(PT)}}(\alpha) = \left(1 + \frac{p_2}{p_1}\right)\bigg\{ & 1 + \frac{p_2}{p_1}[1 - H_{p_2+2}(c_\alpha; \Delta^2_{\mathrm{R(PT)}}(\min))] \\ & + \frac{\Delta^2_{\mathrm{R(PT)}}(\min)}{p_1}[2H_{p_2+2}(c_\alpha; \Delta^2_{\mathrm{R(PT)}}(\min)) - H_{p_2+4}(c_\alpha; \Delta^2_{\mathrm{R(PT)}}(\min))]\bigg\}^{-1}. \end{aligned}$$

The quantity $\Delta^2_{\mathrm{R(PT)}}(\min)$ is the value Δ^2 at which the RWRE value is minimum.

10.4.3 Comparison of RE with SRE and PRSRE

To express the RWRE of SRE and PRSRE, we assume always that $p_2 \geq 3$. We have then

$$\mathrm{RWRE}(\hat{\boldsymbol{\beta}}_n^{\mathrm{R(S)}}; \tilde{\boldsymbol{\beta}}_n) = \left(1 + \frac{p_2}{p_1}\right)\left(1 + \frac{p_2}{p_1} - \frac{(p_2-2)^2}{p_1}\mathbb{E}[\chi_{p_2}^{-2}(\Delta^2)]\right)^{-1}. \tag{10.65}$$

It is a decreasing function of Δ^2. At $\Delta^2 = 0$, its value is $\left(1 + \frac{p_2}{p_1}\right)\left(1 + \frac{2}{p_1}\right)^{-1}$ and when $\Delta^2 \to \infty$, its value goes to 1. Hence, for $\Delta^2 \in \mathbb{R}^+$,

$$1 \leq \left(1 + \frac{p_2}{p_1}\right)\left(1 + \frac{p_2}{p_1} - \frac{(p_2-2)^2}{p_1}\mathbb{E}[\chi_{p_2}^{-2}(\Delta^2)]\right)^{-1} \leq \left(1 + \frac{p_2}{p_1}\right)\left(1 + \frac{2}{p_1}\right)^{-1}.$$

Also,

$$\begin{aligned}\text{RWRE}(\hat{\boldsymbol{\beta}}_n^{\text{R(S+)}};\tilde{\boldsymbol{\beta}}_n) = \left(1+\frac{p_2}{p_1}\right)\Bigg(1+\frac{p_2}{p_1} - \frac{(p_2-2)^2}{p_1}\mathbb{E}[\chi_{p_2}^{-2}(\Delta^2)]\\ -\frac{p_2}{p_1}\mathbb{E}[(1-(p_2-2)\chi_{p_2+2}^{-2}(\Delta^2))^2 I(\chi_{p_2+2}^2(\Delta^2) < (p_2-2))]\\ +\frac{\Delta^2}{p_1}\{2\mathbb{E}[(1-(p_2-2)\chi_{p_2+2}^{-2}(\Delta^2))I(\chi_{p_2+2}^2(\Delta^2) < (p_2-2))]\\ -\mathbb{E}[(1-(p_2-2)\chi_{p_2+4}^{-2}(\Delta^2))^2 I(\chi_{p_2+4}^2(\Delta^2) < (p_2-2))]\}\Bigg)^{-1}.\end{aligned} \tag{10.66}$$

So that,

$$\text{RWRE}(\hat{\boldsymbol{\beta}}_n^{\text{R(S+)}};\tilde{\boldsymbol{\beta}}_n) \geq \text{RWRE}(\hat{\boldsymbol{\beta}}_n^{\text{R(S)}};\tilde{\boldsymbol{\beta}}_n) \geq 1 \qquad \forall \Delta^2 \in \mathbb{R}^+.$$

We also provide a graphical representation (Figure 10.1) of RWRE of the estimators for $p_1 = 3$ and $p_2 = 7$.

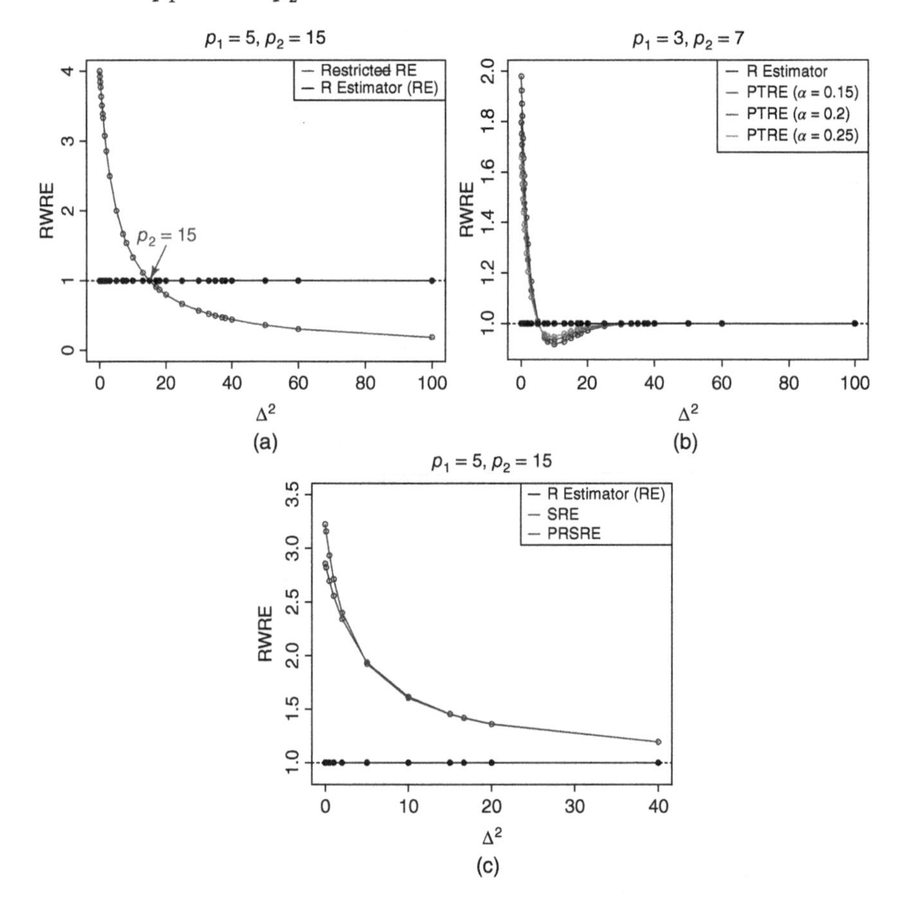

Figure 10.1 RWRE for the restricted, preliminary test, Stein-type, and its positive-rule R-estimators.

10.4.4 Comparison of RE and Restricted RE with RRRE

First, we consider the weighted L_2-risk difference of RE and RRRE given by

$$\sigma^2(p_1+p_2)-\sigma^2 p_1-\sigma^2\frac{p_2\Delta^2}{p_2+\Delta^2}=\sigma^2 p_2\left(1-\frac{\Delta^2}{p_2+\Delta^2}\right)$$
$$=\frac{\sigma^2 p_2^2}{p_2+\Delta^2}>0,\ \forall\ \Delta^2\in\mathbb{R}^+. \quad (10.67)$$

Hence, RRRE outperforms the RE uniformly. Similarly, for the restricted RE and RRRE, the weighted L_2-risk difference is given by

$$\sigma^2(\text{tr}\boldsymbol{C}_{11}^{-1}\boldsymbol{C}_{11\cdot 2}+\Delta^2)-\left(\sigma^2 p_1+\frac{\sigma^2 p_2\Delta^2}{p_2+\Delta^2}\right)$$
$$=\sigma^2\left\{[\text{tr}\boldsymbol{C}_{11}^{-1}\boldsymbol{C}_{11\cdot 2}-p_1]+\frac{\Delta^4}{p_2+\Delta^2}\right\}$$
$$=\sigma^2\left(\frac{\Delta^4}{p_2+\Delta^2}-\text{tr}(\boldsymbol{M}_0)\right). \quad (10.68)$$

If $\Delta^2=0$, then (10.68) is negative. The restricted RE outperforms RRRE at this point. Solving the equation

$$\frac{\Delta^4}{p_2+\Delta^2}=\text{tr}(\boldsymbol{M}_0), \quad (10.69)$$

we get

$$\Delta_0^2=\frac{1}{2}\,\text{tr}(\boldsymbol{M}_0)\left\{1+\sqrt{1+\frac{4p_2}{\text{tr}(\boldsymbol{M}_0)}}\right\}. \quad (10.70)$$

If $0\le\Delta^2\le\Delta_0^2$, then the restricted RE outperforms the RRRE, and if $\Delta^2\in(\Delta_0^2,\infty)$, RRRE performs better than the restricted RE. Thus, neither restricted RE nor RRRE outperforms the other uniformly.

In addition, the RWRE of RRRE versus RE equals

$$\text{RWRE}(\hat{\boldsymbol{\beta}}_n^{\text{RRR}}(k_{\text{opt}}):\tilde{\boldsymbol{\beta}}_n^{\text{R}})=\frac{p_1+p_2}{p_1+\frac{p_2\Delta^2}{p_2+\Delta^2}}=\left(1+\frac{p_2}{p_1}\right)\left(1+\frac{p_2\Delta^2}{p_1(p_2+\Delta^2)}\right)^{-1}, \quad (10.71)$$

which is a decreasing function of Δ^2 with maximum $\left(1+\frac{p_2}{p_1}\right)$ at $\Delta^2=0$ and minimum 1 as $\Delta^2\to\infty$. So,

$$1\le\left(1+\frac{p_2}{p_1}\right)\left(1+\frac{p_2}{p_1\left(1+\frac{p_2}{\Delta^2}\right)}\right)^{-1}\le 1+\frac{p_2}{p_1};\quad \forall\Delta^2\in\mathbb{R}^+.$$

10.4.5 Comparison of RRRE with PTRE, SRE, and PRSRE

Here, the weighted L_2-risk difference of PTRE and RRRE is given by

$$\sigma^2[p_2 - p_2 1 - H_{p_2+2}(c_\alpha;\Delta^2) + \Delta^2\{2H_{p_2+2}(c_\alpha;\Delta^2) - H_{p_2+4}(c_\alpha;\Delta^2)\}] - \frac{\sigma^2 p_2 \Delta^2}{p_2+\Delta^2}$$
$$= \sigma^2\left[\frac{p_2}{p_2+\Delta^2} - \{p_2 H_{p_2+2}(c_\alpha;\Delta^2) - \Delta^2(2H_{p_2+2}(c_\alpha;\Delta^2) - H_{p_2+4}(c_\alpha;\Delta^2))\}\right] \geq 0. \tag{10.72}$$

Since the first term is a decreasing function of Δ^2 with a maximum value p_2 at $\Delta^2 = 0$ and tends to 0 as $\Delta^2 \to \infty$. The second function in brackets is also decreasing in Δ^2 with maximum $p_2 H_{p_2+2}(c_\alpha;0)$ at $\Delta^2 = 0$ which is less than p_2, and the function tends to 0 as $\Delta^2 \to \infty$. Hence, (10.72) is nonnegative for $\Delta^2 \in \mathbb{R}^+$. Hence, the RRRE uniformly performs better than the PTRE.

Similarly, we show the RRE uniformly performs better than the SRE, i.e. the weighted L_2 risk of $\hat{\boldsymbol{\beta}}_n^{\text{RRR}}(k_{\text{opt}})$ and $\hat{\boldsymbol{\beta}}_n^{\text{R(S)}}$ is given by

$$R(\hat{\boldsymbol{\beta}}_n^{\text{RRR}}(k_{\text{opt}}) : \mathbf{C}_{11\cdot2}, \mathbf{C}_{22\cdot1}) \leq R(\hat{\boldsymbol{\beta}}_n^{\text{R(S)}} : \mathbf{C}_{11\cdot2}, \mathbf{C}_{22\cdot1}), \quad \forall\, \Delta^2 \in \mathbb{R}^+. \tag{10.73}$$

The weighted L_2-risk difference of SRE and RRRE is given by

$$\sigma^2\left[\begin{array}{l} \frac{p_2^2}{p_2+\Delta^2} - p_2\{\mathbb{E}[\chi^{-2}_{p_2+2}(\Delta^2)] + \Delta^2\mathbb{E}[\chi^{-4}_{p_2+2}(\Delta^2)]\} \\ \qquad - (p_2^2-4)\mathbb{E}[\chi^{-4}_{p_2+4}(\Delta^2)] \end{array}\right] \geq 0 \quad \forall \Delta^2 \in \mathbb{R}^+.$$

Since the first function decreases with a maximum value p_2 at $\Delta^2 = 0$, the second function decreases with a maximum value $1(\leq p_2)$ and tends to 0 as $\Delta^2 \to \infty$. Hence, the two functions are one below the other and the difference is nonnegative for $\Delta^2 \in \mathbb{R}^+$.

Next, we show that the weighted L_2 risk (WL_2R) of the two estimators may be ordered as

$$R(\hat{\boldsymbol{\beta}}_n^{\text{RRR}}(k_{\text{opt}}) : \mathbf{C}_{11\cdot2}, \mathbf{C}_{22\cdot1}) \leq R(\hat{\boldsymbol{\beta}}_n^{\text{R(S+)}} : \mathbf{C}_{11\cdot2}, \mathbf{C}_{22\cdot1}), \quad \forall\, \Delta^2 \in \mathbb{R}^+.$$

Note that

$$R(\hat{\boldsymbol{\beta}}_n^{\text{R(S+)}}; \mathbf{C}_{11\cdot2}, \mathbf{C}_{22\cdot1}) = R(\hat{\boldsymbol{\beta}}_n^{\text{R(S)}}; \mathbf{C}_{11\cdot2}, \mathbf{C}_{22\cdot1}) - \text{R}^*, \tag{10.74}$$

where R^* is defined by Eq. (9.70).

Thus, we find that the WL_2R difference is given by

$$\begin{aligned} &R(\hat{\boldsymbol{\beta}}_n^{\text{R(S+)}} : \mathbf{C}_{11\cdot2}, \mathbf{C}_{22\cdot1}) - R(\hat{\boldsymbol{\beta}}_n^{\text{RRR}}(k_{\text{opt}}) : \mathbf{C}_{11\cdot2}, \mathbf{C}_{22\cdot1}) \\ &= \{R(\hat{\boldsymbol{\beta}}_n^{\text{R(S)}} : \mathbf{C}_{11\cdot2}, \mathbf{C}_{22\cdot1}) - R(\hat{\boldsymbol{\beta}}_n^{\text{RRR}}(k_{\text{opt}}) : \mathbf{C}_{11\cdot2}, \mathbf{C}_{22\cdot1})\} - \text{R}^* \\ &= \{\cdot\} \text{ is negative} - \text{R}^*(\text{non-negative}) \leq 0. \end{aligned} \tag{10.75}$$

Hence, the RRE uniformly performs better than the PRSRE.

10.4.6 Comparison of RLASSO with RE and Restricted RE

First, note that if p_1 coefficients $|\delta_j| > \eta\sqrt{C^{jj}}$ and p_2 coefficients are zero in a sparse solution, the lower bound of the weighted L_2 risk is given by $\sigma^2(p_1 + \Delta^2)$. Thereby, we compare all estimators relative to this quantity. Hence, the WL_2R difference between RE and RLASSO is given by

$$\sigma^2(p_1 + p_2) - \sigma^2(p_1 + \Delta^2 - \mathrm{tr}(\boldsymbol{M}_0)) = \sigma^2[(p_2 + \mathrm{tr}(\boldsymbol{M}_0)) - \Delta^2]. \tag{10.76}$$

Hence, if $\Delta^2 \in (0, p_2 + \mathrm{tr}(\boldsymbol{M}_0))$, the RLASSO performs better than the RE; while if $\Delta^2 \in (p_2 + \mathrm{tr}(\boldsymbol{M}_0), \infty)$, the RE performs better than the RLASSO. Consequently, neither the RE nor the RLASSO performs better than the other uniformly.

Next, we compare the restricted RE and RLASSO. In this case, the WL_2R difference is given by

$$\sigma^2(p_1 + \Delta^2 - \mathrm{tr}(\boldsymbol{M}_0)) - \sigma^2(p_1 + \Delta^2) = -\sigma^2(\mathrm{tr}(\boldsymbol{M}_0)) < 0. \tag{10.77}$$

Hence, the RRE uniformly performs better than the RLASSO. If $\mathrm{tr}(\boldsymbol{M}_0) = 0$, RLASSO and RRE are L_2-risk equivalent. If the RE estimators are independent, then $\mathrm{tr}(\boldsymbol{M}_0) = 0$. Hence, RLASSO satisfies the oracle properties.

10.4.7 Comparison of RLASSO with PTRE, SRE, and PRSRE

We first consider the PTRE versus RLASSO. In this case, the WL_2R difference is given by

$$\begin{aligned}
&R(\hat{\boldsymbol{\beta}}_n^{\mathrm{R(PT)}}(\alpha); \boldsymbol{C}_{11\cdot 2}, \boldsymbol{C}_{22\cdot 1}) - R(\hat{\boldsymbol{\beta}}_n^{\mathrm{MRL}}(\lambda); \boldsymbol{C}_{11\cdot 2}, \boldsymbol{C}_{22\cdot 1}) \\
&\quad = \sigma^2[p_2(1 - H_{p_2+2}(c_\alpha; \Delta^2)) - \Delta^2\{1 - 2H_{p_2+2}(c_\alpha; \Delta^2) + H_{p_2+4}(c_\alpha; \Delta^2)\}]. \\
&\quad \geq \sigma^2 p_2(1 - H_{p_2+2}(c_\alpha; 0)) \geq 0, \quad \text{if } \Delta^2 = 0.
\end{aligned} \tag{10.78}$$

Hence, the RLASSO outperforms the PTRE when $\Delta^2 = 0$. But, when $\Delta^2 \neq 0$, then the RLASSO outperforms the PTRE for

$$0 \leq \Delta^2 \leq \frac{p_2[1 - H_{p_2+2}(c_\alpha; \Delta^2)]}{1 - 2H_{p_2+2}(c_\alpha; \Delta^2) + H_{p_2+4}(c_\alpha; \Delta^2)}. \tag{10.79}$$

Otherwise, PTRE outperforms the modified RLASSO estimator. Hence, neither PTRE nor the modified RLASSO estimator outperforms the other uniformly.

Next, we consider SRE and PRSRE versus the RLASSO. In these two cases, we have the WL_2R differences given by

$$\begin{aligned}
&R(\hat{\boldsymbol{\beta}}_n^{\mathrm{R(S)}}; \boldsymbol{C}_{11\cdot 2}, \boldsymbol{C}_{22\cdot 1}) - R(\hat{\boldsymbol{\beta}}_n^{\mathrm{MRL}}(\lambda); \boldsymbol{C}_{11\cdot 2}, \boldsymbol{C}_{22\cdot 1}) \\
&\quad = \sigma^2[p_1 + p_2 - (p_2 - 2)^2\mathbb{E}[\chi_{p_2+2}^{-2}(\Delta^2)] - (p_1 + \Delta^2)] \\
&\quad = \sigma^2[p_2 - (p_2 - 2)^2\mathbb{E}[\chi_{p_2+2}^{-2}(\Delta^2)] - \Delta^2]
\end{aligned} \tag{10.80}$$

and from (10.74),

$$R(\hat{\beta}_n^{\mathrm{R(S+)}}; C_{11\cdot2}, C_{22\cdot1}) - R(\hat{\beta}_n^{\mathrm{MRL}}(\lambda); C_{11\cdot2}, C_{22\cdot1})$$
$$= R(\hat{\beta}_n^{\mathrm{R(S)}}; C_{11\cdot2}, C_{22\cdot1}) - R(\hat{\beta}_n^{\mathrm{MRL}}(\lambda); C_{11\cdot2}, C_{22\cdot1}) - \mathrm{R}^*, \tag{10.81}$$

where R^* is given by (9.70). Hence, the modified RLASSO estimator outperforms the SRE as well as the PRSRE in the interval

$$0 \le \Delta^2 \le p_2 - (p_2-2)^2\mathbb{E}[\chi_{p_2}^{-2}(\Delta^2)]. \tag{10.82}$$

Thus, neither SRE nor the PRSRE outperforms the modified RLASSO estimator uniformly.

10.4.8 Comparison of Modified RLASSO with RRRE

Here, the weighted L_2-risk difference is given by

$$R(\hat{\beta}_n^{\mathrm{MRL}}(\lambda); C_{11\cdot2}, C_{22\cdot1}) - R(\hat{\beta}_n^{\mathrm{RRR}}(k_{\mathrm{opt}}); C_{11\cdot2}, C_{22\cdot1})$$
$$= \sigma^2\left[(p_1+\Delta^2) - \left(p_1 + \frac{p_2\Delta^2}{p_2+\Delta^2}\right)\right] = \frac{\sigma^2\Delta^2}{p_2+\Delta^2} \ge 0. \tag{10.83}$$

Hence, the RRRE outperforms the modified RLASSO estimator, uniformly.

10.5 Summary and Concluding Remarks

In this section, we discuss the contents of Tables 10.1–10.10 presented as confirmatory evidence of the theoretical findings of the estimators.

First, we note that we have two classes of estimators, namely, the traditional rank-based PTEs and Stein-type estimators and the penalty estimators. The restricted R-estimators play an important role due to the fact that LASSO belongs to the class of restricted estimators. We have the following conclusion from our study.

(i) Since the inception of the RRE by Hoerl and Kennard (1970), there have been articles comparing RRE with PTE and Stein-type estimators. We have now definitive conclusion that the RRE dominates the RE, PTRE, and Stein-type estimators uniformly. See Tables 10.1 and 10.2 and graphs thereof in Figure 10.1. The RRRE ridge estimator dominates the modified RLASSO estimator uniformly for $\Delta^2 > 0$, while they are L_2-risk equivalent at $\Delta^2 = 0$. The RRRE ridge estimator does not select variables, but the modified RLASSO estimator does.

(ii) The restricted R- and modified RLASSO estimators are competitive, although the modified RLASSO estimator lags behind the restricted R-estimator, uniformly. Both estimators outperform the URE, PTRE, SRE, and PRSRE in a subinterval of $[0, p_2]$ (see Tables 10.1 and 10.2).

Table 10.1 RWRE for the estimators for $p_1 = 3$ and $p_2 = 7$.

			RRE					PTRE					
			ρ^2					α					
	Δ^2	RE	0.1	0.2	0.8	0.9	MRLASSO	0.15	0.2	0.25	SRE	PRSRE	RRRE
	0	1.0000	3.7179	3.9415	5.0593	5.1990	3.3333	1.9787	1.7965	1.6565	2.0000	2.3149	3.3333
	0.1	1.0000	3.5845	3.7919	4.8155	4.9419	3.2258	1.9229	1.7512	1.6194	1.9721	2.2553	3.2273
	0.5	1.0000	3.1347	3.2920	4.0374	4.1259	2.8571	1.7335	1.5970	1.4923	1.8733	2.0602	2.8846
	1	1.0000	2.7097	2.8265	3.3591	3.4201	2.5000	1.5541	1.4499	1.3703	1.7725	1.8843	2.5806
$\Delta_0^2(\rho^2 = 0.1)$	1.62	1.0000	2.3218	2.4070	2.7829	2.8246	2.1662	1.3920	1.3164	1.2590	1.6739	1.7315	2.3185
	2	1.0000	2.1318	2.2034	2.5143	2.5483	2.0000	1.3141	1.2520	1.2052	1.6231	1.6597	2.1951
$\Delta_0^2(\rho^2 = 0.2)$	2.03	1.0000	2.1181	2.1887	2.4952	2.5287	1.9879	1.3085	1.2474	1.2014	1.6194	1.6545	2.1863
	3	1.0000	1.7571	1.8054	2.0091	2.0308	1.6667	1.1664	1.1302	1.1035	1.5184	1.5245	1.9608
$\Delta_0^2(\rho^2 = 0.8)$	3.23	1.0000	1.6879	1.7324	1.9191	1.9388	1.6042	1.1404	1.1088	1.0857	1.4983	1.5006	1.9187
$\Delta_0^2(\rho^2 = 0.9)$	3.33	1.0000	1.6601	1.7031	1.8832	1.9023	1.5791	1.1302	1.1004	1.0787	1.4903	1.4911	1.9020
	5	1.0000	1.3002	1.3264	1.4332	1.4442	1.2500	1.0088	1.0018	0.9978	1.3829	1.3729	1.6901
p_2	7	1.0000	1.0318	1.0483	1.1139	1.1205	1.0000	0.9419	0.9500	0.9571	1.3005	1.2910	1.5385
$p_2 + \text{tr}(\boldsymbol{M}_0)(\rho^2 = 0.1)$	7.31	1.0000	1.0001	1.0155	1.0770	1.0832	0.9701	0.9362	0.9458	0.9541	1.2907	1.2816	1.5208
$p_2 + \text{tr}(\boldsymbol{M}_0)(\rho^2 = 0.2)$	7.46	1.0000	0.9851	1.0001	1.0596	1.0656	0.9560	0.9337	0.9440	0.9528	1.2860	1.2771	1.5126
$p_2 + \text{tr}(\boldsymbol{M}_0)(\rho^2 = 0.8)$	8.02	1.0000	0.9334	0.9468	1.0000	1.0054	0.9073	0.9262	0.9389	0.9493	1.2700	1.2620	1.4841
$p_2 + \text{tr}(\boldsymbol{M}_0)(\rho^2 = 0.9)$	8.07	1.0000	0.9288	0.9421	0.9947	1.0000	0.9029	0.9256	0.9385	0.9490	1.2686	1.2606	1.4816
	10	1.0000	0.7879	0.7975	0.8349	0.8386	0.7692	0.9160	0.9338	0.9473	1.2250	1.2199	1.4050
	13	1.0000	0.6373	0.6435	0.6677	0.6700	0.6250	0.9269	0.9458	0.9591	1.1788	1.1766	1.3245
	15	1.0000	0.5652	0.5701	0.5890	0.5909	0.5556	0.9407	0.9576	0.9690	1.1571	1.1558	1.2865
	20	1.0000	0.4407	0.4437	0.4550	0.4561	0.4348	0.9722	0.9818	0.9877	1.1201	1.1199	1.2217
	30	1.0000	0.3059	0.3073	0.3127	0.3132	0.3030	0.9967	0.9981	0.9989	1.0814	1.0814	1.1526
	50	1.0000	0.1898	0.1903	0.1924	0.1926	0.1887	1.0000	1.0000	1.0000	1.0494	1.0494	1.0940
	100	1.0000	0.0974	0.0975	0.0981	0.0981	0.0971	1.0000	1.0000	1.0000	1.0145	1.0145	1.0480

Table 10.2 RWRE for the estimators for $p_1 = 7$ and $p_2 = 13$.

			RRE					PTRE					
			ρ^2					α					
	Δ^2	RE	0.1	0.2	0.8	0.9	MRLASSO	0.15	0.2	0.25	SRE	PRSRE	RRRE
	0	1.0000	2.9890	3.0586	3.3459	3.3759	2.8571	1.9458	1.7926	1.6694	2.2222	2.4326	2.8571
	0.1	1.0000	2.9450	3.0125	3.2909	3.3199	2.8169	1.9146	1.7658	1.6464	2.2017	2.3958	2.8172
	0.5	1.0000	2.7811	2.8413	3.0876	3.1132	2.6667	1.8009	1.6683	1.5627	2.1255	2.2661	2.6733
	1	1.0000	2.6003	2.6528	2.8664	2.8883	2.5000	1.6797	1.5648	1.4739	2.0423	2.1352	2.5225
	2	1.0000	2.3011	2.3421	2.5070	2.5238	2.2222	1.4910	1.4041	1.3365	1.9065	1.9434	2.2901
$\Delta_0^2(\rho^2 = 0.1)$	2.14	1.0000	2.2642	2.3039	2.4633	2.4795	2.1878	1.4688	1.3854	1.3205	1.8899	1.9217	2.2628
$\Delta_0^2(\rho^2 = 0.2)$	2.67	1.0000	2.1354	2.1707	2.3116	2.3259	2.0673	1.3937	1.3218	1.2663	1.8323	1.8488	2.1697
	3	1.0000	2.0637	2.0966	2.2278	2.2410	2.0000	1.3535	1.2878	1.2375	1.8005	1.8100	2.1192
$\Delta_0^2(\rho^2 = 0.8)$	4.19	1.0000	1.8378	1.8639	1.9669	1.9772	1.7872	1.2352	1.1885	1.1534	1.7014	1.6962	1.9667
$\Delta_0^2(\rho^2 = 0.9)$	4.31	1.0000	1.8173	1.8427	1.9433	1.9534	1.7677	1.2251	1.1801	1.1463	1.6925	1.6864	1.9533
	5	1.0000	1.7106	1.7332	1.8219	1.8307	1.6667	1.1750	1.1385	1.1114	1.6464	1.6370	1.8848
	7	1.0000	1.4607	1.4771	1.5411	1.5474	1.4286	1.0735	1.0554	1.0425	1.5403	1.5289	1.7316
	10	1.0000	1.1982	1.2092	1.2517	1.2559	1.1765	0.9982	0.9961	0.9952	1.4319	1.4243	1.5808
	15	1.0000	2.0533	2.0685	2.0958	2.0966	1.8182	1.0934	1.0694	1.0529	2.1262	2.1157	2.3105
p_2	13	1.0000	1.0156	1.0236	1.0539	1.0568	1.0000	0.9711	0.9767	0.9811	1.3588	1.3547	1.4815
$p_2 + \mathrm{tr}(\boldsymbol{M}_0)(\rho^2 = 0.1)$	13.31	1.0000	1.0000	1.0077	1.0370	1.0399	0.9848	0.9699	0.9760	0.9807	1.3526	1.3488	1.4732
$p_2 + \mathrm{tr}(\boldsymbol{M}_0)(\rho^2 = 0.2)$	13.46	1.0000	0.9925	1.0000	1.0289	1.0318	0.9775	0.9694	0.9757	0.9805	1.3496	1.3459	1.4692
$p_2 + \mathrm{tr}(\boldsymbol{M}_0)(\rho^2 = 0.8)$	14.02	1.0000	0.9655	0.9727	1.0000	1.0027	0.9514	0.9679	0.9749	0.9801	1.3391	1.3358	1.4550
$p_2 + \mathrm{tr}(\boldsymbol{M}_0)(\rho^2 = 0.9)$	14.07	1.0000	0.9631	0.9702	0.9974	1.0000	0.9490	0.9678	0.9748	0.9801	1.3381	1.3349	1.4537
	15	1.0000	0.9220	0.9285	0.9534	0.9558	0.9091	0.9667	0.9745	0.9802	1.3221	1.3195	1.4322
	20	1.0000	0.7493	0.7536	0.7699	0.7715	0.7407	0.9743	0.9820	0.9870	1.2560	1.2553	1.3442
	30	1.0000	0.5451	0.5473	0.5559	0.5567	0.5405	0.9940	0.9964	0.9977	1.1809	1.1808	1.2446
	50	1.0000	0.3528	0.3537	0.3573	0.3576	0.3509	0.9999	1.0000	1.0000	1.1136	1.1136	1.1549
	100	1.0000	0.1875	0.1877	0.1887	0.1888	0.1869	1.0000	1.0000	1.0000	1.0337	1.0337	1.0808

Table 10.3 RWRE of the R-estimators for $p = 10$ and different Δ^2-values for varying p_1.

Estimators	$p_1 = 2$	$p_1 = 3$	$p_1 = 5$	$p_1 = 7$	$p_1 = 2$	$p_1 = 3$	$p_1 = 5$	$p_1 = 7$
	$\Delta^2 = 0$				$\Delta^2 = 1$			
RE	1.0000	1.0000	1.0000	1.0000	1.0000	1.0000	1.0000	1.0000
RRE ($\rho^2 = 0.1$)	5.6807	3.7105	2.1589	1.4946	3.6213	2.7059	1.7755	1.3003
RRE ($\rho^2 = 0.2$)	6.1242	3.9296	2.2357	1.5291	3.7959	2.8204	1.8271	1.3262
RRE ($\rho^2 = 0.8$)	9.4863	5.0497	2.5478	1.6721	4.8660	3.3549	2.0304	1.4325
RRE ($\rho^2 = 0.9$)	10.1255	5.1921	2.5793	1.6884	5.0292	3.4171	2.0504	1.4445
RLASSO	5.0000	3.3333	2.0000	1.4286	3.3333	2.5000	1.6667	1.2500
PTRE ($\alpha = 0.15$)	2.3441	1.9787	1.5122	1.2292	1.7548	1.5541	1.2714	1.0873
PTRE ($\alpha = 0.2$)	2.0655	1.7965	1.4292	1.1928	1.6044	1.4499	1.2228	1.0698
PTRE ($\alpha = 0.25$)	1.8615	1.6565	1.3616	1.1626	1.4925	1.3703	1.1846	1.0564
SRE	2.5000	2.0000	1.4286	1.1111	2.1364	1.7725	1.3293	1.0781
PRSRE	3.0354	2.3149	1.5625	1.1625	2.3107	1.8843	1.3825	1.1026
RRRE	5.0000	3.3333	2.0000	1.4286	3.4615	2.5806	1.7143	1.2903
	$\Delta^2 = 5$				$\Delta^2 = 10$			
RE	1.0000	1.0000	1.0000	1.0000	1.0000	1.0000	1.0000	1.0000
RRE ($\rho^2 = 0.1$)	1.4787	1.2993	1.0381	0.8554	0.8501	0.7876	0.6834	0.5991
RRE ($\rho^2 = 0.2$)	1.5069	1.3251	1.0555	0.8665	0.8594	0.7970	0.6909	0.6046
RRE ($\rho^2 = 0.8$)	1.6513	1.4324	1.1204	0.9107	0.9045	0.8346	0.7181	0.6257
RRE ($\rho^2 = 0.9$)	1.6698	1.4437	1.1264	0.9155	0.9100	0.8384	0.7206	0.6280
RLASSO	1.4286	1.2500	1.0000	0.8333	0.8333	0.7692	0.6667	0.5882
PTRE ($\alpha = 0.15$)	1.0515	1.0088	0.9465	0.9169	0.9208	0.9160	0.9176	0.9369
PTRE ($\alpha = 0.2$)	1.0357	1.0018	0.9530	0.9323	0.9366	0.9338	0.9374	0.9545
PTRE ($\alpha = 0.25$)	1.0250	0.9978	0.9591	0.9447	0.9488	0.9473	0.9517	0.9665
SRE	1.5516	1.3829	1.1546	1.0263	1.3238	1.2250	1.0865	1.0117
PRSRE	1.5374	1.3729	1.1505	1.0268	1.3165	1.2199	1.0843	1.0114
RRRE	1.9697	1.6901	1.3333	1.1268	1.5517	1.4050	1.2000	1.0744
	$\Delta^2 = 20$				$\Delta^2 = 60$			
RE	1.0000	1.0000	1.0000	1.0000	1.0000	1.0000	1.0000	1.0000
RRE ($\rho^2 = 0.1$)	0.4595	0.4406	0.4060	0.3747	0.1619	0.1595	0.1547	0.1499
RRE ($\rho^2 = 0.2$)	0.4622	0.4435	0.4086	0.3768	0.1622	0.1599	0.1551	0.1503
RRE ($\rho^2 = 0.8$)	0.4749	0.4549	0.4180	0.3849	0.1638	0.1613	0.1564	0.1516
RRE ($\rho^2 = 0.9$)	0.4764	0.4561	0.4188	0.3858	0.1640	0.1615	0.1566	0.1517
RLASSO	0.4545	0.4348	0.4000	0.3704	0.1613	0.1587	0.1538	0.1493
PTRE ($\alpha = 0.15$)	0.9673	0.9722	0.9826	0.9922	1.0000	1.0000	1.0000	1.0000
PTRE ($\alpha = 0.2$)	0.9783	0.9818	0.9890	0.9954	1.0000	1.0000	1.0000	1.0000
PTRE ($\alpha = 0.25$)	0.9850	0.9877	0.9928	0.9971	1.0000	1.0000	1.0000	1.0000
SRE	1.1732	1.1201	1.0445	1.0053	1.0595	1.0412	1.0150	1.0017
PRSRE	1.1728	1.1199	1.0444	1.0053	1.0595	1.0412	1.0150	1.0017
RRRE	1.2963	1.2217	1.1111	1.0407	1.1039	1.0789	1.0400	1.0145

Table 10.4 RWRE of the R-estimators for $p = 15$ and different Δ^2-values for varying p_1.

Estimators	$p_1 = 2$	$p_1 = 3$	$p_1 = 5$	$p_1 = 7$	$p_1 = 2$	$p_1 = 3$	$p_1 = 5$	$p_1 = 7$
	$\Delta^2 = 0$				$\Delta^2 = 1$			
RE	1.0000	1.0000	1.0000	1.0000	1.0000	1.0000	1.0000	1.0000
RRE ($\rho^2 = 0.1$)	9.1045	5.9788	3.4558	2.3953	5.6626	4.2736	2.8084	2.0654
RRE ($\rho^2 = 0.2$)	9.8493	6.3606	3.6039	2.4653	5.9410	4.4650	2.9055	2.1172
RRE ($\rho^2 = 0.8$)	15.1589	8.0862	4.0566	2.6637	7.5344	5.2523	3.1928	2.2620
RRE ($\rho^2 = 0.9$)	16.1151	8.2786	4.0949	2.6780	7.7633	5.3329	3.2165	2.2723
RLASSO	7.5000	5.0000	3.0000	2.1429	5.0000	3.7500	2.5000	1.8750
PTRE ($\alpha = 0.15$)	2.8417	2.4667	1.9533	1.6188	2.1718	1.9542	1.6304	1.4021
PTRE ($\alpha = 0.2$)	2.4362	2.1739	1.7905	1.5242	1.9277	1.7680	1.5192	1.3354
PTRE ($\alpha = 0.25$)	2.1488	1.9568	1.6617	1.4462	1.7505	1.6288	1.4325	1.2820
SRE	3.7500	3.0000	2.1429	1.6667	3.1296	2.5964	1.9385	1.5495
PRSRE	4.6560	3.5445	2.3971	1.8084	3.4352	2.7966	2.0402	1.6081
RRRE	7.5000	5.0000	3.0000	2.1429	5.1220	3.8235	2.5385	1.9014
	$\Delta^2 = 5$				$\Delta^2 = 10$			
RE	1.0000	1.0000	1.0000	1.0000	1.0000	1.0000	1.0000	1.0000
RRE ($\rho^2 = 0.1$)	2.2555	1.9970	1.6056	1.3318	1.2875	1.1989	1.0458	0.9223
RRE ($\rho^2 = 0.2$)	2.2983	2.0378	1.6369	1.3531	1.3013	1.2134	1.0590	0.9325
RRE ($\rho^2 = 0.8$)	2.5034	2.1876	1.7244	1.4109	1.3646	1.2651	1.0950	0.9596
RRE ($\rho^2 = 0.9$)	2.5282	2.2015	1.7313	1.4149	1.3720	1.2697	1.0977	0.9614
RLASSO	2.1429	1.8750	1.5000	1.2500	1.2500	1.1538	1.0000	0.8824
PTRE ($\alpha = 0.15$)	1.2477	1.1936	1.1034	1.0338	0.9976	0.9828	0.9598	0.9458
PTRE ($\alpha = 0.2$)	1.1936	1.1510	1.0793	1.0235	0.9948	0.9836	0.9665	0.9568
PTRE ($\alpha = 0.25$)	1.1542	1.1200	1.0619	1.0166	0.9936	0.9850	0.9721	0.9653
SRE	2.0987	1.8655	1.5332	1.3106	1.6727	1.5403	1.3382	1.1948
PRSRE	2.0783	1.8494	1.5227	1.3038	1.6589	1.5294	1.3315	1.1909
RRRE	2.6733	2.2973	1.8000	1.4885	1.9602	1.7742	1.5000	1.3107
	$\Delta^2 = 20$				$\Delta^2 = 60$			
RE	1.0000	1.0000	1.0000	1.0000	1.0000	1.0000	1.0000	1.0000
RRE ($\rho^2 = 0.1$)	0.6928	0.6663	0.6162	0.5711	0.2433	0.2400	0.2331	0.2264
RRE ($\rho^2 = 0.2$)	0.6968	0.6708	0.6208	0.5750	0.2438	0.2405	0.2338	0.2270
RRE ($\rho^2 = 0.8$)	0.7146	0.6863	0.6329	0.5852	0.2459	0.2425	0.2355	0.2285
RRE ($\rho^2 = 0.9$)	0.7166	0.6876	0.6339	0.5859	0.2462	0.2427	0.2356	0.2287
RLASSO	0.6818	0.6522	0.6000	0.5556	0.2419	0.2381	0.2308	0.2239
PTRE ($\alpha = 0.15$)	0.9660	0.9675	0.9720	0.9779	1.0000	1.0000	1.0000	1.0000
PTRE ($\alpha = 0.2$)	0.9761	0.9774	0.9810	0.9854	1.0000	1.0000	1.0000	1.0000
PTRE ($\alpha = 0.25$)	0.9828	0.9839	0.9866	0.9900	1.0000	1.0000	1.0000	1.0000
SRE	1.3731	1.3034	1.1917	1.1091	1.1318	1.1082	1.0689	1.0389
PRSRE	1.3720	1.3026	1.1912	1.1089	1.1318	1.1082	1.0689	1.0389
RRRE	1.5184	1.4286	1.2857	1.1798	1.1825	1.1538	1.1053	1.0669

Table 10.5 RWRE of the R-estimators for $p = 20$ and different Δ^2-values for varying p_1.

Estimators	$p_1 = 2$	$p_1 = 3$	$p_1 = 5$	$p_1 = 7$	$p_1 = 2$	$p_1 = 3$	$p_1 = 5$	$p_1 = 7$
	$\Delta^2 = 0$				$\Delta^2 = 1$			
RE	1.0000	1.0000	1.0000	1.0000	1.0000	1.0000	1.0000	1.0000
RRE ($\rho^2 = 0.1$)	12.9872	8.5010	4.9167	3.4047	7.8677	5.9631	3.9460	2.9092
RRE ($\rho^2 = 0.2$)	14.0913	9.0423	5.1335	3.5107	8.2590	6.2241	4.0844	2.9863
RRE ($\rho^2 = 0.8$)	21.5624	11.4369	5.7360	3.7651	10.3661	7.2732	4.4569	3.1684
RRE ($\rho^2 = 0.9$)	22.8744	11.6969	5.7922	3.7803	10.6590	7.3774	4.4908	3.1791
RLASSO	10.0000	6.6667	4.0000	2.8571	6.6667	5.0000	3.3333	2.5000
PTRE ($\alpha = 0.15$)	3.2041	2.8361	2.3073	1.9458	2.4964	2.2738	1.9310	1.6797
PTRE ($\alpha = 0.2$)	2.6977	2.4493	2.0693	1.7926	2.1721	2.0143	1.7602	1.5648
PTRE ($\alpha = 0.25$)	2.3469	2.1698	1.8862	1.6694	1.9413	1.8244	1.6295	1.4739
SRE	5.0000	4.0000	2.8571	2.2222	4.1268	3.4258	2.5581	2.0423
PRSRE	6.2792	4.7722	3.2234	2.4326	4.5790	3.7253	2.7140	2.1352
RRRE	10.0000	6.6667	4.0000	2.8571	6.7857	5.0704	3.3684	2.5225
	$\Delta^2 = 5$				$\Delta^2 = 10$			
RE	1.0000	1.0000	1.0000	1.0000	1.0000	1.0000	1.0000	1.0000
RRE ($\rho^2 = 0.1$)	3.0563	2.7190	2.2051	1.8390	1.7324	1.6186	1.4214	1.2598
RRE ($\rho^2 = 0.2$)	3.1135	2.7720	2.2477	1.8695	1.7507	1.6372	1.4390	1.2740
RRE ($\rho^2 = 0.8$)	3.3724	2.9625	2.3561	1.9393	1.8297	1.7019	1.4827	1.3060
RRE ($\rho^2 = 0.9$)	3.4028	2.9797	2.3656	1.9433	1.8386	1.7076	1.4864	1.3079
RLASSO	2.8571	2.5000	2.0000	1.6667	1.6667	1.5385	1.3333	1.1765
PTRE ($\alpha = 0.15$)	1.4223	1.3625	1.2595	1.1750	1.0788	1.0592	1.0254	0.9982
PTRE ($\alpha = 0.2$)	1.3324	1.2864	1.2058	1.1385	1.0580	1.0429	1.0169	0.9961
PTRE ($\alpha = 0.25$)	1.2671	1.2306	1.1660	1.1114	1.0437	1.0319	1.0114	0.9952
SRE	2.6519	2.3593	1.9363	1.6464	2.0283	1.8677	1.6170	1.4319
PRSRE	2.6304	2.3416	1.9237	1.6370	2.0082	1.8513	1.6059	1.4243
RRRE	3.3824	2.9139	2.2857	1.8848	2.3729	2.1514	1.8182	1.5808
	$\Delta^2 = 20$				$\Delta^2 = 60$			
RE	1.0000	1.0000	1.0000	1.0000	1.0000	1.0000	1.0000	1.0000
RRE ($\rho^2 = 0.1$)	0.9283	0.8946	0.8309	0.7729	0.3250	0.3207	0.3122	0.3036
RRE ($\rho^2 = 0.2$)	0.9335	0.9002	0.8369	0.7782	0.3256	0.3215	0.3130	0.3044
RRE ($\rho^2 = 0.8$)	0.9555	0.9195	0.8515	0.7901	0.3282	0.3239	0.3150	0.3062
RRE ($\rho^2 = 0.9$)	0.9580	0.9211	0.8527	0.7908	0.3285	0.3241	0.3152	0.3063
RLASSO	0.9091	0.8696	0.8000	0.7407	0.3226	0.3175	0.3077	0.2985
PTRE ($\alpha = 0.15$)	0.9747	0.9737	0.9731	0.9743	1.0000	1.0000	1.0000	1.0000
PTRE ($\alpha = 0.2$)	0.9813	0.9808	0.9808	0.9820	1.0000	1.0000	1.0000	1.0000
PTRE ($\alpha = 0.25$)	0.9860	0.9857	0.9859	0.9870	1.0000	1.0000	1.0000	1.0000
SRE	1.5796	1.4978	1.3623	1.2560	1.2078	1.1811	1.1344	1.0957
PRSRE	1.5775	1.4961	1.3612	1.2553	1.2078	1.1811	1.1344	1.0957
RRRE	1.7431	1.6408	1.4737	1.3442	1.2621	1.2310	1.1765	1.1309

Table 10.6 RWRE of the R-estimators for $p = 30$ and different Δ^2-values for varying p_1.

Estimators	$p_1 = 2$	$p_1 = 3$	$p_1 = 5$	$p_1 = 7$	$p_1 = 2$	$p_1 = 3$	$p_1 = 5$	$p_1 = 7$
	$\Delta^2 = 0$				$\Delta^2 = 1$			
RE	1.0000	1.0000	1.0000	1.0000	1.0000	1.0000	1.0000	1.0000
RRE ($\rho^2 = 0.1$)	22.3753	14.6442	8.4349	5.8203	12.8002	9.8339	6.5821	4.8740
RRE ($\rho^2 = 0.2$)	24.3855	15.6276	8.7997	6.0133	13.4302	10.2680	6.8022	5.0086
RRE ($\rho^2 = 0.8$)	37.2461	19.5764	9.7789	6.3861	16.5880	11.8373	7.3729	5.2646
RRE ($\rho^2 = 0.9$)	39.3232	20.0631	9.8674	6.4129	16.9876	12.0125	7.4229	5.2828
RLASSO	15.0000	10.0000	6.0000	4.2857	10.0000	7.5000	5.0000	3.7500
PTRE ($\alpha = 0.15$)	3.7076	3.3671	2.8451	2.4637	2.9782	2.7592	2.4060	2.1337
PTRE ($\alpha = 0.2$)	3.0508	2.8314	2.4758	2.2002	2.5244	2.3765	2.1279	1.9271
PTRE ($\alpha = 0.25$)	2.6089	2.4579	2.2032	1.9968	2.2107	2.1051	1.9219	1.7688
SRE	7.5000	6.0000	4.2857	3.3333	6.1243	5.0891	3.8037	3.0371
PRSRE	9.5356	7.2308	4.8740	3.6755	6.8992	5.6069	4.0788	3.2054
RRRE	15.0000	10.0000	6.0000	4.2857	10.1163	7.5676	5.0323	3.7696
	$\Delta^2 = 5$				$\Delta^2 = 10$			
RE	1.0000	1.0000	1.0000	1.0000	1.0000	1.0000	1.0000	1.0000
RRE ($\rho^2 = 0.1$)	4.7273	4.2535	3.5048	2.9537	2.6439	2.4888	2.2123	1.9792
RRE ($\rho^2 = 0.2$)	4.8104	4.3328	3.5663	3.0026	2.6697	2.5158	2.2366	2.0011
RRE ($\rho^2 = 0.8$)	5.1633	4.5899	3.7172	3.0928	2.7751	2.6004	2.2951	2.0407
RRE ($\rho^2 = 0.9$)	5.2015	4.6159	3.7298	3.0991	2.7861	2.6087	2.2999	2.0435
RLASSO	4.2857	3.7500	3.0000	2.5000	2.5000	2.3077	2.0000	1.7647
PTRE ($\alpha = 0.15$)	1.7190	1.6538	1.5384	1.4395	1.2353	1.2113	1.1675	1.1287
PTRE ($\alpha = 0.2$)	1.5642	1.5158	1.4286	1.3524	1.1806	1.1622	1.1285	1.0984
PTRE ($\alpha = 0.25$)	1.4532	1.4159	1.3478	1.2873	1.1418	1.1273	1.1007	1.0768
SRE	3.7621	3.3545	2.7588	2.3444	2.7433	2.5307	2.1934	1.9380
PRSRE	3.7497	3.3433	2.7492	2.3362	2.7120	2.5044	2.1742	1.9237
RRRE	4.8058	4.1558	3.2727	2.7010	3.2022	2.9134	2.4706	2.1475
	$\Delta^2 = 20$				$\Delta^2 = 60$			
RE	1.0000	1.0000	1.0000	1.0000	1.0000	1.0000	1.0000	1.0000
RRE ($\rho^2 = 0.1$)	1.4053	1.3603	1.2733	1.1925	0.4890	0.4834	0.4720	0.4604
RRE ($\rho^2 = 0.2$)	1.4126	1.3683	1.2813	1.2004	0.4899	0.4845	0.4731	0.4616
RRE ($\rho^2 = 0.8$)	1.4416	1.3930	1.3003	1.2145	0.4933	0.4875	0.4756	0.4637
RRE ($\rho^2 = 0.9$)	1.4445	1.3953	1.3018	1.2155	0.4937	0.4878	0.4759	0.4638
RLASSO	1.3636	1.3043	1.2000	1.1111	0.4839	0.4762	0.4615	0.4478
PTRE ($\alpha = 0.15$)	1.0081	1.0040	0.9969	0.9912	0.9999	0.9999	0.9999	1.0000
PTRE ($\alpha = 0.2$)	1.0047	1.0018	0.9968	0.9928	1.0000	1.0000	1.0000	1.0000
PTRE ($\alpha = 0.25$)	1.0027	1.0006	0.9970	0.9942	1.0000	1.0000	1.0000	1.0000
SRE	1.9969	1.8948	1.7215	1.5804	1.3630	1.3320	1.2759	1.2268
PRSRE	1.9921	1.8907	1.7185	1.5782	1.3630	1.3320	1.2759	1.2268
RRRE	2.1951	2.0705	1.8621	1.6951	1.4224	1.3876	1.3247	1.2698

Table 10.7 RWRE values of estimators for $p_1 = 5$ and different values of p_2 and Δ^2.

		RRE			PTRE					
		ρ^2			α					
p_2	RE	0.8	0.9	MRLASSO	0.15	0.2	0.25	SRE	PRSRE	RRRE
					$\Delta^2 = 0$					
7	1.0000	4.0468	4.1621	3.3333	2.6044	1.9787	1.6565	2.0000	2.3149	3.3333
17	1.0000	10.2978	10.5396	6.6667	4.3870	2.8361	2.1698	4.0000	4.7722	6.6667
27	1.0000	18.3009	18.7144	10.0000	5.7507	3.3671	2.4579	6.0000	7.2308	10.0000
37	1.0000	28.9022	29.4914	13.3333	6.8414	3.7368	2.6477	8.0000	9.6941	13.3333
57	1.0000	64.9968	66.2169	20.0000	8.4968	4.2281	2.8887	12.0000	14.6307	20.0000
					$\Delta^2 = 0.5$					
7	1.0000	3.2295	3.3026	2.8571	2.2117	1.7335	1.4923	1.8733	2.0602	2.8846
17	1.0000	8.0048	8.1499	5.7143	3.7699	2.5211	1.9784	3.6834	4.1609	5.7377
27	1.0000	13.7847	14.0169	8.5714	4.9963	3.0321	2.2657	5.4994	6.2907	8.5938
37	1.0000	20.9166	21.2239	11.4286	5.9990	3.3986	2.4608	7.3167	8.4350	11.4504
57	1.0000	41.5401	42.0259	17.1429	7.5598	3.8998	2.7153	10.9521	12.7487	17.1642
					$\Delta^2 = 1$					
7	1.0000	2.6869	2.7373	2.5000	1.9266	1.5541	1.3703	1.7725	1.8843	2.5806
17	1.0000	6.5476	6.6443	5.0000	3.3014	2.2738	1.8244	3.4258	3.7253	5.0704
27	1.0000	11.0584	11.2071	7.5000	4.4087	2.7592	2.1051	5.0891	5.6069	7.5676
37	1.0000	16.3946	16.5829	10.0000	5.3306	3.1162	2.3008	6.7542	7.5071	10.0662
57	1.0000	30.5579	30.8182	15.0000	6.7954	3.6170	2.5624	10.0860	11.3392	15.0649
					$\Delta^2 = 5$					
7	1.0000	1.1465	1.1556	1.2500	1.0489	1.0088	0.9978	1.3829	1.3729	1.6901
17	1.0000	2.6666	2.6824	2.5000	1.6731	1.3625	1.2306	2.3593	2.3416	2.9139
27	1.0000	4.2854	4.3074	3.7500	2.2360	1.6538	1.4159	3.3545	3.3433	4.1558
37	1.0000	6.0123	6.0376	5.0000	2.7442	1.8968	1.5652	4.3528	4.3574	5.4019
57	1.0000	9.8282	9.8547	7.5000	3.6311	2.2844	1.7942	6.3514	6.4081	7.8981

Table 10.8 RWRE values of estimators for $p_1 = 7$ and different values of p_2 and Δ^2.

		RRE			PTRE					
		ρ^2			α					
p_2	RE	0.8	0.9	MRLASSO	0.15	0.2	0.25	SRE	PRSRE	RRRE
					$\Delta^2 = 0$					
3	1.0000	2.0072	2.0250	1.4286	1.3333	1.2292	1.1626	1.1111	1.1625	1.4286
13	1.0000	4.1385	4.1573	2.8571	2.4147	1.9458	1.6694	2.2222	2.4326	2.8571
23	1.0000	6.8094	6.8430	4.2857	3.3490	2.4637	1.9968	3.3333	3.6755	4.2857
33	1.0000	10.2779	10.3259	5.7143	4.1679	2.8587	2.2279	4.4444	4.9176	5.7143
53	1.0000	21.7368	21.7965	8.5714	5.5442	3.4283	2.5371	6.6667	7.4030	8.5714
					$\Delta^2 = 0.5$					
3	1.0000	1.8523	1.8674	1.3333	1.2307	1.1485	1.1015	1.0928	1.1282	1.3462
13	1.0000	3.7826	3.7983	2.6667	2.2203	1.8009	1.5627	2.1255	2.2661	2.6733
23	1.0000	6.1542	6.1815	4.0000	3.0831	2.2860	1.8742	3.1754	3.4159	4.0057
33	1.0000	9.1562	9.1942	5.3333	3.8441	2.6619	2.0987	4.2270	4.5705	5.3386
53	1.0000	18.4860	18.5298	8.0000	5.1338	3.2133	2.4055	6.3313	6.8869	8.0050
					$\Delta^2 = 1$					
3	1.0000	1.7196	1.7326	1.2500	1.1485	1.0873	1.0564	1.0781	1.1026	1.2903
13	1.0000	3.4830	3.4963	2.5000	2.0544	1.6797	1.4739	2.0423	2.1352	2.5225
23	1.0000	5.6140	5.6367	3.7500	2.8534	2.1337	1.7688	3.0371	3.2054	3.7696
33	1.0000	8.2554	8.2862	5.0000	3.5625	2.4906	1.9856	4.0350	4.2840	5.0185
53	1.0000	16.0828	16.1162	7.5000	4.7733	3.0225	2.2877	6.0331	6.4531	7.5174
					$\Delta^2 = 5$					
3	1.0000	1.0930	1.0983	0.8333	0.8688	0.9169	0.9447	1.0263	1.0268	1.1268
13	1.0000	2.1324	2.1374	1.6667	1.3171	1.1750	1.1114	1.6464	1.6370	1.8848
23	1.0000	3.2985	3.3063	2.5000	1.7768	1.4395	1.2873	2.3444	2.3362	2.7010
33	1.0000	4.6206	4.6302	3.3333	2.2057	1.6701	1.4360	3.0520	3.0505	3.5267
53	1.0000	7.8891	7.8973	5.0000	2.9764	2.0491	1.6701	4.4745	4.4980	5.1863

Table 10.9 RWRE values of estimators for $p_2 = 3$ and different values of p_1 and Δ^2.

		RRE			PTRE					
		ρ^2			α					
p_1	RE	0.8	0.9	MRLASSO	0.15	0.2	0.25	SRE	PRSRE	RRRE
					$\Delta^2 = 0$					
7	1.0000	2.0384	2.0631	1.4286	1.3333	1.2292	1.1626	1.1111	1.1625	1.4286
17	1.0000	1.3348	1.3378	1.1765	1.1428	1.1028	1.0752	1.0526	1.0751	1.1765
27	1.0000	1.2136	1.2150	1.1111	1.0909	1.0663	1.0489	1.0345	1.0489	1.1111
37	1.0000	1.1638	1.1642	1.0811	1.0667	1.0489	1.0362	1.0256	1.0362	1.0811
57	1.0000	1.1188	1.1190	1.0526	1.0435	1.0321	1.0239	1.0169	1.0238	1.0526
					$\Delta^2 = 0.5$					
7	1.0000	1.8080	1.8274	1.3333	1.2307	1.1485	1.1015	1.0928	1.1282	1.3462
17	1.0000	1.2871	1.2898	1.1429	1.1034	1.0691	1.0483	1.0443	1.0602	1.1475
27	1.0000	1.1879	1.1892	1.0909	1.0667	1.0450	1.0317	1.0291	1.0394	1.0938
37	1.0000	1.1462	1.1467	1.0667	1.0492	1.0334	1.0236	1.0217	1.0292	1.0687
57	1.0000	1.1081	1.1083	1.0435	1.0323	1.0220	1.0156	1.0144	1.0193	1.0448
					$\Delta^2 = 1$					
7	1.0000	1.6244	1.6401	1.2500	1.1485	1.0873	1.0564	1.0781	1.1026	1.2903
17	1.0000	1.2427	1.2452	1.1111	1.0691	1.0418	1.0274	1.0376	1.0488	1.1268
27	1.0000	1.1632	1.1645	1.0714	1.0450	1.0275	1.0181	1.0248	1.0320	1.0811
37	1.0000	1.1292	1.1296	1.0526	1.0334	1.0205	1.0135	1.0185	1.0238	1.0596
57	1.0000	1.0976	1.0978	1.0345	1.0220	1.0136	1.0090	1.0122	1.0158	1.0390
					$\Delta^2 = 5$					
7	1.0000	0.8963	0.9011	0.8333	0.8688	0.9169	0.9447	1.0263	1.0268	1.1268
17	1.0000	0.9738	0.9753	0.9091	0.9298	0.9567	0.9716	1.0130	1.0132	1.0596
27	1.0000	0.9974	0.9984	0.9375	0.9521	0.9707	0.9809	1.0086	1.0088	1.0390
37	1.0000	1.0092	1.0096	0.9524	0.9636	0.9778	0.9856	1.0064	1.0066	1.0289
57	1.0000	1.0204	1.0206	0.9677	0.9754	0.9851	0.9903	1.0043	1.0044	1.0191

Table 10.10 RWRE values of estimators for $p_2 = 7$ and different values of p_1 and Δ^2.

		RRE			PTRE					
		ρ^2			α					
p_1	RE	0.8	0.9	MRLASSO	0.15	0.2	0.25	SRE	PRSRE	RRRE
					$\Delta^2 = 0$					
3	1.0000	5.0591	5.1915	3.3333	2.6044	1.9787	1.6565	2.0000	2.3149	3.3333
13	1.0000	1.7676	1.7705	1.5385	1.4451	1.3286	1.2471	1.3333	1.3967	1.5385
23	1.0000	1.4489	1.4506	1.3043	1.2584	1.1974	1.1522	1.2000	1.2336	1.3043
33	1.0000	1.3296	1.3304	1.2121	1.1820	1.1411	1.1100	1.1429	1.1655	1.2121
53	1.0000	1.2279	1.2281	1.1321	1.1144	1.0898	1.0707	1.0909	1.1046	1.1321
					$\Delta^2 = 0.5$					
3	1.0000	4.0373	4.1212	2.8571	2.2117	1.7335	1.4923	1.8733	2.0602	2.8846
13	1.0000	1.6928	1.6954	1.4815	1.3773	1.2683	1.1975	1.3039	1.3464	1.4851
23	1.0000	1.4147	1.4164	1.2766	1.2234	1.1642	1.1235	1.1840	1.2071	1.2784
33	1.0000	1.3078	1.3086	1.1940	1.1587	1.1183	1.0899	1.1319	1.1476	1.1952
53	1.0000	1.2155	1.2156	1.1215	1.1005	1.0759	1.0582	1.0842	1.0938	1.1222
					$\Delta^2 = 1$					
3	1.0000	3.3589	3.4169	2.5000	1.9266	1.5541	1.3703	1.7725	1.8843	2.5806
13	1.0000	1.6240	1.6265	1.4286	1.3166	1.2169	1.1562	1.2786	1.3066	1.4414
23	1.0000	1.3822	1.3837	1.2500	1.1909	1.1349	1.0990	1.1700	1.1854	1.2565
33	1.0000	1.2868	1.2876	1.1765	1.1367	1.0979	1.0725	1.1223	1.1329	1.1808
53	1.0000	1.2033	1.2034	1.1111	1.0871	1.0632	1.0472	1.0783	1.0849	1.1137
					$\Delta^2 = 5$					
3	1.0000	1.4331	1.4436	1.2500	1.0489	1.0088	0.9978	1.3829	1.3729	1.6901
13	1.0000	1.2259	1.2272	1.1111	1.0239	1.0044	0.9989	1.1607	1.1572	1.2565
23	1.0000	1.1671	1.1682	1.0714	1.0158	1.0029	0.9993	1.1017	1.0996	1.1576
33	1.0000	1.1401	1.1407	1.0526	1.0118	1.0022	0.9994	1.0744	1.0729	1.1137
53	1.0000	1.1139	1.1141	1.0345	1.0078	1.0015	0.9996	1.0484	1.0474	1.0730

(iii) The lower bound of L_2 risk of HTE and the modified RLASSO estimator is the same and independent of the threshold parameter. But the upper bound of L_2 risk is dependent on the threshold parameter.
(iv) Maximum of RWRE occurs at $\Delta^2 = 0$, which indicates that the RE underperforms all estimators for any value of (p_1, p_2). Clearly, RRE outperforms all estimators for any (p_1, p_2) at $\Delta^2 = 0$. However, as Δ^2 deviates from 0, the rank-based PTE and the Stein-type estimator outperform URE, RRE, and the modified RLASSO estimator (see Tables 10.1 and 10.2).
(v) If p_1 is fixed and p_2 increases, the RWRE of all estimators increases (see Tables 10.7 and 10.8).
(vi) If p_2 is fixed and p_1 increases, the RWRE of all estimators decreases. Then, for p_2 small and p_2 large, the modified RLASSO, PTRE, SRE, and PRSRE are competitive (see Table 10.9 and 10.10).
(vii) The PRSRE always outperforms the SRE (see Tables 10.1–10.10).

Finally, we present the RWRE formula from which we prepared our tables and figures, for a quick summary.

$$\text{RWRE}(\hat{\boldsymbol{\beta}}_n^{\text{R(R)}}; \tilde{\boldsymbol{\beta}}_n^{\text{R}}) = \left(1 + \frac{p_2}{p_1}\right)\left(1 - \frac{\text{tr}(\boldsymbol{M}_0)}{p_1} + \frac{\Delta^2}{p_1}\right)^{-1}$$

$$\text{RWRE}(\hat{\boldsymbol{\beta}}_n^{\text{MRL}}(\lambda); \tilde{\boldsymbol{\beta}}_n^{\text{R}}) = \left(1 + \frac{p_2}{p_1}\right)\left(1 + \frac{\Delta^2}{p_1}\right)^{-1}$$

$$\text{RWRE}(\hat{\boldsymbol{\beta}}_n^{\text{RRR}}(k_{\text{opt}}); \tilde{\boldsymbol{\beta}}_n^{\text{R}}) = \left(1 + \frac{p_2}{p_1}\right)\left(1 + \frac{p_2\Delta^2}{p_1(p_2 + \Delta^2)}\right)^{-1}$$

$$\begin{aligned}\text{RWRE}(\hat{\boldsymbol{\beta}}_n^{\text{R(PT)}}(\alpha); \tilde{\boldsymbol{\beta}}_n^{\text{R}}) &= \left(1 + \frac{p_2}{p_1}\right)\left\{1 + \frac{p_2}{p_1}(1 - H_{p_2+2}(c_\alpha; \Delta^2))\right.\\ &\quad \left. + \frac{\Delta^2}{p_1}(2H_{p_2+2}(c_\alpha; \Delta^2) - H_{p_2+4}(c_\alpha; \Delta^2))\right\}^{-1}\end{aligned}$$

$$\text{RWRE}(\hat{\boldsymbol{\beta}}_n^{\text{R(S)}}(\alpha); \tilde{\boldsymbol{\beta}}_n^{\text{R}}) = \left(1 + \frac{p_2}{p_1}\right)\left\{1 + \frac{p_2}{p_1} - \frac{1}{p_1}(p_2 - 2)\mathbb{E}[\chi_{p_2}^{-2}(\Delta^2)]\right\}^{-1}$$

$$\begin{aligned}\text{RWRE}(\hat{\boldsymbol{\beta}}_n^{\text{R(S+)}}(\alpha); \tilde{\boldsymbol{\beta}}_n^{\text{R}}) &= \left(1 + \frac{p_2}{p_1}\right)\left\{1 + \frac{p_2}{p_1} - \frac{1}{p_1}(p_2 - 2)\mathbb{E}[\chi_{p_2}^{-2}(\Delta^2)]\right.\\ &\quad - \frac{p_2}{p_1}\mathbb{E}[(1 - (p_2 - 2)\chi_{p_2+2}^{-2}(\Delta^2))^2 I(\chi_{p_2+2}^2(\Delta^2) < (p_2 - 2))]\\ &\quad + \frac{\Delta^2}{p_1}[2\mathbb{E}[(1 - (p_2 - 2)\chi_{p_2+2}^{-2}(\Delta^2))I(\chi_{p_2+2}^2(\Delta^2) < (p_2 - 2))]\\ &\quad \left. - \mathbb{E}[(1 - (p_2 - 2)\chi_{p_2+4}^{-2}(\Delta^2))^2 I(\chi_{p_2+4}^2(\Delta^2) < (p_2 - 2))]]\right\}^{-1}.\end{aligned}$$

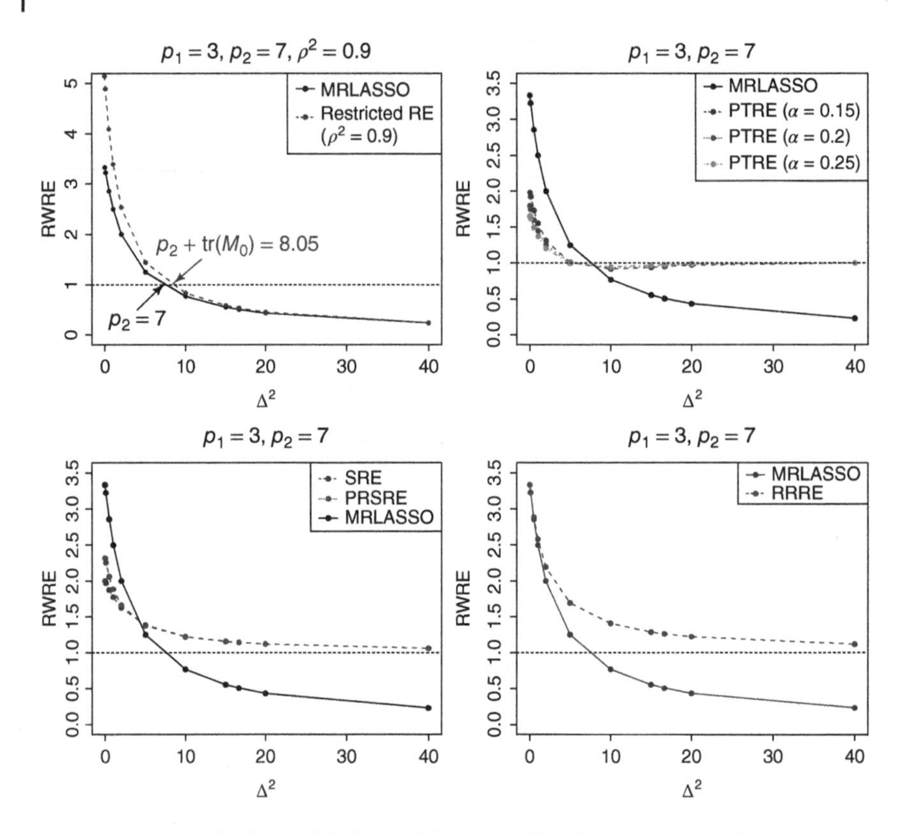

Figure 10.2 RWRE for the modified RLASSO (MRLASSO), ridge, restricted, preliminary test and the Stein-type and its positive rule estimators.

Now, we describe Table 10.1. This table presents RWRE of the seven estimators for $p_1 = 3, p_2 = 7$, and $p_1 = 7, p_2 = 13$ against Δ^2-values using a sample of size $n = 100$, the $\boldsymbol{X}$ matrix is produced. Using the model given by Eq. (10.62) for chosen values, $\rho^2 = 0.1, 0.2$ and $0.8, 0.9$. Therefore, the RWRE value of RLSE has four entries – two for low correlation and two for high correlation. Some Δ^2-values are given as p_2 and $p_2 + \text{tr}(\boldsymbol{M}_0)$ for chosen ρ^2-values. Now, one may use the table for the performance characteristics of each estimator compared to any other.

Tables 10.2–10.3 give the RWRE values of estimators for $p_1 = 2, 3, 5$, and 7 for $p = 10, 15, 20$, and 30.

Table 10.4 gives the RWRE values of estimators for $p_1 = 3$ and $p_2 = 5, 15, 25, 35$, and 55, and also for $p_1 = 7$ and $p_2 = 3, 13, 23, 33$, and 53. Also, Table 10.5 presents the RWRE values of estimators for $p_2 = 3$ and $p_1 = 7, 17, 27, 37$, and 57 as well as $p_2 = 7$ and $p_1 = 3, 13, 23, 33$, and 53 to see the effect of p_2 variation on RWRE (Figures 10.2 and 10.3).

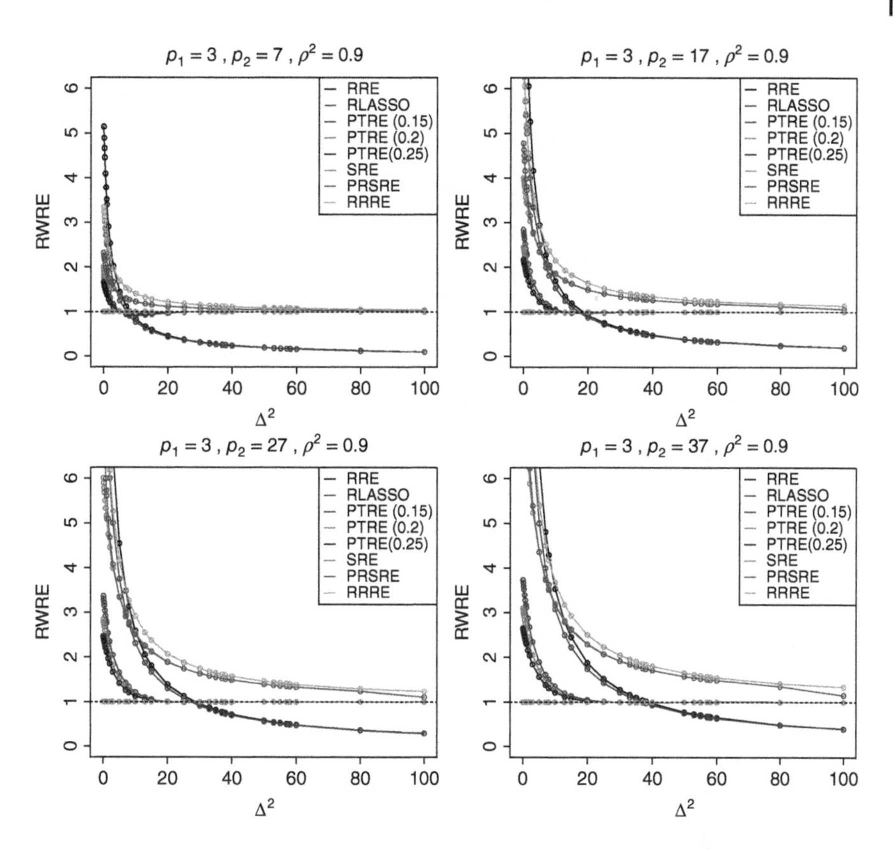

Figure 10.3 RWRE of R-estimates of a function of Δ^2 for $p_1 = 3$, $\rho^2 = 0.9$, and different p_2.

Problems

10.1 Consider model (10.1) and prove that

$$\lim_{n\to\infty} P\left(\sup_{\|\omega\|<k} \left\| \mathbf{L}_n\left(\boldsymbol{\beta} + \frac{\omega}{\sqrt{n}}\right) - \mathbf{L}_n(\boldsymbol{\beta}) + \gamma\omega \right\| > \varepsilon \right) = 0.$$

10.2 Prove that

$$\sqrt{n}(\tilde{\boldsymbol{\beta}}_n^{\mathrm{R}} - \boldsymbol{\beta}) \xrightarrow{\mathcal{D}} \mathcal{N}_p(\mathbf{0}, \eta^2 \boldsymbol{I}_p),$$

where

$$\eta^2 = \frac{A_\varphi^2}{\gamma^2}, \quad \gamma = \int_0^1 \varphi(u) \left\{ -\frac{f'(F^{-1}(u))}{f(F^{-1}(u))} \right\} \mathrm{d}u.$$

10.3 For the model (10.1), show that the test of the null-hypothesis $\mathcal{H}_0$: $\boldsymbol{\beta}_2 = \mathbf{0}$ vs. $\mathcal{H}_A$: $\boldsymbol{\beta}_2 \neq \mathbf{0}$, is given by

$$\mathcal{L}_n = A_n^{-2}\{\hat{L}_{n(2)}^\top \boldsymbol{C}_{n22\cdot 1}\hat{L}_{n(2)}\}, \quad \hat{L}_{n(2)} = L_{n(2)}(\hat{\boldsymbol{\beta}}_{1n}, \mathbf{0}),$$

where $L_n(b_1, b_2) = (L_{n(2)}^\top(b_1, b_2), L_{n(2)}^\top(b_1, b_2))^\top$, $\boldsymbol{C}_{n22\cdot 1} = \boldsymbol{C}_{n22} - \boldsymbol{C}_{n21}\boldsymbol{C}_{n11}^{-1}\boldsymbol{C}_{n12}$, and

$$A_n^2 = \frac{1}{n-1}\sum_{i=1}^n (a_n(i) - \overline{a}_n)^2, \quad \overline{a} = \frac{1}{n}\sum_{i=1}^n a_n(i).$$

10.4 Show that the ADR of $\hat{\boldsymbol{\beta}}_n^{\mathrm{R(PT)}}(\alpha)$ is given by

$$\begin{aligned}&\eta^2[p_1 + p_2(1 - H_{p_2+2}(\chi^2_{p_2}(\alpha); \Delta^2))\\&\quad + \Delta^2(2H_{p_2+2}(\chi^2_{p_2}(\alpha); \Delta^2) - H_{p_2+4}(\chi^2_{p_2}(\alpha); \Delta^2))],\end{aligned}$$

where $H_\nu(c; \Delta^2)$ is the c.d.f. of chi-square distribution with ν DF and noncentrality parameter Δ^2 evaluated at c.

10.5 Prove that the ADR of $\hat{\boldsymbol{\beta}}_n^{\mathrm{R(S)}}$ is given by

$$\eta^2[p_1 + p_2 - (p_2 - 2)^2\mathbb{E}[\chi_{p_2}^{-2}(\Delta^2)]],$$

where

$$\mathbb{E}[\chi_{p_2}^{-2\nu}(\Delta^2)] = \int_0^\infty x^{-2\nu}\mathrm{d}H_{p_2}(x; \Delta^2).$$

10.6 Verify that ADR of $\hat{\boldsymbol{\beta}}_n^{\mathrm{S+}}$ is given by

$$\begin{aligned}&\eta^2[p - (p_2 - 2)^2\mathbb{E}[\chi_{p_2}^{-2}(\Delta^2)]]\\&\quad - \eta^2[p_2\mathbb{E}[(1 - (p_2 - 2)\chi_{p_2+2}^{-2}(\Delta^2))^2 I(\chi^2_{p_2+2}(\Delta^2) < p_2 - 2)]\\&\quad + \Delta^2\{2\mathbb{E}[(1 - (p_2 - 2)\chi_{p_2+2}^{-2}(\Delta^2))I(\chi^2_{p_2+2}(\Delta^2) < p_2 - 2)]\\&\quad - \mathbb{E}[(1 - (p_2 - 2)\chi_{p_2+4}^{-2}(\Delta^2))I(\chi^2_{p_2+4}(\Delta^2) < p_2 - 2)]\}]\\&= \eta^2[(p_1 + p_2) - (p_2 - 2)^2\mathbb{E}[\chi_{p_2}^{-2}(\Delta^2)] + \mathrm{R}^*],\end{aligned}$$

where R^* is given by (9.70).

10.7 Prove that the PTRE dominates the RE whenever

$$\Delta^2 \leq \frac{p_2 H_{p_2+2}(\chi^2_{p_2}(\alpha); \Delta^2)}{2H_{p_2+2}(\chi^2_{p_2}(\alpha); \Delta^2) - H_{p_2+4}(\chi^2_{p_2}(\alpha); \Delta^2)};$$

Otherwise, RE outperforms the PTRE in the given interval.

10.8 Show that modified RLASSO estimator dominates over PTRE whenever

$$0 \leq \Delta^2 \leq \frac{p_2[1 - H_{p_2+2}(\chi^2_{p_2}(\alpha); \Delta^2)]}{1 - 2H_{p_2+2}(\chi^2_{p_2}(\alpha); \Delta^2) + H_{p_2+4}(\chi^2_{p_2}(\alpha); \Delta^2)};$$

Otherwise, PTRE will dominate the modified RLASSO estimator.

10.9 Derive the ADB and ADR of the shrinkage R-estimators in (10.29), (10.30), and (10.36).

10.10 Derive the ADB and ADR of the shrinkage R-estimators in (10.37), (10.38), and (10.39).

10.11 Consider a real data set, where the design matrix elements are moderate to highly correlated, then find the efficiency of the estimators using unweighted risk functions. Find parallel formulas for the efficiency expressions and compare the results with that of the efficiency using the weighted risk function. Are the two results consistent?

11

High-Dimensional Ridge Regression

In biological, medical, bioinformatics, chemometrics, and many other fields, we involve high-dimensional data, where the number of features (variables p) is (much) larger than the number of samples n. In DNA microarray studies, e.g. identifying a set of candidate genes that are most likely related to the outcome in the experiment, we analyze thousands of genes simultaneously, while a limited number of samples are available. In computer vision and human face recognition, eigenvectors (known as eigenfaces) over the high-dimensional vector space are used to make a low-dimensional representation of face images. In functional magnetic resonance images (fMRI), we deal with hundreds of thousands of measurements (volumetric elements or "voxels" within the brain) sampled at hundreds of time points. Other examples include spatiotemporal data, financial data, ecological data, and so on. In such studies, traditional statistical methods fail to be applied. Hence, developing new statistical methods that can deal with the cases where the number of unknown parameters is much larger than the sample size, is of interest.

In the context of multiple linear models, it is challenging to have a least squares estimator (LSE) in high dimension. As it is outlined in Section 1.6, the projection of $\boldsymbol{\beta}$ onto the row space of $\boldsymbol{X}$ seems to be a plausible method, whereas the ridge regression estimator (RRE) is very practical. It always exists and can be used in high-dimensional settings. However, it is of interest to see what kind of properties this estimator possesses, and under what conditions. In this chapter, we review two important cases where the RRE is used in a high-dimensional setting.

Specifically, consider the multiple linear model with coefficient vector, $\boldsymbol{\beta} = (\beta_1, \dots, \beta_p)^\top$ given by

$$\boldsymbol{Y} = \boldsymbol{X}\boldsymbol{\beta} + \boldsymbol{\epsilon}, \tag{11.1}$$

where $\boldsymbol{Y} = (y_1, \dots, y_n)^\top$ is a vector of n responses, $\boldsymbol{X} = (\boldsymbol{x}_1, \dots, \boldsymbol{x}_n)^\top$ is an $n \times p$ design matrix, $\boldsymbol{x}_i \in \mathbb{R}^p$ is ith predictor, and $\boldsymbol{\epsilon}$ is an n-vector of unobserved

Theory of Ridge Regression Estimation with Applications, First Edition.
A.K. Md. Ehsanes Saleh, Mohammad Arashi, and B.M. Golam Kibria.

errors. Particular distributional assumptions may be provided on ϵ as we progress. Under a high-dimensional setting, we assume $p > n$ and particularly consider two scenarios in general:

(i) $p \to \infty$, n is fixed.
(ii) p can grow with n at an almost exponential rate.

Scenario (i) is termed *large p, fixed n* and is more general than scenario (ii). For our purpose, we shall follow the analyses in Luo (2010, 2012) for scenario (i) and Gao et al. (2017) for scenario (ii).

Dicker (2016) studied the minimum property of the ridge regression and derived its asymptotic risk for the growing dimension, i.e. $p \to \infty$.

11.1 High-Dimensional RRE

In this section, we adopt scenario (i) and study the asymptotic properties of a high-dimensional RRE for a diverging number of variables, when the sample size is fixed. Unlike the previous chapters, since $p \to \infty$ and n is fixed, we denote the tuning parameter by k_p and the solution to the optimization problem

$$\min_{\beta}\{(\boldsymbol{Y} - \boldsymbol{X}\boldsymbol{\beta})^\top(\boldsymbol{Y} - \boldsymbol{X}\boldsymbol{\beta}) + k_p\|\boldsymbol{\beta}\|\}$$

is given by

$$\hat{\boldsymbol{\beta}}^{\mathrm{RR}}(k_p) = (\boldsymbol{X}^\top\boldsymbol{X} + k_p\boldsymbol{I}_p)^{-1}\boldsymbol{X}^\top\boldsymbol{Y}.$$

Let $\mathbb{E}(\boldsymbol{\epsilon}) = \boldsymbol{0}$, $\mathbb{E}(\boldsymbol{\epsilon}\boldsymbol{\epsilon}^\top) = \sigma_p^2\boldsymbol{I}_n$. In this section, we derive the asymptotic bias and variance of $\hat{\boldsymbol{\beta}}^{\mathrm{RR}}(k_p)$ as $p \to \infty$. Following Luo (2010), assume the following regularity conditions hold:

(A1) $1/k_p = o(1)$ and $\sigma_p = O(k_p^{1/2})$. There exists a constant $0 \le \delta < 0.5$ such that components of $\boldsymbol{X}$ is $O(k_p^{\delta})$. Further, let a_p be a sequence of positive numbers satisfying $a_p = o(1)$ and $\sqrt{p} = o(a_p k_p^{1-2\delta})$.
(A2) For sufficiently large p, there is a vector $\boldsymbol{b}$ such that $\boldsymbol{\beta} = \boldsymbol{X}^\top\boldsymbol{X}\boldsymbol{b}$ and there exists a constant $\varepsilon > 0$ such that each component of the vector $\boldsymbol{b}$ is $O(1/p^{1.5+\epsilon})$ and $k_p = o(p^{\epsilon}a_p)$.

An example satisfying (A2) is $k_p = \sqrt{p}$ and $\varepsilon = 0.5 - \delta$.

Let $\hat{\boldsymbol{\beta}}^{\mathrm{RR}}(k_p) = (\hat{\beta}_1^{\mathrm{RR}}(k_p), \ldots, \hat{\beta}_p^{\mathrm{RR}}(k_p))^\top$.

Theorem 11.1 *Assume (A1). Then,* $\mathrm{Var}(\hat{\boldsymbol{\beta}}_i(k_p)) = o(1)$ *for all* $i = 1, \ldots, p$.

Proof: By definition

$$\begin{aligned}
\mathrm{Var}(\hat{\boldsymbol{\beta}}^{\mathrm{RR}}(k_p)) &= \sigma_p^2(\boldsymbol{X}^\top\boldsymbol{X} + k_p\boldsymbol{I}_p)^{-1}\boldsymbol{X}^\top\boldsymbol{X}(\boldsymbol{X}^\top\boldsymbol{X} + k_p\boldsymbol{I}_p)^{-1} \\
&= \frac{\sigma_p^2}{k_p}\left(\frac{\boldsymbol{X}^\top\boldsymbol{X}}{k_p} + \boldsymbol{I}_p\right)^{-1}\frac{\boldsymbol{X}^\top\boldsymbol{X}}{k_p}\left(\frac{\boldsymbol{X}^\top\boldsymbol{X}}{k_p} + \boldsymbol{I}_p\right)^{-1} \\
&= \frac{\sigma_p^2}{k_p}\left[\left(\frac{\boldsymbol{X}^\top\boldsymbol{X}}{k_p} + \boldsymbol{I}_p\right)^{-1}\left(\frac{\boldsymbol{X}^\top\boldsymbol{X}}{k_p} + \boldsymbol{I}_p - \boldsymbol{I}_p\right)\left(\frac{\boldsymbol{X}^\top\boldsymbol{X}}{k_p} + \boldsymbol{I}_p\right)^{-1}\right] \\
&= \frac{\sigma_p^2}{k_p}\left[\left(\frac{\boldsymbol{X}^\top\boldsymbol{X}}{k_p} + \boldsymbol{I}_p\right)^{-1} - \left(\frac{\boldsymbol{X}^\top\boldsymbol{X}}{k_p} + \boldsymbol{I}_p\right)^{-1}\left(\frac{\boldsymbol{X}^\top\boldsymbol{X}}{k_p} + \boldsymbol{I}_p\right)^{-1}\right] \\
&= \frac{\sigma_p^2}{k_p}\left[\left(\frac{\boldsymbol{X}^\top\boldsymbol{X}}{k_p} + \boldsymbol{I}_p\right)^{-1} - \left(\frac{\boldsymbol{X}^\top\boldsymbol{X}}{k_p}\frac{\boldsymbol{X}^\top\boldsymbol{X}}{k_p} + \boldsymbol{I}_p + 2\frac{\boldsymbol{X}^\top\boldsymbol{X}}{k_p}\right)^{-1}\right].
\end{aligned}$$

By (A1), $\boldsymbol{X}^\top\boldsymbol{X}/k_p + \boldsymbol{I}_p \to \boldsymbol{I}_p$ and $(\boldsymbol{X}^\top\boldsymbol{X}/k_p)(\boldsymbol{X}^\top\boldsymbol{X}/k_p) + \boldsymbol{I}_p + 2(\boldsymbol{X}^\top\boldsymbol{X}/k_p) \to \boldsymbol{I}_p$ as $p \to \infty$. Hence, $\mathrm{Var}(\hat{\boldsymbol{\beta}}^{\mathrm{RR}}(k_p)) \to 0$ and the proof is complete. □

Theorem 11.2 *Assume (A1) and (A2). Further suppose $\lambda_{ip} = O(k_p)$, where $\lambda_{ip} > 0$ is the ith eigenvalue of $\boldsymbol{X}^\top\boldsymbol{X}$. Then, $b(\hat{\boldsymbol{\beta}}_i(k_p)) = o(1)$ for all $i = 1, \ldots, p$.*

Proof: Let $\boldsymbol{\Gamma}$ be an orthogonal matrix such that

$$\boldsymbol{\Gamma}^\top\boldsymbol{X}^\top\boldsymbol{X}\boldsymbol{\Gamma} = \begin{bmatrix} \boldsymbol{\Lambda}_{n\times n} & \boldsymbol{0}_{n\times(p-n)} \\ \boldsymbol{0}_{(p-n)\times n} & \boldsymbol{0}_{(p-n)\times(p-n)} \end{bmatrix}, \tag{11.2}$$

where $\boldsymbol{\Lambda}_{n\times n}$ is a diagonal matrix with elements λ_{ip}, $i = 1, \ldots, n$. Then, we have

$$\begin{aligned}
b(\hat{\boldsymbol{\beta}}^{\mathrm{RR}}(k_p)) &= \mathbb{E}(\hat{\boldsymbol{\beta}}^{\mathrm{RR}}(k_p)) - \boldsymbol{\beta} \\
&= (\boldsymbol{X}^\top\boldsymbol{X} + k_p\boldsymbol{I}_p)^{-1}\boldsymbol{X}^\top\boldsymbol{X}\boldsymbol{\beta} - \boldsymbol{\beta} \\
&= (\boldsymbol{X}^\top\boldsymbol{X} + k_p\boldsymbol{I}_p)^{-1}(\boldsymbol{X}^\top\boldsymbol{X} + k_p\boldsymbol{I}_p - k_p\boldsymbol{I}_p)\boldsymbol{\beta} - \boldsymbol{\beta} \\
&= -k_p(\boldsymbol{X}^\top\boldsymbol{X} + k_p\boldsymbol{I}_p)^{-1}\boldsymbol{\beta} \\
&= -\left(\frac{\boldsymbol{X}^\top\boldsymbol{X}}{k_p} + \boldsymbol{I}_p\right)^{-1}\boldsymbol{\beta} \\
&= -\left[\boldsymbol{\Gamma}\left(\frac{\boldsymbol{\Gamma}^\top\boldsymbol{X}^\top\boldsymbol{X}\boldsymbol{\Gamma}}{k_p} + \boldsymbol{I}_p\right)\boldsymbol{\Gamma}^\top\right]^{-1}\boldsymbol{\beta} \\
&= -\boldsymbol{\Gamma}\boldsymbol{A}\boldsymbol{\Gamma}^\top\boldsymbol{\beta},
\end{aligned}$$

where $\boldsymbol{A} = ((\boldsymbol{\Gamma}^\top\boldsymbol{X}^\top\boldsymbol{X}\boldsymbol{\Gamma}/k_p) + \boldsymbol{I}_p)^{-1}$ is a diagonal matrix with $k_p/(k_p + \lambda_{ip})$, for $i = 1, \ldots, n$ as first n diagonal elements, and the rest $(p - n)$ diagonal elements all equal to 1. Under (A2),

$$\begin{aligned}\boldsymbol{\Gamma}^\top\boldsymbol{\beta} &= \boldsymbol{\Gamma}^\top\boldsymbol{X}^\top\boldsymbol{X}\boldsymbol{\Gamma}\boldsymbol{\Gamma}^\top\boldsymbol{b}\\ &= \begin{bmatrix}\boldsymbol{\Lambda}_{n\times n} & \boldsymbol{0}_{n\times(p-n)}\\ \boldsymbol{0}_{(p-n)\times n} & \boldsymbol{0}_{(p-n)\times(p-n)}\end{bmatrix}\boldsymbol{\Gamma}^\top\boldsymbol{b}.\end{aligned}$$

Hence,

$$\begin{aligned}b(\hat{\beta}_i^{\mathrm{RR}}(k_p)) &= O\left(\frac{p^{-1.5-\epsilon}p\lambda_{ip}k_p}{k_p+\lambda_{ip}}\right)\\ &= O(p^{-\epsilon-0.5}k_p)\\ &= o(p^{-\epsilon}k_p)\\ &= o(1) \quad \text{for all } i=1,\ldots,n.\end{aligned}$$

□

Using Theorems 11.1 and 11.2, it can be verified that the ridge regression is mean squared error (MSE) consistent for $\boldsymbol{\beta}$ as $p \to \infty$.

The following result, according to Luo (2012), gives the asymptotic distribution of the RRE as $p \to \infty$.

Theorem 11.3 *Assume $1/k_p = o(1)$ and for sufficiently large p, there exists a constant $\delta > 0$ such that each component of $\boldsymbol{\beta}$ is $O(1/p^{2+\delta})$. Let $k_p = o(p^\delta)$, $\lambda_{ip} = o(k_p)$. Further, suppose that $\epsilon \sim \mathcal{N}_n(0, \sigma^2\boldsymbol{I}_n)$, $\sigma^2 > 0$. Then,*

$$k_p(\hat{\boldsymbol{\beta}}_p(k_p) - \boldsymbol{\beta}) \xrightarrow{D} \mathcal{N}(\boldsymbol{0}, \sigma^2\boldsymbol{X}^\top\boldsymbol{X}) \qquad as \quad p \to \infty.$$

In the following section, we develop shrinkage estimators based on the RRE. We use the result of Arashi and Norouzirad (2016).

11.2 High-Dimensional Stein-Type RRE

In order to develop shrinkage estimators, specifically the Stein-type estimator using the high-dimensional ridge, one needs a prior information $\boldsymbol{\beta} = \boldsymbol{\beta}_0$. However, the correctness of this information, say, is under suspicion and must be tested. Using the result of Luo and Zuo (2011), one can develop a test statistic for testing the null-hypothesis $\mathcal{H}_o : \boldsymbol{\beta} = \boldsymbol{\beta}_0$ for our high-dimensional case, i.e. growing dimension p and fixed sample size n. Here, we consider $\epsilon \sim \mathcal{N}_n(0, \sigma_p^2\boldsymbol{I}_n)$, where σ_p^2 grows as $p \to \infty$.

For our purpose, we use the following regularity condition:

(A3) $1/k_p = o(1)$. For sufficiently large p, there is a vector $\boldsymbol{b}^*$ such that $\boldsymbol{\beta} = \boldsymbol{X}^\top\boldsymbol{X}\boldsymbol{b}^*$. Further, there exists a constant $\varepsilon > 0$ such that each component of $\boldsymbol{b}^*$ is $O(1/p^{1+\varepsilon})$. Moreover, $p^{-\varepsilon}k_p = o(1)$ and $\sigma_p = o(k_p^{1/2})$

Theorem 11.4 *Assume (A3) and* $p^{-\frac{\varepsilon}{2}}k_p/\sigma_p = o(1)$, $\lambda_{ip} = o(k_p)$ *for all* $i = 1, \ldots, n$. *Further, suppose that the p covariates are uncorrelated. Then,*

$$\frac{k_p}{\sigma_p}(\hat{\boldsymbol{\beta}}^{\rm RR}(k_p) - \boldsymbol{\beta}) \xrightarrow{\mathcal{D}} \mathcal{N}_p(\mathbf{0}, {\rm Diag}(\boldsymbol{X}^\top\boldsymbol{X}))$$

Proof: Let $\boldsymbol{\Gamma} = (\gamma_{ij})$ be an orthogonal matrix such that

$$\boldsymbol{\Gamma}^\top\boldsymbol{X}^\top\boldsymbol{X}\boldsymbol{\Gamma} = \begin{bmatrix} \boldsymbol{\Lambda}_{n\times n} & \mathbf{0}_{n\times(p-n)} \\ \mathbf{0}_{(p-n)\times n} & \mathbf{0}_{(p-n)\times(p-n)} \end{bmatrix}.$$

Then, it is easy to see that

$$\text{Cov}(\hat{\boldsymbol{\beta}}^{\rm RR}(k_p) - \boldsymbol{\beta}) = \frac{\sigma_p^2}{k_p}[\boldsymbol{\Gamma}(A - A^2)\boldsymbol{\Gamma}^\top],$$

$$A = \left(\frac{\boldsymbol{X}^\top\boldsymbol{X}}{k_p} + \boldsymbol{I}_p\right)^{-1}.$$

Since $\frac{\sigma_p^2}{k_p}\frac{k_p\lambda_{ip}}{(k_p+\lambda_{ip})^2}$ is the ith diagonal element of $\frac{\sigma_p^2}{k_p}(A - A^2)$, we have

$$\text{Var}(\hat{\beta}_j^{\rm RR}(k_p)) = \frac{\sigma_p^2}{k_p^2}\sum_{i=1}^n \gamma_{ij}^2\frac{k_p^2\lambda_{ip}}{(k_p + \lambda_{ip})^2}.$$

Under $\lambda_{ip} = o(k_p)$, $k_p^2/(k_p + \lambda_{ip})^2 = o(1)$ for all $i = 1, 2, \ldots, n$. So

$$\begin{aligned}\sum_{i=1}^n \gamma_{ij}^2\frac{k_p^2\lambda_{ip}}{(k_p + \lambda_{ip})^2} &\to \sum_{i=1}^n \gamma_{ij}^2\lambda_{ip} \\ &= j\text{th diagonal element of } \boldsymbol{X}^\top\boldsymbol{X} \\ &= \text{Diag }(\boldsymbol{X}^\top\boldsymbol{X})_j, \quad j = 1, \ldots, p.\end{aligned}$$

Hence,

$$\begin{aligned}\lim_{p\to\infty}\frac{k_p^2}{\sigma_p^2}\text{ Var}(\hat{\beta}_j(k_p)) &= \lim_{p\to\infty}\sum_{i=1}^n \gamma_{ij}^2\frac{k_p^2\lambda_{ip}}{(k_p + \lambda_{ip})^2} \\ &= \text{Diag }(\boldsymbol{X}^\top\boldsymbol{X})_j.\end{aligned}$$

Given that the p covariates are uncorrelated, we deduce

$$\lim_{p\to\infty}\frac{k_p^2}{\sigma_p^2}\text{ Cov}(\hat{\boldsymbol{\beta}}^{\rm RR}(k_p)) = \text{Diag}(\boldsymbol{X}^\top\boldsymbol{X}).$$

The result follows using Theorem 11.3. □

Now, we may design a test statistic using the result of Theorem 11.4 for testing $\mathcal{H}_o: \boldsymbol{\beta} = \boldsymbol{\beta}_0$ as

$$G_{n,p} = \frac{p(\boldsymbol{Y} - \boldsymbol{X}\hat{\boldsymbol{\beta}}^{\rm RR}(k_p))^\top(\boldsymbol{Y} - \boldsymbol{X}\hat{\boldsymbol{\beta}}^{\rm RR}(k_p))}{k_p^2(\hat{\boldsymbol{\beta}}^{\rm RR}(k_p) - \boldsymbol{\beta}_0)^\top[\text{Diag}(\boldsymbol{X}^\top\boldsymbol{X})]^{-1}(\hat{\boldsymbol{\beta}}^{\rm RR}(k_p) - \boldsymbol{\beta}_0)}.$$

Luo and Zuo (2011) showed that under $\mathcal{H}_o$, $G_{n,p}$ converges in distribution to χ_n^2 as $p \to \infty$.

Using the test statistic $G_{n,p}$, we define a high-dimensional Stein-type RRE (SE, again under high-dimensional setup) with the form

$$\hat{\boldsymbol{\beta}}^{\mathrm{RR(S)}}(k_p) = \boldsymbol{\beta}_0 + \left(1 - \frac{c}{G_{n,p}}\right)(\hat{\boldsymbol{\beta}}^{\mathrm{RR}}(k_p) - \boldsymbol{\beta}_0)$$
$$= \hat{\boldsymbol{\beta}}^{\mathrm{RR}}(k_p) - \frac{c}{G_{n,p}}(\hat{\boldsymbol{\beta}}^{\mathrm{RR}}(k_p) - \boldsymbol{\beta}_0),$$

where $c > 0$ is the shrinkage parameter. As in Arashi and Norouzirad (2016), we take $c = n - 2$, $n \geq 3$.

Depending on the amount of shrinkage that is determined by value of the test statistic $G_{n,p}$ and the constant c, some coefficients might be zero. If $cG_{n,p}^{-1} > 1$, the shrinkage factor $(1 - cG_{n,p}^{-1})$ will be negative, causing some coefficients to have their signs reversed. This is called as overshrinkage in our setup. To avoid such inconsistency and overshrinking, we define a convex combination of $\hat{\boldsymbol{\beta}}^{\mathrm{RR}}(k_p)$ and $\hat{\boldsymbol{\beta}}^{\mathrm{RR(S)}}(k_p)$, namely, the high-dimensional positive-rule Stein-type RRE (PRSE) as

$$\hat{\boldsymbol{\beta}}^{\mathrm{RR(S+)}}(k_p) = \boldsymbol{\beta}_0 + \left(1 - \frac{c}{G_{n,p}}\right)^+ (\hat{\boldsymbol{\beta}}^{\mathrm{RR}}(k_p) - \boldsymbol{\beta}_0)$$
$$= \hat{\boldsymbol{\beta}}^{\mathrm{RR(S)}}(k_p) - \left(1 - \frac{c}{G_{n,p}}\right) I(G_{n,p} \leq c)(\hat{\boldsymbol{\beta}}^{\mathrm{RR}}(k_p) - \boldsymbol{\beta}_0),$$

where $a^+ = \max\{a, 0\}$.

The proposed PRSE outperforms the SE, which is demonstrated here.

Theorem 11.5 *Assume* $\boldsymbol{\beta}_0 = \mathbf{0}$. *The* $\hat{\boldsymbol{\beta}}^{\mathrm{RR(S+)}}(k_p)$ *outperforms* $\hat{\boldsymbol{\beta}}^{\mathrm{RR(S)}}(k_p)$ *in the risk sense.*

Proof: Consider the difference in risk given by

$$D = R(\hat{\boldsymbol{\beta}}^{\mathrm{RR(S+)}}(k_p)) - R(\hat{\boldsymbol{\beta}}^{\mathrm{RR(S)}}(k_p))$$
$$= -\mathbb{E}\left[\left(1 - \frac{c}{G_{n,p}}\right)^2 I(G_{n,p} < c)\hat{\boldsymbol{\beta}}^{\mathrm{RR}^\top}(k_p)\hat{\boldsymbol{\beta}}^{\mathrm{RR}}(k_p)\right]$$
$$+ 2\mathbb{E}\left[\left(1 - \frac{c}{G_{n,p}}\right) I(G_{n,p} < c)\hat{\boldsymbol{\beta}}^{\mathrm{RR}^\top}(k_p)(\hat{\boldsymbol{\beta}}^{\mathrm{RR}}(k_p) - \boldsymbol{\beta})\right]$$
$$< 0$$

since for values $G_{n,p} < c$, $1 - cG_{n,p}^{-1} < 0$ and the expected value of a positive random variable is always positive. The proof is complete. □

Table 11.1 Model fit characteristics for the proposed high-dimensional estimators: riboflavin data example.

Method	RRE	SE	PRSE
RSS	9.3621	8.3529	4.9318
R^2	0.8421	0.8591	0.9168

11.2.1 Numerical Results

In this section, we proceed with our analysis by giving a real example about application of the proposed estimation method to the riboflavin production data set. A Monte Carlo study is also conducted to support our results.

11.2.1.1 Example: Riboflavin Data

Here, we again consider the real example studies in Section 7.6.1; however, we assume a multiple linear model. This assumption does not alter our analysis, since we only want to illustrate the superiority of the PRSE. Table 11.1 shows a summary of the results.

In Table 11.1, the RSS and R^2, respectively, are the residual sum of squares and coefficient of determination of the model, i.e.

$$\text{RSS} = \sum_{i=1}^{n} (y_i - \hat{y}_i)^2, \ \hat{y}_i = \boldsymbol{x}_i \hat{\boldsymbol{\beta}}$$
$$R^2 = 1 - \text{RSS}/S_{yy}.$$

From Table 11.1, it is clear that the proposed high-dimensional positive-rule Stein-type RRE has better performance compared to the high-dimensional RRE and SE.

We further plotted the RSS vs. the ridge parameter for these three named estimators. The result is depicted in Figure 11.1. Surprisingly, for all ridge parameters in the specified range, the PRSE is superior. In conclusion, independent of the choice of k_p, the PRSE is always the best.

11.2.1.2 Monte Carlo Simulation

To examine the performance of the proposed estimators, we perform a Monte Carlo simulation. To achieve different degrees of collinearity, following McDonald and Galarneau (1975) and Gibbons (1981) the explanatory variables were generated using the following scheme for $n = 75$ and different values $p = 120,160$:

$$x_{ij} = (1 - \gamma^2)^{\frac{1}{2}} z_{ij} + \gamma z_{ip}, \qquad i = 1, \dots, n, \quad j = 1, \dots, p,$$

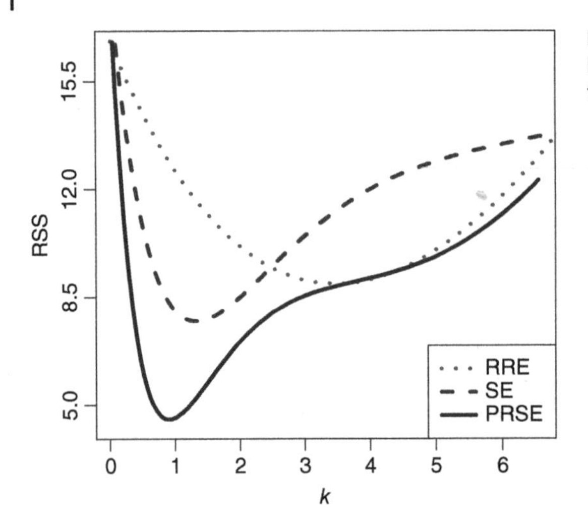

Figure 11.1 Estimation of RSS based on the ridge parameter for riboflavin data example.

where z_{ij} are independent standard normal pseudorandom numbers and γ is specified so that the correlation between any two explanatory variables is given by γ^2. These variables are then standardized so that $\boldsymbol{X}^\top\boldsymbol{X}$ and $\boldsymbol{X}^\top\boldsymbol{y}$ are in correlation forms. Three different sets of correlation corresponding to $\gamma = 0.60, 0.75$, and 0.90 are considered. Then the observations for the dependent variable are determined by

$$\boldsymbol{y} = \boldsymbol{X}_1\boldsymbol{\beta}_1 + \boldsymbol{X}_2\boldsymbol{\beta}_2 + \boldsymbol{\epsilon}, \tag{11.3}$$

with $\boldsymbol{X}_1 \sim \mathcal{N}_{p_1}(\boldsymbol{2}, 4 \times \boldsymbol{I}_{p_1})$, $p_1 = 0.75p$, and $n = 75$. To accommodate sparsity in our model, we generated $\boldsymbol{X}_2$ as $\mathcal{N}_{p_2}(\boldsymbol{0}, 0.01 \times \boldsymbol{I}_{p_2})$, $p_2 = p - p_1$. Further, we assumed $\boldsymbol{\epsilon} \backsim \mathcal{N}_n(\boldsymbol{0}, \sigma_p^2\boldsymbol{I}_n)$, $\sigma_p = 1/p$.

The Monte Carlo simulation is performed with $M = 10^3$ replications.

The relative efficiencies of the aforementioned estimators with respect to RRE are estimated as

$$\text{REff}(\hat{\boldsymbol{\beta}}_i^*) = \frac{\frac{1}{M}\sum_{m=1}^{M}\left\|\boldsymbol{y}^{(m)} - \boldsymbol{X}^{(m)}\hat{\boldsymbol{\beta}}_1^{\text{RR(m)}}\right\|_2^2}{\frac{1}{M}\sum_{m=1}^{M}\left\|\boldsymbol{y}^{(m)} - \boldsymbol{X}^{(m)}\hat{\boldsymbol{\beta}}_i^{*\text{(m)}}\right\|_2^2}, \quad i = 2, 3,$$

where $(\boldsymbol{X}^{(m)}, \boldsymbol{y}^{(m)})$ stands for the generated sample in the mth iteration and $\hat{\boldsymbol{\beta}}_i^{*(m)}$ is one of the $\hat{\boldsymbol{\beta}}^{\text{RR}}(k_p)$, $\hat{\boldsymbol{\beta}}^{\text{RR(S)}}(k_p)$, and $\hat{\boldsymbol{\beta}}^{\text{RR(S+)}}(k_p)$ estimators obtained in the mth iteration.

Figures 11.2–11.4 show the estimated risks of the proposed estimators for different values of γ and p. In these graphs, since p is specified in each case, instead of k_p we used shrinkage parameter k. Superior performance of the PRSE is apparent in all cases.

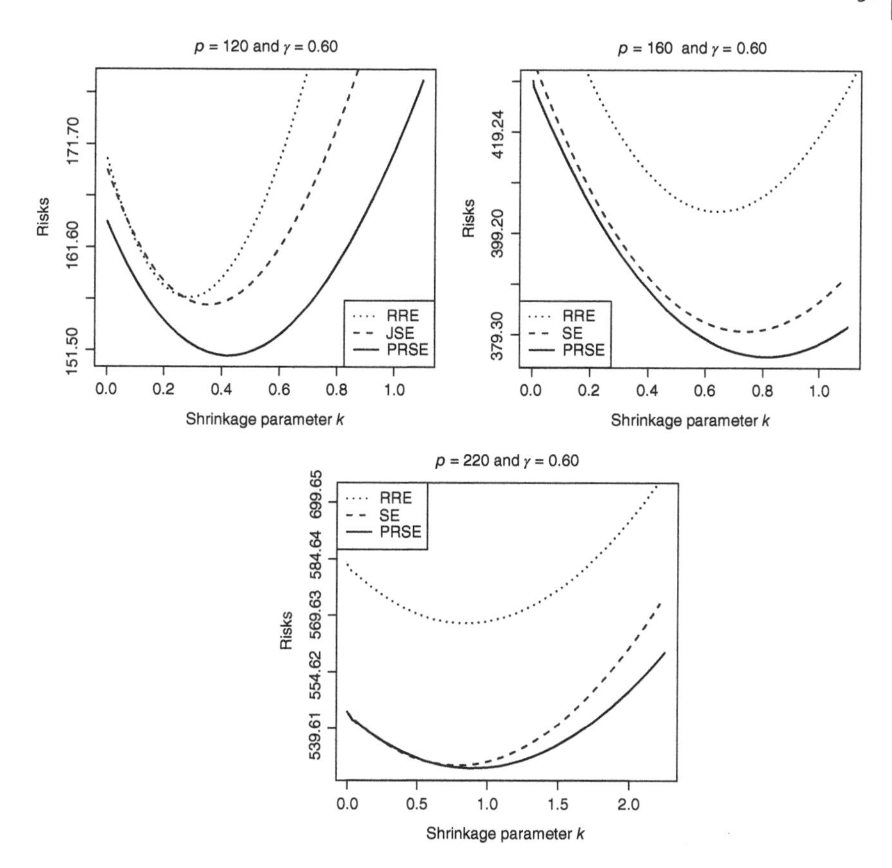

Figure 11.2 Estimated risks for the estimators of model (11.3), for $\gamma = 0.60$ and different values of p.

Table 11.2 gives the values of REff in the Monte Carlo simulation study. Both SE and PRSE dominate the high-dimensional RRE. Moreover, the PRSE is always superior.

11.3 Post Selection Shrinkage

In this section, we follow the scenario (ii), where unlike in Section 11.1, there is a relation between p and n in the high-dimensional setting. In other words, we shall assume the number of variables p can grow exponentially faster than the number of samples n.

Gao et al. (2017) argued between model complexity and model prediction and developed a platform for an interesting statistical inference of model selection in high-dimensional analysis. They specifically designed a method consisting

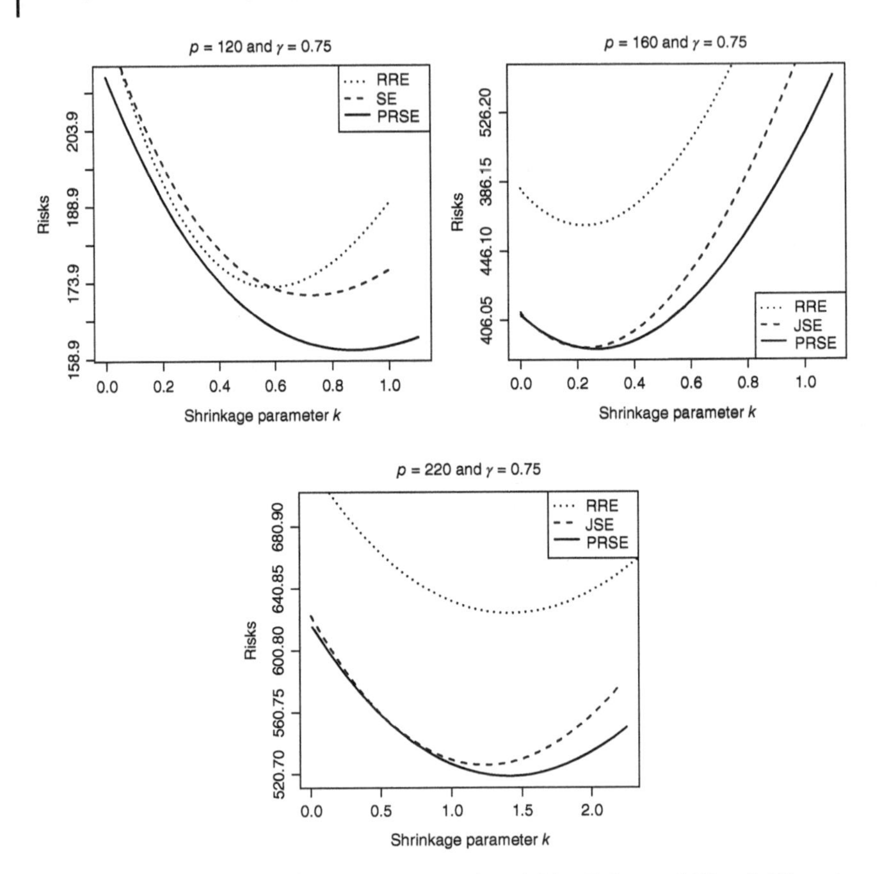

Figure 11.3 Estimated risks for the estimators of model (11.3), for $\gamma = 0.75$ and different values of p.

of three steps to combine the impact of both strong and weak signals by a post selection shrinkage estimator (PSE) in a high-dimensional regression model, in the context of scenario (ii).

In the first step, their method selects predictors with strong signals using the least absolute shrinkage and selection operator (LASSO) (or adaptive LASSO of Zou 2006); and in the second step, a ridge regression estimate of regression coefficients for the variables not selected in the first step is computed. The third step is devoted to combining the results of the two steps using a shrinkage procedure. In what follows, we specifically explain their study for practical sake in post selection analysis.

Consider the multiple linear model

$$Y = X\beta + \epsilon$$

as given by (11.1), in which $p = p_n$, to show the dimension p grows by n.

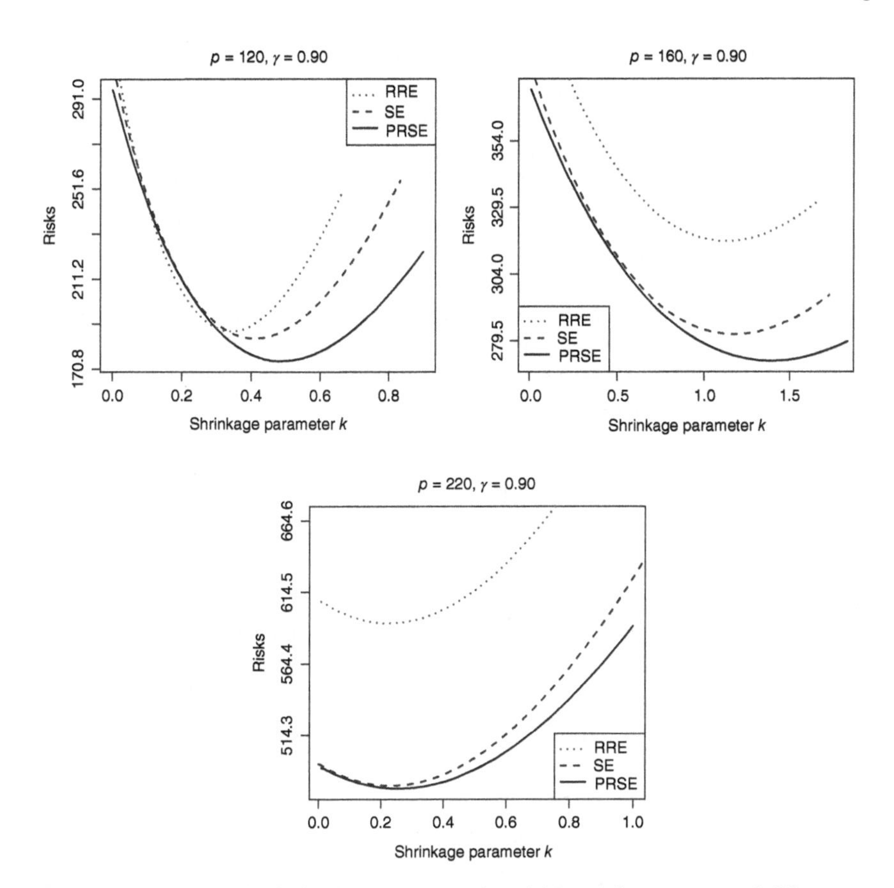

Figure 11.4 Estimated risks for the estimators of model (11.3), for $\gamma = 0.90$ and different values of p.

Table 11.2 REff values of the proposed high-dimensional estimators relative to high-dimensional RRE.

	p								
	120	**160**	**220**	**120**	**160**	**220**	**120**	**160**	**220**
Method		$\gamma = 0.60$			$\gamma = 0.75$			$\gamma = 0.09$	
SE	1.0139	1.0513	1.0700	1.0219	1.1202	1.2237	1.0268	1.1859	1.1882
PRSE	1.0447	1.0546	1.0714	1.0938	1.1238	1.2394	1.1241	1.2343	1.1931

Let $\boldsymbol{\beta}^* = (\beta_1^*, \ldots, \beta_{p_n}^*)^\top$ be the true coefficient vector in the model. For any subset $S \subset \{1, \ldots, p_n\}$ with cardinal value $|S|$, denote $\boldsymbol{\beta}_S^*$ a sub-vector of $\boldsymbol{\beta}^*$ indexed by S. Further, suppose we can divide the index set $\{1, \ldots, p_n\}$ into three disjoint subsets: S_1, S_2, and S_3. In particular, S_1 includes indexes of nonzero $\boldsymbol{\beta}_i$s which are moderately large and easily detected; S_3 includes indexes with only zero coefficients; S_2, being the intermediate, includes indexes of those nonzero β_j with weak but nonzero effects. Fan (2017) commented on such well-separated regimes of regression coefficients, and pointed that this can be too ideal in various applications. However, their contribution is worthwhile to be considered for practical post selection analysis. Indeed, they proposed a data adaptive post selection shrinkage strategy to improve the risk of the LASSO-type estimators in high-dimensional settings.

11.3.1 Notation and Assumptions

For the purpose of post selection analysis, we adhere to the standard notation in Gao et al. (2017) and give some definitions.

Partition the true vector parameter as $\boldsymbol{\beta}^* = (\boldsymbol{\beta}_1^{*\top}, \boldsymbol{\beta}_2^{*\top}, \boldsymbol{\beta}_3^{*\top})^\top$ according to the partitioning of the index set. In a similar manner, we have $\boldsymbol{X} = (\boldsymbol{Z}, \boldsymbol{X}_3)$, $\boldsymbol{Z} = (\boldsymbol{X}_1, \boldsymbol{X}_2)$. Letting $p_k = |S_k|$, $k = 1, 2, 3$, the total dimension $\sum_{j=1}^{3} p_j = p_n$ can be very large, such that $q = p_1 + p_2 \leq n$ and $\boldsymbol{\Sigma}_n = n^{-1}\boldsymbol{Z}^\top\boldsymbol{Z}$ is nonsingular. In case $\boldsymbol{\Sigma}_n$ is singular, we use the Moore–Penrose generalized inverse.

Let

$$\begin{aligned}
\boldsymbol{\Sigma}_{n11} &= \frac{1}{n}\boldsymbol{X}_1^\top\boldsymbol{X}_1, \quad \boldsymbol{\Sigma}_{n22} = \frac{1}{n}\boldsymbol{X}_2^\top\boldsymbol{X}_2,\\
\boldsymbol{\Sigma}_{n12} &= \frac{1}{n}\boldsymbol{X}_1^\top\boldsymbol{X}_2, \quad \boldsymbol{\Sigma}_{n21} = \frac{1}{n}\boldsymbol{X}_2^\top\boldsymbol{X}_1,\\
\boldsymbol{\Sigma}_{n22\cdot1} &= \frac{1}{n}\boldsymbol{X}_2^\top\boldsymbol{X}_2 - \boldsymbol{X}_2^\top\boldsymbol{X}_1(\boldsymbol{X}_1^\top\boldsymbol{X}_1)^{-1}\boldsymbol{X}_1^\top\boldsymbol{X}_2,\\
\boldsymbol{\Sigma}_{n11\cdot2} &= \frac{1}{n}\boldsymbol{X}_1^\top\boldsymbol{X}_1 - \boldsymbol{X}_1^\top\boldsymbol{X}_2(\boldsymbol{X}_2^\top\boldsymbol{X}_2)^{-1}\boldsymbol{X}_2^\top\boldsymbol{X}_1.
\end{aligned}$$

Further, let $\boldsymbol{U} = (\boldsymbol{X}_2, \boldsymbol{X}_3)$ and $\boldsymbol{M}_1 = \boldsymbol{I}_n - \boldsymbol{X}_1(\boldsymbol{X}_1^\top\boldsymbol{X}_1)^{-1}\boldsymbol{X}_1^\top$. Then, the $(p_n - p_1) \times (p_n - p_1)$ dimensional matrix $\boldsymbol{U}^\top\boldsymbol{M}_1\boldsymbol{U}$ is singular with rank $k_n \geq 0$. Denote all its eigenvalues by $\varrho_{1n} \leq \ldots \leq \varrho_{k_n n}$.

Gao et al. (2017) provided some signal strength and model error assumptions. We list them here to be referred to later.

Signal Strength Regularity Conditions

(A1) There exists a positive constant c_1, such that $|\beta_j^*| > c_1\sqrt{(\log p_n)/n}$ for all $j \in S$.

(A2) The parameter vector $\boldsymbol{\beta}^*$ satisfies $\|\boldsymbol{\beta}_{S_2}^*\|_2^2 \sim n^\tau$ for some $0 < \tau < 1$ and $\boldsymbol{\beta}_j^* \neq \boldsymbol{0}$ for all $j \in S_2$.

(A3) $\hat{\boldsymbol{\beta}}_j^* = \boldsymbol{0}$ for all $j \in S_3$.

Random Error and Model Sparsity Regularity Conditions

(B1) $\epsilon_i \sim \mathcal{N}(0, \sigma^2)$.
(B2) $\varrho_{1n}^{-1} = O(n^{-\eta})$, where $\tau < \eta \leq 1$ for τ in (A2).
(B3) $\log(p_n) = O(n^{\nu})$ for $0 < \nu < 1$.
(B4) There exists a positive definite matrix $\boldsymbol{\Sigma}$ such that $\lim_{n\to\infty}\boldsymbol{\Sigma}_n = \boldsymbol{\Sigma}$, where eigenvalues of $\boldsymbol{\Sigma}$ satisfy $0 < \rho_1 < \rho_{\boldsymbol{\Sigma}} < \rho_2 < \infty$.

11.3.2 Estimation Strategy

Consider the partitioning $\boldsymbol{X} = (\boldsymbol{X}_{S_1}|\boldsymbol{X}_{S_2}|\boldsymbol{X}_{S_3})$, where $\boldsymbol{X}_S$ is the sub-matrix and consists of vector indexed by $S \subset \{1, \ldots, p_n\}$. Let $\hat{S}_1 \subset \{1, \ldots, p_n\}$ index an active subset from the following penalized least squares (PLS) estimator.

$$\hat{\boldsymbol{\beta}}_n^{\text{PLS}} = \text{argmin}_{\boldsymbol{\beta}}\{\|\boldsymbol{Y} - \boldsymbol{X}\boldsymbol{\beta}\|_2^2 + \lambda \mathbf{1}_{p_n}^{\top}|\boldsymbol{\beta}|\},$$

where $\mathbf{1}_{p_n} = (1, \ldots, 1)^{\top}$ and $\boldsymbol{\beta} = (\beta_1, \ldots, \beta_{p_n})^{\top}$ for which $\hat{\boldsymbol{\beta}}_{jn}^{\text{PLS}} = \mathbf{0}$ iff. $j \notin \hat{S}_1$, with $\hat{\boldsymbol{\beta}}_n^{\text{PLS}} = (\hat{\boldsymbol{\beta}}_{1n}^{\text{PLS}}, \ldots, \hat{\boldsymbol{\beta}}_{p_n n}^{\text{PLS}})^{\top}$ (the adaptive LASSO is used in Gao et al. (2017) as a penalty function).

Following Belloni and Chernozhukov (2013), a restricted estimator is then defined by

$$\hat{\boldsymbol{\beta}}_{\hat{S}_1}^{\text{RE}} = (\boldsymbol{X}_{\hat{S}_1}^{\top}\boldsymbol{X}_{\hat{S}_1})^{-1}\boldsymbol{X}_{\hat{S}_1}^{\top}\boldsymbol{Y}.$$

When S_1 and S_2 are not separable, one can select the important subset $\hat{S}_1$, such that $\hat{S}_1 \subseteq S_1$ for a large enough λ, or $S_1 \subset \hat{S}_2 \subset S_1 \cup S_2$ for a smaller λ.

Gao et al. (2017) argued that although $\hat{\boldsymbol{\beta}}_{\hat{S}_1}^{\text{RE}}$ is more efficient than $\hat{\boldsymbol{\beta}}_{\hat{S}_1}^{\text{PLS}}$, the prediction risk of $\hat{\boldsymbol{\beta}}_{\hat{S}_1}^{\text{RE}}$ can still be high because many weak signals in S_2 are ignored in $\hat{\boldsymbol{\beta}}_{\hat{S}_1}^{\text{RE}}$; and, hence, they proposed to improve the risk performance of $\hat{\boldsymbol{\beta}}_{\hat{S}_1}^{\text{RE}}$ by picking up some information from $\hat{S}_1^c$, a complement subset of the selected candidate submodel using the weighted ridge (WR) estimator.

More specifically, they designed the following post selection shrinkage estimators algorithm:

Algorithm Post selection shrinkage estimation algorithm

Step 1. Obtain a subset $\hat{S}_1$ and construct $\hat{\boldsymbol{\beta}}_{\hat{S}_1}^{\text{RE}}$.
Step 2. Obtain a post selection WR estimator, $\hat{\boldsymbol{\beta}}_n^{\text{WR}} = (\hat{\boldsymbol{\beta}}_{\hat{S}_1}^{\text{WR}}, \hat{\boldsymbol{\beta}}_{\hat{S}_1^c}^{\text{WR}})$, using a ridge procedure based on $\hat{S}_1$ selected from Step 1.
Step 3. Obtain a positive-rule Stein-type shrinkage estimator (PRSE) by shrinking $\hat{\boldsymbol{\beta}}_{\hat{S}_1}^{\text{WR}}$ from Step 2 in the direction of $\hat{\boldsymbol{\beta}}_{\hat{S}_1}^{\text{RE}}$ from Step 1.

Interestingly, the post selection WR estimator in Step 2 can handle three scenarios simultaneously:

(1) The sparsity in high-dimensional data analysis,
(2) The strong correlation among covariates,
(3) The jointly weak contribution from some covariates.

Based on this algorithm, the Stein-type shrinkage estimator and its positive-rule are respectively defined by

$$\hat{\boldsymbol{\beta}}_{\hat{S}_1}^{\mathrm{RR(S)}} = \hat{\boldsymbol{\beta}}_{\hat{S}_1}^{\mathrm{WR}} - ((\hat{s}_2 - 2)\hat{T}_n^{-1})(\hat{\boldsymbol{\beta}}_{\hat{S}_1}^{\mathrm{WR}} - \hat{\boldsymbol{\beta}}_{\hat{S}_1}^{\mathrm{RE}}), \quad \hat{s}_2 = |\hat{S}_2|$$

and

$$\hat{\boldsymbol{\beta}}_{\hat{S}_1}^{\mathrm{RR(S+)}} = \hat{\boldsymbol{\beta}}_{\hat{S}_1}^{\mathrm{RE}} + (1 - (\hat{s}_2 - 2)\hat{T}_n^{-1})I(\hat{T}_n > \hat{s}_2 - 2)(\hat{\boldsymbol{\beta}}_{\hat{S}_1}^{\mathrm{WR}} - \hat{\boldsymbol{\beta}}_{\hat{S}_1}^{\mathrm{RE}}),$$

where the elements of the post selection WR estimator $\hat{\boldsymbol{\beta}}_{\hat{S}_1}^{\mathrm{WR}}$ are obtained as

$$\hat{\beta}_j^{\mathrm{WR}}(r_n, a_n) = \begin{cases} \tilde{\beta}_j(r_n) & ; \quad j \in \hat{S}_1 \\ \tilde{\beta}_j(r_n) I(\tilde{\beta}_j(r_n) > a_n) & ; \quad j \in \hat{S}_1^c \end{cases}$$

with $a_n = c_1 n^{-\alpha}$, $0 < \alpha \leq 1/2$ and some $c_1 > 0$,

$$\tilde{\boldsymbol{\beta}}(r_n) = \arg\min_{\boldsymbol{\beta}} \{\|\boldsymbol{Y} - \boldsymbol{X}\boldsymbol{\beta}\|_2^2 + r_n \|\boldsymbol{\beta}_{\hat{S}_1^c}\|_2^2\}, \quad r_n > 0$$

$$\hat{S}_2 = \{j \in \hat{S}_1^c : \hat{\beta}_j^{\mathrm{WR}}(r_n, a_n) \neq 0\}, \quad \text{and}$$

$$\hat{T}_n = \frac{1}{\sigma^2}(\hat{\boldsymbol{\beta}}_{\hat{S}_2}^{\mathrm{WR}})^\top (\boldsymbol{X}_{\hat{S}_2}^\top \mathcal{M}_{\hat{S}_1} \boldsymbol{X}_{\hat{S}_2}) \hat{\boldsymbol{\beta}}_{\hat{S}_2}^{\mathrm{WR}}$$

with $\mathcal{M}_{\hat{S}_1} = \boldsymbol{I}_n - \boldsymbol{X}_{\hat{S}_1}(\boldsymbol{X}_{\hat{S}_1}^\top \boldsymbol{X}_{\hat{S}_1})^{-1}\boldsymbol{X}_{\hat{S}_1}^\top$. The following result according to Gao et al. (2017) gives the asymptotic distribution of $\hat{\boldsymbol{\beta}}_{S_3^c}^{\mathrm{WR}}$ as $n \to \infty$. Let $\boldsymbol{\Sigma}_n = n^{-1}\boldsymbol{Z}^\top\boldsymbol{Z}$, with $\boldsymbol{Z} = (\boldsymbol{X}_1, \boldsymbol{X}_2)$.

Theorem 11.6 *Let $s_n^2 = \sigma^2 \boldsymbol{d}_n^\top \boldsymbol{\Sigma}_n^{-1} \boldsymbol{d}_n$ for any $(p_{1n} + p_{2n}) \times 1$ vector $\boldsymbol{d}_n$ satisfying $\|\boldsymbol{d}_n\| \leq 1$, and assume (B1)–(B4). Consider a sparse model with signal strength under (A1), (A3), and (A2) with $0 < \tau < 1/2$. Further, suppose a preselected model such that $S_1 \subset \hat{S}_1 \subset S_1 \cup S_2$ is obtained with probability one. Also let $r_n = c_2 a_n^{-2} (\log\log n)^3 \log(n \vee p_n)$ and $\alpha < \{(\eta - \nu - \tau)/3, 1/4 - \tau/2\}$, then, as $n \to \infty$, we have*

$$\sqrt{n} s_n^{-1} \boldsymbol{d}_n^\top (\hat{\boldsymbol{\beta}}_{S_3^c}^{\mathrm{WR}} - \boldsymbol{\beta}_{S_3^c}^*) \xrightarrow{D} \mathcal{N}(0, 1).$$

11.3.3 Asymptotic Distributional L_2-Risks

In this section, we give the ADL$_2$R (ADR) of the post selection shrinkage estimators. For this purpose, assume $\hat{S}_1 = S_1$ and both S_2 and S_3 are known in advance (see Gao et al. 2017 for details).

Unlike previous chapters, here, the ADR of any estimator $\hat{\boldsymbol{\beta}}_{1n}$ is defined by

$$\text{ADR}(\hat{\boldsymbol{\beta}}_{1n}) = \lim_{n\to\infty} \mathbb{E}\{[\sqrt{n}s_{1n}^{-1}\boldsymbol{d}_{1n}^\top(\hat{\boldsymbol{\beta}}_{1n} - \boldsymbol{\beta}_1^*)]^2\},$$

where $s_{1n}^2 = \sigma^2\boldsymbol{d}_{1n}^\top\boldsymbol{\Sigma}_{n11\cdot2}^{-1}\boldsymbol{d}_{1n}$, with p_{1n}-dimensional vector $\boldsymbol{d}_{1n}$, satisfying $\|\boldsymbol{d}_{1n}\| \le 1$.

Let $\boldsymbol{\delta} = (\delta_1, \dots, \delta_{p_{2n}})^\top \in \mathbb{R}^{p_{2n}}$, and

$$\Delta_{\boldsymbol{d}_{1n}} = \frac{\boldsymbol{d}_{1n}^\top(\boldsymbol{\Sigma}_{n11}^{-1}\boldsymbol{\Sigma}_{n12}\boldsymbol{\delta}\boldsymbol{\delta}^\top\boldsymbol{\Sigma}_{n21}\boldsymbol{\Sigma}_{n11}^{-1})\boldsymbol{d}_{1n}}{\boldsymbol{d}_{1n}^\top(\boldsymbol{\Sigma}_{n11}^{-1}\boldsymbol{\Sigma}_{n12}\boldsymbol{\Sigma}_{n22\cdot1}^{-1}\boldsymbol{\Sigma}_{n21}\boldsymbol{\Sigma}_{n11}^{-1})\boldsymbol{d}_{1n}}.$$

Further, let

$$c = \lim_{n\to\infty} \frac{\boldsymbol{d}_{1n}^\top\boldsymbol{\Sigma}_{n11}^{-1}\boldsymbol{d}_{1n}}{\boldsymbol{d}_{1n}^\top\boldsymbol{\Sigma}_{n11\cdot2}^{-1}\boldsymbol{d}_{1n}}, \quad \boldsymbol{d}_{2n} = \sigma^2\boldsymbol{\Sigma}_{n21}\boldsymbol{\Sigma}_{n11}^{-1}\boldsymbol{d}_{1n}, \quad s_{2n}^2 = \boldsymbol{d}_{2n}^\top\boldsymbol{\Sigma}_{n22\cdot1}^{-1}\boldsymbol{d}_{2n}.$$

We have the following result.

Theorem 11.7 *Under the assumptions of Theorem 11.6, except (A2) is replaced by $\beta_j^* = \delta_j/\sqrt{n}$ for $j \in S_2$, with $|\delta_j| < \delta_{\max}$ for some $\delta_{\max} > 0$, we have*

$$\text{ADR}(\hat{\boldsymbol{\beta}}_{1n}^{\text{WR}}) = 1,$$

$$\text{ADR}(\hat{\boldsymbol{\beta}}_{1n}^{\text{RE}}) = 1 - (1-c)(1-\Delta_{1n}),$$

$$\text{ADR}(\hat{\boldsymbol{\beta}}_{1n}^{\text{S}}) = 1 - \mathbb{E}[g_1(\boldsymbol{z}_2 + \boldsymbol{\delta})],$$

$$\text{ADR}(\hat{\boldsymbol{\beta}}_{1n}^{\text{RR(S+)}}) = 1 - \mathbb{E}[g_2(\boldsymbol{z}_2 + \boldsymbol{\delta})],$$

where

$$g_1(\boldsymbol{x}) = \lim_{n\to\infty}(1-c)\frac{p_{2n}-2}{\boldsymbol{x}^\top\boldsymbol{\Sigma}_{n22\cdot1}\boldsymbol{x}}\left[2 - \frac{\boldsymbol{x}^\top((p_{2n}+2)\boldsymbol{d}_{2n}\boldsymbol{d}_{2n}^\top)\boldsymbol{x}}{s_{2n}^2\boldsymbol{x}^\top\boldsymbol{\Sigma}_{n22\cdot1}\boldsymbol{x}}\right],$$

$$g_2(\boldsymbol{x}) = \lim_{n\to\infty} \frac{p_{2n}-2}{\boldsymbol{x}^\top \boldsymbol{\Sigma}_{n22\cdot1}\boldsymbol{x}} \left[(1-c)\left(2 - \frac{\boldsymbol{x}^\top((p_{2n}+2)\boldsymbol{d}_{2n}\boldsymbol{d}_{2n}^\top)\boldsymbol{x}}{s_{2n}^2 \boldsymbol{x}^\top \boldsymbol{\Sigma}_{n22\cdot1}\boldsymbol{x}}\right)\right]$$
$$\times I(\boldsymbol{x}^\top \boldsymbol{\Sigma}_{n22\cdot1}\boldsymbol{x} \geq p_{2n}-2)$$
$$+ \lim_{n\to\infty} [(2 - s_{2n}^{-2}\boldsymbol{x}^\top \boldsymbol{d}_{2n}\boldsymbol{d}_{2n}^\top \boldsymbol{x})(1-c)] I(\boldsymbol{x}^\top \boldsymbol{\Sigma}_{n22\cdot1}\boldsymbol{x} \leq p_{2n}-2).$$

Under normality assumption for the error distribution, Gao et al. (2017) studied the asymptotic properties of the post selection shrinkage estimators $\hat{\boldsymbol{\beta}}_{\hat{S}_1}^{\text{RR(S)}}$ and $\hat{\boldsymbol{\beta}}_{\hat{S}_1}^{\text{RR(S+)}}$. They numerically illustrated that as p_n gets larger, both the restricted estimator $\hat{\boldsymbol{\beta}}_{\hat{S}_1}^{\text{RE}}$ and adaptive LASSO become worse than $\hat{\boldsymbol{\beta}}_{\hat{S}_1}^{\text{WR}}$. However, the post selection PRSE still performs better that the post selection WR. Therefore, $\hat{\boldsymbol{\beta}}_{\hat{S}_1}^{\text{RR(S+)}}$ provides a protection of the adaptive LASSO in the case that it loses its efficiency. The advantages of $\hat{\boldsymbol{\beta}}_{\hat{S}_1}^{\text{RR(S+)}}$ over the LASSO is more obvious. Also, it is much more robust and at least as good as the WR estimator.

11.4 Summary and Concluding Remarks

In this chapter, we discussed the high-dimensional ridge estimator for estimating the regression parameters of the regression models, with diverging number of variables. Asymptotic performance of the high-dimensional ridge regression estimator studies, when $p \to \infty$. Then, the high-dimensional Stein-type estimator developed and its performance analyzed for the riboflavin data. Further, the post selection shrinkage methodology was reviewed and we gave the asymptotic properties of the post shrinkage estimators. In conclusion, the high-dimensional shrinkage estimators performed well, comparatively.

Problems

11.1 Find the MSE of the estimator, $\hat{\boldsymbol{\beta}}^{\text{RR}}(k_p)$.

11.2 Prove Theorem 11.3.

11.3 Prove the test statistic $G_{n,p}$ converges to χ_n^2, as $p \to \infty$.

11.4 Define the preliminary test estimator using $\hat{\boldsymbol{\beta}}(k_p)$ and derive its asymptotic properties.

11.5 Assume $1/k_p = o(1)$ and for sufficiently large p, there exists a constant $\delta > 0$ such that each component of $\boldsymbol{\beta}$ is $O(1/p^{2+\delta})$. Let $k_p = o(p^{\delta})$, $\lambda_{ip} = o(k_p)$. Further, suppose that $\epsilon \sim \mathcal{N}_n(0, \sigma^2 \boldsymbol{I}_n)$, $\sigma^2 > 0$. Then, show that $k_p(\hat{\boldsymbol{\beta}}_p(k_p) - \boldsymbol{\beta}) \xrightarrow{\mathcal{D}} \mathcal{N}(\mathbf{0}, \sigma^2 \boldsymbol{X}^\top \boldsymbol{X}) \quad \text{as} \quad p \to \infty$.

11.6 Verify

$$(\boldsymbol{X}_{S_1^c}^\top \boldsymbol{M}_1 \boldsymbol{X}_{S_1^c} + r_n \boldsymbol{I}_{q_n})^{-1} \boldsymbol{X}_{S_1^c}^\top \boldsymbol{M}_1 \boldsymbol{X}_{1n} \boldsymbol{\beta}_{10} = \mathbf{0},$$

where $q_n = p_{2n} + p_{3n}$.

11.7 Prove Theorem 11.6.

11.8 Prove Theorem 11.7.

11.9 Under the assumptions of Theorem 11.7, for $0 < \|\boldsymbol{\delta}\|^2 \leq 1$, show

$$\text{ADR}(\hat{\boldsymbol{\beta}}_{1n}^{\text{RR(S+)}}) \leq \text{ADR}(\hat{\boldsymbol{\beta}}_{1n}^{\text{S}}) \leq \text{ADR}(\hat{\boldsymbol{\beta}}_{1n}^{\text{WR}}).$$

12

Applications: Neural Networks and Big Data

This book thus far has focused on detailed comparisons of ridge regression, least absolute shrinkage and selection operator (LASSO), and other variants to examine their relative effectiveness in different regimes. In this chapter, we take it a step further and examine their use in emerging applications of advanced statistical methods. We study the practical implementation of neural networks (NNs), which can be viewed as an extension of logistic regression (LR), and examine the importance of penalty functions in this field. It should be noted that the ridge-based L_2-penalty function is widely, and almost exclusively, used in neural networks today; and we cover this subject here to gain an understanding as to how and why it is used. Furthermore, it is worth noting that this field is becoming more and more important and we hope to provide the reader with valuable information about emerging trends in this area.

This chapter concerns the practical application of penalty functions and their use in applied statistics. This area has seen a rapid growth in the past few years and will continue to grow in the foreseeable future. There are three main reasons why applied statistics has gained increasing importance and given rise to the areas of machine learning, statistical learning, and data science. The first is the advance of compute power. Today, computers are faster and have vectored capabilities to carry out matrix–vector operations quickly and efficiently. Second, the amount of data being generated in the digital era has ushered in a new world where big data sets are readily available, and big data analytics are dominating research in many institutions. The ability to accumulate and store terabytes of information is routinely carried out today in many companies and academic institutions, along with the availability of large data sets for analysis from many open sources on the Internet. And, third, there has been significant algorithmic advances in neural networks and deep learning when handling large data sets. As a result, there is a tremendous amount of activity both in industry and academia in applied statistics today.

Theory of Ridge Regression Estimation with Applications, First Edition.
A.K. Md. Ehsanes Saleh, Mohammad Arashi, and B.M. Golam Kibria.

12.1 Introduction

Our objective in this chapter is to build upon Chapter 8, which covered logistic regression, to better understand the mechanics of neural networks. As stated in that chapter, logistic regression is well suited to binary classification problems and probability prediction, or the likelihood that an event will occur. A similar set of applications exist for neural networks. In fact, we will find that neural networks are essentially built using more than one layer of logistic regression, and deep learning can be viewed (in an oversimplified way) as multiple layers of logistic regression. The meaning of layers and networks will become clear later in this chapter, but the basic idea is that complex relationships between the independent variables can be realized through the use of multilayered logistic networks. In fact, it is accurate to state that most of the recent work in this area falls in the category of multilevel logistic regression. However, the terms neural networks and deep learning are much more provocative and captivating than multilevel logistic regression, so they are now commonly used to describe work in this area.

Work on neural networks (NNs) began in earnest by computer scientists in the late 1980s and early 1990s in the pursuit of artificial intelligence systems (cf. Haykin 1999). It was touted as a way to mimic the behavior and operation of the brain, although that connection is not completely embraced today. There was much excitement at the time regarding the potential of NNs to solve many unsolvable problems. But computers were slow and the era of big data had not arrived, and so NNs were relegated to being a niche technology and lay dormant for many years. During that time, compute power increased significantly and the Internet emerged as a new communication system that connected computers and data to users around the world.

As major Internet companies began to collect volumes of data on consumers and users, it became increasingly clear that the data could be mined for valuable information to be used for sales and advertising, and many other purposes such as predicting the likelihood of certain outcomes. A variety of software tools were developed to micro-target customers with goods and services that they would likely purchase. Alongside this trend was the reemergence of NNs as an effective method for image recognition, speech recognition, and natural language processing. Other advancements in supervised and unsupervised learning (James et al. 2013) were developed in the same period. These methods were rebranded as machine learning. This term caught the attention of the media and is used to describe almost all applied statistical methods in use today.

The key difference between the early work on NNs and its ubiquitous presence today is the availability of big data. To be more specific, consider a data set with p variables $x_{1i}, x_{2i}, \ldots, x_{pi}$, where i is the sample number. Let n be the number of samples in a sample set $s_i = \{(x_i, y_i) | i = 1, \ldots, n\}$. We observe a recent trend of dramatic increases in the values of p and n. Several decades ago, p was small, typically in the range of 10–40, while the sample size n would range from several hundred to a few thousand. Both disk space and compute power severely limited the scope and applicability of NNs. Today, the value of p is usually very large and can easily run into hundreds, or thousands, and even higher. The sample size n can be in millions or billions. It can be static data sitting on a server disk or real-time data collected as users purchase goods and services. In the latter case, it will continue to grow endlessly so the problem of obtaining large data sets is gone. On the other hand, trying to extract important information from the data, managing large data sets, handling missing elements of the data, and reducing the footprint of data sets are all real issues under investigation currently.

It is useful and instructive to understand the relationship between big data and neural networks before proceeding further. So far, we know that NNs are simply multiple layers of logistic regression. If we have one layer, it is standard logistic regression. If we have two or three layers it is a shallow NN. If there are many layers of logistic regression, it is called a deep NN. In the context of NNs, there is a simple way to view its applicability to a given data set. Figure 12.1 illustrates the different regimes of the sampled data aspect ratios $n \times p$ and the different forms of neural networks suited to the data type. In the case of a few tens of variables, and a few hundred samples, a single layer of standard logistic

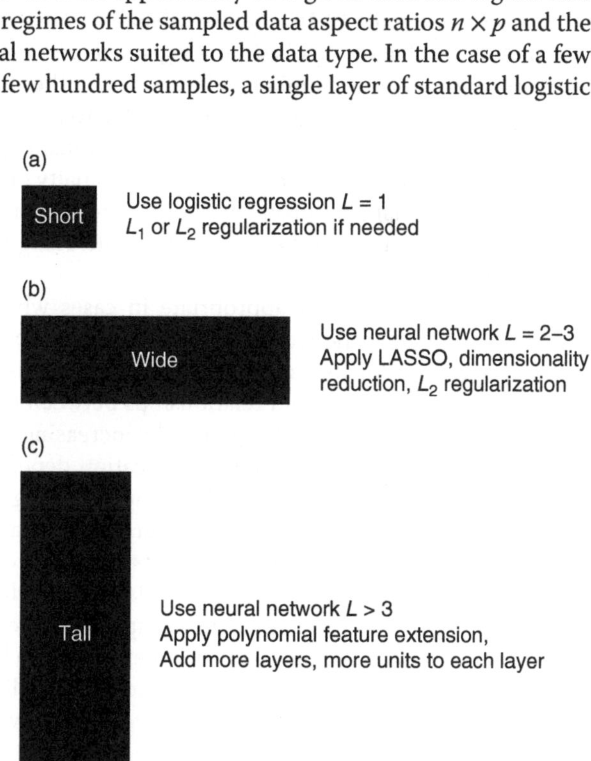

Figure 12.1 Data set aspect ratios and suitable methods. (a) Very small p, very small n = logistic regression. (b) Large p, small n = shallow neural network. (c) Smaller p, very large n = deep neural network.

regression would be sufficient for a desired accuracy level. Accuracy in this context refers to how good a given logistic model is at predicting outcomes.

If the data set has a high dimensionality, that is, p is large relative to n, then a *shallow* NN would suffice to obtain good accuracy. This can be viewed as one or two layers of logistic regression. These layers are also referred to as *hidden layers* in the neural network. On the other hand, if p is small relative to n, then many hidden layers are needed to obtain the desired accuracy. The more samples that are used, the higher the number of layers that are needed to improve accuracy. These are called *deep* neural networks.

Figure 12.1a indicates that standard logistic regression ($L = 1$) is suitable when both p and n are relatively small. If $n > p$, we may run into problems of under-fitting the data. In that case, we could use polynomial expansion of the variables (create x_i^2 terms and $x_i x_j$ cross terms, etc.) to increase the number of variables. On the other hand, if we have $n < p$, we may consider using LASSO or ridge regression, using L_1 penalty or L_2 penalty, respectively, to reduce the possibility of over-fitting the data. The key point here is that these penalty functions are used to avoid over-fitting the data for small data sets with $p > n$.

As shown in Figure 12.1b, the use of shallow NNs is advised in the case of $p \gg n$. Again, a shallow NN is simply logistic regression with a modest number of layers. Typically, the number of layers, L, is either 2 or 3. Therefore, 1 or 2 hidden layers of logistic regression are needed to obtain good prediction accuracy. The data set itself presents a high-dimensionality problem and is prone to over-fitting. For reasons cited earlier, the L_1 penalty or L_2 penalty is needed. If over-fitting wide data is a problem, there are techniques that can be applied to the data to reduce the number of variables, including LASSO and dimensionality reduction techniques.

The use of deep NNs is appropriate in cases where $n \gg p$, as shown in Figure 12.1c. A deep network implies a large number of logistic regression layers. The reason for its use is that such a data set is prone to under-fitting. However, a deep NN will find relationships between the variables and tend to reduce the effects of under-fitting, thereby increasing prediction accuracy. The number of hidden layers needed is application dependent, but usually $L > 3$ defines a deep NN. Note that all three methods (logistic regression, shallow NNs, deep NNs) will produce good accuracy on small data sets, but shallow and deep NNs will be needed for data sets with high dimensionality, and only deep NNs will deliver good accuracy on data sets with a very large number of samples. The price to pay is that the computational expense of NNs is much higher than standard LR due to the number of hidden layers used. The reasons for this will become clear in the detailed sections to follow that provide specific information about the cases discussed before along with experimental results and analysis.

12.2 A Simple Two-Layer Neural Network

In this section, we describe practical aspects of the implementation of logistic regression which will lead us to the discussion on neural networks. The mathematical details have been described in previous chapters, so in this chapter we address mainly the practical aspects and implementation of the methods. Any description of statistical learning algorithms in software involves the use of several tuning parameters (sometimes called hyper-parameters) that are adjusted when applied to specific use cases. These hyper-parameters are identified in the sections to follow along with their purpose and typical ranges.

12.2.1 Logistic Regression Revisited

Logistic regression is well suited to binary classification problems and probability prediction (Hosmer and Lemeshow 1989). We begin with a short review of the basics of logistic regression in the context of neural networks. Recall that $Y_i \in \{0, 1\}$ is the dichotomous dependent variable and $\boldsymbol{x}_i = \{1, x_{1i}, x_{2i}, \dots, x_{pi}\}^\top$ is a $p + 1$-dimensional vector of independent variables for the ith observation. Then the conditional probability of $Y_i = 1$ given x_i is

$$\tilde{P}(Y_i = 1|x_i) = \pi(x_i) = \left(1 + \exp\{-\boldsymbol{\beta}^\top \boldsymbol{x}_i\}\right)^{-1}, \tag{12.1}$$

where $\boldsymbol{\beta} = (\beta_0, \beta_1, \dots, \beta_p)^\top$ is the $p + 1$-vector regression parameter of interest. The first step is to decompose this equation into two steps. In the first step, we declare an intermediate variable called the logit, z.

$$z = \ln \frac{\pi(x_i)}{1 - \pi(x_i)} = \boldsymbol{\beta}^\top \boldsymbol{x}_i = \beta_0 + \beta_1 x_1 + \cdots + \beta_p x_p. \tag{12.2}$$

In the second step, we obtain the predicted value $\tilde{p}$ using the logistic function (specifically, the sigmoid function) represented as $\pi(x)$ but with a variable substitution using the logit z.

$$\tilde{p} = \pi(z) = \frac{1}{1 + \exp\{-z\}}. \tag{12.3}$$

There is a way to view these two steps that will become very useful, and necessary, for the study of neural networks. We represent this two-step sequence of operations in terms of a dataflow graph with the input $\boldsymbol{x}$ and the output $\tilde{p}$. The diagram of the dataflow from input to output is very useful when implementing neural networks, so we will start with a very simple version of the dataflow graph. Such a diagram is shown in Figure 12.2. It illustrates a specific example: the case of three independent variables (after setting $\beta_0 = 0$) and one dependent variable.

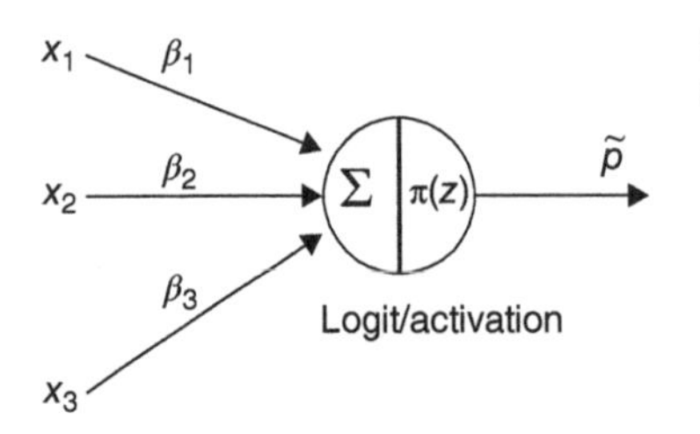

Figure 12.2 Computational flow graph for logistic regression.

This figure is simply a restatement of (12.1) using (12.2) and (12.3) to construct a dataflow diagram. Each input x is multiplied by its respective edge weights ($\boldsymbol{\beta}$) and summed together to form z. Then z is applied to the logistic function $\pi(z)$ to produce the predicted value $\tilde{p}$, called the output value. The reason for this representation will become clear shortly as it is used in representing the computational graphs associated with neural networks. However, for the moment, consider (12.1) and Figure 12.2 to be identical to one another for the case of three inputs, one output, and one sample point.

It is also instructive to examine the logistic function in more detail. This form of the cumulative distribution function (in particular, the sigmoid function) is used to ensure that the result lies between 0 and 1. Without it, we would revert to simple linear regression, which defeats the whole purpose of logistic regression. Instead, this highly nonlinear function is used to compute the probability that the dependent variable $\tilde{p}$ is 1. In neural networks, the logistic function is referred to as the activation function and it operates on logits, z. While the activation function can take many different forms, a graph of the logistic function is shown in Figure 12.3. Note that for large positive values of z, the curve asymptotically approaches one, while for large negative values it approaches 0.

The actual (observed) value of the dependent variable y for any given input $\boldsymbol{x}$ is either 0 or 1. However, the predicted value $\tilde{p}$ lies between 0 and 1, but rarely reaches either value for reasons that should be clear from (12.3). For binary classification, the predicted output is typically compared to 0.5. If $\tilde{p} > 0.5$, it is taken as 1; whereas if $\tilde{p} < 0.5$, it is taken as 0. For the general case in (12.1), the output value represents the probability of a particular event occurring so it is taken directly as the probability rather than being quantized as done in binary classification.

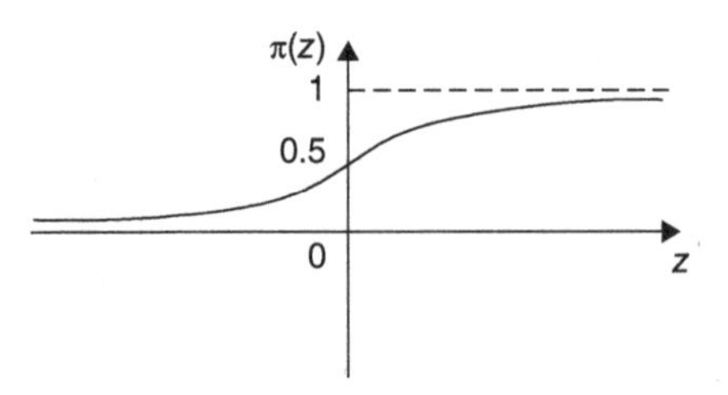

Figure 12.3 Logistic function used in logistic regression.

To find the values of $\boldsymbol{\beta}$, (12.1) is solved numerically by optimizing $\boldsymbol{\beta}$ relative to a loss function. For the one-sample case under consideration thus far, the loss function is given by

$$J = \text{loss}(y, \tilde{p}) = -y\log(\tilde{p}) - (1-y)\log(1-\tilde{p}). \tag{12.4}$$

This is commonly referred to as the *cross-entropy loss* function. The function, while seemingly complex, relies on the fact that y can only be 0 or 1. If $y = 0$, then the first term drops out and the loss is given by the second term, which simplifies to $\log(1 - \tilde{p})$. This is a useful measure since we require that $\tilde{p} = 0$ to produce a loss of 0. If $y = 1$, then the second term drops out and the loss is given by $\log(\tilde{p})$. Again, we require that $\tilde{p} = 1$ in order to reduce the loss to 0. Therefore, the loss function in (12.4) has the desired characteristics for use with logistic regression.

The numerical optimization technique used in logistic regression is usually gradient descent, an iterative method which requires the partial derivatives of J in (12.4) with respect to the parameters in the $\boldsymbol{\beta}$ vector. We can apply the chain rule here to obtain the needed gradients

$$\frac{\partial J}{\partial \beta_i} = \frac{\partial J}{\partial \tilde{p}}\frac{\partial \tilde{p}}{\partial \beta_i}. \tag{12.5}$$

Then, computing the derivative of the first term with respect to $\tilde{p}$

$$\frac{\partial J}{\partial \tilde{p}} = -\frac{y}{\tilde{p}} + \frac{1-y}{1-\tilde{p}}, \tag{12.6}$$

and the second term with respect to each parameter, β_i, $i = 1, \ldots, p$, which involves the derivative of the logistic function

$$\frac{\partial \tilde{p}}{\partial \beta_i} = \frac{\exp\{-z\}\boldsymbol{x}_i}{(1+\exp\{-z\})^2} = \tilde{p}(1-\tilde{p})\boldsymbol{x}_i. \tag{12.7}$$

The last step is the update equation for each β_i to compute the new parameter values

$$\beta_i^{t+1} = \beta_i^{\top} - \alpha\frac{\partial J}{\partial \beta_i^{\top}} = \beta_i^{\top} - \alpha(\tilde{p} - y)\boldsymbol{x}_i, \tag{12.8}$$

where α is a tuning parameter for gradient descent that controls the rate of descent, also called the *learning rate*. The goal is to drive the gradient terms toward 0 with each successive iteration. These steps are all associated with the case of $n = 1$, that is, for one sample of the training set.

For $n > 1$, it is straightforward to develop matrix equations. Assume we have n samples. Then, the entire set of samples $\boldsymbol{X}$ is an $n \times p$ matrix. The equations can be formulated as follows. The logit values are computed using an initial set of $\boldsymbol{\beta}$ values as

$$\boldsymbol{Z} = \boldsymbol{X\beta}. \tag{12.9}$$

The predicted values are computed using the logistic function

$$\tilde{P} = \pi(\boldsymbol{Z}). \tag{12.10}$$

Next, the loss is determined using the cross-entropy loss function

$$J = -\frac{1}{n}\sum_{i=1}^{n}(Y\log\tilde{P} + (1-Y)\log(1-\tilde{P})). \tag{12.11}$$

Finally, the $\boldsymbol{\beta}$ values are updated at each iteration of gradient descent using

$$\boldsymbol{\beta}^{t+1} = \boldsymbol{\beta}^{\top} - \frac{\alpha}{n}\boldsymbol{X}^{\top}(Y - \tilde{P}). \tag{12.12}$$

Here, we seek to drive the second term to zero within a given accuracy tolerance using numerical techniques, and specifically we desire that, at convergence,

$$\boldsymbol{X}^{\top}(Y - \pi(\boldsymbol{x})) = 0. \tag{12.13}$$

The number of iterations needed to reach an acceptable level of accuracy is controlled by the parameter α. If α is small, the rate of convergence can be slow, which implies a large number of iterations. If α is too high, it could lead to nonconvergence. A suitable α value can be obtained by sweeping α over a small range and selecting the appropriate value. Note that the numerical computations will be relatively cheap during each iteration since it only requires calculations of derivatives that have closed-form expressions. Therefore, logistic regression is inherently a fast method relative to neural networks, as we see in the sections to follow.

12.2.2 Logistic Regression Loss Function with Penalty

We mentioned earlier that logistic regression is suitable for small problems with relatively small values of p and n. However, in cases where $p > n$, we may have the problem of over-fitting the data. This was illustrated earlier in Figure 12.1 in the case of a "wide" data set. If particular, we may have a data set with a large number of variables, $\boldsymbol{x}_i$, such that $\tilde{P}$ matches Y at the sample points but is otherwise inaccurate due to the lack of a sufficient number of samples, n. In such cases, a suitable penalty function is applied to the loss function to reduce the degree of over-fitting. A penalty function based on LASSO is called an L_1-penalty function and a penalty function based on ridge regression is referred to as an L_2-penalty function.

$$L_1(Y,\tilde{P}) = -\frac{1}{n}\sum_{i=1}^{n}(Y\log\tilde{P} + (1-Y)\log(1-\tilde{P})) + \frac{\lambda}{n}\|\boldsymbol{\beta}\|_1, \tag{12.14}$$

$$L_2(Y,\tilde{P}) = -\frac{1}{n}\sum_{i=1}^{n}(Y\log\tilde{P} + (1-Y)\log(1-\tilde{P})) + \frac{\lambda}{2n}\|\boldsymbol{\beta}\|_2^2, \tag{12.15}$$

where

$$\|\boldsymbol{\beta}\|_1 = \sum_{i=1}^{p} |\beta_i|, \tag{12.16}$$

and

$$|\boldsymbol{\beta}\|_2^2 = \boldsymbol{\beta}^\top \boldsymbol{\beta} \tag{12.17}$$

Note the use of the tuning parameter λ to control the effect of the penalty function, or sometimes referred to as the degree of regularization. If $\lambda = 0$, the penalty function is removed. When $\lambda > 0$, the level of regularization increases accordingly. When gradient descent is used, the partial derivative terms must be adjusted reflecting the addition of the penalty term mentioned.

For L_1, the update equation is given by

$$\boldsymbol{\beta}^{t+1} = \boldsymbol{\beta}^t - \frac{\alpha}{n}[\boldsymbol{X}^\top(Y - \tilde{P}) + \lambda \text{sgn}(\boldsymbol{\beta}^t)] \tag{12.18}$$

In practice, the L_2 penalty is typically used in most machine learning applications. The update equation for the L_2 penalty is given by

$$\boldsymbol{\beta}^{t+1} = \boldsymbol{\beta}^\top - \frac{\alpha}{n}[\boldsymbol{X}^\top(Y - \tilde{P}) + \lambda \boldsymbol{\beta}^t] \tag{12.19}$$

12.2.3 Two-Layer Logistic Regression

The simplest form of neural networks is a two-layer logistic regression model. To illustrate its structure, consider the simplification of Figure 12.2 in Figure 12.4 where the logit and activation functions have been merged into one unit called a neuron. This diagram has the usual inputs and a new output a. This will reduce the complexity of neural network schematics because we make use of several of these units to construct a neural network.

A two-layer logic regression model is the simplest neural network. Its structure is depicted in Figure 12.5. Each circle is a neuron that performs logistic regression.

Rather than directly affecting the output, the input variables are converted to a set of intermediate variables, a_1, a_2, a_3, and a_4 using a layer of intermediate units. Each of these units includes both logit and activation functions.

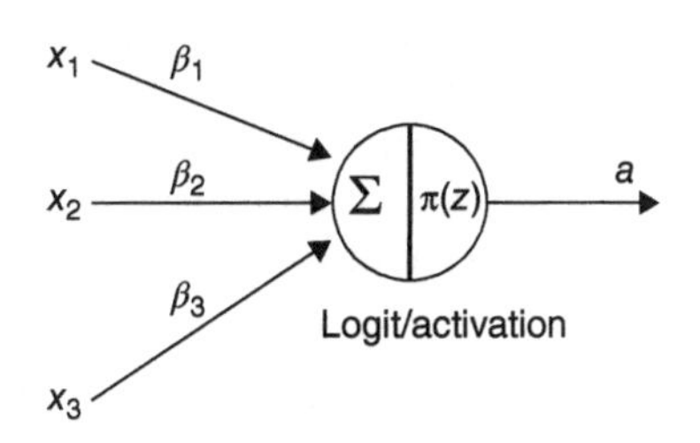

Figure 12.4 Simplified flow graph for logistic regression.

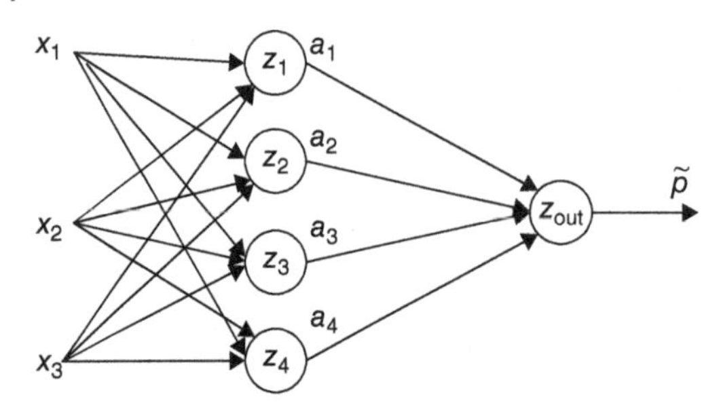

Figure 12.5 Two-layer neural network.

The same holds true for the output unit. The layer with four units is referred to as a *hidden layer*. There are four units in the hidden layer and one unit in the output layer. Therefore, this two-layer neural network would be considered as a shallow neural network because it has only one hidden layer and one output layer.

The sequence of steps to determine $\tilde{P}$ in a two-stage network is as follows:

$$\begin{aligned} \boldsymbol{Z}_1 &= \boldsymbol{\beta}_{\text{matrix}} \boldsymbol{X}^{\mathsf{T}} \\ A_1 &= \pi(Z_1) \\ z_{\text{out}} &= \boldsymbol{\beta}_{\text{vector}}^{\mathsf{T}} A_1 \\ \tilde{P} &= \pi(z_{\text{out}}). \end{aligned} \tag{12.20}$$

In each stage, we determine the logits and then apply the logistic function to obtain the outputs. To illustrate the details of this process for one sample, Figure 12.6 shows the computations as the input is forward propagated to produce the predicted output. It provides information about the benefits and limitations of neural networks. In particular, the hidden layer allows for more complex relationships between the inputs to be formed since there is a fully connected set of links between the input and the hidden layer. This implies that a neural network can deliver a higher accuracy for a given data set. However, it comes at a price. There is a matrix–vector multiplication required in each iteration of gradient descent, and so the computational run time is much higher than logistic regression.

To reduce this time, the implementation can be vectorized to take advantage of special computer instructions tailored to this type of operation. But the overall computation time is still higher than logistic regression. In addition, there is a gradient calculation step required to update the $\boldsymbol{\beta}_{\text{matrix}}$ and $\boldsymbol{\beta}_{\text{vector}}$ terms, although this represents only a marginal increase in the computation time.

The application of the penalty functions for neural networks proceeds in the same way as that for logistic regression, shown in (12.16) and (12.17) with the following changes to the loss function, assuming k units in the hidden layer:

$$\|\beta\|_1 = \sum_{i=1}^{k} |\beta_i|_{\text{vector}} + \sum_{j=1}^{p} \sum_{i=1}^{p} |\beta_{ij}|_{\text{matrix}}, \tag{12.21}$$

1st layer

$$\begin{bmatrix} z_1 \\ z_2 \\ z_3 \\ z_4 \end{bmatrix} = \overbrace{\begin{bmatrix} \beta_{11} & \beta_{12} & \beta_{13} \\ \beta_{21} & \beta_{22} & \beta_{23} \\ \beta_{31} & \beta_{32} & \beta_{33} \\ \beta_{41} & \beta_{42} & \beta_{43} \end{bmatrix}}^{\beta_{\text{matrix}}} \begin{bmatrix} x_1 \\ x_2 \\ x_3 \end{bmatrix} \qquad \begin{bmatrix} a_1 \\ a_2 \\ a_3 \\ a_4 \end{bmatrix} = \begin{bmatrix} \pi(z_1) \\ \pi(z_2) \\ \pi(z_3) \\ \pi(z_4) \end{bmatrix}$$

2nd layer

$$\begin{bmatrix} z_{\text{out}} \end{bmatrix} \quad \overbrace{\begin{bmatrix} \beta_1 & \beta_2 & \beta_3 & \beta_4 \end{bmatrix}}^{\beta_{\text{vector}}} \begin{bmatrix} a_1 \\ a_2 \\ a_3 \\ a_4 \end{bmatrix} \qquad \tilde{p} = \pi(z_{\text{out}})$$

Figure 12.6 Detailed equations for a two-layer neural network.

$$\|\boldsymbol{\beta}\|_2^2 = |\boldsymbol{\beta}^{\mathrm{T}}\boldsymbol{\beta}|_{\text{vector}} + \sum_{j=1}^{k}\sum_{i=1}^{p} |\beta_{ij}|^2_{\text{matrix}}. \tag{12.22}$$

The gradient update equations for $\boldsymbol{\beta}_{\text{matrix}}$ and $\boldsymbol{\beta}_{\text{vector}}$ can be easily derived for the two-layer neural network and is left as an exercise for the reader.

12.3 Deep Neural Networks

A natural next step to take with neural networks is to increase the number of layers. In doing so, we create deep neural networks. A simple four-layer neural network is illustrated in Figure 12.7. It has three hidden layers and one output layer. The architecture of this neural network is symmetric, but actual deep networks can be highly asymmetric. That is, the number of units in each layer may vary considerably, and the number of layers used depends on the characteristics of the application and the associated data set.

A key point to mention upfront is that deep neural networks are best suited to "tall" data sets, as shown earlier in Figure 12.1c. In particular, if $n \gg p$, then deep neural networks should be investigated to address the problem. In effect, these architectures are intended to address the problem of under-fitting. Specifically, big data in the "tall" format is prone to under-fitting. Therefore, ridge and

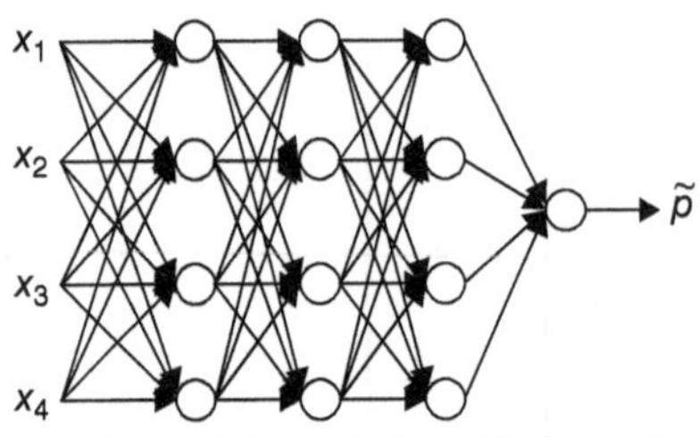

Figure 12.7 A four-layer neural network.

LASSO regularization will not be as effective for these types of data sets. In fact, we seek to increase the number variables by introducing more layers and more units in each layer to counteract under-fitting. In short, the use of regularization in the form of a penalty function should not be the first choice to improve the accuracy of these types of networks.

A few other comments are appropriate for deep learning on big data. First, the logistic function is rarely used in such networks, except in the final layer. An alternative activation function, called the *relu* function, is used instead. This function can be viewed as a highly simplified logistic function, but is linear in nature. Its use is motivated by the fact that the logistic function gradient approaches zero for large positive or negative values of z. This is clear from a quick glance at Figure 12.3 shown earlier. These near-zero gradients present major convergence problems in cases when large values of z (positive or negative) arise in the network.

The more commonly used relu(z) function is shown in Figure 12.8.

It can be represented as follows

$$\text{relu}(z) = \begin{cases} 0 & \text{if } z < 0 \\ z & \text{if } z > 0. \end{cases} \tag{12.23}$$

Accordingly, the derivative of the relu function is given by

$$\frac{\text{drelu(z)}}{\text{dz}} = \begin{cases} 0 & \text{if } z < 0 \\ 1 & \text{if } z > 0. \end{cases} \tag{12.24}$$

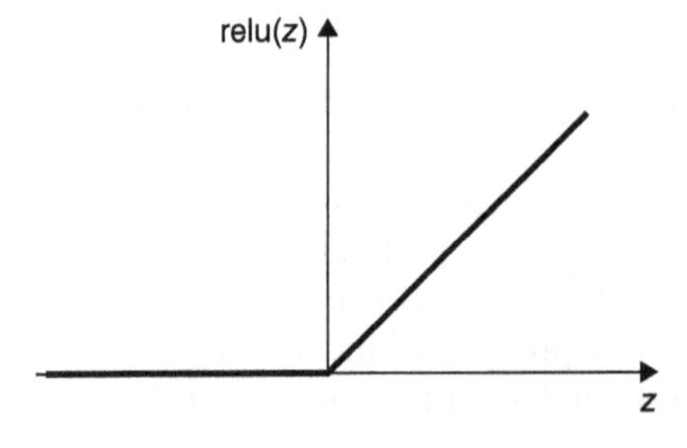

Figure 12.8 The relu activation function used in deep neural networks.

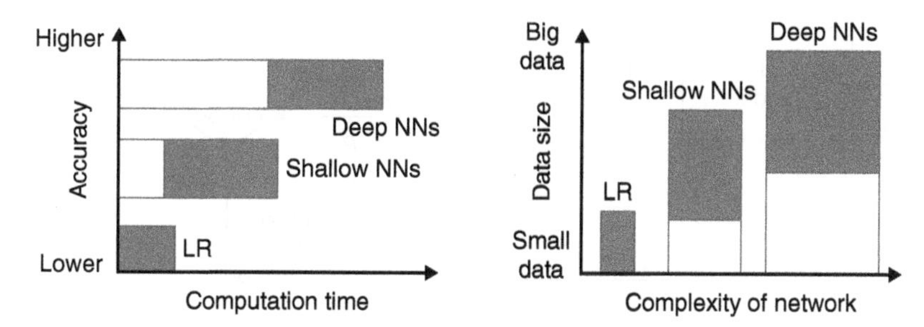

Figure 12.9 Qualitative assessment of neural networks.

The equations presented earlier would be updated using the relu function in place of the standard logistic function to obtain the required equations.

Finally, deep neural networks are much more computationally expensive than logistic regression or shallow neural networks for obvious reasons. This is depicted qualitatively in Figure 12.9. Specifically, there are many more matrix–vector operations due to the number of hidden layers and units in a deep network. They also require many more iterations to reach acceptable levels of accuracy. However, deep NNs can provide higher accuracy for big data sets, whereas logistic regression is more suitable for small data sets.

12.4 Application: Image Recognition

12.4.1 Background

Statistical learning methods can be categorized into two broad areas: supervised learning and unsupervised learning. In the case of supervised learning, the "right answer" is given for a set of sample points. Techniques include predictive methods such as linear regression, logistic regression, and neural networks, as described in this book. It also includes classification methods such as binary classification where each sample is assigned to one of two possible groups, as is described shortly. On the other hand, in unsupervised learning, the "right answer" is not given so the techniques must discover relationships between the sample points. Well-known methods in this category include K-means clustering, anomaly detection, and dimensionality reduction (cf. James et al. 2013).

In this section, we describe an application of supervised learning for binary classification using logistic regression, shallow neural networks, and deep neural networks. We build regression-based models using training data and then validate the accuracy using test data. The accuracy is determined by the number of correct predictions made on samples in the test set. The experimental setup is shown in Figure 12.10.

The sequence of steps is as follows. First, the supervised training data $(X_{\text{train}}, Y_{\text{train}})$ is applied to the selected regression-based model, either logistic

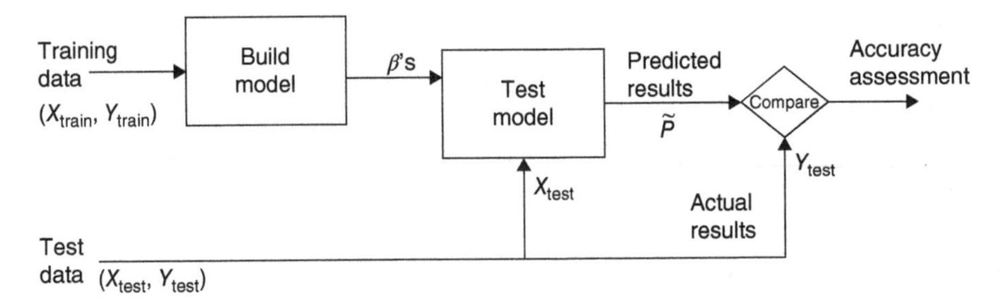

Figure 12.10 Typical setup for supervised learning methods.

regression, shallow NN or deep NN, in order to compute the $\boldsymbol{\beta}$ parameters. The accuracy of the model generated using the training data is usually quite high when measured on the same training data. Typically, it is 80–100% depending on the application. This can be misleading since the goal of the optimization is to closely match the training data and, in fact, a close match is produced. But that does not necessarily translate to the same accuracy on new data that is not in the training set. Therefore, it must also be evaluated on an independent set of test data (X_{test}, Y_{test}) to properly assess its accuracy. This is carried out by applying X_{test} to the derived model and then comparing the predicted result $\tilde{P}$ against Y_{test}.

When given an initial data set, it is important to divide it into two sets, one for training and another for testing. For small data sets, 70% is used for training and 30% is used for testing. As a practical matter, there is usually a cross-validation test set used during the development of the model to improve the model and then the final model evaluation is performed using the test set. Furthermore, on big data sets, 98% may be used for training, while 1% is used for cross-validation and 1% for testing. Here we will consider only the training set and test set to simplify the presentation and use roughly 70% to train and 30% to test the models.

12.4.2 Binary Classification

In order to compare the accuracy and speed of logistic regression, shallow neural networks and deep neural networks, a suitable application must be selected. One such application is binary classification of images, which is a simple form of image recognition. The problem of binary classification involves the task of deciding whether a set of images belongs to one group or another. The two groups are assigned the labels of 1 or 0. A label of 1 indicates that it belongs in the target group whereas a label of 0 signifies that it does not.

Specifically, if we are given a set of images of cats and dogs, we could label the cat pictures as 1 and the non-cat pictures (i.e. pictures of dogs) as 0. Our training set has 100 images, with 50 cats and 50 dogs, and we label each one

Table 12.1 Test data input, output, and predicted values from a binary classification model.

i	x_{1i}	x_{2i}	x_{3i}	x_{pi}	Y	$\tilde{P}$	
1	0.306	0.898	0.239	—	1	0.602 = 1	tp
2	0.294	0.863	0.176	—	1	0.507 = 1	tp
3	0.361	0.835	0.125	—	1	0.715 = 1	tp
4	0.310	0.902	0.235	—	0	0.222 = 0	tn
5	0.298	0.871	0.173	—	1	0.624 = 1	tp
6	0.365	0.863	0.122	—	1	0.369 = 0	fp
7	0.314	0.741	0.235	—	1	0.751 = 1	tp
8	0.302	0.733	0.173	—	0	0.343 = 0	tn
9	0.369	0.745	0.122	—	0	0.698 = 1	fn
10	0.318	0.757	0.239	—	0	0.343 = 0	tn

appropriately and use this to train a logistic regression model or a neural network. After training, we test the model on a set of 35 new images and evaluate the accuracy of the model. We also keep track of the amount of computer time to determine the trade-off between accuracy and speed of the different methods.

Table 12.1 shows representative results from this type of analysis to set the context for experiments to be described in later sections. Column 1 indicates the test sample number, where we have listed only 10 samples of the test set. The next p columns are the input values $x_{1i}, x_{2i}, \ldots, x_{pi}$ associated with each image. Only the first few data values are shown here. The method used to convert images into these x values is described in the next section. The next column in the table is Y, which is the correct label (0 or 1) for the image. And, finally, $\tilde{P}$ is the value predicted by the model, which will range between 0 and 1 since it is the output of a logistic function. This value is compared to 0.5 to quantize it to a label of either 0 or 1, which is also provided in the table.

To assess the accuracy of each model, we could simply take the correct results and divide it by the total number of test samples to determine the success rate. However, this type of analysis, although useful as a first-order assessment, may be misleading if most of the test data are cat pictures. Then by guessing 1 in all cases, a high accuracy can be obtained, but it would not be a true indication of the accuracy of the model. Instead, it is common practice to compute the F_1 score as follows. First, we categorize each sample based on whether or not the model predicted the correct label of 1 or 0. There are four possible scenarios, as listed in Table 12.2.

Table 12.2 Interpretation of test set results.

Y	$\tilde{P}$	Interpretation
1	1	tp = true positive
0	0	tn = true negative
1	0	fp = false positive
0	1	fn = false negative

For example, if the correct value is 1 and the quantized model predicts 1, it is referred to as a true positive (tp). Similarly a correct value of 0 and a predicted value of 0 is called a true negative (tn). On the other hand, incorrect predictions are either false positives (fp) or false negatives (fn). We can simply categorize each result and count up the numbers in each category. Then, using this information, we compute the recall and precision, which are both standard terms in binary classification (cf. James et al. 2013). Finally, the F_1 score is computed using these two quantities to assess the overall effectiveness of the model.

In particular, the recall value is given by

$$\text{recall} = \frac{\text{tp}}{\text{tp} + \text{fn}}, \tag{12.25}$$

and the precision value is given by

$$\text{precision} = \frac{\text{tp}}{\text{tp} + \text{fp}}. \tag{12.26}$$

Then, the F_1 score is computed using

$$F_1 = \frac{2 \times \text{precision} \times \text{recall}}{\text{precision} + \text{recall}}. \tag{12.27}$$

For the example of Table 12.1 using only the 10 samples shown, the recall = $5/(5+1)$, the precision = $5/(5+1)$, and the F_1 score = 0.84. The optimal value of F_1 score is 1, so this would be a relatively good F_1 score.

12.4.3 Image Preparation

In order to carry out binary classification, images must be converted into values. The basic idea is to partition images into picture elements (pixels) and assign a number to each pixel based on the intensity and color of the image in that pixel. A set of images were prepared in this manner to train and test models for logistic regression, shallow neural networks, and deep neural networks. Each image was preprocessed as shown in Figure 12.11. First, the image was divided into pixels as illustrated in the figure. For simplicity, the case of 4×4

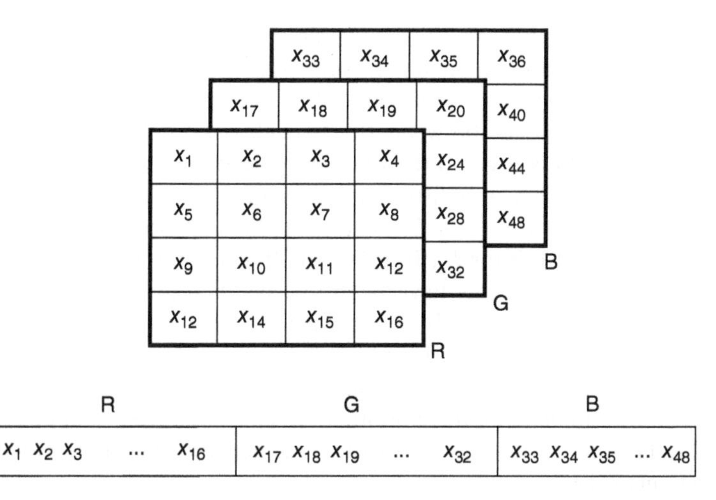

Figure 12.11 Preparing the image for model building.

pixels is shown here to produce a total of 16 pixels. However, there are three possible color combinations (RGB = red, green, blue) for each pixel resulting in a total of $4 \times 4 \times 3 = 48$ pixels in this example. Next, the values are "unrolled" to create one long $\boldsymbol{x}$ vector. In the case to be presented in the experimental results shortly, images were actually divided into $64 \times 64 \times 3 = 12\ 288$ pixels to obtain higher resolution for image recognition purposes.

Each image was assigned the proper observed Y value of $1(=\text{cat})$ or $0(=\text{dog})$. A total of 100 images were used to build models, and 35 images were used to test the model. Each model was generated first without a penalty function and then with the ridge and LASSO penalty functions. To illustrate the expected effect of ridge or LASSO penalty on the data, a simple case of two independent variables is shown in Figure 12.12. In the actual case, there are 12 288 independent

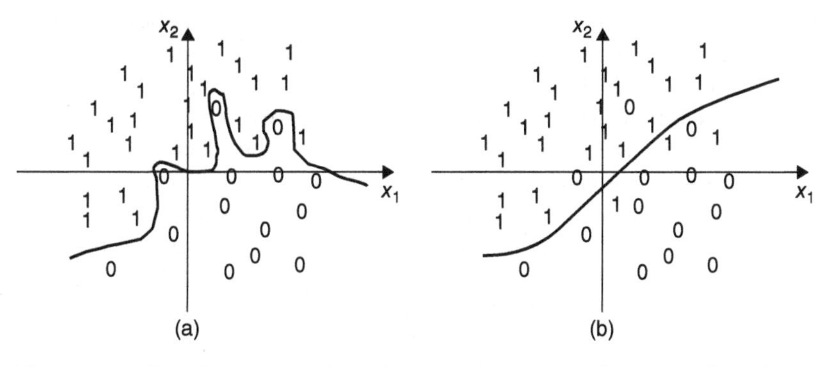

Figure 12.12 Over-fitting vs. regularized training data. (a) Over-fitting. (b) Effect of L_2 penalty.

variables as mentioned, but two variables are sufficient to describe conceptually the effect of a penalty function in binary classification.

In the figure, the 1's represent cats, while 0's represent dogs. The model defines the decision boundary between the two classes. In Figure 12.12a, neither ridge nor LASSO is used so the decision boundary is some complex path through the data. This over-fitted situation produces 100% accuracy on the training data. With the L_1 or L_2 penalty applied, a smoother curve is expected, as in Figure 12.12b for L_2. In this case, the training data accuracy is below 100% because some cats have been misclassified as dogs and vice versa. On the other hand, it will likely be more accurate on test data since the decision boundary is more realistic. The inherent errors are due to the fact that some cat images may appear to look like dogs, while dog pictures may appear to look like cats. But, in general, the smoother curve is preferred as opposed to the over-fitted result. In fact, this is why the ridge penalty function is widely used in neural networks.

12.4.4 Experimental Results

A set of $n = 100$ images of cats and dogs (50 each) were used as training data for a series of regression models. An additional 35 images were used as test data. As mentioned, for each sample image, there are $p = 12\ 288$ independent variables (the "unrolled" pixels) and one dependent variable (1 = cat, 0 = dog). Clearly, this is a very wide data set where over-fitting would be a significant problem. Therefore, some type of penalty function is needed in conjunction with the loss function, as described in earlier sections. Numerical methods were used to determine the $\boldsymbol{\beta}$ parameters for each regression model by minimizing the loss function using gradient descent, with a step size of $\alpha = 0.01$. The solution space was assumed to be convex and so a simple gradient descent was sufficient to minimize the loss value and obtain the corresponding $\boldsymbol{\beta}$ values. The number of iterations of gradient descent was initially set to 3000. In the first run of each model, the tuning parameter λ was set to 0 (i.e. no penalty function). Then it was varied between 0.1 and 10.0 for both L_1-penalty and L_2-penalty functions. The results are shown in Tables 12.3–12.6.

Table 12.3 contains the results for logistic regression (LR) using the L_2-penalty function; as λ is increased from 0.0 to 10.0, the effect of the penalty function increases. This can be understood by examining the columns labeled test accuracy and F_1 score. Initially, the test accuracy increases as λ is increased until an optimal value of the test accuracy is achieved. The optimal result is obtained for $\lambda = 0.4$ and this case delivers a test accuracy of 74.3% along with the highest F_1 score of 0.77. This implies that 74.3% of the 35 test images were correctly identified as a cat or dog, after training on the 100 sample images with 100% accuracy. However, increasing the value of λ further does not improve the test accuracy and, in fact, it begins to degrade at very high values.

Table 12.3 Results for L_2-penalty (ridge) using LR.

L_2-penalty parameter	Train accuracy (%)	Testing accuracy (%)	F_1 score	Runtime (s)
$\lambda = 0.0$	100	68.6	0.72	14.9
$\lambda = 0.2$	100	68.6	0.72	15.2
$\lambda = 0.4$	100	74.3	0.77 optimal	16.6
$\lambda = 1.0$	100	71.4	0.74	15.3
$\lambda = 2.0$	100	71.4	0.74	12.5
$\lambda = 4.0$	100	68.6	0.72	15.7
$\lambda = 10.0$	100	60.0	0.63	15.4

Table 12.4 Results for L_1-penalty (LASSO) using LR.

L_1-penalty parameter	Train accuracy (%)	Testing accuracy (%)	F_1 score	Runtime (s)
$\lambda = 0.0$	100	68.6	0.72	14.8
$\lambda = 0.1$	100	71.4	0.75 optimal	14.9
$\lambda = 0.2$	100	62.9	0.70	15.3
$\lambda = 0.4$	100	60.0	0.65	15.3
$\lambda = 0.8$	97	60.0	0.65	15.4
$\lambda = 1.0$	57	57.1	0.61	15.7

Table 12.5 Results for L_2-penalty (ridge) using two-layer NN.

L_2-penalty parameter	Train accuracy (%)	Testing accuracy (%)	F_1 score	Runtime (s)
$\lambda = 0.0$	100	77	0.79	184
$\lambda = 0.1$	100	77	0.79	186
$\lambda = 0.2$	100	77	0.79	187
$\lambda = 1.0$	100	80	0.82	186
$\lambda = 2.0$	100	82	0.84 Optimal	192
$\lambda = 4.0$	100	77	0.79	202
$\lambda = 10.0$	100	65	0.68	194

Table 12.6 Results for L_2-penalty (ridge) using three-layer NN.

L_2-penalty parameter	Train accuracy (%)	Testing accuracy (%)	F_1 score	Runtime (s)
$\lambda = 0.0$	100	77	0.79	384
$\lambda = 0.2$	100	80	0.82 optimal	386
$\lambda = 0.4$	100	80	0.82	387
$\lambda = 1.0$	100	77	0.79	386
$\lambda = 2.0$	100	77	0.79	392
$\lambda = 4.0$	100	77	0.79	402

Table 12.4 indicates that the L_1 penalty has difficulty in improving the results beyond those obtained from the L_2 penalty. Inherently, LASSO seeks to zero out some of the parameters, but since the inputs are all pixels of images, no variables can be eliminated. Another effect is due to slow convergence of gradient descent when the L_1-penalty is used. One can increase the number of iterations beyond 3000 in this particular case, but it becomes computationally more expensive without delivering improvements over ridge. For these and other reasons, the L_2-penalty function is widely used in applications with characteristics similar to image recognition rather than the L_1-penalty function.

The next model was built using a two-layer neural network with one hidden layer with five units using only the L_2-penalty function, since the L_1-penalty function produced very poor results. The five hidden units used the relu(z) function of Figure 12.8, while the output layer used the logistic function $\pi(z)$ of Figure 12.3. The results are given in Table 12.5. To obtain the desired accuracy, 12 000 iterations were required. Note that the accuracy of the model has improved greatly relative to logistic regression but the runtime has also increased. This is expected based on Figure 12.9 shown previously. In this case, each iteration requires operations on a $12\,288 \times 5$ matrix so the runtime is much higher. However, the test accuracy is now 82% with a corresponding F_1 score of 0.84.

The final set of results are shown in Table 12.6 for a three-layer neural network. There are two hidden layers with ten and five hidden units, respectively. All 15 hidden units used the relu(z) function and, as before, the output function employed the logistic function $\pi(z)$. While not strictly a deep neural network, it provides enough information to assess the accuracy and speed characteristics of deep networks. The best test accuracy of the different cases is 80% with an F_1 score of 0.82, which is lower than the two-layer case, but the overall accuracy of all cases is higher. However, it is clearly more computationally expensive than the other models. This is due to the fact that it now operates on a 12 288 × 10 matrix on each iteration of gradient descent. Therefore, it requires a higher run-time. In this case, a total of 10 000 iterations were used to obtain the results.

To summarize, the L_2-penalty function is more effective on logistic and neural networks than the L_1-penalty function for image recognition. These penalty functions are best suited to wide data sets where over-fitting is the main problem. It is very useful in shallow neural networks, but less effective as more layers are used in the network. One should note that the results are controlled to a large extent by a set of tunable hyper-parameters: α, λ the number of layers, the number of hidden units in each layer, and the number of iterations used to build the model. Furthermore, additional improvements can be obtained by increasing the size of the sample set. An important part of using neural networks is to properly determine these (and other) tunable hyper-parameters to deliver optimal results. In any case, it is clear that the ridge penalty function is a critical part of neural networks and worth pursuing as future research.

12.5 Summary and Concluding Remarks

In this chapter, we used logistic regression in neural networks, from the importance of penalty functions. As we outlines in Section 12.1, we dealt with the practical application of penalty functions and their use in applied statistics. In conclusion, we found the L_2-penalty function is more effective on logistic and neural networks than the L_1-penalty function.

Problems

12.1 Derive the updated equations for $\boldsymbol{\beta}_{\text{vector}}$ and $\boldsymbol{\beta}_{\text{matrix}}$ for a two-layer neural network that is similar in form to (12.19) assuming the logistic function is used in both layers.

12.2 Another possible activation function (that is, a function that would play the role of the logistic function) is the relu(z) function. Derive the updated equations for a two-layer neural network if the relu(z) function is used in the first layer and the logistic function is used in the second layer.

12.3 Derive the gradient equation for the tanh(z) activation function where

$$\tanh(z) = \frac{\exp\{x\} - \exp\{-x\}}{\exp\{x\} + \exp\{-x\}}.$$

12.4 The figure depicts a four-layer neural network. Based on Figure 12.6, what are the sizes of the respective $\boldsymbol{\beta}$ matrices or vector for each stage? (Hint: the matrix for layer 1 is 5×4 since the first layer has five hidden units and the input has four variables).

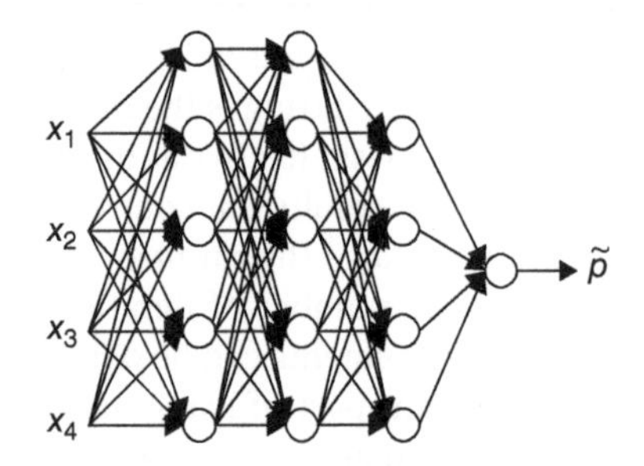

12.5 Derive equations similar to (12.20) for the deep neural network shown. You can assume that the logistic function is used in every layer.

References

Ahmed, S.E. and Saleh, A.K.M.E. (1988). Estimation strategy using a preliminary test in some univariate normal models. *Soochow Journal of Mathematics* 14: 135–165.

Ahsanullah, M. and Saleh, A.K.M.E. (1972). Estimation of intercept in a linear regression model with one dependent variable after a preliminary test on the regression coefficient. *International Statistical Review* 40: 139–145.

Akdeniz, F. and Ozturk, F. (1981). The effect of multicollinearity: a geometric view. *Communications de la Faculte des Sciences de l'Universite d'Ankara* 30: 17–26.

Akdeniz, F. and Tabakan, G. (2009). Restricted ridge estimators of the parameters in semiparametric regression model. *Communications in Statistics - Theory and Methods* 38 (11): 1852–1869.

Alkhamisi, M. and Shukur, G. (2008). Developing ridge parameters for SUR model. *Communications in Statistics - Theory and Methods* 37 (4): 544–564.

Anderson, T.W. (1984). *An Introduction to Multivariate Statistical Analysis.* New York: Wiley.

Arashi, M. (2012). Preliminary test and Stein estimators in simultaneous linear equations. *Linear Algebra and its Applications* 436 (5): 1195–1211.

Arashi. M. and Norouzirad, M. (2016). Steinian shrinkage estimation in high dimensional regression. In: *13th Iranian Statistics Conference*. Kerman, Iran: Shahid Bahonar University of Kerman.

Arashi, M. and Roozbeh, M. (2016). Some improved estimation strategies in high-dimensional semiparametric regression models with application to the Riboflavin production data. *Statistical Papers*. doi: 10.1007/s00362-016-0843-y.

Arashi, M. and Tabatabaey, S.M.M. (2009). Improved variance estimation under sub-space restriction. *Journal of Multivariate Analysis* 100: 1752–1760.

Arashi, M. and Valizadeh, T. (2015). Performance of Kibria's methods in partial linear ridge regression model. *Statistical Papers* 56 (1): 231–246.

Arashi, M., Saleh, A.K.M.E, and Tabatabaey, S.M.M. (2010). Estimation of parameters of parallelism model with elliptically distributed errors. *Metrika* 71: 79–100.

Theory of Ridge Regression Estimation with Applications, First Edition.
A.K. Md. Ehsanes Saleh, Mohammad Arashi, and B.M. Golam Kibria.

Arashi, M., Roozbeh, M., and Niroomand, H.A. (2012). A note on Stein type shrinkage estimator in partial linear models. *Statistics* 64 (5): 673–685.

Arashi, M., Janfada, M., and Norouzirad, M. (2015). Singular ridge regression with stochastic constraints. *Communication in Statistics - Theory and Methods* 44: 1281–1292.

Arashi, M., Kibria, B.M.G., and Valizadeh, T. (2017). On ridge parameter estimators under stochastic subspace hypothesis. *The Journal of Statistical Computation and Simulation* 87 (5): 966–983.

Asar, M., Arashi, M., and Wu, J. (2017). Restricted ridge estimator in the logistic regression model. *Communications in Statistics - Simulation and Computation* 46 (8): 6538–6544.

Aslam, M. (2014). Performance of Kibria's method for the heteroscedastic ridge regression model: some Monte Carlo evidence. *Communications in Statistics - Simulation and Computation* 43 (4): 673–686.

Avalos, M., Grandvalet, Y., and Ambroise, C. (2007). Parsimonious additive models. *Computational Statistics and Data Analysis* 51: 2851–2870.

Aydin, D., Yuzbasi, B., and Ahmed, S.E. (2016). Modified ridge type estimator in partially linear regression models and numerical comparisons. *Journal of Computational and Theoretical Nanoscience* 13 (10): 7040–7053.

Baltagi, B.H. (1980). On seemingly unrelated regressions with error components. *Econometrica* 48: 1547–1551.

Bancroft, T. (1964). Analysis and inference for incompletely specified models involving the use of preliminary test(s) of significance. *Biometrics* 57: 579–594.

Beaver, W.H. (1966). Financial ratios as predictors of failure. *Journal of Accounting Research* 4 (3): 71–111.

Belloni, A. and Chernozhukov, V. (2013). Least squares after model selection in high-dimensional sparse models. *Bernoulli* 19: 521–547.

Breiman, L. (1996). Heuristics of instability and stabilization in model selection. *The Annals of Statistics* 24: 2350–2383.

Buhlmann, P., Kalisch, M., and Meier, L. (2014). High-dimensional statistics with a view toward applications in biology. *Annual Review of Statistics and its Applications* 1: 255–278.

Chandrasekhar, C.K., Bagyalakshmi, H., Srinivasan, M.R., and Gallo, M. (2016). Partial ridge regression under multicollinearity. *Journal of Applied Statistics* 43 (13): 2462–2473.

Dempster, A.P., Schatzoff, M., and Wermuth, N. (1977). A simulation study of alternatives to ordinary least squares. *Journal of the American Statistical Association* 72: 77–91.

Dicker, L.H. (2016). Ridge regression and asymptotic minimum estimation over spheres of growing dimension. *Bernoulli* 22: 1–37.

Donoho, D.L. and Johnstone, I.M. (1994). Ideal spatial addaption by wavelet shrinkage. *Biometrika* 81 (3): 425–455.

Draper, N.R. and Nostrand, R.C.V. (1979). Ridge regression and James-Stein estimation: review and comments. *Technometrics* 21: 451–466.

Fallah, R., Arashi, M., and Tabatabaey, S.M.M. (2017). On the ridge regression estimator with sub-space restriction. *Communication in Statistics - Theory and Methods* 46 (23): 11854–11865.

Fan, J. (2017). Discusion of "post selection shrinkage estimation for high dimensional data anlysis". *Applied Stochastic Model in Business and Industry* 33: 121–122.

Fan, J. and Li, R. (2001). Variable selection via nonconcave penalized likelihood and its oracle properties. *Journal of the American Statistical Association* 96: 1348–1360.

Farrar, D.E. and Glauber, R.R. (1967). Multicollinearity in regression analysis: the problem revisited. *Review of Economics and Statistics* 49 (1): 92–107.

Foschi, P., Belsley, D.A., and Kontoghiorghes, E.J. (2003). A comparative study of algorithms for solving seemingly unrelated regressions models. *Computational Statistics and Data Analysis* 44: 3–35.

Frank, I.E. and Friedman, J.H. (1993). A statistical view of some chemometrics regression tools. *Technometrics* 35: 109–135.

Gao, J.T. (1995). Asymptotic theory for partially linear models. *Communication in Statistics - Theory and Methods* 22: 3327–3354.

Gao, J. (1997). Adaptive parametric test in a semi parametric regression model. *Communication in Statistics - Theory and Methods* 26: 787–800.

Gao, X., Ahmed, S.E., and Feng, Y. (2017). Post selection shrinkage estimation for high dimensional data analysis. *Applied Stochastic Models in Business and Industry* 33: 97–120.

Gibbons, D.G. (1981). A simulation study of some ridge estimators. *Journal of the American Statistical Association* 76: 131–139.

Golub, G.H., Heath, M., and Wahba, G. (1979). Generalized cross-validation as a method for choosing a good ridge parameter. *Technometrics* 21 (2): 215–223.

Grandvalet, Y. (1998). Least absolute shrinkage is equivalent to quadratic penalization. In: *ICANN'98*, Perspectives in Neural Computing, vol. 1 (ed. L. Niklasson, M. Boden, and T. Ziemske), 201–206. Springer.

Grob, J. (2003). Restricted ridge estimator. *Statistics and Probability Letters* 65: 57–64.

Gruber, M.H.J. (1998). *Improving Efficiency by Shrinkage: The James-Stein and Ridge Regression Estimators*. New York: Marcel Dekker.

Gruber, M.H.J. (2010). *Regression Estimators*, 2e. Baltimore, MD: Johns Hopkins University Press.

Gunst, R.F. (1983). Regression analysis with multicollinear predictor variables: definition, detection, and effects. *Communication in Statistics - Theory and Methods* 12 (19): 2217–2260.

Hansen, B.E. (2016). The risk of James-Stein and lasso shrinkage. *Econometric Reviews* 35: 1456–1470.

Hardle, W., Liang, H., and Gao, J. (2000). *Partially Linear Models*. Heidelberg: Springer Physica-Verlag.

Haykin, S. (1999). *Neural Networks and Learning Machines*, 3e. Pearson Prentice-Hall.

Hefnawy, E.A. and Farag, A. (2013). A combined nonlinear programming model and Kibria method for choosing ridge parameter regression. *Communications in Statistics - Simulation and Computation* 43 (6): 1442–1470.

Hoerl, A.E. (1962). Application of ridge analysis to regression problems. *Chemical Engineering Progress* 58 (3): 54–59.

Hoerl, A.E. and Kennard, R.W. (1970). Ridge regression: biased estimation for non-orthogonal problems. *Technometrics* 12: 55–67.

Hoerl, A.E., Kennard, R.W., and Baldwin, K.F. (1975). Ridge regression: some simulations. *Communications in Statistics - Theory and Methods* 4: 105–123.

Hosmer, D.W. and Lemeshow, J.S. (1989). *Applied Logistic Regression*, 2e. Wiley.

Huang, C.C.L., Jou, Y.J., and Cho, H.J. (2016). A new multicollinearity diagnostic for generalized linear models. *Journal of Applied Statistics* 43 (11): 2029–2043.

Ismail, B. and Suvarna, M. (2016). Estimation of linear regression model with correlated regressors in the presence of auto correlation. *International Journal of Statistics and Applications* 6 (2): 35–39.

Jamal, N. and Rind, M.Q. (2007). Ridge regression: a tool to forecast wheat area and production. *Pakistan Journal of Statistics and Operation Research* 3 (2): 125–134.

James, W. and Stein, C. (1961). Estimation with quadratic loss. In: *Proceeding Berkeley Symposium on Mathematical Statistics and Probability*, vol. 1, 361–379. University of California.

James, G., Witten, D., Hastie, T., and Tibshirani, R. (2013). *An Introduction to Statistical Learning: with Applications in R*. Springer.

Judge, G.G. and Bock, M.E. (1978). *The Statistical Implications of Pre-test and Stein-Rule Estimators in Econometrics*. Amsterdam: North-Holland Publishing Company.

Jureckova, J. (1971). Non parametric estimate of regression coefficients. *The Annals of Mathematical Statistics* 42: 1328–1338.

Jureckova, J. and Sen, P.K. (1996). *Robust Statistical Procedures: Asymptotic and Interrelations*. New York: Wiley.

Kaciranlar, S., Sakallioglu, S., Akdeniz, F. et al. (1999). A new biased estimator in linear regression and a detailed analysis of the widely-analyzed data set on Portland cement. *Sankhya The Indian Journal of Statistics* 61: 443–456.

Khalaf, G. and Shukur, G. (2005). Choosing ridge parameters for regression problems. *Communications in Statistics - Theory and Methods* 34 (5): 1177–1182.

Kibria, B.M.G. (2003). Performance of some new ridge regression estimators. *Communications in statistics - Simulation and Computations* 32 (2): 419–435.

Kibria, B.M.G. (2012). Some Liu and Ridge type estimators and their properties under the Ill-conditioned Gaussian linear regression model. *Journal of Statistical Computation and Simulation* 82 (1): 1–17.

Kibria, B.M.G. and Banik, S. (2016). Some ridge regression estimators and their performances. *Journal of Modern Applied Statistical Methods* 15 (1): 206–238.

Kibria, B.M.G. and Saleh, A.K.M.E. (2012). Improving the estimators of the parameters of a probit regression model: a ridge regression approach. *Journal of Statistical Planning and Inference* 14 (2): 1421–1435.

Knight, K. and Fu, W.J. (2000). Asymptotic for Lasso-type estimators. *Annals of Statistics* 28: 1356–1378.

Kontoghiorghes, E.J. (2000). Inconsistencies and redundancies in SURE models: computational aspects. *Computational Economics* 16 (1-2): 63–70.

Kontoghiorghes, E.J. (2004). Computational methods for modifying seemingly unrelated regressions models. *Journal of Computational and Applied Mathematics* 162 (1): 247–261.

Kontoghiorghes, E.J. and Clarke, M.R.B. (1995). An alternative approach for the numerical solution of seemingly unrelated regression equations models. *Computational Statistics and Data Analysis* 19 (4): 369–377.

Lawless, J.F. and Wang, P.A. (1976). simulation study of ridge and other regression estimators. *Communications in Statistics - Theory and Methods* 5: 307–323.

Liang, H. and Hardle, W. (1999). Large sample theory of the estimation of the error distribution for semi parametric models. *Communications in Statistics - Theory and Methods* 28 (9): 2025–2036.

Luo, J. (2010). The discovery of mean square error consistency of ridge estimator. *Statistics and Probability Letters* 80: 343–347.

Luo, J. (2012). Asymptotic efficiency of ridge estimator in linear and semi parametric linear models. *Statistics and Probability Letters* 82: 58–62.

Luo, J. and Zuo, Y.J. (2011). A new test for large dimensional regression coefficients. *Open Journal of Statistics* 1: 212–216.

McDonald, G.C. and Galarneau, D.I. (1975). A Monte Carlo evaluation of ridge-type estimators. *Journal of the American Statistical Association* 70 (350): 407–416.

Mansson, K., Shukur, G., and Kibria, B.M.G. (2010). On some ridge regression estimators: a Monte Carlo simulation study under different error variances. *Journal of Statistics* 17 (1): 1–22.

Maronna, R.A. (2011). Robust ridge regression for high-dimensional data. *Technometrics* 53 (1): 44–53.

Marquardt, D.W. and Snee, R.D. (1975). Ridge regression in practice. *The American Statistician* 29 (1): 3–20.

Martin, D. (1977). Early warning of bank failure: a logit regression approach. *Journal of Banking and Finance* 1 (3): 249–276.

Montgomery, D.C., Peck, E.A., and Vining, G.G. (2012). *Introduction to Linear Regression Analysis*. 5e. Hoboken, NJ: Wiley.

Müller, M. and Rönz, B. (1999). Credit Scoring Using Semiparametric Methods, Discussion Papers, Interdisciplinary Research Project 373: Quantification and Simulation of Economic Processes, No. 1999,93, Humboldt-Universität Berlin http://nbnresolving.de/urn:nbn:de:kobv:11-10046812.

Muniz, G. and Kibria, B.M.G. (2009). On some ridge regression estimators: an empirical comparison. *Communications in Statistics - Simulation and Computation* 38 (3): 621–630.

Muniz, G., Kibria, B.M.G., Mansson, K., and Shukur, G. (2012). On developing ridge regression parameters: a graphical investigation. *Statistics and Operations Research Transactions* 36 (2): 115–138.

Najarian, S., Arashi, M., and Kibria, B.M. (2013). A simulation study on some restricted ridge regression estimators. *Communication in Statistics - Simulation and Computation* 42 (4): 871–890.

Norouzirad, M. and Arashi, M. (2017). Preliminary test and Stein-type shrinkage ridge estimators in robust regression. *Statistical Papers*. doi: 10.1007/s00362-017-0899-3.

Norouzirad, M., Arashi, M., and Ahmed, S. E. (2017). Improved robust ridge M-estimation. *Journal of Statistical Computation and Simulation* 87 (18): 3469–3490.

Ozturk, F. and Akdeniz, F. (2000). Ill-conditioning and multicollinearity. *Linear Algebra and its Applications* 321: 295–305.

Puri, M.L. and Sen, P.K. (1986). *Nonparametric Methods in General Linear Models*. New York: Wiley.

Raheem, E., Ahmed, S.E., and Doksum, K. (2012). Absolute penalty and shrinkage estimation in partially linear models. *Computational Statistics and Data Analysis* 56 (4): 874–891.

Roozbeh, M. (2015). Shrinkage ridge estimators in semi parametric regression models. *Journal of Multivariate Analysis* 136: 56–74.

Roozbeh, M. and Arashi, M. (2013). Feasible ridge estimator in partially linear models. *Journal of Multivariate Analysis* 116: 35–44.

Roozbeh, M. and Arashi, M. (2016a). New ridge regression estimator in semi-parametric regression models. *Communications in Statistics - Simulation and Computation* 45 (10): 3683–3715.

Roozbeh, M. and Arashi, M. (2016b). Shrinkage ridge regression in partial linear models. *Communications in Statistics - Theory and Methods* 45 (20): 6022–6044.

Roozbeh, M., Arashi, M., and Gasparini, M. (2012). Seemingly unrelated ridge regression in semi parametric models. *Communications in Statistics - Theory and Methods* 41: 1364–1386.

Sakallioglu, S. and Akdeniz, F. (1998). Generalized inverse estimator and comparison with least squares estimator. *Communication in Statistics - Theory and Methods* 22: 77–84.

Saleh, A. (2006). *Theory of Preliminary Test and Stein-type Estimation with Applications*. New York: Wiley.
Saleh, A.K.M.E. and Kibria, B.M.G. (1993). Performances of some new preliminary test ridge regression estimators and their properties. *Communications in Statistics - Theory and Methods* 22: 2747–2764.
Saleh, A.K.M.E. and Sen, P.K. (1978). Non-parametric estimation of location parameter after a preliminary test on regression. *Annals of Statistics* 6: 154–168.
Saleh, A.K.M.E. and Sen, P.K. (1985). On shrinkage R-estimation in a multiple regression model. *Communication in Statistics -Theory and Methods* 15: 2229–2244.
Saleh, A.K.M.E., Arashi, M., and Tabatabaey, S.M.M. (2014). *Statistical Inference for Models with Multivariate t-Distributed Errors*. Hoboken, NJ: Wiley.
Saleh, A.K.M.E., Arashi, M., Norouzirad, M., and Kibria, B.M.G. (2017). On shrinkage and selection: ANOVA modle. *Journal of Statistical Research* 51 (2): 165–191.
Sarkar, N. (1992). A new estimator combining the ridge regression and the restricted least squares methods of estimation. *Communication in Statistics - Theory and Methods* 21: 1987–2000.
Sen, P.K. and Saleh, A.K.M.E. (1979). Non parametric estimation of location parameter after a preliminary test on regression in multivariate case. *Journal of Multivariate Analysis* 9: 322–331.
Sen, P.K. and Saleh, A.K.M.E. (1985). On some shrinkage estimators of multivariate location. *Annals of Statistics* 13: 172–281.
Sen, P.K. and Saleh, A.K.M.E. (1987). On preliminary test and shrinkage M-estimation in linear models. *Annals of Statistics* 15: 1580–1592.
Shao, J. and Deng, X. (2012). Estimation in high-dimensional linear models with deterministic design matrices. *Annals of Statistics* 40 (2): 812–831.
Shi, J. and Lau, T.S. (2000). Empirical likelihood for partially linear models. *Journal of Multivariate Analysis* 72 (1): 132–148.
Speckman, P. (1988). Kernel smoothing in partial linear models. *Journal of the Royal Statistical Society: Series B (Statistical Methodology)* 50: 413–436.
Stamey, T., Kabalin, J., McNeal, J. et al. (1989). Prostate specific antigen in the diagnosis and treatment of adenocarcinoma of the prostate II: radical prostatectomy treated patients. *The Journal of Urology* 16: 1076–1083.
Stein, C. (1956). Inadmissibility of the usual estimator for the mean of a multivariate normal distribution. *Proceedings of the 3rd Berkeley Symposium on Mathematical Statistics and Probability*, pp. 1954–1955.
Tam, K. and Kiang, M. (1992). Managerial applications of neural networks: the case of bank failure predictions. *Management Science* 38 (7): 926–947.
Tibshirani, R. (1996). Regression shrinkage and selection via the lasso. *Journal of the Royal Statistical Society: Series B (Statistical Methodology)* 58 (1): 267–288.
Tikhonov, A.N. (1963). Translated in "solution of incorrectly formulated problems and the regularization method". *Soviet Mathematics* 4: 1035–1038.

Wan, A.T.K. (2002). On generalized ridge regression estimators under collinearity and balanced loss. *Applied Mathematics and Computation* 129: 455–467.

Wang, X., Dunson, D., and Leng, C. (eds.) (2016). No penalty no tears: least squares in high dimensional linear models. *Proceedings of the 33rd International Conference on Machine Learning (ICML 2016).*

Woods, H., Steinnour, H.H., and Starke, H.R. (1932). Effect of composition of Portland cement on heat evolved during hardening. *Industrial and Engineering Chemistry* 24: 1207–1241.

Yuzbasi, B. and Ahmed, S.E. (2015). Shrinkage ridge regression estimators in high-dimensional linear models. In: *Proceedings of the 9th International Conference on Management Science and Engineering Management, Advances in Intelligent Systems and Computing*, vol. 362 (ed. J. Xu, S. Nickel, V. Machado, and A. Hajiyev). Berlin, Heidelberg: Springer-Verlag.

Zhang, M., Tsiatis, A.A., and Davidian, M. (2008). Improving efficiency of inferences in randomized clinical trials using auxiliary covariates. *Biometrics* 64 (3): 707–715.

Zou, H. (2006). The adaptive lasso and its oracle properties. *Journal of the American Statistical Association* 101: 1418–1429.

Zou, H. and Hastie, T. (2005). Regularization and variable selection via the elastic net. *Journal of the Royal Statistical Society: Series B (Statistical Methodology)* 67: 301–320.

Index

Theory of Ridge Regression Estimation with Applications, First Edition.
A.K. Md. Ehsanes Saleh, Mohammad Arashi, and B.M. Golam Kibria.